What became of the rhetorical promises of the early Royal Society? How was it that the experimental philosophy of the early Newtonians was regarded by some as destructive of British society and by others as its savior? Why did numerous entrepreneurs in the eighteenth century come to see Newton's disciples as the means to ensure success in the wild world of technical adventures? In the convergence of social and intellectual issues, these are some of the questions *The Rise of Public Science* seeks to answer by exploring the social attitudes toward the claims and activities of the natural philosophers in Britain from the Restoration to the first stage of industrialization.

Situating science amidst the social and intellectual transformation of the times, Stewart considers the religious conflicts, political changes, and technological demands that influenced the development of natural philosophy by the early eighteenth century. Through Newton's disciples, he explores the reasons for the popularity of science and its controversial links with a variety of technical and entrepreneurial ventures on the verge of the Industrial Revolution.

The Rise of Public Science

The Rise of Public Science

Rhetoric, Technology, and Natural Philosophy in Newtonian Britain, 1660–1750

Larry Stewart
University of Saskatchewan

CAMBRIDGE
UNIVERSITY PRESS

Published by the Press Syndicate of the University of Cambridge
The Pitt Building, Trumpington Street, Cambridge CB2 1RP
40 West 20th Street, New York, NY 10011-4211, USA
10 Stamford Road, Oakleigh, Victoria 3166, Australia

First published 1992

Printed in the United States of America

Library of Congress Cataloging-in-Publication Data
Stewart, Larry
The rise of public science : rhetoric, technology, and natural philosophy in Newtonian Britain / by Larry Stewart.
p. cm.
Includes bibliographical references and index.
ISBN 0-521-41700-7
1. Science—Great Britain—History—17th century. 2. Science—Great Britain—History—18th century. 3. Technology—Great Britain—History—17th century. 4. Technology—Great Britain—History—18th century. I. Title.
Q127.07S74 1992
809.41'09'032—dc20 91-44172
CIP

A catalog record for this book is available from the British Library

ISBN 0-521-417007-7 hardback

FOR UTA

Contents

List of illustrations

Preface

This book is a step in the recognition of the public world of modern science. This then is, explicitly, the product of the culture of discourse that explodes the private solitude of ideas, for it owes much to those who have offered their comments, their enthusiasm, and their encouragement over the years of rummaging in archives and the agonies of trying to make it all sound convincing.

It has been my good fortune to have had the doors opened to a succession of archives and libraries where I have been met with unfailing patience and not a little amusement. It would be a remarkable slight if I did not especially acknowledge assistance of those to whom I must have been a particular nuisance: the Royal Society; the Royal Astronomical Society; the William Andrews Clarke Memorial Library; the Huntington Library and Mary of the chocolates; McMaster University Library; the Master and Fellows of Trinity College, Cambridge for the papers in the Wren Library; the syndics of the University Library, Cambridge; the Spalding Gentlemen's Society; the Carlisle Record Office; the Northumberland Record Office; the Gateshead Public Library; Sir John Conant for documents held in the Leicestershire Record Office; Mr. James Stirling of Garden and the Scottish Record Office, Edinburgh; the British Library; and the Public Record Office, London. I am particularly thankful to the Richard Landon and the staff of the Thomas Fisher Rare Book Library, University of Toronto, who have made possible much of what follows.

Over the past decade when this book took shape, I received much support from friends and colleagues. Trevor Levere and George Grinnell will be amused at how much of their influence is in this book. I especially want to thank Peter Millard, Ray Stephanson, Hugh Johnson, and Graham Cummings for their opinions on an early draft of the manuscript, and to Simon Schaffer and Jan Golinski for many insightful comments that have helped give this work a more precise focus. John Parry has been a willing sounding board longer than either of us care to remember. In spite of computers, Jean Horosko, Joan Aschim, Jackie Fraser, and Lisa Stewart were still able to help me put all of this together.

The Social Science and Humanities Research Council of Canada and, against great odds, the University of Saskatchewan, have supported the research for this book. This is remarkable when the humanities are generally regarded as a luxury by those who have come to believe, as one colleague has written, that "the university means business." But I am, nonetheless, grateful for the finer edge such an attitude to research has forced on my work on the origins of the early Industrial Revolution. The myopia of the current academic world, and the pathetically antiintellectual attitude of many governments, makes the survival of the humanities precarious. I recall one young student from the College of Commerce at the University of Saskatchewan whose exasperation in my course of European history once bubbled over. He asked, in front of a class of about 120 undergraduates, "Why are we taking all this stuff for anyways; it's all out of date?" My stunned silence was soon filled by gales of laughter from his fellow students. For him, the past was a foreign country. To those university administrators who would agree, this book is extended so that they too may comprehend their past.

It is difficult to thank all those who have offered their comments and their assistance. I include here the readers and editors of the *Journal of British Studies, Isis, Medical History* and the *Journal of the History of Ideas* where early versions of some of my arguments were published. Many will see their influence in the following pages, and they can take any credit that may be due. Quite simply, whatever its flaws, this is deliberately a work of the acknowledgment of those historians who have preceded me and especially of those who, in an ongoing debate, would reexamine a past that has pushed us from private laboratories to a wider, now intensely, political sanction. This is territory that is well mapped out by Simon Schaffer, Steven Shapin, Margaret and James Jacob, Michael Hunter, and Jan Golinski, to name only a few. To them I owe many of the questions I have faced during writing and the interminable revision. But the notion that something so esoteric and seemingly unattainable as science might actually have attracted those who clamored to have entry to an otherwise private world, the rules of which were policed by the likes of the Royal Society, meant that the creation of avenues of access was fundamentally a political comment on the production of knowledge.

But do we need another look at the science of the early modern world

– especially in this age when science has become an adjunct of industry and of government, when contemplation is overrun by unambiguous commerce? This only readers can determine. But it does seem that our own material culture is surely the legacy of the popular acceptance of the science and of the mechanization whose history concerns me. Such studies as the one proposed here continue to fill the book shops, and readers surface, even when science is no longer regarded as our unambiguous savior. It is these very ambiguities that impose an interest upon a present public.

Introduction

This book was written as an exploration into the reception of science in the modern world. I have set out in the seventeenth and eighteenth centuries to examine a world much like our own, where intellectual life is not always isolated from the seduction of adventure, where even enlightened doctrines of reason and reasonableness cannot sensibly be divorced from the personal rationalizations of their proponents. At the outset, I want to be clear that this book is concerned more with attitudes and actions than with causes. It seeks to explore the convergence of intellect and industry. For this reason, we begin with the rhetoric of the early Royal Society and the lingering influence of natural philosophy not in abstruse and complex metaphysics, but in the manifest doctrine of the public interest. In the eighteenth century seen here, what rhetoric proposes the public disposes. I have suggested a notion of popular science that adopted the presumption of benefits – and thereby provided legitimacy to the new science. Thus it was the public that would increasingly be the arbiter of the value of natural philosophy in the first half of the eighteenth century.

In an effort to reconstruct the world of public science, I have sought to avoid the idealisms of intellectual history as much as possible. This is one reason for my concern with rhetoric and its reception in the seventeenth and eighteenth centuries. In so doing, this book attempts to reach into the perceptions and the attitudes of a world that has long fascinated me. Those seeking a grand platonist revival or a stunningly meaningful sociological re-creation of the eighteenth century will not find it here. This book is devoted more to sympathies and to actions, to notions and sometimes to ideals that were wrapped in rhetoric – that naturally implies we should pay some attention to the listeners.

There was after all, in western Europe at least, a great struggle of conscience over the meaning of authority in its immense variety of literate, political, and social forms. This was a world infected with a sense of urgency – where the pressures for change would build not only to political revolution, but also to a dynamic of social and economic transformation that ultimately made new political relationships both possible and necessary. In this process, science would play a critical role by helping to redefine authority and legitimacy. It did so by establishing a forceful and credible experimentalism – but, far more important, by likewise creating a public science to which many might obtain entry. Readers, listeners, observers could prove as important as authors and orators. In these terms, it is less the creation of scientific facts that matters, rather than the manufacture of credibility.

There are large philosophical, social, and historiographical issues here and I have chosen not to treat them separately but within the context that gives them meaning. I was especially struck by these matters when visiting an exhibition of the paintings of Joseph Wright of Derby. His portraits of Erasmus Darwin, done at the end of the eighteenth century, are undoubtedly impressive and humane. Much less so, it seems to me, are his portrayals of groups huddled about the orrery or the experiments of the vacuum pump. They now seem, in effect, too obvious, too direct. But that is also their significance. Wright does speak to the historian in unique ways for his group of anxious witnesses is suffused with a light that knowledge might bring. Think of the glow as a scientific halo – no saints here, nor mangers, just artisans and curious children. It is not necessary, of course, for his audience to watch and occasionally they don't. But this light is exceptional for its intimacy and its focus, and for the way it emanates in the siting of knowledge in the workshop or the iron forge. Wright has attempted to capture a sense from the last half of the eighteenth century that the natural world was both accessible and could be tamed. That had been the message of the previous century.

To decipher this message, I have sought to recognize the powerful myth of the scientist as a man of adventure – fundamentally out of my own frustration at the bifurcation of the history of science into externalist and internalist cliches. We need to pay more attention to the range of activity inherent to natural philosophy. Although this book acknowledges the force of rhetoric, the emphasis is overwhelmingly placed on the *activities* of natural philosophers as they responded to the various technical, epistemological, religious, and sociopolitical contingencies they had to face as *practitioners* of natural philosophy.[1] Those artificial structures that

[1] Barry Barnes, *T. S. Kuhn and Social Science* (New York, 1982), pp. 114–20; Michael

isolate an intellectual life from one of action are, it seems to me, obstinately class-ridden. In the nineteenth century, defense of class meant restraint on the spread of knowledge.[2] Nature, not technology, was the proper preserve of gentlemen. There was an early demarcation between scientists and grimy mechanics that has since been transformed into a history that is, in the broadest sense, antiintellectualist – and magnifies a mirage of distance to separate the modern academic from the artisan. It is one of the ironies of the modern world, where technology is often applied science, that science might be regarded as master of simple artisans or of mere purveyors of instruments or machines. Perhaps this mastery was an image made easy by class distinction, but it was not ever thus. Where the early instrument makers fit in the rise of popular science must surely give us pause. Science, as Joseph Wright tried to show his viewers, was increasingly a means to probe nature, not to master men or reinforce social barriers.

The authority we now attach to science was the result of an intense debate in the seventeenth and eighteenth centuries that industrialism and social power seemed once to have settled. My particular concern is with the pressing question of the legitimation of science and with the presumption of beneficent mechanization that forms part of what has been described as "the grammar of industrial society." If we now look back beyond two centuries of industrial culture, this grammar resonates through the Restoration and the Glorious Revolution and in the "metaphysic of mechanization" that transformed the Western world in the eighteenth century.[3]

The universe was changing, England's Restoration gentry knew – some adapted, and some looked on in grim dismay. The fragile Restoration promise of political and social stability was soon in tatters. Even after the Glorious Revolution, the country was drifting from one crisis to another, later ones perhaps even manufactured by Robert Walpole and his Whig cronies. The Protestant Succession lived on the knife-edge of collapse and Hanoverian stability was sullied with a deep despair of corruption under the Robinocracy. At stake was the legitimacy of a world-view – in politics, as in nature. The terrible uncertainty in those same

Nerlich, *Ideology of Adventure. Studies in Modern Consciousness, 1100–1750*, trans. Ruth Crowley, Volume II (Minneapolis, 1987), pp. 269–83. The discursive power of scientific language in the eighteenth century was suggested in Michael Foucault, *The Order of Things. An Archaeology of the Human Sciences* (London, 1970), pp. 86–8, 206–7.

2 Martin J. Wiener, *English Culture and the Decline of the Industrial Spirit, 1850–1980* (Cambridge, 1981), pp. 16–21. Cf. Morris Berman, *Social Change and Scientific Organization* (Ithaca, 1978), pp. xx–xxi.

3 The phrase was that of Oskar Kokoschka. Quoted in Berman, *Social Change and Scientific Organization*, p. xviii.

winds of Revolution still haunted the Newtonian heretic William Whiston many years later. His memory magnified perhaps by time, he claimed to have heard the French guns whose distant thunder from across the marshes and the sea off Beachy-Head in Sussex had reached the windows of his Cambridge college in 1690.[4] Science could not remain immune to such circumstances. Unlike Whiston in the early eighteenth century, non-juring Roger North withdrew from Whigs, Newtonians and anti-Trinitarians to self-imposed internal exile at Rougham in Norfolk. There North sneered against the Newtonian universe, calling it a "whim piece of machinery" of the "cosmografye of the attractors."[5] This sad derivation of his sorry Stuart sympathies, rising and falling with the gravity of Jacobite plots, had denied him the validity to which Newton's apostles could aspire. Even the *Principia* symbolized a revolution in Church and State. The political moment made a universe with which Robert Walpole, North's Norfolk antagonist, would be far more at ease. Walpole secured power, whereas North shunned it.

The division of political, and hence social, power was a reflection of a revolution in the public mind – which was to be mesmerized, in the eighteenth century, by rampant capitalism and the insistence of technological promise. There was nothing progressive in this mentality; it was a swamp of opportunism, full of uncertainties and confusing directions. In seventeenth-century England, the force of machines – especially in the broader search for economic power – was still elusive. Yet, the discursive centers of this culture came to include public experiments along with powerful fountains and belching steam engines that would transfix a public. For Daniel Defoe, even the untutored, unlettered mechanic might have access to a public culture of science and technology.[6] Even so, no magic elixir made machines more acceptable in popular opinion. By 1744, the natural philosopher John Theophilus Desaguliers, son of Huguenot hostility to the Sun King, had helped tie the ideology of experiment to the force of the machine. Yet his life was overwhelmed by the contradictions produced by mechanization. The gouty Desaguliers grumbled at the end of this life that

[4] Millicent Barton Rex, *University Representation in England, 1604–1690* (London, 1954), pp. 302–20. Whiston recalled hearing the guns about a hundred miles away in 1690 while the battle raged. This, purportedly, was what gave him his method for discovering longitude. William Whiston and Humphry Ditton, *A New Method For Discovering the Longitude Both at Sea and Land, Humbly Proposed to the Consideration of the Publick* (London, 1714), p. 19.

[5] Peter T. Millard, "An Edition of Roger North's Life of Dr. John North with a Critical Introduction" (D.Phil. thesis, Oxford University, 1969), pp. 92–4.

[6] Terry Eagleton, *The Function of Criticism. From The Spectator to Post-Structuralism* (London, 1984), pp. 14, 23–4.

I have known some Persons of Fortune, who being covetous, but unwilling to own it, full of the Notion of the Difference between Theory and Practice (or excus'd by it) have hearken'd to the most ignorant Pretenders, who have call'd themselves Men of Practice, in order to save Money, when they have wanted Water to be raised, which they suppos'd those who were eminent for their knowledge that way would expect; and thought that Work done well enough and very cheap might be sufficient, as if raising of Water requir'd no more Skill than Hedging and Ditching. Thus many People employ the Apothecary to save the Charge of the Physician.[7]

There was a dissonance here, in the allure and contradictions of industrialism, that, in Max Weber's words, would ultimately "determine man's fate until the last ton of fossilized coal is burnt."[8] The assertion of the legitimacy of science has become contingent on its utility. This notion was vigorously propounded by Desaguliers, whose career provides much of the focus of this book. Equally active in the Royal Society and Exchange Alley, Desaguliers' aggressive campaign on behalf of natural philosophers demonstrated how clearly credibility and usefulness would merge in the eighteenth-century rhetoric of technology. For Desaguliers, only the proponents of the true philosophy could navigate between the Scylla and Charybdis of patents and false projects. The lure of true, Newtonian, philosophy was ultimately to be emphasized by the legitimation of the doctrine of industry in the eighteenth century.

If it is true that science has not always been an intellectual preserve uncontaminated by the rabble, then it is futile to write a history of scientific ideas simply from the texts. It is on this proposition that this book is founded. Widespread acceptance in the eighteenth century of natural philosophy as worthwhile, even useful, suggests that Thomas Birch's distinction of the humane Robert Boyle from "the vulgar herds of chemists, naturalists and philosophers" was a late lament against the

[7] J.T. Desaguliers, *A Course of Experimental Philosophy*, II (London, 1744), p. 416. A good example of the significance of the lecturers in technological change can be seen in the career of the engineer, John Grundy (ca. 1697–1760). The Spalding Gentlemen's Society recorded how "Mr. Grundy communicated his proposals for draining lands that lie near the sea, showing the necessity of mathematical and philosophical knowledge thereto required, approved by J. T. Desaguliers, D.D., F.R.S." Raymond James Evans, "The Diffusion of Science: The Geographical Transmission of Natural Philosophy in the English Provinces, 1660–1760" (Ph.D. thesis, Cambridge University, 1982), p. 243. On Grundy, see Taylor, *Hanoverian England*, p. 122; and Jacob, "Scientific Culture in the Early English Enlightenment: Mechanisms, Industry, and Gentlemanly Facts," in Alan Charles Kors and Paul J. Korshin, *Anticipations of the Enlightenment in England, France, and Germany* (Philadelphia, 1987), p. 149.

[8] Max Weber, *The Protestant Ethic and the Spirit of Capitalism*, trans. Talcott Parsons (New York, 1958), p. 181, quoted in the especially useful Marshall Berman, *All that is Solid Melts into Air. The Experience of Modernity* (New York, 1988), p. 27.

hurricane of popular knowledge.[9] One might suspect that Birch longed for the old Baconian vision of philosopher-priest. But he was out of touch with a new public world of philosophy, where commerce and public utility would ultimately overwhelm the confessional and the priest. Natural philosophy has long occupied a space – as Steven Shapin has reminded us – that sought to escape mere rhetorical images of success. And it has had its social place at least since Galileo determined to demand one.[10] The rapid multiplication of the sites of science meant the emergence, in coffee houses and in country towns, of what I have chosen to call public science. Access was constrained solely by the flexibility of the purse.

Thus, the present book is about the place science occupied in the minds and in the social reality of preindustrial Britain. Such is an exploration barely begun. Occasionally, we will find ourselves in busy alleys as much as in churches, in fetid prisons, and among the company of coffeehouses, at Court and in the bleak lead mines of Lanarkshire or on the sullen moors of Northumberland. In all of these was done the worldly business of experimental philosophy. These were the venues of the eighteenth-century philosophers. It was these men who gave the texts, whether of Isaac Newton or of Samuel Clarke, their place in the library of a remote country house or read while crossing a wind-lashed sea to Holland. Their bindings now foxed, pages worm-stained, these texts once had a place. It is their place beyond the library and their purpose among the itinerant engineers or the listeners of a public lecture that will concern us here.

Structure

The legitimation of science required more than texts. So the polemic of science took an odd – and exceedingly important – turn. The power and the fragility of words were recognized as soon as the Royal Society emerged from the Restoration. Bishop Thomas Sprat's *History of the*

[9] Thomas Birch, ed.,"The Life of the Honourable Robert Boyle," in *The Works of the Honourable Robert Boyle* (London, 1772), Vol. I, cl.

[10] Steven Shapin, "The House of Experiment in Seventeenth-Century England," *Isis*, LXXIX (1988), 373–404; Shapin, "Robert Boyle and Mathematics: Reality, Representation, and Experimental Practice," *Science in Context*, II (1988), 23–58, esp. 42–3, 50–2; Pietro Redondi, *Galileo Heretic*, trans. Raymond Rosenthal (Princeton, 1987), pp. 51–9; Elizabeth L. Eisenstein, *The Printing Press as an Agent of Change. Communications and Cultural Transformations in Early-Modern Europe* (Cambridge, 1980), pp. 477–8, 529–30, 679–80; and Mario Biagioli, "Galileo's System of Patronage," *History of Science*, XXVIII (March 1990), 1–62. Cf. Margaret C. Jacob, *The Cultural Meaning of the Scientific Revolution* (Philadelphia, 1987), pp. 19–31. I am much indebted here to Steven Shapin and Simon Schaffer, "Making Newton: On Interpreting Scientific Texts," typescript.

Royal Society of 1667 was a book written, ironically, against words. Sprat understood that material reality must overcome rhetoric, that natural philosophy could not survive wrapped in a cocoon of words that claimed more than nature could be made to surrender. The search for the secrets of nature had once been bound tight to the dreams of metaphysicians. The antischolastic rhetoric of the Royal Society, and the example of its predecessor the Accademia del Cimento, was really a quest for a more authentic and legitimate nature. The book of nature must replace the libraries of all authorities. This was the first basis for making nature the ground of a far-reaching epistemology; there was little enough in words, except to make men enemies. Such was a refrain that echoed throughout the eighteenth century. Indeed, I would maintain that the deliberate employment of scientific apparatus in *public* spaces – even to the point of reproducing, in texts, lists of experiments to be done – was an effort to establish an epistemology of common experience.[11] In effect, this effort was one important link between the enthusiasm of Thomas Sprat and its fruition among the public philosophers of the eighteenth century.

Rhetoric, in whatever guise, gave rise to an unease over the whims of the mob. This was a concern that could not sit well in a world so recently recovered from civil strife. As Brian Vickers' study of rhetoric points out, efforts to influence the many have long fueled the suspicion of democracy, of popular prejudices, and of the "flattery and corruption" that made "the masses" or "the vulgar" more pejorative words than might be necessary. This fluidity of meaning, in reference to fears of a popular culture, was also rhetorical and served a political purpose.[12] Just as or-

[11] See J. T. Desaguliers, *Physico-Mechanical Lectures. OR, An Account of What is Explain'd and Demonstrated in the Course of Mechanical and Experimental Philosophy . . . Wherein the Principles of Mechanics, Hydrostatics and Opticks, are demonstrated and explain'd by a great Number of Experiments. Design'd for the Use of all such as have seen, or may see Course of Experimental Philosophy* (London,1717), pp. 75–8; and William Vream, Pneumatical Instrument Maker, *A Description of the Air-Pump. According to the late Mr. Hawksbee's best and last Improvements; With the manner of making Fifty of the most curious Experiments upon it: The Figures of the Air-Pump-Glasses, and all the Machines belonging to it, being curiously engraven in Copper Plates* (London, 1717). See also Pietro Redondi, "Theology and Epistemology in the Scientific Revolution," in William R. Shea, ed., *Revolutions in Science. Their Meaning and Relevance* (Canton, 1988), pp. 108–9; Jan V. Golinski, "Robert Boyle: Scepticism and Authority in Seventeenth-Century Chemical Discourse," in Andrew E. Benjamin, Geoffrey N. Cantor, and John R. R. Christie, eds., *The Figural and the Literal. Problems of Language in the History of Science and Philosophy, 1630–1800* (Manchester, 1987), pp. 65, 77. See also Murray Cohen, *Sensible Words. Linguistic Practice in England, 1640–1785* (Baltimore and London, 1977), Chapter I; and Desaguliers, *Course of Experimental Philosophy*, second edition (London, 1745), I, 1.

[12] Brian Vickers, *In Defence of Rhetoric* (Oxford, 1988), pp. 87–9.

atorical skill was transformed by the printed world, so too was the literal extended by the drama of the experimentalists. The creation of a public place for natural philosophy had to face the issue not merely of the skill of argument, but also of the composition of its audience. In the Restoration, this caused difficulties for both Sprat and the Royal Society.

In rhetoric, I assume both an argument and an audience. The argument prompted by purveyors of natural philosophy was one that was increasingly linked to assertions of the public good – whether that good be religious or economic. Moreover, it must be clear that experimental natural philosophy required an audience. Even experimental texts in the eighteenth century assumed as much; the lists of experiments to be done were fundamentally extensions of experiments already completed. It was not merely the replications that mattered, but the expansion of audience as well.[13] The rhetoric of science took on the new form of public disclosure that made the secrets of the natural world available to all prepared to subscribe to lectures or sometimes to support the collection plate at sermons. Hence, public science is a notion that is constructed out of the intent of the experimental method of the late seventeenth and early eighteenth centuries and from the purpose of its practitioners. We cannot then be confined by the rhetorical disputes of the texts themselves; rather it will be necessary to move beyond increasingly barren debates to seek the ground of sanction for science in the eighteenth century.

The common discourse over natural philosophy was constructed on two important themes explored by Sprat in 1667 and unwittingly reinforced by his critics like Henry Stubbe. Sprat had set the agenda – as indeed he has for this book. Part I examines the crucial relationship between religion and the new philosophy within the power of the text, even if this led to the dry paths of the very metaphysics Sprat was loath to explore. It is worthwhile emphasizing here that the force of natural theology at the end of the seventeenth century insisted on the prominence of providence. It is a remarkable fact that a kind of economic providentialism emerged in the seventeenth century that seemed to attenuate private interest in the service of the public good. Providence, to put this in a current light, promised wealth might tickle down if not upon the un-

[13] See, for example, *A Catalogue of the Experiments in Mr. Desaguliers's Course* [dated December 14, 1713], British Library, Shelf mark C.112.f.7; J. T. Desaguliers, *Physico-Mechanical Lectures, OR, An Account of what is Explain'd and Demonstrated in the Course of Mechanical and Experimental Philosophy...* (London, 1717), pp. 75ff.; and the lists of courses of experiments appended to William Vream, *A Description of the Air-Pump, According to the late Mr. hawksbee's best and last Improvements...* (London, 1717), pp. 1–17; and to Peter Shaw and Francis Hawksbee, *An Essay for Introducing a Portable Laboratory...* (London, 1731), pp. 27ff.

washed at least upon the worthy. This we will explore at length for, I believe, providence proved crucial in demonstrating the importance of natural philosophers – if their authority could be established, they might provide an effective prophylaxis against the ravages of projectors.

If Thomas Sprat had hoped to avoid censure, he was hopelessly mistaken. His *History* was immediately the target of Stubbe's censure upon a Society he prayed would have no successor – lest "they were making way for the introducing of something else besides a New Philosophy."[14] Stubbe's hints at heterodoxy were precisely what Sprat's virtuosi sought anxiously to avoid.[15] There were great social issues here to which natural philosophers might hold the key.

Heterodoxy haunted the new philosophy. It was here that the texts, and the sermons turned into texts, demonstrated their own futility. Their range remained narrowly focused on the troubles of the Established Church. Part I, therefore, explores the agenda of the natural theologians. Their efforts, while understandable enough in the defense of scripture, swam in the same sectarian swamp that produced civil unrest and endangered Church and King. As in all sectarian rancor, it was easy for a few disputants to narrow the ground where salvation was possible. The later presumptions of doctrinal innovation amongst Newton's disciples, like William Whiston and Samuel Clarke, and accusations of deliberate misrepresentations of scripture and of each other, settled absolutely nothing.[16] Contrary to what might be expected of these largely clerical efforts to seek public approval, the textual enterprise suitably appeared shallow and suspect.

A great deal of recent attention has been paid to the rhetorical efforts to obtain assent for the scientific enterprise. The link between the methods of science and the texts that assert discoveries was a complex process that united a public audience with experiment and credibility.[17] Part II is about

[14] Henry Stubbe, *Legends No Histories: Or, A Specimen of Some Animadversions Upon the History of the Royal Society* (London, 1670), pp. 16–21.

[15] Tho. Sprat, *The History of the Royal Society of London, For the Improving of Natural Knowledge* (London, 1667), pp. 25–7.

[16] See Edward Wells, *Remarks on Dr Clarke's Introduction to his Scripture-Doctrin of the Trinity* (Oxford, 1713), pp. 38–9.

[17] J. R. R. Christie and J. V. Golinski, "The Spreading of the Word: New Directions in the Historiography of Chemistry 1600–1800," *History of Science*, XX (1982), 235–66; G. Nigel Gilbert and Michael Mulkay, "Experiments Are the Key. Participant's Histories and Historians' Histories of Science," *Isis*, LXXV (March 1984), 105–25, and the reply by Steven Shapin, "Talking History: Reflections on Discourse Analysis," *ibid.*, 125–8; Steven Shapin and Simon Schaffer, *Leviathan and the Air Pump: Hobbes, Boyle, and the Experimental Life* (Princeton, 1985); Peter Dear, "'Totius in verba.' Rhetoric and Authority in the Early Royal Society," *Isis*, LXXVI (June 1985), 145–61; Brian Vickers,

the transition to public knowledge. Restoration assertions of the connection between the defense of religion and the usefulness of natural philosophy were difficult to prove. In one respect, this was fundamentally a problem of epistemology, perhaps of the relation between reason and revelation, but more critically of the establishment of a convincing, and occasionally dramatic, experimental method – especially when consent was sought through published reports and public lectures. Here the books of experiments and apparatus were to have an important place. The *Philosophical Transactions*, no less than Sprat's *History*, were efforts to create a form of knowledge that would obtain wider consent than the limits of the laboratory could easily allow.

This is obviously more than an issue of scientific community. It is rather a question of the force of method, and of the production of what sociologists of science have called the artifactual – or the mechanical manufacture of philosophical facts within the laboratory. Experiment was everything. Such a process became truly revolutionary through the expansion of the sites of knowledge, thereby making possible the alteration of public attitudes to both natural philosophy and technology. Experimental demonstrations, say, of a vacuum or of electricity, also helped to render scientific specialization irrelevant. Consequently, reservations about the artificial production of natural events – whether from Hobbes or the later Hutchinsonians – reflected an essential distrust of the democratization of natural philosophy and, thus, of nature. This was partly the result of recognizing the consequences of new forces ready to embrace new philosophy. The most important achievement in natural philosophy in the eighteenth century was a burgeoning public interest. Nature, once marginally controlled by law and by the lord's bailiffs, had entered the realm of public property. The production of knowledge, of scientific

"The Royal Society and English Prose Style: A Reassessment," in Brian Vickers and Nancy S. Struever, *Rhetoric and the Pursuit of Truth. Language and Change in the Seventeenth and Eighteenth Centuries* (Los Angeles, 1985), pp. 3–76; Bruno Latour, *Science in Action* (Cambridge, Mass., 1987), Chapter 1; Steven Shapin, "Robert Boyle and Mathematics: Reality, Representation, and Experimental Practice," *Science in Context*, II (1988), 23–58; Simon Schaffer, "Glass Works: Newton's Prisms and the Uses of Experiment," in David Gooding, Trevor Pinch, and Simon Schaffer, eds., *The Uses of Experiment. Studies in the Natural Sciences* (Cambridge, 1989), pp. 67–104; and Geoffrey Cantor, "The Rhetoric of Experiment," in *ibid.*, pp. 159–80. For a recent study of rhetoric and Sprat's *History*, see Alan Gross, *The Rhetoric of Science* (Cambridge, Mass., 1990), pp. 165–8. After this work was completed I encountered Yaron Ezrahi, *The Descent of Icarus. Science and the Transformation of Contemporary Democracy* (Cambridge, Mass., 1990), which explores the consequences of the achievement of public science.

facts, increasingly depended upon the victory of a market model of public competition and consumption.[18]

This model exploded with manifold force in the eighteenth century; the unification of the rhetoric of science with the language of power was the means by which science emerged into public consciousness. Without experimentation, without the structures of instruments and laboratories, a public science was inconceivable. It is worth noting the recent remark of Geoffrey Cantor: "Scientific narrative in general and experimental reports in particular are *rhetorical*, in the accepted sense of the term, since they aim to persuade or influence. This makes scientific discourse a discourse of power. This point is frequently overlooked. . . . Far from being rhetoric free, modern scientific prose has become the most potent instrument of persuasion in our culture" – especially, one might add, when it moved beyond the text into public lectures.[19]

The enthusiasm of improvement entranced the early apologists of the Royal Society. Part III deals with this transformation – from the dissipation of religiosity to the affirmation of experiment and of mechanical power. This was based, increasingly, on the acknowledgment of the value of particular interests demonstrated in an enthusiasm for technical innovation. The discursive defense of the philosophers and their Society was also frequently founded on the technical promises Sprat assumed inherent in the new philosophy. The florid prose of Joseph Glanvill, recently installed as rector of Bath, recommended the pleasures of philosophy and of mechanics, "fruitfull in works, not words," to the sons of the wealthy who might improve their estates. By the turn of the century, it was clear patent seekers had been listening; petitioners threatened to overwhelm the Royal Society.

Henry Oldenburg, Secretary to the Society, expressed some nervousness that Glanvill's polemics might raise excessive expectations, although he was encouraged by the intention to defend experimental philosophers "by representing the advantages of this way of trial, both *for light and use*, above that of former times."[20] But more than the gentry and a landed

[18] Karin D. Knorr-Cetina, "The Ethnographic Study of Scientific Work: Towards a Constructivist Interpretation of Science," in Knorr-Cetina and Michael Mulkay, eds., *Science Observed. Perspectives on the Social Study of Science* (London, 1983), pp. 117–20, 128–32. See also Bruno Latour, "Give Me a Laboratory and I will Raise the World," *ibid.*, esp. pp. 153–6; Steven Shapin and Simon Schaffer, *Leviathan and the Air-Pump. Hobbes, Boyle, and the Experimental Life* (Princeton, 1985), Chapters VII and VIII; and Latour, *Science in Action* (Cambridge, Mass., 1987), pp. 103ff.

[19] Geoffrey Cantor, "The Rhetoric of Experiment," p. 161.

[20] My italics; H. Oldenburg to R. Boyle, October 1, 1667, in Thomas Birch, ed., *The*

oligarchy might benefit. The broader run of men might be persuaded that "there is worthier Imployment for them, than tying knots in bulrushes; and that they may be better accommodated in a well-built House, than in a Castle in the Air."[21] This refrain proved equally problematical. Nevertheless, Sprat's defense clearly recognized that the survival of the Royal Society and of experimental philosophy could not be determined by the broadsides endlessly evoked from the imaginations and anguish of churchmen and pamphleteers.

The impact of the texts would not be confined by the edges of literacy. And, in the seventeenth century, it was not books alone that would reveal the secrets of nature. But how was public knowledge to be formed if not by books? Access to the Royal Society was equally perplexing. As Michael Hunter has shown, there was enormous ambivalence in some quarters to public access frequently leading to calls for secrecy in meetings and in 1674 even for the exclusion of "troublesome and prejudicial" visitors.[22] Yet there was also some recognition of the inadequacy of the printed word – in the injunction of the Committee for Mechanicks "to take notice of what is practised and not found in Books. . . ."[23] In the gathering of information, this may have seemed obvious, but the process might also be reversed. While artisans and mechanics might disclose the secrets of trades, they were also potentially beneficiaries of such exchanges. Inherent in the search for improvement was the assumption of the diffusion of science as well.

The role of natural philosophy would require the sorting out of the place of a public audience. The difficulty in doing so was revealed in the institutional tensions of the Royal Society – over the distance between a scientific elite and vulgar mechanics. Partly this would depend upon the Fellows' image of their role, partly upon the esoteric nature of some of

Works of Robert Boyle, VI, 243. The phrase was repeated the following year in Joseph Glanvill, *Plus Ultra: Or, the Progress and Advancement of Knowledge Since the Days of Aristotle* (London, 1668), p. 52. See also Glanvill, *Scepsis scientifica: Or, Confest Ignorance, the way to Science; In an Essay of the Vanity of Dogmatizing, And Confident Opinion. With a Reply to the Exceptions of the learned Thomas Albius* (London, 1665), "Dedication" to the Royal Society; and Glanvill, *Scirei tuum nihil est: Or, The Authors Defense of the Vanity of Dogmatizing* (London, 1665), pp. 48–9; Sprat, *History*, p. 349; and Oldenburg to Boyle, December 10, 1664, in Thomas Birch, ed., *The Works of Robert Boyle*, VI, 185. Cf. B. C. Southgate, "'Forgotten and Lost': Some Reactions to Autonomous Science in the Seventeenth Century," *Journal of the History of Ideas*, L (April-June 1989), 260–2.

21 Glanvill, *Plus Ultra*, p. 91.

22 Michael Hunter, *Establishing the New Science. The Experience of the Early Royal Society* (Woodbridge, 1989), p. 218.

23 Hunter, *Establishing the New Science*, p. 116 (October 17, 1664).

their interests. Yet by the early eighteenth century, if Newton was to prove utterly unintelligible to the majority of readers, he had legions of translators and expositors. Various digests of Newton's philosophy, numerous displays of mechanical and philosophical principles, and public lectures galore gave an entirely new voice to the market in natural philosophy. But it was to be a market defined in large measure by religion and by the conquest of nature in the syllabus of Thomas Sprat.

This book affirms Sprat's purpose. It also adopts the ground surveyed by Sprat's contemporary John Evelyn in his 1667 ironic view of *Public Employment, and An Active Life, With all its Appanages... Preferr'd to Solitude*. Evelyn might have had a political purpose, to isolate from the active philosophy of a Royal Society those such as the Fifth Monarchy Men, those "highest affected with solitude and devout enthusiasme... whilst they breathe nothing save ruin and destruction." Solitude might induce such dementia. In contrast, he asserted, "the wisest men are not made in chambers and closets crowded with shelves, but by habitude and active conversations."[24] The debate between private and public, between vice and virtue, would have implications for the practice of natural philosophy. I have accepted that the *place* of discourse proved crucial to the creation of a public science. This book, thus, seeks out an experiential culture as prelude to a broadly literate one – where technique and nature were to be united in the assertion that the world was for man's use and not merely for his amusement. It is about the pleasures and problems of a contemplative life, but also about the efforts of many whose later interest in Newton's philosophy helped to break the bounds of what were once seen as the limits of nature.

The fleeting glimpses we catch of scientific lecturers near the Royal Exchange or in the salons of Bath tell us little of the dynamism of Newtonian Britain. We are forced to recognize that by the early eighteenth century, science had become part of the frame of a culture. So natural philosophy could not remain a narrow academic, intellectual issue or one confined to the miasma of metaphysical speculation; rather it reflected more prescient economic and social tensions within Britain after the Revolution. The fashions and fallacies *ex libri* were increasingly anemic in the face of mechanics. Making science public was itself an intellectual

[24] J. E. Esq. F.R.S., *Public Employment, and An Active Life, With all its Appanages, Such as Fame, Command, Riches, Conversation, &c. Preferr'd to Solitude*, in William Upcott, ed., *The Miscellaneous Writings of John Evelyn, Esq. F.R.S.* (London, 1825), pp. 501–52, esp. 523, 537. On Evelyn's intentions, see Brian Vickers, ed., *Public and Private Life in the Seventeenth Century: The Mckenzie-Evelyn Debate* (Delmar, New York, 1986), esp. introduction.

revolution that preceded the force of industry. Natural philosophers were increasingly determined to make themselves heard. On an empirical level, Newtonianism entered the world of practical men through the efforts of a few individuals who refused to limit themselves to the disputes with Leibniz. The points of contact were initially found in the coffeehouses that brought entrepreneurs and financiers together with the wits and politicians of London and Westminster. But the transition induced other tensions.

From Galileo to Newton, there had been an aversion to the vulgar sort. Not for Joseph Glanvill the "Transient view" of vulgar opinion; there was nothing quite so common as a commonweal.[25] Having survived the rebellion and the Restoration, Glanvill took pride in the antifanatical divines of the Royal Society, who "had no ambition to be cry'd up by the common Herd."[26] But from the origins of the Royal Society, there had been ambivalence, if not outright confusion, over popular knowledge. The sociopolitics of a commonwealth of learning was one factor in the failure of the Society's program for a History of Trades.[27] There was much suspicion and dismay over public knowledge of the natural world that compounded a distrust of what, if anything, the vulgar might offer.[28]

Yet the influential Robert Boyle argued continually that philosophers might "meliorate the Inventions of illiterate Tradesmen" as much as they might depend on the apparatus of instrument makers.[29] By the eighteenth century, it was increasingly evident that not all philosophers shared the suspicion of the masses. In 1738, Robert Walpole's good friend Robert

[25] Joseph Glanvill, *The Vanity of Dogmatizing: Or Confidence in Opinions Manifest in a Discourse of the Shortness and Uncertainty of our Knowledge, And its Causes: With some Reflexions on Peripateticism; and An Apology for Philosophy* (London, 1661), pp. 64–6, 76–7, 246.

[26] Glanvill, *Essays on Several Subjects in Philosophy and Religion* (London, 1676), p. 42.

[27] Michael Hunter and Paul B. Wood, "Towards Solomon's House: Rival Strategies for Reforming the Early Royal Society," *History of Science*, XXIV (1986), 49–108; and J. V. Golinski, "A Noble Spectacle. Phosphorous and the Public Cultures of Science in the Early Royal Society," *Isis*, LXXX (1989), esp. pp. 11–15, 38–9.

[28] Keith Thomas, *Man and the Natural World. Changing Attitudes in England 1500–1800* (London, 1983), pp. 78–86; see the commentary in Alan Macfarlane, *The Culture of Capitalism* (Oxford, 1987), pp. 79–83.

[29] Robert Boyle, *Some Considerations Touching the Usefulness of Experimental Natural Philosophy: Proposed in a familiar Discourse to a Friend, By Way of Invitation to the Study of it*, second edition (Oxford, 1664), p. 20; Boyle, *Some Considerations Touching the Usefulness of Experimental Natural Philosophy*, second tome (Oxford, 1671), "The Second Section." "Of its Usefulnesse to the Empire of Man over inferior Creatures," pp. 2, 4, 30–1; "Of the Usefulness of Mechanical Disciplines to Natural Philosophy," p. 1; Boyle, *The Excellency of Theology, Compared with Natural Philosophy* (1674) in Birch, *Works of Robert Boyle*, IV, 116–17. Note the similar comment by Desaguliers in *A Course of Experimental Philosophy*, second edition (London, 1745), I, 70–1.

Smith wrote forcefully about the importance of popular optics, with proofs drawn from experiments, especially for those familiar with mechanical devices.[30] As Benjamin Martin was also to show, the literate and the curious were not to be quarantined from philosophy by the portals of Crane Court. The power and significance of the market were precisely what overcame the isolation of the crowd from the prestige of the Royal Society – that crowd, at least, willing to pay to be part of the anointed. Thus, Parts II and III of this book explore the community of science and technology where the realm of public science was to be defined, largely, by the efforts of Newton's disciples.

Constructing public science

By the time of Newton's death in 1727, his work was available to a broad range of the reading public.[31] However, even with an increase in literacy, the public was hardly likely to approach the formidable *Principia*, and certainly not without a guide.[32] The greater opportunity provided for exposure to Newton's experimental science through the publication in 1704 of his *Opticks* in English had a considerable effect. In addition, the wider glorification of Newton through poetry can be traced to the non-mathematical suggestions of the *Opticks* as well as to the active principles of the *Principia*. Newton's *Opticks*, however, which lent itself more readily to the mathematically naive, also led to a program of the replication of controversial experiments. Simon Schaffer has recently shown that Newton's experimental method and its difficulties were intimately involved with the achievement of consensus in the scientific community.[33] But we must consider that the formation of consensus over the philosopher's lasting reputation in the wider society was one significant objective of Newton's popular disciples. All of this was to contribute to the deification of Newton. This will bring us, inevitably, to consider the

[30] Jan V. Golinski, "Robert Boyle: scepticism and authority," p. 66; and Jacob, *The Cultural Meaning of the Scientific Revolution*, pp. 19–31. Robert Smith, *A Compleat System of Opticks in Four Books, viz. A Popular, A Mathematical, a Mechanical, and a Philosophical Treatise* (Cambridge, 1738), "Preface." See also Desaguliers, *A Course of Experimental Philosophy*, second edition, corrected (London, 1745), I, viii–ix.

[31] I. Bernard Cohen, *Introduction to Newton's Principia* (Cambridge, Mass., 1978), pp. 138, 284.

[32] R. M. Wiles, "Middle-Class Literacy in Eighteenth-Century England: Fresh Evidence," in R. F. Brissenden, ed., *Studies in the Eighteenth Century* (Toronto, 1968), pp. 49–65; David Cressy, *Literacy and the Social Order. Reading and Writing in Tudor and Stuart England* (Cambridge, 1980), Chapters I–III, *passim.*

[33] Simon Schaffer, "Glass Works: Newton's Prisms and the Uses of Experiment," esp. p. 92.

intellectual tension that necessarily occurs between the text, even when based on an experimental epistemology, and the society that must interpret it.[34]

Thomas Sprat's *History* in 1667 had acknowledged the force of social sanction. And there was more than a tacit recognition of this social power when Descartes was later overcome by gravitation. Such were the warnings of Richard Marsh, a Fellow of St. John's College, Cambridge, in 1699, when he preached about the vanity of fashions in philosophy. By the end of the century, Marsh complained that because of new speculations, "Men are turning Levellers in Religion, as they were of old in Government."[35] Such fashions were to be a powerful force in the growth of scientific education. Edward Wells, the rector of Cotesbach, Leicestershire, later stated as much. Wells was the author of numerous works of instruction, including *The Young Gentleman's Astronomy* in 1718 in which it was explained how the universe was now seen to be preserved by gravity alone.[36] Wells' reluctant Newtonian acknowledgments reflected the vogue in natural philosophy, or at least the adoption of the principles that were perceived to provide a successful science.

There was no neutrality in such discussions. That is why the public lectures became so critically important; they revealed a Newtonian universe that was largely empty, populated only in small part by passive matter and tended by active powers. In this case, it was not necessary to enter into debates of metaphysical principles that would be barely understood, if at all. Here, in the lectures, was the evidence of attraction as Voltaire's "great Spring by which all Nature is mov'd" of which even Newton "plainly foresaw that its very Name wou'd offend."[37] It could

[34] See Steven Shapin, "Social Uses of Science," in G. S. Rousseau and Roy Porter, *Ferment of Knowledge. Studies in the Historiography of Eighteenth-Century Science* (Cambridge, 1980), pp. 93–105; Marjorie Hope Nicolson, *Newton Demands the Muse: Newton's Opticks and Eighteenth Century Poets* (Princeton, 1966), p. 15; and Steven Shapin and Simon Schaffer, *Leviathan and the Air-Pump*, esp. Chapter VIII.

[35] Richard Marsh, *The Vanity and Danger of Modern Theories. A Sermon Preach'd at St. Marys Church in Cambridge, On Sunday the 13th Day of August, 1699* (Cambridge, 1699), pp. 5–6, esp. p. 8. Cited in Geoffrey Bowles, "The Place of Newtonian Explanation in English Popular Thought, 1687–1727," (D.Phil. thesis, Oxford University, 1977), p. 313. See also John Gascoigne, *Cambridge in the Age of the Enlightenment. Science, Religion and Politics from the Restoration to the French Revolution* (Cambridge, 1989), p. 166. Such comments were not new and were often repeated. Southgate, "'Forgotten and Lost': Some Reactions to Autonomous Science in the Eighteenth Century," pp. 257–8.

[36] Edward Wells, *The Young Gentleman's Astronomy*, second edition (London, 1718), pp. 13–14; *D.N.B.*, LX, 227.

[37] "Ainsi viola l'attraction qui est le grand ressort qui fait mouvoir toute le nature." In Voltaire, *Lettres ecrites de Londres sur les Anglois et autres sujets*, XV, "Sur l'attraction"

not be otherwise that science in the early eighteenth century made the transition from metaphysics to a social and intellectual presence inherited from notions of the frame of the world or of the public good in natural theology. How else could one account for the delights of *The Spectator* in 1712?

> There is something very engaging to the Fancy, as well as to our Reason, in the treatise of metals, minerals, plants, and meteors. But when we survey the whole Earth at once, and the several planets that lie within its neighbourhood, we are filled with a pleasing astonishment, to see so many worlds hanging one above another, and sliding round their axes in such an amazing pomp and solemnity. If, after this, we contemplate those wild fields of Aether, that reach in height as far as from Saturn to the fixed stars, and run abroad almost to an infinitude, our Imagination finds its capacity filled with so immense a prospect, and puts itself upon the stretch to comprehend it. . . . [38]

Whether the *Spectator* merely reflects an imperfect understanding of Newton, a residual Cartesianism, or the poetic mystification of nature is impossible to tell. What it does reveal, however, is that musings about natural philosophy had become the subject of public contemplation. What remains is the central issue of the form that interest in Newtonianism would express in the social world.

A comprehension of the power of Newton's natural philosophy, therefore, expanded beyond the colleges, or Crane Court, or even beyond the subculture of instrument makers in Fleet Street. As Roy Porter has shown, the social fluidity and the commerce of rationality in enlightenment England rested on the presumptions of the concrete, the practical, and the entertaining. The efforts of those like Joseph Addison and Richard Steele in the coffeehouses, the *Spectator* could easily reflect, forced "Philosophy out of Closets and Libraries, Schools and Colleges."[39] The sanction of natural philosophy came to rest in a far wider community than literacy

(Basle, 1734), p. 131; written between 1728 and 1730 and published first in England as *Letters Concerning the English Nation* (London, 1733), pp. 136–7; Simon Schaffer, "Natural Philosophy and Public Spectacle in the Eighteenth Century," *History of Science*, XXI (1983), 4–6.

[38] *The Spectator* (July 2, 1712); and Edward A. Bloom and Lillian D. Bloom, *Joseph Addison's Sociable Animal* (Providence, 1971), *passim*. This reference to the ether in 1712 is curious because Newton was not to explore the issue at length until the new queries were added to the Opticks in 1717. Cf. Arnold Thackray, *Atoms and Powers. An Essay on Newtonian Matter-Theory and the Development of Chemistry* (Cambridge, Mass., 1970), pp. 26–8.

[39] Roy Porter, "The Enlightenment in England," in Porter and M. Teich, eds., *The Enlightenment in National Context* (Cambridge, 1981), pp. 5–14. Cf. John H. Brooke, "Why did the English mix their Science and their Religion?" in Sergio Rossi, ed., *Science and Imagination in XVIIIth-Century British Culture* (Milan, 1987), pp. 58–66.

alone might lead us to expect; the liberty of the coffeehouses may have been one reason, just as certainly as the rising cult of money was another.

As Newton's apostles like John Theophilus Desaguliers and John Harris reveal, there was a Newtonianism in the first decades of the eighteenth century that soon proved aggressively practical in its outlook. If we do not accept this, we are reduced to claiming the success of Newton rested solely on the merit of his natural philosophy. Even if we were to suppose praise followed brilliance, such a notion tells us nothing of why Newton's science captivated eighteenth-century readers and the coffeehouse patrons who could hardly have been expected to judge such merit. There was nothing at all obvious about the *Principia*. In order to appreciate the social impact of natural philosophy, at any point, it is necessary to admit that a wider society exists – in this case, beyond the Royal Society.[40]

This means we must pay more heed to Newton's expositors and translators than has been the case. To be sure, this defines his disciples as men of the second or third rank, never quite reaching, if they ever intended, to the pedestal upon which the greats have been placed. This, however, distorts the victory of Newtonianism. Newton's prism experiments entered the realm of folklore, and his universal gravitation and attractions were recognized throughout a broad social spectrum, precisely because of the efforts of his apostles. Thus, to understand how Newton's philosophy came to predominate, and what role it played in the eighteenth century, it is essential to appreciate the agenda of his popularizers as well. Only by doing so will it be possible to grasp the forces of the marketplace. Newton's disciples frequently embraced the experimental method as a means to explain Newton's principles, and to lay claim to that method of philosophical speculation, but to do so in a manner that was resolutely dramatic and entertaining. By this means, Newton's philosophy entered a public space.

It also meant that the shape of Newtonianism changed to fit the nature of public interest. A prism experiment might be as remarkable as an eclipse, but it also must have meaning – especially, in this instance, to those who had little contact with the initiates of the Royal Society. It was a consequence of the agendas of both Newton's group of apostles and their audiences that a great emphasis was placed on the utility of the principles of true philosophy. Hence, Newtonianism was soon represented in the public world as holding the keys to the solution to a wide

[40] Cf. Margaret 'Espinasse, "The Decline and Fall of Restoration Science," in Charles Webster, ed., *The Intellectual Revolution of the Seventeenth Century* (London, 1974), pp. 364–8.

range of obstacles in mechanics, mining, hydraulics, and various technical enterprises. Thus, we need to know how the mathematical practitioner, the instrument maker, the trader, and the entrepreneur, and sometimes those who were all of these, could be captivated by the descriptions and demonstrations, and perhaps the promises, of Newton's followers.[41] It was their adoption of experimentalism that made public philosophy possible. Seeing truly did become believing.

If the reader has come this far, I want to be absolutely clear about my purpose. My intention is an interrogation of the Newtonian social and intellectual universe. I first wish to show that the disputes over the origin of the world or over Christian doctrine were increasingly desperate, but with a Byzantine intensity perhaps only the devout or the demented could really appreciate. The veil of reason soon decayed to reveal far too many metaphysical torments for most to care in a world that was to be defined essentially by the forces of matter and a grasping materialism. Matter had its attractions in both of these senses – hence the second purpose of this book. Religious tensions and natural theology might continue to inform the charged atmosphere of college politics in the eighteenth century. Yet, while assiduously turning every stone in the violence over church gospel and the structure of the world, the real action for many philosophers had moved elsewhere.

It is crucial to move beyond rhetoric and rationalization into the actions of the proponents of the new philosophy. Thus, the archeology of popular science is an exploration of the ground upon which the natural philosophy of the Restoration sought to unify the ideal with the actual, and explicitly to demonstrate the connection of nature with the public good. In the world of the early Royal Society and of Restoration Anglicanism, conquest and credibility were intimately tied together. The particles of matter and a material social world revealed a mutual attraction. It may very well be the case that there is a great anonymity in public philosophy, in the "joint-stock company," as Roland Barthes describes the bourgeoisie.[42] Yet it remains possible for the historian to examine the intentions of those who produced this public culture in the seventeenth and eighteenth centuries.

Natural theologians encouraged the continued assertion of the doctrine of improvement. Improvement became not so much an ideology as an epistemological conclusion – and, as we shall see, it frequently blurred

[41] Cf. G. S. Rousseau, "Science books and their readers in the eighteenth century," in Isabel Rivers, ed., *Books and their Readers in Eighteenth-Century England* (Leicester, 1982), pp. 208–17.

[42] Roland Barthes, *Mythologies*, trans. Annette Lavers (London, 1989), pp. 149–52.

the boundaries between the philosopher and the projector. The continued promotion of experiment from Boyle and Newton to their apostles essentially enhanced the credibility of conquest as a reasonable and possible achievement. Thus, the world of the public lecturer and experimenter was prescient with meaning for the acceptance of natural philosophy and, through its practitioners, of the legitimacy of the domination and manipulation of nature upon which a materialist society came to rest. The industrialization of eighteenth-century Britain was as much a function of this attitude as a response to economic or technical factors.

This book, then, is about the formation of science as culture. It is not about science as an additional factor to the social world, still less about social factors merely intruding on an otherwise objective and rational activity. To understand this may prove something of a problem. As Michael Fores has recently argued, the convergence of notions of the utility of science and rational, scientific behavior has established a mythology, a "secular theology" that has included Newton in the establishment of the edifice of Reason.[43] But why should this have been the case? The clues clearly reside in Newton's disciples themselves. As Samuel Johnson incisively pointed out, "Every new System of Nature gives Birth to a Swarm of Expositors, whose Business it is to explain and illustrate it, and who can hope to exist no longer than the Founder of their Sect preserves his Reputation."[44] He was thinking expressly of the voluminous studies of experiments and natural philosophy, and the opportunism, we are about to explore.

[43] Michael Fores, "Constructed Science and the Seventeenth Century 'Revolution,' " *History of Science*, XXII (1984), esp. pp. 237–9.

[44] Samuel Johnson, *The Rambler*, no. 106 (Saturday, March 23, 1751).

PART I

For light

But Life did never to one Man allow
Time to Discover Worlds, and Conquer too;
Nor can so short a Line sufficient be
To fathome the vast depths of Natures Sea; . . .

– Abraham Cowley (1608–67) on Bacon
"To the Royal Society"
in Thomas Sprat, *History of the Royal Society* (1667)

1

Deeds, not words

When the Royal Society adopted the motto "*Nullius in verba*" it sought to make a public declaration. This was fostered, of course, by the anxiety over sectarian violence forcing its way into its deliberations; natural phenomena must be allowed to speak for themselves.[1] More importantly, the Society was also declaring that it was not the words, hypotheses, or theories that mattered in natural philosophy. The observance of nature and the probing of her by experiment became a more defensible position than repetition of interminable disputes between sceptics and dogmatists. The texts of the controversialists or of the scholars could no longer be given preeminence. Nature was now to have its public place.

The early Royal Society

Science has long been a matter of much public dispute. No sooner were early scientific cabals formed than they were subject to vilification and controversy. When, shortly after the group received its charter, Bishop Thomas Sprat wrote his *History of the Royal Society*, it turned out to be not so much a history as a polemic. This was a case of necessity. In the charged atmosphere of the Restoration, the Royal Society was forced to defend itself against assertions that its origins and intentions were less than orthodox. Thus, Sprat's *History*, which was begun in 1663, although not published until 1667, prepared the Society for a vigorous defense of natural philosophy.

[1] Thomas Sprat, *History of the Royal Society* (London, 1667), pp. 111–15; Joseph Glanvill, *Scirei tuum nihil est; Or, the Author's Defence of the Vanity of Dogmatizing* (London, 1665), p. 3. See Michael Hunter, *Establishing the New Science. The Experience of the Early Royal Society* (Woodbridge, 1989), pp. 17, 36, 93.

The Royal Society was soon trapped in a sea of suspicion in which the tides of Restoration politics ran very swiftly. In May 1662, Simon Patrick, later successively Bishop of Chichester and Ely, received a letter from Oxford in which his correspondent complained that the followers of mechanical philosophy were given to excessive theological latitude. Perhaps it was the case that the Fellows avoided the necessity of searching out the theological divisions that might separate their members, but such deliberate evasion could easily be turned against them. Patrick had reason to be anxious about what was reported by a friend in Oxford concerning the advocates of latitude:

> On the one side I hear them represented as a party very dangerous to the King and Church, and seeking to undermine them both: on the other side I cannot hear what their particular opinions or practices are, that bear any such dangerous aspect; to say the truth, I can meet with nothing distinct concerning them, but that they had their rise at Cambridge, and are followers for the most part of the new Philosophy, wherewith they have so poysoned that Fountain that there are like to issue out very unwholsome streams throughout the whole Kingdome.[2]

It was these aspersions that provoked Thomas Sprat into what Michael Hunter has recently called "as much a confession of faith as a factual record."[3] The vigorous defense of natural philosophy and rationalism was cloaked in a catalogue of the activities of the Society whose experiments would, Sprat argued, benefit

> the wonted Courses of Education; to the Principles, and instruction of the minds of Men in general; to the Christian Religion, to the Church of England; to all Manual Trades; to Physic; to the Nobility, and Gentry; and the Universal Interest of the whole Kingdom.[4]

Sprat's case was only slightly disingenuous, based as it was on a broad utilitarianism. This was an acknowledged debt to Baconianism but not necessarily to its radical or Puritan form that had carried English science to the creation of the Royal Society.[5] Sprat neatly wove together the

[2] S. P. of Cambridge, *A Brief Account of the new Sect of Latitude-Men Together with some Reflections upon the New Philosophy. In answer to a letter to a Friend at Oxford* (London, 1662), p. 3. See Gordon Rupp, *Religion in England, 1688–1791* (Oxford, 1986), pp. 29ff.; John Gascoigne, *Cambridge in the Age of the Enlightenment. Science, Religion and Politics from the Restoration to the French Revolution* (Cambridge, 1989), pp. 41–3, 48–50, 63–5; and Glanvill, *Plus Ultra* (1668), "Preface."

[3] Michael Hunter, *Science and Society in Restoration England* (Cambridge, 1981), p. 29; and Hunter, *Establishing the New Science*, p. 48.

[4] Sprat, *History*, pp. 322–3.

[5] Charles Webster, *The Great Instauration: Science Medicine and Reform, 1626–1660* (New York, 1976), p. 492; and Antonio Perez-Ramos, *Francis Bacon's Idea of Science and the Maker's Knowledge Tradition* (Oxford, 1988), pp. 13–15.

strands of private interest, from men of traffic to Churchmen and Charles' Court, and argued that all could benefit from the examinations of nature in which the Royal Society was engaged. As others saw it, this was precisely the problem.

Sprat's *History* evoked from the physician-pamphleteer Henry Stubbe a *Censure upon Certain Passages Contained in the History of the Royal Society*. Stubbe's intentions have been a matter of some puzzlement and, recently, of much debate. He has often been viewed as a conservative in the pay of the rival Royal College of Physicians, and as such, he resurrected and displayed the religious apprehension that Sprat had attempted to set to rest. On the other hand, James Jacob has argued that Stubbe was really a radical supporter of Hobbes, whose attack on Sprat and the Royal Society was intended to undermine the moderate and latitudinarian ideology that many in the Society sought to promote.[6] In this context, the Baconianism of the Royal Society was simply "vulgar"; hence the jaundiced comment of Thomas Hobbes that "not every one that brings from beyond seas a new gin, or other jaunty device, is therefore a philosopher."[7] Sprat's assertion that true natural philosophy should be intended for "Mechanicks, and Artificers" was thus a weakness and not a strength to those who feared the undermining of society and of religion to the benefit of dreaded popery.[8] The popularity of natural philosophy was no agreed preservative of social order. As Stubbe recognized, no effective support for the Society could rest solely on promises it might not keep. Baconian optimism could not guarantee a mine would be drained or iron would be better smelted.

This Sprat understood – as he also knew, given the religious sensibilities of the age, that theological innuendo might prove deeply damaging. Religion, then, would define Sprat's first line of defense, as slippery a place as he admitted it was.[9] By its enemies, the Society was looked upon as destructive of Anglican theology, and thereby subversive of the state. This especially seems to have been Stubbe's view.[10] Thus, those who

[6] James R. Jacob, *Henry Stubbe, Radical Protestantism and the Early Enlightenment* (Cambridge, 1983), pp. 84ff.; cf. Richard S. Westfall, *Science and Religion in Seventeenth Century England* (Ann Arbor, 1973), pp. 22–4; and Dorothy Stimson, *Scientists and Amateurs: A History of the Royal Society* (New York, 1968), pp. 70–89.

[7] Quoted in Hunter, *Science and Society*, p. 89.

[8] Jacob, *Henry Stubbe*, p. 87; Hunter, *Science and Society*, p. 87; and Sprat, *History*, p. 438. See Henry Stubbe, *Legends No Histories: Or, A Specimen of Some Animadversions Upon the History of the Royal Society* (London, 1670), pp. 17–18.

[9] Sprat, *History*, pp. 345ff.

[10] Margaret 'Espinasse, "The Decline and Fall of Restoration Science," in Charles Webster, ed., *The Intellectual Revolution of the Seventeenth Century* (London, 1974), pp. 353–4. See also Jacob, *Henry Stubbe*, esp. Chapter V.

proposed a latitude in matters of religion became the focus not only of fears of atheism, republicanism, and subversion, but of the terror of Catholicism as well. Hobbes haunted the entire debate.[11]

Religious sensibility could readily be exploited by those hostile to the Royal Society. On the other hand, many leading divines of the Church had a great interest in natural philosophy. Sprat, of course, is a good example of this, along with John Wilkins, the Bishop of Chester, and Seth Ward, the Savilian Professor of Astronomy at Oxford. One must also include here one of Stubbe's bitter adversaries, Joseph Glanvill, an original Fellow of the Society and zealous opponent of what he called Stubbe's "malevolent, envious humour against the Royal Society, and its Friends."[12]

The Royal Society was under seige. The Oxford cleric Robert South, who rose as chaplain successively to the Earl of Clarendon and to James, Duke of York, was one implacable enemy. Shortly before Sprat's *History* appeared in August 1667, South warned in a sermon in Westminster Abbey of "a set of fellows [that] get together, and [who] formed into a kind of diabolical society, for the finding out new experiments in vice. . . . " South's effort was widely regarded as an attempt to sink experimentalists with damnations of enthusiasm and of "that profane, atheistical, epicurean rabble, whom the whole nation so rings of. . . . "[13] Likewise, South delivered an Oration in July 1669 in the presence of John Wallis, John Evelyn, Joseph Glanvill, and probably Thomas Sprat, who had been drawn to Oxford for the dedication of the Sheldonian Theatre. Wallis complained that South's performance was composed "of execrations against fanatics, conventicles, comprehension, and new philosophy; damning them *ad inferos, ad gehennam*." This whole Restoration spectrum of discord provided a rich legacy upon which High-Churchmen would later draw.[14]

[11] John Redwood, *Reason, Ridicule and Religion: The Age of Enlightenment in England, 1660–1750* (London, 1976), pp. 70–3; and Quentin Skinner, "Thomas Hobbes and the Nature of the early Royal Society," *Historical Journal*, XII (1969), 217–39.

[12] Cited in Jackson I. Cope, *Joseph Glanvill, Anglican Apologist* (St. Louis, 1956), p. 30; Jacob, *Henry Stubbe*, *passim*. See also Norman Sykes, *From Sheldon to Secker: Aspects of English Church History, 1660–1728* (Cambridge, 1959), p. 155.

[13] Robert South, "The practise of religion enforced by reason," in *Sermons Preached Upon Several Occasions* (Oxford, 1823), I, pp. 373–5. See also John Beale to Henry Oldenburg, June 1, 1667, in A. Rupert Hall and Marie Boas Hall, eds., *The Correspondence of Henry Oldenburg*, III (Wisconsin, 1966), pp. 427, 429n12; Jan V. Golinski, "A Noble Spectacle. Phosphorous and the Public Cultures of Science in the Early Royal Society," *Isis*, LXXX (March 1989), 37–8; and Hunter, *Establishing the New Science*, p. 60.

[14] John Wallis to Robert Boyle, July 17, 1669, in Thomas Birch, ed., *The Works of the Honourable Robert Boyle* (London, 1772), VI, p. 459; E. S. de Beer, ed., *The Diary of*

Given the realities of the political terrain, the Fellows of the Royal Society could hardly afford to be indifferent to such animosity. This was especially the case when the experimental philosophy of the Society was often associated in the minds of critics with mechanical philosophy, epicureanism, and atheism.[15] These theological and philosophical demons were never fully exorcised in the seventeenth century; the later fears of the Newtonians clearly indicated the battle was far from over by the eighteenth century. Few familiar with the bitter controversies surrounding natural philosophy in the 1690s, or with those about the implications of science during the reigns of Anne or the first two Georges, could seriously assume a notion of "the waning role of religion."[16] Natural theology would long remain important. Certainly, the pressure for religious toleration and latitude, which some of the Fellows supported, should not be construed as theological indifference.

If we are to believe Sprat, a sense of a broad spirituality followed from the rational investigation of nature. Thus, natural philosophy and rigid dogmatism were incompatible. But to imply laxity in matters of religion was, in Glanvill's words, a "vulgar conceit."[17] It was foolish, impious, and deceitful, Sprat complained, to conclude that an experimenter "should be prone to deny the essence, and properties of God, the universal Sovereignty of his Dominion and his Providence over the Creation." On the contrary, it was from experiments that the philosopher would have arguments for "the beauty, contrivance, and order of Gods Works: From hence, he will learn to serve him with all reverence, who in all that he has made, consulted Ornament, as well as Use."[18] Here Sprat could not

John Evelyn (Oxford, 1955), III, p. 531 (July 9, 1669). See also Wallis to Oldenburg, July 16, 1669, and Glanvill to Oldenburg, July 19, 1699, in Hall and Hall, eds., *Correspondence of Henry Oldenburg*, VI (Wisconsin, 1969), pp. 129, 137; Irene Simon, *Three Restoration Divines. Barrow, South, Tillotson* (Paris, 1967), I, pp. 243, 251; and Jacob, *Henry Stubbe*, pp. 105, 155.

15 Steven Shapin and Simon Schaffer, *Leviathan and the Air-Pump. Hobbes, Boyle and the Experimental Life* (Princeton, 1985), pp. 290ff.; and Joseph Glanvill, *Philosophia Pia, Or, A Discourse of the Religious Temper, and Tendencies of the Experimental Philosophy Which is posest By the Royal Society* (London, 1671), pp. 1–3.

16 Barbara Shapiro, "Debate: Science, Politics and Religion," *Past and Present*, LXVI (February 1975), 136–8; and Michael Hunter, *Establishing the New Science*, pp. 60–1. Cf. Mulligan, "Civil War Politics, Religion and the Royal Society," in Charles Webster, ed., *The Intellectual Revolution of the Seventeenth Century* (London and Boston, 1974), p. 339; "Debate," pp. 141–2; Mulligan, "Anglicanism, Latitudinarianism and Science in Seventeenth Century England," *Annals of Science*, XXX (June 1973), 217. For debates in college circles, see especially Gascoigne, *Cambridge in the Age of the Enlightenment*, *passim*

17 Joseph Glanvill, *The Vanity of Dogmatizing* (1661), p. 248. In case anyone missed the point, the exact lines were repeated in his *Scepsis Scientifica* (1665), p. 182.

18 Sprat, *History*, pp. 347–9, 373–4.

only cite Bacon but echo the Psalms, the doctrine of Boyle and the cant of Glanvill.[19]

The emphasis on reason and reasonableness amongst some of the early members of the Society was the direct reaction to the violence of the previous generation. Even so, there was a great deal of moral sanctimony in the erastian proposition that the Restoration and the Royal Society were, intellectually at least, the antithesis of the dogmatism of the Interregnum. Sprat very deliberately pointed out that the study of nature "never separates us into mortal Factions; [but] gives us room to differ, without animosity; and permits us, to raise contrary imaginations upon it, without danger of a Civil War."[20] His Restoration readers would not have missed the point.

Nature was presumed to provide a respite from enthusiasm. And the recognition of human fallibility provided for the establishment of moderation as the watchword of those divines who became known as Latitudinarians.[21] This did not necessarily imply heterodoxy, despite the definition of a Latitudinarian as "a Gentleman of a wide swallow."[22] In fact, the true foundations of theology were to be defended by the founders of the Royal Society. Many of the members of the Royal Society were Royalists and staunch Anglicans, as Glanvill happily noticed and Stubbe had to admit.[23] Some were Anglican clerics who joined the Royal Society with the King in 1664. This group stands in marked contrast to those founder fellows, especially those who might be called Latitudinarian, who were all natural philosophers or supporters of the new philosophy, like Sprat, John Wallis, or John Wilkins, who played an important role in

[19] Sprat, *History*, p. 351; Boyle, *The Excellency of Theology, Compared with Natural Philosophy* (London, 1674), p. 73; Glanvill, *Plus Ultra*, p. 52. See also J. Beale to Boyle, October 31, 1666, in Birch, ed., *The Works of Boyle*, VI, p. 419.

[20] Sprat, *History*, pp. 53, 56. This appears to follow from Glanvill, *Scepsis Scientifica* (1665), pp. 72–5, esp. 72–3. See also John Spurr, "'Rational Religion' in Restoration England," *Journal of the History of Ideas*, XLIX (Oct.–Dec., 1988), esp. pp. 563–9; and Barbara J. Shapiro, "Latitudinarianism and Science in Seventeenth-Century England," in Webster, ed., *The Intellectual Revolution*, p. 300.

[21] Shapiro, "Latitudinarianism and Science," pp. 302–3. See Gascoigne, *Cambridge in the Age of the Enlightenment*, pp. 82ff.

[22] Barbara J. Shapiro, *John Wilkins, 1614–1652: An Intellectual Biography* (Berkeley, 1969), p. 160. Cf. John Marshall, "The Ecclesiology of the Latitude-men 1660–1689: Stillingfleet, Tillotson, and 'Hobbism,'" *Journal of Ecclesiastical History*, XXXVI (July 1985), 407–27.

[23] Joseph Glanvill, *Philosophia Pia; Or, A Discourse of the Religious Temper, and Tendencies of the Experimental Philosophy, which is posest by the Royal Society. To which is annext A Recommendation, and Defense of Reason in the Affairs of Religion* (London, 1671), pp. 78–80; Glanvill, *Plus Ultra* (1668), "Preface"; Stubbe, *Legends No Histories*, "Preface to the Reader," and pp. 4–5

the final publication of Sprat's *History*.[24] For the most orthodox Fellows, the rot may well have been in the foundation despite claims to construct ramparts to "serve Religion against the godless Rout" of atheists.[25] Yet the Latitudinarians were to set the tone and direction of scientific investigation for much of the next century.

Sprat had once been a student at Wadham College, Oxford, while John Wilkins was Warden. Under Wilkins' influence, Oxford became the locus of scientific studies during the 1650s. It is often suggested that the group that met in Wilkins' lodgings later went on to form the nucleus of the Royal Society. For a short period in 1659–60, Wilkins became Master of Trinity College, Cambridge, where his moderation and tolerance made him natural ally of those known as the Cambridge Platonists. It was Wilkins who proposed Henry More and Ralph Cudworth as members of the Royal Society, and obtained the living of St. Lawrence Jewry for Benjamin Whichcote. It was Wilkins' influence that led to the appointment of Isaac Barrow, Newton's mentor, as Gresham professor of geometry. But quiet college lawns were not capable of isolating the scholar from the tumult of Anglican revival. Although Wilkins' tenure at Trinity did not survive the Restoration, he became the center of a brilliant Latitudinarian circle in London, which included John Tillotson, the future Archbishop of Canterbury, Thomas Sprat, formerly of Wadham, Simon Patrick, the defender of the Latitudinarians, and Gilbert Burnet, a Fellow of the Royal Society and future Whig and Low-Church bishop.[26]

In philosophy and in politics, "reasonableness" and moderation were the invocations of the Latitudinarians. Apart from the apologia of Thomas Sprat, the closest thing to a manifesto for the war-weary moderates came from Joseph Glanvill, who, while formerly chaplain to the Cromwellian Francis Rous, embraced the Restoration and successively the rectories of Wimbush and of Bath. His conversion was sufficiently successful under the Restoration, at the height of his defense of the Royal Society, to become chaplain to Charles II in 1672. Glanvill had no use for enthusiasts of any stripe. When he stated in 1661 that "The knowledge I teach, is ignorance," he was helping lay the foundations for a new epistemology that would begin with a distaste for dogma and lead to the scepticism of the scientific experimenter.[27] This had profound implications for the great

24 Hunter, *Establishing the New Science*, pp. 50–4.

25 Glanvill, *Plus Ultra* (1668), p. 138; Mulligan, "Civil War Politics," pp. 325, 328–9.

26 Shapiro, *Wilkins*, pp. 121–30, 143–5, 154–5; and Gascoigne, *Cambridge in the age of the Enlightenment*, pp. 41–50. On Sprat and Wilkins, see Hunter, *Establishing the New Science*, pp. 50–3, 67.

27 Jos. Glanvill, M.A., *The Vanity of Dogmatizing*, "Preface"; and Shapiro, *Wilkins*, pp. 227ff.

issues of religion and theology as the Latitudinarians similarly had little use for claims of infallibility. Ten years later, Glanvill was to lay this program clearly before his readers:

> The Real Philosophy tends to the ending of disputes, by taking men off from unnecessary Terms of Art, which very often are occasions of great contests: If things were stated in clear, and plain words, many Controversies would be at an end; and the Philosophy I am recommending, inclines men to define with those that are the simplest and plainest, and thereby also very much promotes the interests both of truth and peace.[28]

Glanvill's earlier rationalization for natural philosophy strikes a chord that would resonate far beyond the battles of the books. It was nature, and not the failure of the peripatetic philosophy, that revealed that the rhetorical power of the text could not survive the scepticism of the new philosophy. Books may begin the reverence for ideas, but "Sun-beams best command themselves."[29] Experimentalism, especially once triumphant in Newton's own rhetoric, meant nature's glories should eclipse endless metaphysical speculation. But there was a basic contradiction in the efforts of the scholars like Sprat or Glanvill who used books to promote the power of nature.

Baconianism

Central to these disputes was the meaning of Baconianism in English science. The debate over man's rehabilitation of nature was, in effect, a conflict over social change. The power of Bacon's vision was confined neither to the period of the formation of the Royal Society nor to the Restoration. As both Sprat and Stubbe had revealed, the apparent Baconian aims of the Society were an essential part of a wider canvas of religious, political, and social policy – a canvas that ultimately would extend into the eighteenth century. The indications were numerous. Without entering into the debate over the Puritan influence on the Royal Society, there can be little doubt that it was the intention of those associated with the early Royal Society to promote useful knowledge. As Charles Webster has argued, the various early seventeenth-century

[28] Joseph Glanvill, *Philosophia Pia*, pp. 93–4.

[29] Joseph Glanvill, *Scepsis Scientifica: OR, Confest Ignorance, the way to Science; In an Essay of the Vanity of Dogmatizing, And Confident Opinion. With a Reply to the Exceptions of the Learned Thomas Albius* (London, 1665), p. 184. For an excellent discussion of the language of the new science, see Brian Vickers, "The Royal Society and English Prose Style: A Reassessment," in Vickers and Nancy C. Struever, eds., *Rhetoric and the Pursuit of Truth* (Los Angeles, 1985), pp. 3–76.

models for a society of natural philosophers came together in the efforts of Boyle, Petty, Oldenburg, Wilkins, and Evelyn in the shape of a Baconian ideology. Thus, the influence of the projectors Samuel Hartlib or Gabriel Plattes came to fruition in the empiricism and improving spirit of those who justified the Royal Society.[30] In the early days of the Society, the adoption of Bacon's idea to establish a history of trades announced a commitment to a base for further experiments and for potential improvement through the exchange of information. Bacon's "refutation of rhetorical culture" proved Sprat's purpose and he adopted for inclusion in the *History* some of the first fruits of the scheme like Petty's work on dyeing. The affirmation of a system of exchange by John Beale and John Houghton linked them directly to a Baconian program that could be traced from Hartlib to Nehemiah Grew and the early Newtonians.[31]

Before 1660, science had not only its promoters, but also some general support either in the Oxford group who met in John Wilkins' chambers, or in the meetings of scientists at Gresham College after 1645 in the rooms of Samuel Foster, Lawrence Rooke, or Christopher Wren. Although the precise origins of the Royal Society may be problematical, what is significant is that many of the Gresham College meetings consisted of those with strong utilitarian interests.[32] By then the link between the Gresham mathematicians and the mercantile and naval community had been cemented by the attempts of Henry Gellibrand and Lawrence Rooke to improve the methods of navigation. One might doubt Gresham's Baconian success in reaching the mass of London tradesmen, mathematicians, and instrument makers.[33] Nevertheless, whatever the origins of the Royal Society, Sprat relied for defense of the new institution upon the kinds of practical interests that were widely accepted. This is not to say he was always circumspect in the choice of evidence. His statement that

[30] Charles Webster, *The Great Instauration. Science, Medicine and Reform, 1626–1660* (New York, 1976), pp. 47, 334.

[31] Paolo Rossi, *Francis Bacon. From Magic to Science*, trans. Sacha Rabinovich (Chicago, 1978), pp. 8–11, 218; Hunter, *Science and Society*, pp. 91–2, 96–9; Walter E. Houghton, Jr., "The History of Trades: Its Relation to Seventeenth-Century Thought," *Journal of the History of Ideas*, II (1941), 37–43; Mayling Stubbs, "John Beale, Philosophical Gardener of Herefordshire. Part II. The Improvement of Agriculture and Trade in the Royal Society (1663–83)," *Annals of Science*, XLVI (1989), 323–63, esp. 352–3; and Stan A. E. Mendyk, "*Speculum Britanniae.*" *Regional Study, Antiquarianism, and Science in Britain to 1700* (Toronto, 1989), pp. 139ff.

[32] Francis R. Johnson, "Gresham College: Precursor of the Royal Society," *Journal of the History of Ideas*, I (1940), 414–21; cf. Webster, *Great Instauration*, pp. 51–2, 222–3.

[33] Johnson, "Gresham College," pp. 427–36; and Mordechai Feingold, *The Mathematician's Apprenticeship. Science, Universities and Society in England, 1560–1640* (Cambridge, 1984), pp. 166ff., 208–9.

the Society had been formed "of all sorts of men, of the Gown, of the Sword, of the Shop, of the Field, of the Court, of the Sea; all mutually assisting each other" smacks more of faith than of reality.[34] Even so, it was in the links with Gresham College and later with Christ's Hospital Mathematical School that the Society came closest to approaching the general leveling effect that was perhaps inherent in Baconianism.[35]

The effect of Baconianism was best reflected in the wider culture, beyond the Society that initially sought to adopt it. The effort to produce solutions to obvious problems like the longitude or to the evaluation of ores had long been evident in the activities of the Gresham professors. Certainly, the search for the longitude at sea by Fellows of the Royal Society was also a quest for credibility.[36] In this respect, the enthusiastic arguments of Sprat represented what predecessors of the Society had already recognized. One might suggest that the formation of the Royal Society was actually a reversal of the kind of personal patronage or assistance from merchant companies that Bacon's utilitarianism encouraged and that Sprat merely assumed would be adopted by the Society. But the mathematics teachers and instrument makers who received their support from the Muscovy Company at the end of the sixteenth century, or the immigrant experts on the drainage of mines or of the fens encouraged by the Commonwealth, were likely to be disappointed by the attitude of the Royal Society.[37] Sprat was not wrong; he was merely overly encouraged. One good example of reality overcoming intentions was in the matter of patents for invention. Patents, of course, are merely limited monopolies for a fixed span of time at the gift of the Crown. For Francis Bacon, patents were fraught with difficulties as they were part of the economic framework from which he and many others so handsomely benefited. And yet, the patents of invention could also be destructive and constrictive. It is remarkable, therefore, that it was Henry Stubbe who appeared as the champion of the tradesmen and guilds he felt were threatened by proposals that the Royal Society be given the power to supervise the award of patents for invention.[38]

[34] Sprat, *History of the Royal Society*, p. 76; see Richard Grassby, "Social Mobility and Business Enterprise in Seventeenth-Century England," in Donald Pennington and Keith Thomas, eds., *Puritans and Revolutionaries* (Oxford, 1978), pp. 362–3.

[35] Hunter, *Science and Society*, pp. 64–5.

[36] Feingold, *Mathematicians' Apprenticeship*, pp. 208–9; and Stephen Pumfrey, "Ideas above his Station: A Social Study of Hooke's Curatorship of Experiments," *History of Science*, XXI (March 1991), 19–20.

[37] Webster, *Great Instauration*, pp. 352–3, 358–9.

[38] James Jacob, *Henry Stubbe*, p. 90; Henry Stubbe, *The Plus Ultra reduced to a Non Plus: OR, A Specimen of some Animadversions upon the Plus Ultra of Mr. Glanvill,*

Experimentalism associated with the usefulness of knowledge was one legacy of Baconianism. It had also been central to Sprat's defense of the Royal Society.[39] One of the best examples of the way in which the Society promoted experiments was the work of Robert Hooke, whose efforts Sprat cited extensively in his *History* and whose own problematic connections with merchant patronage were important to the publication of the famous *Micrographia* in 1665. In his dedication to the merchant Sir John Cutler, Hooke praised him for endowing a lecture "for the promotion of Mechanick Arts":

> This Gentleman has well observ'd, that the Arts of Life have been too long imprison'd in the dark shops of Mechanicks themselves, & there hindered from growth, either by ignorance, or self-interest: and he has bravely freed them from these inconveniences: He hath not only obliged Tradesmen, but Trade itself: he has done a work that is worthy of London, and has taught the chief City of Commerce in the world the right way how Commerce is to be improv'd.[40]

Hooke's magnificent *Micrographia* elevated rhetoric beyond disputation. Its remarkable graphics transformed the discourse of natural philosophy, in Hooke's words, *ad oculum*. Thereby, Hooke linked the issue of credibility to the rhetorical utility of statements that *showed*, rather than told, the scientific facts. This was an important step in visual representation that, in the end, meant natural philosophy would require experimental demonstrations in a form that the learned could not only comprehend but to which an intelligent public could achieve access.[41] Even so, it was a notion that Sir John Cutler failed to appreciate.

The bounty of the parsimonious Cutler proved far more perplexing

wherein sundry Errors of some Virtuosi are discovered, the Credit of the Aristotelians in part Re-advanced; and Enquiries made... (London, 1670), "A Premonition." Cf. Sprat, *History*, p. 75. On the Royal Society and patents, see Christine MacLeod, *Inventing the Industrial Revolution. The English Patent System, 1660–1800* (Cambridge, 1988), esp. pp. 185–8, 204–6.

39 Sprat, *History*, pp. 149–50.

40 Barbara J. Shapiro, *Probability and Certainty in Seventeenth Century England* (Princeton, 1983), pp. 35, 66ff.; Robert Hooke, *Micrographia; or, Some physiological descriptions of minute bodies made by magnifying glasses. With observations and inquiries thereupon* (London, 1665), "Preface"; Sprat, *History*, pp. 129–30; and Oldenburg to Boyle, November 10, 1664, in Birch, ed., *The Works of Boyle*, VI, p. 175. See also Michael Hunter and Paul B. Wood, "Towards Solomon's House: Rival Strategies for Reforming the Early Royal Society," *History of Science*, XXIV (1986), 55–7.

41 John T. Harwood, "Rhetoric and Graphics in *Micrographia*," and Steven Shapin, "Who was Robert Hooke?" in Michael Hunter and Simon Schaffer, eds., *Robert Hooke. New Studies* (Woodbridge, 1989), esp. pp. 121–2, 261–2. See also Stephen Pumfrey, "Ideas above his Station," p. 27. On the utility of facts and public reception, see Ludwick Fleck, *Genesis and Development of a Scientific Fact*, trans. Fred Bradley (Chicago and London, 1979), pp. 111–25.

than Hooke had hoped. Although he was no longer being paid by 1674, Hooke hoped praise might oblige the merchant. Optimism within the Royal Society ultimately soured into angry litigation.[42] Such arrangements between merchants and philosophers were still sufficiently novel as easily to evaporate in disappointment. Crucial to this failure was the ambivalence within the Royal Society to merchants and tradesmen; presumably artisans were not sufficiently conversant with the arcana of gentlemen philosophers. This attitude scuttled the History of Trades of course; and it was much lamented by Hooke and William Petty. It was Hooke who insisted on the unity of theory and practice and who argued that nature could yet be subdued by a new "Cortesian army well Disciplined and regulated though their number be but small."[43]

There was much revealed in this language of conquest. This is not to argue that Hooke's work proved useful, or that indeed much of the scientific activity of the seventeenth century produced practical results. Rather it was the case that the utilitarian aims of Baconians or merchants served to promote the kind of scientific investigations that did take place.[44] As Marie Boas Hall once sensibly pointed out, in the early days of the Society, "hardly a meeting went by without some reference to trades"[45] Yet the controversy over practice could agitate important elements in the Royal Society. Inventions might just as easily induce conflict – and they smelled of projectors. The passion produced in Hooke's dispute with the secretary Henry Oldenburg over the idea of spring-balance watches certainly suggests utilitarian intensity.

On the other hand, the general disinterest in technology among the Fellows by the end of the seventeenth century may be the result of their persuasion that the path to useful knowledge was much more frustrating than they had first believed.[46] But this merely addresses the attitude of the Royal Society and not that of the broader range of merchants or mathematical practitioners. This is one of the difficulties of reliance on a text like Sprat's *History*; it reflects an institutional interest. If the original determination to produce a history of trades soon diminished, this does

[42] Robert Hooke, *An Attempt To prove the Motion of the Earth from Observations* (London, 1674), sigs. A.2–3; and esp. Michael Hunter, *Establishing the New Science*, pp. 303, 306ff.

[43] Hunter, *Establishing the New Science*, pp. 214, 233.

[44] Cf. Richard S. Westfall, "Robert Hooke, Mechanical Technology and Scientific Investigation," in John G. Burke, ed., *The Uses of Science in the Age of Newton* (Berkeley and Los Angeles, 1983), pp. 106–9.

[45] Marie Boas Hall, "Oldenburg, the *Philosophical Transactions* and Technology," in Burke, ed., *The Uses of Science*, p. 27.

[46] Marie Boas Hall, "Oldenburg, the *Philosophical Transactions* and Technology," p. 43.

not mean the scientific interests of entrepreneurs, projectors, or mathematical instrument makers ceased to exist. Nor, for that matter, does it mean that the Fellows ceased to be cognizant of the world of merchants or traders. Admittedly, the recurrent proposals for secrecy suggests a principle of exclusivity. But, in fact, secrecy was suggested as a means to attract artisans to divulge their secrets. The related efforts to put an end to "troublesome and prejudicial" visitors produced its own exclusion crisis from time to time. But this should be contrasted with the abortive proposal to provide cheap broadsides in natural philosophy to cost no more than two pence.[47] Exclusivity and public philosophy proved difficult to reconcile.

Shortly after the turn of the century, Abraham Sharp, who had been trained as an instrument maker and who became assistant to the astronomer John Flamsteed, complained that the Royal Society had become "in appearance, sadly drooping and languishing." At the time, he was hoping some means could be found to support James Hodgson, the mathematics teacher who ultimately joined Christ's Hospital Mathematical School.[48] Although Hodgson had recently become a Fellow of the Society, Sharp's remarks were quite telling. His sarcastic barb that the Society had lost much of its utilitarian zeal so that it was "open to all classes rather in the same way as the law-courts and the Ritz" may explain part of the transformation. The frequently repeated philosophers' fears of the vulgar suggests there were social pressures to which the Society was highly susceptible. Nevertheless, the burgeoning interest of aristocrats and gentry in improvement throughout the next century would indicate no impermeable barrier had been built.[49]

The declining passion in the Royal Society for technological innovation may largely be attributed to a widespread suspicion of promoters and projectors. This was an attitude that was not only difficult to overcome, but also one that was continually reinforced well into the eighteenth century. The longstanding wariness of projectors was crucial to the posture taken by the Royal Society – and perhaps it suggests a widening of a gap between entrepreneurs and the Society itself. Thomas Sprat had

[47] Hunter, *Establishing the New Science*, p. 199.

[48] Abr. Sharp to Flamsteed (October 25, 1704) in William Cudworth, *Life and Correspondence of Abraham Sharp* (London, 1889), pp. 81–2; and E. G. R. Taylor, *The Mathematical Practitioners of Tudor and Stuart England* (Cambridge, 1954), pp. 288–9. On Flamsteed, Sharp, and Hodgson, see Lesley Murdin, *Under Newton's Shadow. Astronomical Practices in the Seventeenth Century* (Bristol and Boston, 1985), *passim.*

[49] Margaret 'Espinasse, "The Decline and Fall of Restoration Science," in Charles Webster, ed., *The Intellectual Revolution of the Seventeenth Century* (London, 1974), pp. 365–8. Cf. Hunter, *Science and Society*, p. 99.

hinted as much in his *History* in 1667. He had argued that the "publick Faith of Experimental Philosophy" would have been undermined had it not been supported by the Fellows' own contributions. Otherwise, a claim on the exchequer of King Charles would simply have resulted in them being condemned "as vain Projectors. So ready is man-kind, to suspect all new undertakings to be Cheats, and Chimeraes; especially when they seem chargeable."[50] Natural philosophy should not be reduced to personal advantage.

The uses of nature

The eighteenth-century Newtonians were to inherit a legacy that was long connected to the rise of science in the seventeenth century and to the formation of the Royal Society. This was a legacy that linked the usefulness of nature with the forces of God in the universe. Thomas Sprat had emphasized the utilitarian advantages inherent in the experimental philosophy and, in the end, such a notion produced a ubiquitous doctrine that clearly crossed political boundaries. Indeed, "promoting the Publick Good" became part of the enduring rhetoric of the rise of science in the seventeenth and eighteenth centuries.[51] Whatever else might be said of a Whig and Latitudinarian ideology, the case for a broad bottomed enthusiasm for invention can likewise be seen in the Tory propagandist and Fellow of the Royal Society, John Houghton. It was Houghton who extolled the virtues of joint stocks and whose support for a state-sponsored improvement of husbandry and trade was also conceived as an argument to shore up the Stuart monarchy. Houghton's Tory utilitarianism is evidence of an incipient ideology that was to spread throughout a large part of the philosophical community – even if its Puritan legacy might hold strong reservations over Houghton's delight in luxury and vanity as engines of demand.[52]

Assertions of the need for improvement were intimately related to the justification for the Royal Society. The claims of Sprat and Houghton also represent a vestigial Baconianism of the first half of the seventeenth

[50] Sprat, *History*, p. 77.
[51] Robert Hooke, *An Attempt To prove the Motion of the Earth from Observations* (London, 1674), sigs. A2–3; Sprat, *History*, pp. 129–30, and esp. Part III.
[52] J. R. Jacob, "Restoration Ideologies and the Royal Society," *History of Science*, XVIII (1980), 28–32; and John Houghton, *A Collection for Improvement of Husbandry and Trade*, V, no. 103 (Friday, July 20, 1694). On Houghton, see Christine Macleod, "The 1690s Patents Boom: Invention or Stock-Jobbing?" *Economic History Review*, 2nd ser. XXXIX (1986), 549–71; and Stubbs, "John Beale, Philosophical Gardener of Herefordshire. Part II," pp. 354ff.

century found, for example, in Gabriel Plattes' unfinished "Treasure House of Nature Unlocked." A similar attitude was evident in the work of Plattes' publisher, Samuel Hartlib, who made much of his own particular mission that "Providence having directed mee by the improvement of several relations unto the Experiences and Observations of Others, I find myself obliged to become a conduit-pipe thereof towards the Public."[53] Hartlib's influence on the first generation of Fellows should not be underestimated. It can be seen, for example, in the program of the Royal Society to develop a history of trades. Although unsuccessful, it nonetheless formed part of the background to Robert Boyle's publication in 1663 of *Some considerations touching the usefulness of experimental natural philosophy.*[54]

Sprat had adopted a widely accepted argument. But there were religious benefits besides. The theological justification residing in providence was a complicated business uniting God's will with the utility of knowledge. Hartlib's vision for a storehouse of knowledge faltered as much on these complexities as through a failure of will on the part of the Fellows. Yet, the providentialism of the Baconians survived in the early Fellows of the Royal Society. As Paul Wood has remarked, Sprat's *History of the Royal Society* "was concerned almost exclusively with the cooperative compilation of natural histories, with the primary aim of utilitarian benefits." Such utilitarianism was evident in the partial histories that Sprat appended to his polemic, but this was aimed as much at a "reformation of philosophy" as at "the increas of *Inventions*. . . . "[55] Although Sprat had to be circumspect about any debt that might be owed the Puritan Baconians, his Restoration doctrine of utility allowed him to retain the lineage of the former Stuart Lord Chancellor.[56] Thus Bacon's providential legacy

[53] Quoted in Charles Webster, *The Great Instauration. Science, Medicine and Reform, 1626–1660* (New York, 1976), p. 472.

[54] Robert Boyle, *Some Considerations Touching the Usefulness of Experimental Natural Philosophy*, second edition (Oxford, 1664), Part I, pp. 5–6; Kathleen H. Ochs, "The Royal Society of London's History of Trades Programme: An Early Episode in Applied Science," *Notes and Records of the Royal Society*, XXXIX (April 1985), 130–2; and, J. R. Jacob, "Restoration, Reformation and the Origins of the Royal Society," *History of Science*, XIII (1975), 161–4. On Hartlib's program of educational reform, see Charles Webster, ed., *Samuel Hartlib and the Advancement of Learning* (Cambridge, 1970), esp. pp. 119ff., "Considerations Tending to the Happy Accomplishment of Englands Reformation," and Stan A. E. Mendyk, "*Speculum Britanniae*," Chapter VII.

[55] P. B. Wood, "Methodology and Apologetics: Thomas Sprat's *History of the Royal Society*," *British Journal for the History of Science*, XIII (1980), 6, 14–15. Also, J. R. Jacob, "Restoration Ideologies and the Royal Society," *History of Science*, XVIII (1980), 26–8.

[56] Webster, *Great Instauration*, p. 492; Webster, "Puritanism, Separatism, and Science," in David C. Lindberg and Ronald L. Numbers, eds., *God and Nature. Historical Essays*

remained intact. Bacon's recommendation that there be compiled a "History of Providence" was not unconnected with the efforts of the numerous divines in their documentation of natural phenomena both during the Protectorate and the Restoration.

Part of that enthusiasm can be sought in the search for dominion over nature.[57] Technological achievement had to contend with a moral as much as a mechanical framework. Such a religious sensibility was evident in John Wilkins, for example, who had recognized a special significance in the limits imposed upon machines. In his *Mathematicall Magick* of 1648, Wilkins asserted that human artifice could not exceed the constraints on strength and motion imposed by providence, else "... men might thereby incourage (with the builders of Babel, or the rebell Gyants) to such bold designs as would not become a created being." This much reprinted and well-known work set out an important moral interaction between mechanical law and human capacity that clearly imposed a special onus on those philosophers who claimed great improvements.[58]

The notion of providence governed the minds that searched diligently for a justification for the existence of a scientific society. Providence formed a topology for seventeenth- and eighteenth-century natural philosophers simply because it enabled them to connect their religious and their practical sensibilities. The idea found prominent place in the works of writers as diverse as Richard Cumberland and Joseph Glanvill in the seventeenth century and Daniel Defoe and Benjamin Franklin in the eighteenth. Glanvill, in his essay "On Modern Improvements of Useful Knowledge," argued in defense of the Royal Society

> That it was observed by the excellent Lord Bacon, and some other ingenious Moderns, That Philosophy, which should be an Instrument to work with, to find out those Aides that Providence hath laid up in Nature to help us against the Inconveniences of this State, and to make such applications of things as may tend to universal benefit....[59]

on the Encounter between Christianity and Science (Berkeley and Los Angeles, 1986), pp. 198–203; and Sprat, *History*, p. 372.

57 Keith Thomas, *Religion and the Decline of Magic. Studies in Popular Beliefs in Sixteenth- and Seventeenth-Century England* (Harmondsworth, 1973), pp. 110–11; and, Webster, *Great Instauration*, pp. 2–3, 506.

58 John Wilkins, *Mathematicall Magick. Or, the Wonders That may be performed by Mechanicall Geometry* (London, 1648), p. 104. See the discussion much later in John Rowning, *A Compendious System of Philosophy: With Notes, Containing the Mathematical Demonstrations, and Some Occasional Remarks*, seventh edition (London, 1772), p. 80.

59 Joseph Glanvill, *Essays on Several Important Subjects in Philosophy and Religion*, ed. Richard H. Popkin (New York and London, 1970), p. 35. See also Christine MacLeod, "Patents for Invention and Technical Change in England, 1660–1753" (Ph.D. thesis, Cambridge University, 1982), p. 329; and Jacob Viner, *The Role of Providence in the*

Glanvill repeated this argument frequently. Although well directed, the force of the position could cut in opposite directions, particularly where the extensive manipulation and exploitation of nature were concerned – as in experiments or in projects that defiled the natural world. Indeed, this continuing dialectic was one reason why innovation could not easily overcome the inertia and active resentment of clerics already overwhelmed by suspicion of philosophers.

Experiment

If, as time went on, the Society was wary even of the notion that experimental philosophy could be of assistance to the mechanic trades, this was not the case of other defenses of experimentation. We can have little doubt that Sprat's *History* reflected the official opinion of the Royal Society, which was under pressure to defend its activities. As Stubbe made plain, inventions were one thing, the politics of religion quite another. Thus, experimentalism was also recommended as a rock upon which to build the Restoration church, as a means of turning back dangerous enthusiasm and of supporting rational religion.[60] This was the position adopted by both Joseph Glanvill and Samuel Parker in their defense of natural philosophy. Parker especially seems to have understood the significance of the new method of "exact Experiments and Observations."[61] Through experiments and observations, the Society could establish the glory of creation while not troubling to descend into matters of theology. A great part of the Society's efforts were directed to this end, to the point of employing individuals to assist the fellows in the performance and demonstration of experiments. But, even this had abuse heaped upon it by Stubbe.[62] In the house of experiment, however, there were many meanings.

Social Order. An Essay in Intellectual History (Princeton, 1976), pp. 41–6, 65–7. On Defoe and providence, see G. A. Starr, *Defoe and Spiritual Autobiography* (Princeton, 1965), esp. pp. 95–6, 119–23; and Simon Schaffer, "Defoe's Natural Philosophy and the Worlds of Credit," in John Christie and Sally Shuttleworth, eds., *Nature Transfigured. Science and Literature, 1700–1900* (Manchester, 1989), pp. 13–44.

60 Sprat, *History*, pp. 362, 374–5; Margery Purver, *The Royal Society: Concept and Creation* (Cambridge, Mass., 1987), pp. 9–19; Steven Shapin and Simon Schaffer, *Leviathan and the Air-Pump. Hobbes, Boyle, and the Experimental Life* (Princeton, 1985), p. 340.

61 Robert Hugh Kargon, *Atomism in England from Hariot to Newton* (Oxford, 1966), p. 111.

62 Marie Boas Hall, "Solomon's House Emergent: The Early Royal Society and Cooperative Research," in Harry Woolf, ed., *The Analytic Spirit: Essays in the History of Science* (Ithaca, 1981), pp. 179–82; Shapiro, *Probability and Certainty*, p. 68; and Jacob, *Henry Stubbe*, pp. 85–6.

Symptomatic as Stubbe's rancor was, his blasts had little real effect. But experiment was also a vehicle capable of retaining the providentialism and empiricism inherent in the Baconian heritage. It was Hooke who had, in a memorandum for William Petty, insisted on the unity of experimental philosophy and practice that was at the heart of the difficulties with Cutler. Cutler's notion of utility was too crude for Hooke to accept.[63] On the other hand, there seems to have been no real decline in experimentation throughout the Restoration, as the *Philosophical Transactions* clearly indicate, even though the line between experimentation and simple empirical natural history was not clearly demarcated. Experimentation helped to secure the empiricism of the natural historians like John Ray or Nehemiah Grew and many others like John Harris whose microscopical observations at the end of the seventeenth century place him in the tradition established best by Robert Hooke.[64] This experimental aspect was strenuously maintained throughout the early history of the Society from Hooke and Boyle in the seventeenth century to John Theophilus Desaguliers and Francis Hauksbee in the eighteenth. And Isaac Newton's optical researches, however problematical, presented what Newton preached in the first edition of the *Principia* in 1687 – that hypotheses had no place in experimental philosophy, so that "particular propositions are inferred from the phenomena, and afterwards rendered general by induction."[65]

This was the view propagated by the first generation of Newton's disciples. John Keill, at Oxford, argued for the necessity of experimental evidence as indeed Stephen Hales and Desaguliers were to establish it in practice.[66] The work on chemical forces by Keill and the physician John Freind, on attraction and repulsion by Desaguliers and Hauksbee, and by Hales on pneumatic chemistry all owed much to the Queries Newton was to add to the subsequent editions of the *Opticks*.[67] By the early

[63] Hunter, *Establishing the New Science*, pp. 293–5.

[64] Webster, *Great Instauration*; Shapiro, *Probability and Certainty*, pp. 52–3; see also, John Harris, "Some Microscopical Observations of the Number of Animalicula seen in Water," *Philosophical Transactions*, IX, no. 220 (1696), 254.

[65] Newton, *Mathematical Principles of Natural Philosophy*, ed. F. Cajori (New York, 1962), Rules of Reasoning IV, p. 400. See Anita M. Pampusch, " 'Experimental,' 'Metaphysical' and 'Hypothetical' Philosophy in Newtonian Methodology," *Centaurus*, XVIII (1974), 289–300.

[66] Arnold Thackray, *Atoms and Powers. An Essay on Newtonian Matter-Theory and the Development of Chemistry* (Cambridge, Mass., 1970), pp. 114ff.; Robert E. Schofield, *Mechanism and Materialism. British Natural Philosophy in An Age of Reason* (Princeton, 1970), pp. 25–30; John Keill, M.D., *An Introduction to Natural Philosophy: or, Philosophical Lectures Read in the University of Oxford, Anno. Dom. 1700* (London, 1720), p. 16.

[67] John Theophilus Desaguliers, *A Course of Experimental Philosophy*, vol. II (London,

eighteenth century, this was the legacy of the experimental doctrine established by the nascent Royal Society. The Newtonian natural philosophy, at least in its rhetoric and often in its practice, was the product of the Restoration polemic.[68] This is acutely revealed in the notes of Thomas Birch,who, a century after Sprat, was to comb the minutes of the Royal Society for his own *History*. In his commonplace book, which dates from 1725 and 1726, Birch copied extracts on various topics in experimental science and natural history. Keill, Harris, Hooke, and Boyle were frequently represented although mostly the source was Newton or a commentary upon Newton. Birch concluded that the Newtonian Philosophy was founded upon "nothing but Observations & Experiments, which are evident to the Sense of all Mankind, & From thence he deduces demonstrative Conclusions; & then again, by ye Assistance of these Conclusions, he explains the Causes of many Phaenomena in Nature."[69] *Nullius in verba.*

Practitioners and projectors

There is some danger in assuming that the members of the Royal Society were of one mind in their attitudes toward technology or experiment. Nor did the Society reflect general views on these issues. John Witty, another of Flamsteed's assistants at the Greenwich Observatory, had argued that experimental philosophy might well provide useful applications for the benefit of society.[70] Certainly, the retreat from the History of Trades was not met with alacrity by all within the Royal Society. This was most certainly true of Hooke, John Harris, and Desaguliers. They were hardly alone even if the Society appeared increasingly distanced from mechanics. As Michael Hunter has shown, expressions of the common or public good became increasingly commonplace. Joseph Moxon, who was elected to the Royal Society in 1678 and who styled himself

1744), p. 403. See also Arthur Quinn, "Repulsive Forces in England, 1706–1744," *Historical Studies in the Physical Sciences*, XIII (1982), 117–26; Alister M. Duncan, "Some Eighteenth Century Explanations of Chemical Attraction," *XIe Congres international d'histoire des sciences*, 1965. Actes (1968), 150–1; and on Hauksbee's research program, W. D. Hackmann, "Scientific Instruments: Models of Brass and Aids to Discovery," in David Gooding, Trevor Pinch, and Simon Schaffer, eds., *The Uses of Experiment. Studies in the Natural Sciences* (Cambridge, 1989), pp. 49–52.

68 Shapiro, *Probability and Certainty*, pp. 55–8.

69 BL., Add. MSS. 4478D, Dr Birch's Common Place Book, f. 107.

70 Geoffrey Bowles, "The Place of Newtonian Explanation in English Popular Thought" (D.Phil. thesis, Oxford University, 1977), pp. 15–16; cf. 'Espinasse, "Decline and Fall of Restoration Science," pp. 366–7. Note also Murdin, *Under Newton's Shadow*, pp. 51, 66; and Webster, *Great Instauration*, pp. 495–6.

"Hydrographer" to King Charles, claimed that the gentlemen of other nations had less contempt for manual operations while England suffered, as though "the excellent Invention of a Mill should be dispis'd, because a blind Horse draws in it."[71] Especially in the conflict with the Dutch, the parliamentarian Andrew Yarranton had argued, England suffered for want of encouragement of mechanic arts.[72] Such views were more important to the course of English science than we might have assumed simply by looking into the Royal Society in general. And yet, even there, the cases of Joseph Moxon, the map seller, and John Houghton, the apothecary and tea merchant elected to the Society in 1680, do not support a policy of separation from the mechanical world.[73] Nor is it necessarily the case that such individuals were out of place in the Royal Society.

Projectors were not foreign to the early Royal Society. Among Houghton's numerous schemes was his publication of *A Collection of Letters for the Improvement of Husbandry and Trade*, which relied to some extent on the early History of Trades. Not only was Houghton a statistician and agricultural reformer, he acted as an estate and employment agent, broker, and vendor of chocolate. He seems to have been only one of several publishers who linked projects with improvements in trade and commerce just as had the early Baconians. Like Houghton, the inveterate projector Charles Povey published his own *General Remarks on Trade*, the profits from which were used to set up a college for the education of poor boys.[74] In the early eighteenth century, Povey's name was linked with the inventor Thomas Savery and Francis Lynn, later of the African Company, as one of the commissioners for the sick and wounded.[75] Povey's interests in commerce and education were not especially unique.

As E. G. R. Taylor has established, the late seventeenth century saw

[71] Joseph Moxon, *Mechanick Exercises*, third edition (London, 1703), Preface; Hunter, *Establishing the New Science*, pp. 77–9, 90–1, 286.

[72] Andrew Yarranton, *England's Improvement By Sea and Land. To Out-do the Dutch without Fighting, To Pay Debts Without Moneys, To set at Work all the Poor of England with the Growth of our Lands* (London, 1677), pp. 61ff. A similar expression of the benefits of the earth "dispos'd for Trafick and Commerce" had been made during the Dutch wars by John Evelyn in *Navigation and Commerce, Their Original and Progress* (London, 1674), pp. 2, 4–5.

[73] On Moxon, see Sarah Tyacke, "Map-Sellers and the London Map Trade c1650–1710," in Helen Wallis and Sarah Tyacke, eds., *My Head Is a Map* (London, 1973), p. 68; and Hunter, *Science and Society*, pp. 96, 109–10. On Houghton, see Hunter, *Establishing the New Science*, pp. 11, 344–5.

[74] R. B. Walker, "Advertising in London Newspapers, 1650–1750," *Business History*, XV (1973), 115–18; Webster, *Great Instauration*, pp. 354–5; 448–9; Larry Neal, "The Rise of a Financial Press: London and Amsterdam, 1681–1810," *Business History*, XXX (April 1988), 165–8. Cf. Hunter, *Science and Society*, pp. 97–8.

[75] Bodleian Library, MS. Rigaud 39, f. 72.

close links between various practitioners, instrument makers, and teachers of mathematics. Besides Samuel Pepys, a governor of Christ's Hospital, Hooke and the astronomers Edmund Halley, Sharp, and Flamsteed all had a connection with and an interest in mathematical education, especially with the intention of improving instruments, navigation, or at least making young boys fit for sea.[76] This practical link with the mathematical teachers continued throughout the early eighteenth century although the improvements might be minimal. But it is an attitude we must address. In the eighteenth century, the mathematics teacher James Atkinson of Cherry Garden Stairs, Rotherithe, was still selling editions of the works of the Gresham Professor, Henry Gellibrand, who had died in 1636.[77] The utilitarian concerns of the founders of the Royal Society continued to find fertile ground both within and beyond the Society.

Edmund Halley was a case in point. This son of a London soap boiler, and friend of Moxon and Flamsteed, had a career based largely on extensive practical interests. His work on tides, compass variation, his excursion to the South Atlantic from 1676 to 1678, and his scientific voyage to investigate magnetic variation in the Atlantic in 1698 financed by the Crown give the impression of a singularity of purpose that would have done the Gresham professors proud.[78] The 1698 venture had been postponed for three years because of the continental wars, but even in the interim, Halley found useful employment in the Chester Mint. This was an unhappy experience for Halley, perhaps because of the postponement of the voyage, still more so because of difficulties of tolerating the internal bickering at the Mint as well as his distance from the center of London science. But he seems to have busied himself by writing to his friends in the Royal Society about various observations such as the lunar eclipse of 1697. Likewise, he found time to inform the Society about the structure of an abandoned windmill and of a simple waterworks powered by an undershot waterwheel used to supply Chester with water.[79]

[76] On Hooke, see J. A. Bennett, "Hooke's Instruments for Astronomy and Navigation," and Michael Wright, "Robert Hooke's Longitude Time keeper," in Michael Hunter and Simon Schaffer, eds., *Robert Hooke. New Studies* (Woodbridge, 1989), Chaps. 1, 3.

[77] E. G. R. Taylor, *The Mathematical Practitioners of Tudor and Stuart England* (Cambridge, 1954), pp. 117–18, 126, 205–6, 254, 348–9, 386; Joyce Brown, "Guild Organization and the Instrument-Making Trade, 1550–1830: the Grocers' and Clockmakers' Companies," *Annals of Science*, XXXVI (1979), 27.

[78] Taylor, *Tudor and Stuart England*, pp. 122–5, 273–4; Norman J. W. Thrower, "Edmund Halley and Thematic Geo-Cartography," in Thrower, ed., *The Compleat Plattmaker. Essays on Chart, Map, and Globe Making in England in the Seventeenth and Eighteenth Centuries* (Berkeley and Los Angeles, 1978), pp. 195–228. Cf. Hunter, *Science and Society*, pp. 132–4

[79] Royal Society, MS. Letter Book, H.3, fols. 51–2; Hunt. Lib., Stowe MS. 57, I, f. 21,

Invention and the Royal Society

No one familiar with the Royal Society in these last years of the century could imagine a complete divorce from usefulness. But we should not overstate the case. Certainly, Hooke and Halley were not dismayed by mechanical projects. Nor,curiously, although he had better reason to be, was the Huguenot Denis Papin. Papin had been assistant to both Hooke and Boyle before he went off to the continent in 1687 to the employ of the Landgrave of Hesse-Cassel. Papin was undeterred by his lack of encouragement by Hooke; perhaps the failure of the scientific or commercial world to beat a path to his weekly demonstration in Blackfriars resulted in his twenty-year venture in the pay of his German patron. This accounts for the generation that passed between his early proposals to the Royal Society for inventions for circulating water or to purify air by distillation of water in vacuo, and the later suggestion in 1708 that he would build a vessel of 80 tons to be powered by a "fire-engine."[80] Papin was too much the projector. Unfortunately, he required an advance of four hundred pounds to begin construction as well as the assurance of a post as inspector of such engines for which he would be paid a salary.[81]

Inventions could often precipitate chaotic disputes. These demonstrate the economic dilemma in which inventors often found themselves and it also posed a problem for those whom they solicited. Papin was soon the focus of a furor with Thomas Savery, who claimed that Papin, while in Germany, must have learned the details of his engine for raising water by fire through the Prince of Hesse-Cassel, who had consulted Savery in 1706. Savery made these charges in writing to John Harris in 1709 when Harris served briefly as Secretary to the Society. Papin, of course, denied any such thing and offered to build the Cassel engine, but only if the Society provided the money to do so.[82] This was precisely the kind of controversy that the Society knew all too frequently surrounded the various contrivances and inventions brought to its attention. And it was extremely difficult to distinguish between disreputable projectors and legitimate inventors who might make claims to a patent. The Society apparently was not entirely certain what to make of Papin and, while

James Brydges to Leibniz, January 3, 1694/5; see John M. Kemble, ed., *State Papers and Correspondence Illustrative of the Social and Political State of Europe from the Revolution to the Accession of the House of Hanover* (London, 1857), pp. 164–5, 239.

80 Alan Smith, "Steam and the City: The Committee of Proprietors of the Invention for Raising Water by Fire, 1715–1735," *Transactions of the Newcomen Society*, XLIX (1977–8), 5–7; Royal Society MSS., Classified Papers, 1660–1740, XVIII, nos. 1–80.

81 Royal Society MSS., Classified Papers, XVIII, no. 66. (Read) February 11, 1707/8.

82 Royal Society MSS., Classified Papers, VI, 61, and XVIII, 80.

refusing him a regular salary for making experiments before their meetings, they nonetheless agreed to some payment for experiments approved by the Council.[83] The precedent, ironically, could be found in its earlier failure to settle its difficulties with Hooke.[84]

The uncertainty of the reaction to Papin suggests that there was hardly anything that we might term a policy to discourage Baconianism within the Royal Society, despite appearances. On the contrary, the ambivalence in the reaction may also mask a resurgence of Baconianism, at least among some of the members of the Society as the careers of Hooke, Halley, and Papin indicate. But this did not solve the problem of projectors to which Sprat had originally pointed. Papin reveals just how difficult an issue this could be. In 1711, he proposed to the Royal Society that it create a "New Inventions Company" on the basis of some of his achievements. Papin apparently felt that this would at least assure him of some patronage. The Company was to be formed as a joint stock from which Papin would then draw a salary but only to one-tenth of the profit generated by his own inventions so as not to be seen "... as one of those chymerical projects which are but to[o] frequent." He proposed to market his recently improved "Hellish Bellows," especially useful in the making of iron plates and wire, along with his new furnace which also consumed the smoke, and his improved digesters. To get the project under way, he suggested the manufacture of salt from sea water to provide good initial profit. All of this was to be based on the encouragement of the Royal Society and the subscriptions of investors to the tune of £1000.[85] We can well imagine the nervousness this evoked. Council apparently saw fit to ignore Papin's offer. It voted him £10 for his experiments made before the Society.[86]

But Papin was hardly defeated. He likely raised suspicions in requesting the approbation of the Society for an invention to determine the longitude at sea.[87] Not long after Papin wrote to Dr. Hans Sloane, the Secretary, desiring to be paid for experiments previously performed for the Society, his income already seriously diminished by "being persuaded that there can be no better imployment then to work for the R.S. which is the same

[83] Royal Society MSS., Council Minutes, II (1682–1727), f. 206, January 12, 1708/9; and f. 108, February 23, 1708/9.

[84] Hunter, *Establishing the New Science*, pp. 330ff.

[85] Royal Society MSS, Classified Papers, XVIII, 68. (Read) April 26, 1711; Journal Book (C) X, 278–279. Papin may have been trying to revive a notion by Abraham Hill from the early days of the Royal Society. See Hunter, *Establishing the New Science*, pp. 89–90.

[86] Royal Society MSS., Council Minutes, II (1682–1727), f. 240, May 30, 1711.

[87] Royal Society MSS., Journal Book (C), X, 321–2. October 28?, 1711.

as to work for the publick good." After listing the services that he had done, Papin added a pointed reminder:

Pray, Sir, give me leave to add here that in the Royal Academy at Paris there are three Pensionarys for Mechanick's which have every one a very good yearly Pension: and besides that there are skillful workmen of all trades pay'd by the King, which are ready at any time to do whatever they are commanded by any of the Pensionary's. Then if you please to take the *Memoires de l'Academie Royale des Sciences* and see what there three Pensionary's do every year and compar it with what I have d'on these 7 months: I hope you will find that I have reason to say that I have d'on as much as can be expected from the most honest man with my little ability's and scarcity of money.[88]

As reasonable as this opinion may have been, Papin must have known that the circumstances of the Royal Society differed greatly from those of the French Academy. And yet the comparison must have stung; it was obvious that skilful experimenters could hardly find their livelihood in the pay of a society that continually had problems collecting dues from its Fellows. Moreover, Papin's circumstances were not exceptional.

Papin's most important successor, John Theophilus Desaguliers, had continual problems in the eighteenth century collecting funds from the Society for the experiments he performed. Desaguliers, also a Huguenot refugee, had been born at La Rochelle, but had come to England with his father at a young age. Educated at Oxford, he took orders in the Anglican Church and this admitted him to a series of preferments from the aristocracy that enabled him to devote much of his time to scientific and technical pursuits. By 1710, he became chaplain to James Brydges, the Earl of Carnarvon, later Duke of Chandos, and in 1717, he obtained a living from William Cowper, Lord High Chancellor. Desaguliers managed to ingratiate himself among the Royal family and thereby kept them in touch with the latest scientific advances. As a Fellow of the Royal Society, he was employed to provide experiments as well as making what must have been a considerable living at giving public lectures.[89] The

[88] Royal Society MSS., Letter Book, P (1), 93. Papin to Hans Sloane, December 31, 1711; cf. Huntington Library, San Marino, MS. HM 1264, Nehemiah Grew, "The Means of a Most Ample Encrease of the Wealth & Strength of England In a few years Humbly Represented to Her Majestie In the 5th Year of Her Reign," (1707–8), p. 87; and Alice Stroup, "Royal Funding of the Parisian Academie Royale des Sciences During the 1690s," *Transactions of the American Philosophical Society*, LXXVII, 4 (1987), pp. 27–30, 57–8.

[89] Bodleian Library, MS. Rawlinson J., f. 3, fols. 29–34. Dated July 8, 1736; see also A. Rupert Hall, "Desaguliers," *Dictionary of Scientific Biography*, IV (New York, 1971), 44–6; and W. R. Hurst, *An Outline of the Career of John Theophilus Desaguliers, M.A., LL.D, F.R.S.* (privately printed, 1928), *passim.*

enormous range of Desaguliers' activities will repay examination; for the moment, it should be emphasized that, like Papin, he too gave considerable time to matters of technical interest. His translations of French technical works like Gauger's *Fires improved*, which he dedicated to Hugh, Earl Cholmondely, and of Marriotte on *The Motion of Water*, dedicated to James Brydges, were minor manifestations of his deep involvement in practical and entrepreneurial activities with which the early eighteenth century abounded.

The existence of a latent Baconianism in the eighteenth century cannot be ignored, whatever the institutional sympathies of the Royal Society. It should be noted that while Desaguliers became a devoted disciple of Newton, he did not endeavor to keep mechanics at arms length. In fact, quite the reverse was the case. Desaguliers maintained close contact with mathematical instrument makers like John Senex as did the experimentalist Francis Hauksbee, the younger, and John Harris.[90] Such relationships served to lay the foundations for the rage of public lectures on science in which the century reveled. This was the case, for example, with Dr. Peter Shaw whose alliance with Hauksbee initially developed into lectures and publications with a commercial leaning. Shaw published editions of the works of Bacon and Boyle as well as his own explorations in chemistry, the practical aims of which, he reminded his readers, were to be based on experimentation.[91] By 1731, Hauksbee and Shaw attempted to market a portable chemical laboratory to "Who ever wou'd proceed further" for "the improvement of Philosophy, Arts, Sciences, Trade and Commerce."[92] To this early version of the salesman's glossy prospectus, they appended a series of schemes for courses in chemistry. Their course in the spring of 1731, to employ their new portable furnace, was to cost

[90] On Senex, see Taylor, *Tudor and Stuart England*, p. 301; David A. Woodward, "English Cartography, 1650–1750," in Thrower, ed., *The Compleat Plattmaker*, pp. 173–4, 182; and Sarah Tyacke, "Map Sellers and the London Map Trade c. 1650–1750," in Wallis and Tyacke, eds., *My Head Is a Map*, p. 76. On Hauksbee, see Taylor, *ibid.*, pp. 296–9, 302. On the experimental tradition and Francis Hauksbee, senior, see Henry Guerlac, "Francis Hauksbee: Experimentateur au profit de Newton," in *Essays and Papers in the History of Modern Science* (Baltimore and London, 1972), p. 109.

[91] Peter Shaw, M.D., *Three Essays in Artificial Philosophy, Or Universal Chemistry* (London, 1731), p. 9; Jan V. Golinski, "Peter Shaw: Chemistry and Communication in Augustan England," *Ambix*, XXX (March 1983), 21–3.

[92] Peter Shaw, M.D., and Francis Hauksbee, *An Essay For Introducing a Portable Laboratory: By Means Whereof all the Chemical Operations Are Commodiously Perform'd, For the Purposes of Philosophy, Medicine, Metallurgy, and a Family* (London, 1731), p. 26. On the development of this apparatus, see Brian Gee, "Amusement Chests and Portable Laboratories: Practical Alternatives to the Regular Laboratory," in Frank A. J. L. James, ed., *The Development of the Laboratory. Essays on the Place of Experiment in Industrial Civilization* (London, 1989), pp. 37–59, esp. 40–1.

a relatively high fee of five guineas. Although this scheme was intended to explain the common chemical operations, its wider purpose was the improvement of arts and manufactures.[93] Shaw had adopted the language of materialism and conquest. For him, as for Hooke, "Knowledge and Power are inseparable." But Shaw went further – arguing that it was possible, with due application, for "persons of ordinary natural abilities . . . to carry on the whole business of philosophy with as much success as men of the highest parts." Thus, Shaw was committed to bringing science out of private shadows and, with Bacon, to "set mankind nearly upon a level, in the business of Philosophy."[94] The election as Fellows of the Royal Society of Francis Hauksbee, Desaguliers, and their associates Henry Beighton and John Senex is evidence that the Society did not always deliberately shun those who had practical interests. In addition, their careers suggest a community that saw connections between experimentalists and mechanics as something to be fostered rather than suppressed.

In this respect, their careers are also representative of a problem – the historical links between science and technology just before the first industrial revolution.[95] A. E. Musson and Eric Robinson have pointed out that the Scientific Revolution and the Industrial Revolution "have both tended to draw attention to themselves and away from the intervening period" of the early eighteenth century.[96] There has been very little done to rectify this since they looked for the missing link. However, citing Desaguliers' evidently disparaging remarks on meager mechanics may

[93] Peter Shaw, M.D., and Francis Hauksbee, *Proposals For a Course of Chemical Experiments: With a View to Practical Philosophy, Arts, Trades, and Business* (dated London, May 12, 1731), p. 2.

[94] B.L., Add. MSS. 4258 [printed]. Peter Shaw, *Proposals for Printing, by Subscription, The Philosophical Works of Francis Bacon . . . in Three Volumes* (dated London, March 25, 1731), pp. 4–5. I owe this reference to Jan Golinski.

[95] The literature on science and industry is as voluminous as it is contentious: Arnold Thackray, "Science and Technology in the Industrial Revolution," *History of Science*, IX (1970), 76–89; A. E. Musson and Eric Robinson, *Science and Technology in the Industrial Revolution* (Toronto, 1966); Paul Mantoux, *The Industrial Revolution in the Eighteenth Century* (rev. edn., London, 1928); Michael W. Flinn, *The Origins of the Industrial Revolution* (London, 1966); T. S. Ashton, *The Industrial Revolution, 1760–1830* (London, 1948); Michael Fores, "Francis Bacon and the Myth of Industrial Science," *History of Technology*, VII (1982), 57–76; Margaret C. Jacob, *The Cultural Meaning of the Scientific Revolution* (Philadelphia, 1987), Chap. V–VI. Cf. J. C. D. Clark, *English Society 1688–1832* (Cambridge, 1985), pp. 64ff., on the existence of an industrial revolution at all.

[96] A. E. Musson and Eric Robinson, *Science and Technology in the Industrial Revolution* (Toronto and Manchester, 1969), p. 30.

have obscured the picture as well because Desaguliers' motives and the context in which they were made makes them less than damning.[97]

The issue, ultimately, was not just the ignorance of workmen or their improvement through acquaintance with philosophy, but rather what the nature of the relationship ought to be. This was to be constrained by factors that were as much social as they were intellectual. When Desaguliers wrote about the mechanics he had encountered, he was nearing the end of a life that had been full of difficulties and great disappointments as well as interminable gout.[98] But this was in the context of industrial and technical fraud perpetrated by projectors he knew far too well. Like the Royal Society, Desaguliers was plagued by them. He thought, as well, that he had been their victim. But we cannot conclude that this was his portrait of all, or even most, mechanics. In fact, he maintained close ties to the instrument makers and mathematical practitioners with whom he often worked.

The relationship between science and technology in the first half of the eighteenth century was complex and certainly more subtle than even Musson and Robinson suggested. The debate over the role science played in England's transformation to industrialization must take into account the kind of encouragement of science that technical enterprises brought about. The failure to understand this relationship as one important to the period of transition is merely the result of wishing away a causal connection.[99]

Few have recognized that the enthusiastic search for technical advantage by the swarms of projectors and entrepreneurs served to provide employment even for those versed in natural philosophy. The public lectures on science in eighteenth-century London that took place largely in coffeehouses were not in quarantine from the commercial interests that coffeehouses existed to serve. And the substantial fee of two or three guineas for a series of lectures hardly indicates that there was little demand.

[97] Musson and Robinson, *Science and Technology*, p. 38; cf. Terry S. Reynolds, *Stronger than a Hundred Men. A History of the Vertical Water Wheel* (Baltimore, 1983), pp. 192–3.

[98] John Theophilus Desaguliers, *A Course of Experimental Philosophy*, II (London, 1744), p. 416.

[99] Cf. A. Rupert Hall, "What did the Industrial Revolution in Britain Owe to Science?" in Neil McKendrick, ed., *Historical Perspectives: Studies in English Thought and Society in Honour of J. H. Plumb* (London, 1974), esp. pp. 143, 146; and Hall, "On Knowing, and Knowing how to . . . ," *History of Technology*, III (1978), 91–103. See also the provocative paper by Michael Fores, "Francis Bacon and the Myth of Industrial Science," in A. Rupert Hall and Norman Smith, eds., *History of Technology*, VII (1982), 57–76.

In the eighteenth century, the consequences of science were not, strictly speaking, literate ones. We must, therefore, move beyond the text to overcome the limited horizons of a literate culture. There was a dialogue that existed between the text and the society in which it had sometimes a broad impact.[100] By the eighteenth century, no single text could provide the evidence of the involvement, of the active, personal life or motives, of any of the Newtonians or of their social purpose. But the popularity of their texts and the success of their efforts to reach a public world demonstrate that the Newtonians were deliberate in their search for a sanction for their philosophy – and, ultimately, for their own actions. No history of the early Newtonians could sensibly ignore this. No historian of the acolytes of Newton can now reasonably retreat into the pleasure of the text. We must force ourselves to recognize that the text had a social place – and its author a life that was not confined to contemplation. Physicotheology would successfully provide a justification of their actions; but if their metaphysics became increasingly etherial, their projects were the affirmation of their place in a universe desperate for stability.

[100] See Dominick LaCapra, "Rethinking Intellectual History and Reading Texts," in LaCapra and Steven L. Kaplan, eds., *Modern Intellectual History. Reappraisals and New Perspectives* (Ithaca and London, 1982), pp. 65–72; and Tony Bennett, "Texts in History: The Determinations of Readings and Their Texts," in Derek Attridge, Geoff Bennington and Robert Young, eds., *Post-Structuralism and the Question of History* (Cambridge, 1987), pp. 63–81.

2

Providence and the Newtonians

The great intellectual upheaval of seventeenth-century England also manufactured a legend – that it had attempted to sweep before it the brutal religiosity that dominated Western Europe for the previous two hundred years.[1] Drawing a distinctness from the beast of Rome, such was the metamorphosis of the Protestant Cause nurtured after the Revocation of the Edict of Nantes in 1685 and by the Glorious Revolution that sent James's priests packing in 1688. Yet "No Popery" hardly meant the end of religious controversy; if anything, it meant more of it. But at least the civic polity was alert – sometimes incessantly so – to the dramatic dangers religious animosity could create.

There has been a tendency to assume the intellectual history of the late seventeenth and early eighteenth centuries can be written as the victory of mechanism and rationality over mystery.[2] Here the rhetoric of en-

[1] Tony Davies, "The Ark in Flames: Science, Language and Education in Seventeenth-Century England," in Andrew E. Benjamin, Geoffrey N. Cantor, and John R. R. Christie, eds., *The Figural and the Literal. Problems of Language in the History of Science and Philosophy, 1630–1800* (Manchester, 1987), pp. 94–6.

[2] Norman Sykes, *Church and State in England in the XVIIIth Century* (Cambridge, 1934), p. 345; and Gerald R. Cragg, *Reason and Authority in the Eighteenth Century* (Cambridge, 1964), pp. 62–6. More recently, G. V. Bennett, *The Tory Crisis in Church and State, 1688–1730* (Oxford, 1975), pp. 16–17; and W. A. Speck, *Stability and Strife: England 1714–1760* (London, 1977), p. 100. A similar variant of the same theme is P. M. Heimann, "Newtonian Natural Philosophy and the Scientific Revolution," *History of Science*, XI (1973), 2–3. Cf. Gerald R. Cragg, *The Church & the Age of Reason 1648–1789* (Harmondsworth, 1974), pp. 73–5; Richard S. Westfall, *Science and Religion in Seventeenth Century England* (Ann Arbor, 1973), *passim*; S. C. Carpenter, *Eighteenth Century Church and People* (London, 1959), p. 116; and, Caroline Robbins, "Faith and Freedom (1677–1729)," *Journal of the History of Ideas*, XXXVI (Jan.–March 1975), 48. For a discussion of the problem, see John Redwood, *Reason, Ridicule and Religion: The Age of Enlightenment in England, 1660–1750* (London, 1976), pp. 16–27; and especially

lightenment slides into historical convention. The whole notion of scientific reason in opposition to mystical revelation is a neat and attractive means of explaining the theological difficulties of the early eighteenth century, the rise of deism, and the expansion of scientific interest. Thus, the Restoration Latitudinarians rebuilding the Church from within were seen as part of the same intellectual wave in which Newtonianism found its first support.[3]

After 1687, theological disputes could not be confined to either Church politics or to metaphysical solitude without a cautious glance at the deep social and political tides that were reshaping the English social landscape. The laws of nature, it had been discovered, had meaning for theologians and for politicians who had need to control those social forces that had already unleashed regicide and civil war. No Restoration intellect could ignore that fact. When natural philosophers urged an examination of the relationship between man and nature, they posed, in some minds, a potential threat to the political order; thus, they had to exercise great care.[4]

Natural theology in the seventeenth century increasingly attempted to reinforce revelation and point to the profusion of the effects of providence in the world. Such glorification was critical to the program Thomas Sprat's *History* had anticipated – although apart from the accumulation of natural curiosities very little was seriously achieved until provoked by the precipitous adventures of the Reverend Thomas Burnet in the 1680s. Burnet had been a student at Clare Hall of John Tillotson, the Archbishop of the Revolution. He likewise had very strong ties with the Cambridge Platonists, having followed Ralph Cudworth to Christ's College. During the Restoration, he associated with a number who were sympathetic to millenarianism and hence who had a consuming passion for biblical

Lotte Mulligan, "'Reason,' 'right reason,' and 'revelation' in mid-seventeenth-century England," in Brian Vickers, ed., *Occult and Scientific Mentalities in the Renaissance* (Cambridge, 1984), pp. 375–401; and John Gascoigne, "From Bentley to the Victorians: The Rise and Fall of British Newtonian Natural Theology," *Science in Context*, II (Autumn 1988), 219–56.

[3] Norman Sykes, *Sheldon to Secker: Aspects of English Church History, 1660–1768* (Cambridge, 1959), pp. 152–3; John Gascoigne, *Cambridge in the Age of the Enlightenment. Science, Religion and Politics from the Restoration to the French Revolution* (Cambridge, 1989), *passim*.

[4] Jonathan Clark's recent assertions about the entrenched power of the Church are meaningful here. The revolution in natural philosophy, which Clark evades, was significant precisely *because* of the presumed threat to the retained stature of the Church as an institution and as an intellectual force. Cf. J. C. D. Clark, *Revolution and Rebellion. State and Society in England in the Seventeenth and Eighteenth Centuries* (Cambridge, 1986), pp. 37–9.

prophecy.[5] It was largely to Burnet that late Stuart England owed the eruption of a violent controversy that established that natural investigations of revelation would not be allowed to escape the rigorous scrutiny of the High-Church Anglicans.[6]

Thomas Burnet

Burnet's *Sacred Theory* first appeared after his consultation with Newton. Burnet had been Senior Proctor at Cambridge when Newton received his M.A. in 1668, and in 1680, he sought Newton's advice about the Mosaic account of creation. In Burnet's version, the earth was created as a perfect sphere without the evident imperfections and asymmetries that provoked the rhapsodies of poets. These blemishes had arisen, he argued, as the result of the natural causes associated with the Flood of Noah when the crust of the earth had broken up and subterranean waters had spilled out, bringing about the present geological variation. Burnet's naturalistic explanations were primarily mechanistic and Cartesian for they had recourse to vortices. However, when Newton replied to Burnet's speculations, he presented an alternative hypothesis of the "pressure of ye vortex or of ye Moon upon ye Waters."[7] Like many of the natural theologians who followed him, Burnet had difficulty with the scriptural account of the sufficiency of rainfall in order to produce a general flood. His objections rested on his inability to see the possibility of the creation of such a deluge of new water as he was convinced that "the Universe being full, if any more was created, then there must be as much annihilated at the same time to make room for it; for Bodies cannot penetrate one anothers dimension. . . ."[8] The destruction of the plenum and its ultimate re-

[5] Margaret C. Jacob, *The Newtonians and the English Revolution, 1689–1720* (Ithaca, 1976), pp. 107–8; and David Charles Kubrin, "Providence and the Mechanical Philosophy: The Creation and Dissolution of the World in Newtonian Thought. A Study of the Relations of Science and Religion in Seventeenth Century England" (Ph.D thesis, Cornell University, 1969), pp. 50–1. On the Cambridge Platonists, see John Gascoigne, "The Universities and the Scientific Revolution: The Case of Newton and Restoration Cambridge," *History of Science*, XXIII (1985), esp. 408–13; and Gascoigne, *Cambridge in the age of the Enlightenment*, pp.41–2, 47–8

[6] [George Hickes], *Some Discourses Upon Dr. Burnet and Dr. Tillotson; Occasioned by the late Funeral Sermon of the Former upon the Later* (London, 1695).

[7] *The Correspondence of Isaac Newton*, Vol. II, ed. H. W. Turnbull (Cambridge, 1960), p. 319. Newton to Burnet, December 24, 1680. See also Ernest Tuveson, "Swift and the World-Makers," *Journal of the History of Ideas*, XI (1950), 54–74; and Redwood, *Reason, Ridicule, and Religion*, p. 120.

[8] Thomas Burnet, *The Theory of the Earth; Containing an Account of the Original of the Earth, and all the Changes which it hath already undergone, or is to undergo Till the Consummation of all Things* (second edition, 1691; reprint edn. with introduction by Basil Willey, London, 1965), p. 34.

creation as the waters receded was "to make very bold with the Deity" as far as Burnet was concerned. The key to Burnet's alternative hypothesis was the recognition that no such creation of new waters could have occurred. This issue was so problematic that it may be regarded at least as one of the motives for the recurrent meteorological diaries of countless eighteenth-century naturalists.

For Burnet, the source of the deluge had to be other than a new creation. Miracle was to be given assistance. The clue lay in his view of the perfection of the antediluvian earth. The destruction of the crust would lead to the creation of the present uneven form of the earth, while suggesting the breaking up of the great "womb of the Earth," thus giving birth to the deluge of Noah. When Moses said *The fountains of the great Abyss were broken up*, Burnet took this to mean the destruction of the crust of the Earth thereby releasing great subterranean waters to inundate the entire planet.[9]

Burnet's efforts to explain in naturalistic terms the creation and dissolution of the world were exceptionally provocative in an atmosphere that was intensely sensitive to any scheme to slip the bounds of orthodoxy. From the beginnings of his search into the scriptural version of the creation of the earth and the deluge, Burnet was well aware of the hazards: "Tis a dangerous thing to ingage the authority of Scripture in disputes about the Natural World, in opposition to Reason; lest Time, which brings all things to light, should discover that to be evidently false which we had made Scripture to assert. . . ."[10] What might be evidently false depended partly on the interpretation of natural evidence that was, of itself, much in dispute. But like most Latitudinarians, Burnet could not believe "that any truth concerning the Natural World can be an Enemy to Religion; for Truth cannot be an Enemy to Truth, God is not divided against himself."[11] This was exactly the doctrine of Tillotson. Moreover, while anxiously awaiting the proper license for the publication of his *Theory of the Earth*, Burnet was fully aware of the political difficulties it might occasion. In the summer of 1688, he wrote to a future President of the Royal Society, Sir Robert Southwell, pointing out that the delay of approval may be from the fear in ". . . such a conjuncture of time as this is, when all Lyes at stake, to quarrel their friends about speculations."[12]

[9] Burnet, *Theory of the Earth*, pp. 64–71.

[10] Burnet, *Theory of the Earth*, p. 16. The first English edition was 1684. For one acceptance of Burnet at his word, see Stephen Jay Gould, *Time's Arrow Time's Cycle. Myth and Metaphor in the Discovery of Geological Time* (Cambridge, Mass., 1987), pp. 27–9.

[11] Burnet, *Theory of the Earth*, p. 16.

[12] B.L., Add. MSS. 10039, f. 63. Burnet to Sir Robert Southwell, July 10, 1688. See Jacob, *Newtonians and the English Revolution*, pp. 112, 118–19.

The problem was that not all of his readers accepted the tradition of the Latitudinarians that natural law was itself providential.[13] Hostility was his reward. The rigidly orthodox ignored Burnet's protestations that even on occasions where natural philosophy might have helped to understand the mechanism of the deluge, it had happened "at a time appointed by Divine Providence."[14] Ideology would decide the issue. The crux of the matter was opposing views of the relationship between natural law and revelation. No amount of recourse to reason was likely to help his case.

In 1685, Herbert Croft, the Bishop of Hereford and formerly chaplain to Charles I, took up this very issue in an examination of Burnet's theory. He was aghast that man should "by his shallow weak Reason examine and determine the great and wonderful Works of God? God forbid." The essential problem for Croft, as it had earlier been for John Eachard, was that "This way of philosophizing all from natural causes, will make the whole world turn scoffers."[15] Croft was one of the first to sound an alarm that echoed from pulpit to Convocation.

Burnet's work came to public notice at a particularly inopportune time. The Anglican clergy were on the alert for deism and atheism, which they sensed behind the speculations of natural philosophers. Where Sprat had been nervous, Burnet turned to frontal assault. After the Revolution, Burnet foolishly turned his attention to revelations yet to come to pass, specifically the final conflagration. The result was suitably spectacular. In 1689, when the last two books of his *Theory* appeared in English, Burnet proposed that the dissolution of the earth at the day of judgment could be accounted for by the release of the fire reserved in the earth since the Flood.[16] Apart from the question of the miraculous, Burnet's views seem merely unconventional rather than dangerous. They might, after all, be taken to coincide nicely with Tillotson's refutation of the Aristotelian notion of the eternity of the world or with the Newtonian idea of the degeneration of the universe. Such notions were to become commonplace.[17] But for Burnet's ventures the issue remained; flexibility in such matters smacked too much of deism.

[13] Cf. Westfall, *Science and Religion*, p. 204.

[14] Burnet, *Theory of the Earth*, p. 65.

[15] Herbert Croft, *Some Animadversions upon a book Intituled the Theory of the Earth* (London, 1685), cited in Michael Macklem, *The Anatomy of the World: Relations between Natural and Moral Law from Donne to Pope* (Minneapolis, 1958), p. 27; and in Redwood, *Reason, Ridicule and Religion*, p. 122.

[16] Burnet, *Theory of the Earth*, p. 50.

[17] For example, Henry Pemberton, in response to a challenge by Leibniz to the supposed imperfection in the design of the Creator, was to reply that Newton's view was more like the orthodox notion of an impending day of judgment. Pemberton drew the analogy

It is possible Burnet was emboldened by his success under the patronage of Archbishop Tillotson, through whom he was appointed Clerk of the Closet to William III. In 1692, Burnet published his *Archaeologiae Philosophicae*, which was the occasion of the tempest that had been brewing throughout the previous decade. Victim of the High-Church he was to be, but he was to bear much of the responsibility himself. His *Archaeologiae* went much further than his previous speculations and directly dismissed the Mosaic account of creation. Adam and Eve's fall was treated as a fiction. Thus, protest soon became abuse. Burnet was forced to retire from his post at Court and return to the Mastership of the Charterhouse, where he was to see Addison and Steele as pupils. His boldness had shattered his career. Any ambition there may have been to follow Tillotson to Canterbury dissolved when he was passed over in favor of the far less controversial Tenison.[18] There was a limit to the resilience of Anglican belief.

There were, in effect, two problems with the approach of a natural theologian like Burnet. First, there was the serious and self-defeating proposition that revelation could be explained by way of physical evidence. The orthodox were obviously loath to concede a notion that, as the mathematician Richard Sault later warned, had been erected "... upon the Ruins of Eternal Reason and Religion."[19] Second, among the naturalists themselves, there was enough doubt about the material manifestations of creation to fuel an increasingly bitter debate. This, of course, also gave philosophers an opportunity to claim that Burnet's theological heterodoxy was not sustained by natural philosophy.

The consequent disputes about the relationship between revelation and the use of reason in the explication of either scripture or nature ultimately served to disguise the essence of the seventeenth-century problem.[20] Now to see the conflict over natural philosophy as one of reason against revelation is simply to assume that there was substance to High Church claims of "Church-in-danger" or to accept that other variant – of rational science slaying the superstitious dragons of revelation and religious en-

with the bodies of animals that are "designed to last but a small space of time," but that nevertheless "shews the unlimited wisdom of its author" whose purpose we cannot pretend to know. But that was when the battles of natural theologians had been fought and accommodations found in the English Church. Henry Pemberton, *View of Sir Isaac Newton's Philosophy* (London, 1728), pp.180–1; and Kubrin, "Providence and the Mechanical Philosophy," pp. 7, 51–2.

[18] Kubrin, "Providence and the Mechanical Philosophy," pp. 143–6.

[19] [Richard Sault], *A Supplement to the Athenian Oracle: Being a Collection of the Remaining Questions and Answers in the Old Athenian Mercuries* (London, 1710), p. 430.

[20] Cf. Joseph M. Levine, *Dr. Woodward's Shield. History, Science, and Satire in Augustan England* (Berkeley and Los Angeles, 1977), Chapter III; and John Redwood, *Reason, Ridicule and Religion, passim.*

thusiasm. Neither view seems to me to do more than to provide historical reality with the convenient gloss of rhetoric.

The view that Christian revelation was under broad assault was a belief that was as amorphous as it was popular. It was not exceptional to find the Cambridge don John Eachard complaining, perhaps facetiously, about the contempt for the clergy, or John Norris, the Rector of Bemerton near Salisbury, infuriated by free-thinking John Toland's *Christianity Not Mysterious*.[21] Deists like Toland became fair game in the orthodox hunts of the eighteenth century, but so too was John Locke, on whose epistemology it has often been assumed Toland rested his arguments.[22] Locke was regarded by many as a Socinian or an atheist. Just before the bitter High-Church storm broke in 1709, John Byrom, an inventor of shorthand, was warned by his father to avoid those controversies "as a snare of the devil, thrown among sharp wits and ingenuous youth to oppose their reason to revelation, and because they cannot apprehend reason, to make them sceptics. . . ."[23] The proper palliative to Locke and the rest, Byrom's father suggested, was the newly popular *Cosmologia Sacra* of Nehemiah Grew. But this in itself was an interesting suggestion that betrayed little awareness of the lineage to which Grew belonged or indeed why reason was presumed to undermine revelation. Even Grew's beautifully illustrated work disguised more prescient notions of providence than it evidently revealed. Grew would come to represent, in his work and in his life, the Restoration legacy of improvement, providence, and the public good.

Richard Cumberland

Influential Restoration thinkers maintained a wide audience in the early eighteenth century. Richard Cumberland was one seventeenth-century

[21] John Norris, *An Account of Reason & Faith: In Relation to the Mysteries of Christianity* (London, 1697), pp. 9, 323. See also [J. Eachard], *The Grounds & Occasions of the Contempt of the Clergy and Religion Enquired into* (London, 1672); Robert South, *Sermons Preached upon Several Occasions* (Oxford, 1679), pp. 234–41; and Gascoigne, *Cambridge in the Age of the Enlightenment*, pp. 44–5, 116. The fact that apologists like Glanvill shared this view of the state of the clergy ought to be a caution against seeing the defenders of natural philosophy as an early breed of liberal rationalists. See Glanvill, *Seasonable Reflections and Discourses In Order to the Conviction, & Cure Of the Scoffing, & Infidelity Of a Degenerate Age* (London, 1676), pp. 32–3.

[22] See Margaret C. Jacob, "John Toland and the Newtonian Ideology," *Journal of the Warburg and Courtauld Institutes*, XXXII (1969), 311–12; and Jacob, *The Newtonians and the English Revolution, 1689–1720* (Ithaca, 1976), pp. 212–16. On Toland's debt to Tillotson, see Robert E. Sullivan, *John Toland and the Deist Controversy: A Study in Adaptations* (Cambridge, Mass., 1982), pp. 119–20.

[23] Edward Byrom to John Byrom, September 16, 1709, in Richard Parkinson, ed., *The Private Journal and Literary Remains of John Byrom*, Vol. I, Pt. 1, printed for the Chetham Society, XXXII (1854), p. 7.

divine who continued to have a broad following among those later attracted to Newton's philosophy. In 1672, Cumberland had published one of the most popular and influential attacks on Hobbes after the Restoration. His *De Legibus Naturae* had an immediate following, and its influence among Whiggish circles became more apparent after his elevation as Bishop of Peterborough in 1691 and its translation by Locke's friend James Tyrrell as *A Brief Disquisition of the Laws of Nature*. The work was enlarged by Tyrrell in a second edition in 1701, and it continued to be reprinted in the eighteenth century as part of the swelling armory against materialism.[24]

Cumberland was one of those who sought the foundation of ethics in the common good. And it is instructive that the Bishop and his followers saw public morality in relation to the providence of the natural world.[25] Providence and its powers would become crucial to the Newtonian comprehension of the universe – and, most important, providence would also provide the justification for the ideology of intervention in nature that the eighteenth century would make real. In the century after the Restoration, there evolved a social transformation justified jointly by Newtonian science and providential principles. This is imperfectly reflected in Whiggish interests in the management of property and its link with the public good.[26] It is also revealed in providence (both as a political and theological principle) as a reason for the Revolution itself.

In the long run, the public place of science was to be determined partly by these theological concerns. Religious and theological issues were not only central to Newton's personal universe. Between the Restoration and the Revolution, theology was to prove crucial to public debates over natural philosophy. Thus, the popular appeal of Newtonianism (and the benefits some presumed might result) can be regarded as a partial answer to a question posed in 1672 by Richard Cumberland:

[24] Caroline Robbins, *The Eighteenth-Century Commonwealthman* (Cambridge, Mass., 1959), pp. 76–8; J. W. Gough, "James Tyrrell, Whig Historian and friend of John Locke," *Historical Journal*, XIX (1976), 607; and Linda Kirk, *Richard Cumberland and Natural Law. Secularisation of Thought in Seventeenth-Century England* (Cambridge, 1987), pp. 88–96.

[25] Robbins, *Eighteenth-Century Commonwealthman*, p. 77; J. A. W. Gunn, *Politics and Public Interest in the Seventeenth Century* (London, 1969), p. 278; and Jacob Viner, *The Role of Providence in the Social Order. An Essay in Intellectual History* (Princeton, 1976), pp. 66–70. For Cumberland's scientific reputation, see John R. Millburn, "William Stukeley and the Early History of the Orrery," *Annals of Science*, XXXI (1974) 514–18; and W. C. Lukis, *The Family Memoirs of the Rev. William Stukeley M.D. and the Antiquarian and other Correspondence of William Stukeley, Roger & Samuel Gale*, Vol. III, printed for the Surtees Society, LXXX (1887), p. 463.

[26] Richard Ashcraft, *Revolutionary Politics & Locke's Two Treatises of Government* (Princeton, 1986), pp. 218–21, 256–83.

Does not the Nature of Things, and consequently God its Author, powerfully persuade and command an Endeavour to promote the common Good of Mankind, by every Indication they give, that it is both a possible Effect, and the greatest; and also more closely united with the private Happiness of every one, than any other Effect which we can foresee as possible, and by making us in some degree to promote it necessarily, even then when we give way to our natural Affections, and oppose it to the utmost of out Power? Is it not evident, that he acts most agreeably to practical Reason, and to the imprinted Ideas of the Causes of both publick and private Good, who promotes the first Attempts of corporeal Nature, and exalts them to a greater height by the additional Force of the Human Mind?[27]

It may be going too far to assume, as did the early Newtonians, that the public good might be better served by the utilization of Newton's philosophy. But the search for dominion over nature made Newtonianism a powerful ally to philosophers seeking to sell their expertise. This was obviously the attitude assumed by later Newtonian engineers like John Theophilus Desaguliers, Henry Beighton, Martin Triewald, and others. Either by revealing false projects or by assisting in attempts to drain fens or make flooded mines workable, the public interest could be advanced. This, therefore, made a moral argument, precisely as Cumberland had intended against Hobbes, of the necessity and utility of benevolence.[28] Such was the real force behind the concept of improvement in the early eighteenth century – if nature might be enhanced, then all society would benefit – provided that it was the greater good and not purely private self-interest that drove improvers. Thus, natural philosophy was made moral philosophy as well.

The connection of nature and its uses with morality had critical consequences in the development of Western society. This was the fundamental link between religion and utility emphasized by Sprat.[29] The notion of dominion over nature had early obtained a religious sanction that a search of Scripture could readily assert. Hence, later complaints by nineteenth- and twentieth-century romantics about the alienation of science from nature and about demythologizing nature were really also complaints about Christian ethics. It was the Judeo-Christian legacy that made utilization of nature socially legitimate – and, therefore, industrialism possible. I doubt the later achievement of secularism had much to

27 Richard Cumberland, *A Treatise of the Laws of Nature. Made English from the Latin by John Maxwell, M.A. Prebendary of Connor, and Chaplain to his Excellency the Lord Carteret, Lord Lieutenant of Ireland* (London, 1727), p. 120.
28 Cumberland, *A Treatise of the Laws of Nature*, pp. 242–3.
29 Sprat, *History*, pp. 74–6.

do with it; providentialism was a strong enough force to make nature worth exploring and improving.[30]

Revelation and Revolution

Providentialism was an increasingly powerful ally in sanctioning *human* intervention in the natural world. And the idea of providence was so closely linked to the fate of the Protestant Cause that the success of William's descent upon England in 1688 "raised the prestige of the idea far above anything it had enjoyed before."[31] Thus, history and nature were entwined by forces beyond human comprehension – and only partly revealed in Scripture. It was not merely the brazen Thomas Burnet who had precipitated the controversy over revelation and the natural world; it was the coincident shifting of the political axis of Western Europe that gave force and urgency to the evidence of God's Providence. Politics, theology, and nature were not to be disentangled.

Reason and revelation were political dynamite, especially as the Revolution had to be justified. The affirmation of an actual providence in the natural world had been the mission of seventeenth-century natural theology. It is not surprising that the Latitudinarian theologians who were later to embrace Newton's ideas would inherit the same criticism aimed at the men of the "wide swallow." But there was nothing monolithic about the Latitudinarians. There was much divergence of opinion in the interpretation of Scripture among natural theologians, some of them early followers of Newton. Even so, to determined defenders of the faith, this was proof that natural philosophy tended to the destruction of orthodoxy and confirmed the alleged link between reason, science, and rampant irreligion.

In view of the potential conflict between the natural theologians and the High Church, it may seem odd that naturalists were not more reluctant to engage in scriptural interpretation. But there was a real dilemma for supporters of natural philosophy. As Sprat and others seem to have sensed, to have ignored the natural evidence of biblical revelation would

[30] Cf. Lynn White, Jr., *Machina Ex Deo: Essays in the Dynamism of Western Culture* (Cambridge, Mass., 1968), pp. 84–94; Herbert N. Schneidau, *Sacred Discontent. The Bible and Western Tradition* (Baton Rouge, 1976), pp. 24–5, 181–2; and on the link between secularism and ideology, see J. C. D. Clark, *Revolution and Rebellion. State and Society in England in the Seventeenth and Eighteenth Centuries* (Cambridge, 1986), pp. 97–102.

[31] Gerald M. Straka, *Anglican Reaction to the Revolution of 1688* (Madison, 1962), p. 66; W. A. Speck, *Reluctant Revolutionaries. Englishmen and the Revolution of 1688* (Oxford, 1989), pp. 84–5.

have been to allow there was a conflict. On the other hand, the Latitudinarians went to great pains to deny any antagonism between revelation and discoveries made by reason. It was this attitude that led to the efforts of the unfortunate Thomas Burnet.

Circumstances such as these were tailor-made for natural theologians. The threat to the literal truth of Scripture could not go unchallenged – although the tide of criticism launched against Burnet contained a large degree of apologetics for the speculations of a variety of clerics. By the 1690s, John Ray, Nehemiah Grew, John Woodward, John Harris, John Keill, and William Whiston, among others, were tripping over each other in their rush to prove Scripture was reinforced by natural philosophy. The Reverend William Nicolson's witty remark about "Dr. Burnet's roasted egg, Dr. Woodward's hasty pudding, or Mr. Whiston's snuff of a comet" was not the only bemused comment by an age overrun with speculation. Even so, Nicolson managed to keep on reasonable terms with most of the combatants as they outdid each other to establish the compatibility of nature and Scripture.[32]

The credence of nature rested in the highly politicized battles of physicotheologians. Like Anglican doctrine, nature too had been politicized by the Revolution. The sorry flight of James II was a whimper that presaged an end to the English century of calamity. To replace James with a Dutch Calvinist may seem hardly much more than a constitutional spasm following a palace revolt. But the effect was to assault the chronic virtue of the clerics and their pathological addiction to religious crises. In this context, it is clear the Revolution and its natural theology set loose energies that may have been as difficult to manage as older sectarian squabbles. But in the Revolution Settlement at least there was the possibility of permitting the nascent forces of commerce and politics to unite in a transformation that would ultimately eclipse the power of the Sun King across the water. In this way would the English Reformation come to fruition and the belief in a Protestant destiny be fulfilled. Its justification lay in God's providence.

The language of conquest united Scripture with the mechanical philosophy. The reassertion of providence had deep and far-reaching consequences for Newton's disciples. These were the heirs of Hooke as well. But Thomas Burnet had gone too far. To seem to undermine providence eroded not only authority and orthodoxy, but the credibility of man in

[32] Nicolson to Edward Lhwyd, quoted in Lester M. Beattie, *John Arbuthnot: Mathematician and Satirist* (Cambridge, 1935), p. 203; Shapiro, *Probability and Certainty*, pp. 55–6; and Clyve Jones and Geoffrey Holmes, eds., *The London Diaries of William Nicolson Bishop of Carlisle, 1702–1718* (Oxford, 1985), pp. 14–15.

the conquest of the natural world. We must understand that the Newtonian acceptance of the providential united natural philosophy with the taming of nature. It was Newton's growing army of adherents who would set its mind to forming a system of the world where the production of scientific facts could be tied to the public good.

John Ray and the variety of nature

To define man's place in nature was the business of Latitude. To seek a course out of the quick currents of sectarianism was to be the task of natural theology. It is, therefore, important to recognize that the natural theologians of the late seventeenth century were more than enthusiasts for design in nature. Theirs was no simple-minded exaltation of design as proof of the existence of God.[33]

One of the best efforts at the validation of the human uses of nature lay in the works of John Ray. Ray's long life (1627–1705) and his connections with numerous other naturalists made him a man of considerable stature far beyond the Royal Society. Having abjured the oaths under the Restoration, he retired from the anguish of the Church and turned to a contemplative life of botany at his home in Essex. Although removed from London and the Society, he was drawn to the controversies of the world makers, ultimately experiencing the considerable wrath of Woodward and his ally, John Harris. A much gentler figure, Ray was no match for their continual vituperation, although his reputation outlasted them both. Naive as it may seem, it was constantly Ray's hope that the works of all the naturalists might "... in some measure clear up those difficulties wherewith the original of those bodies [fossils] is entangled, & reconcile all to the novity of the World & Scripture history of the Creation & Deluge."[34] Here lay the latitudinarian vision of the founders of the Royal Society.

The importance of the virulent debates over the vestiges of Creation went far beyond disagreements over the plasticity of nature or the finer points of speciation. It has not been sufficiently recognized that these disputes, which were often bitter, represented the formation of an increasingly consistent picture of how man might intervene in the natural

33 Cf. Shapiro, *Probability and Certainty*, pp. 93–4; and cf. Katherine B. Collier, *Cosmogonies of Our Fathers* (New York, 1934), pp. 92–9, 143–8.

34 Robert W. T. Gunther, ed., *Further Correspondence of John Ray*, Ray Society, no. 114 (London, 1928), p. 252. Ray to Edward Lhwyd, Aug. 16, 1694; pp. 270–1. Ray to Lhwyd, March 15, 1697. Also, Charles E. Raven, *John Ray Naturalist. His Life and Works* (Cambridge, 1942), pp. 449–51.

world. It was not a question of empiricism overcoming teleology; far from it. It was rather the adoption of a teleology that made virtue of the expropriation of the natural world. In the early days of the Royal Society, this had been Boyle's doctrine.[35] It is the case that there are few explicit ideological manifestos we can now unearth – except if practice is evidence of purpose. Such is the weakness of relying on the text as proof of an ideology then more accustomed to civil war and religious conflict than to the contemplation of nature. But it can come as no surprise that the late seventeenth-century controversialists frequently were inclined to consider the uses of the minerals or vegetation they examined. Among the proponents of natural philosophy, a literate tradition, nevertheless, occasionally provided confirmation of their practices. Text and purpose might occasionally overlap in the exploitation of the earth.

John Ray spent most of his life in a dedicated examination of nature from his country home at Black Notley. In the end, and somewhat reluctantly, Ray was drawn into the turmoil that Burnet's *Theory of the Earth* had engendered – even though as late as 1691 he still had not seen it and thought that "by wt I have heard of it I think it needs no great confutation."[36] Nevertheless, during 1690, Ray had begun work on what turned out to be the first of many editions of his *Wisdom of God Manifested in his Works of Creation*. The response was so encouraging that he was quickly busy on his *Miscellaneous Discourses concerning the Dissolution and Changes of the World*, a work that he dedicated to Tillotson, the recently installed Archbishop of Canterbury.[37] Thus, at the very moment Richard Bentley's Boyle Lectures began, Ray ventured nervously into battle while the orthodox loudly sang psalms of revelation.

It is one of the more important, and now ignored, aspects of the seventeenth-century naturalists that they were disposed to do more than merely reinforce the miraculous in Scripture. They were also inclined to accept nature as the creation of a benevolent God for the uses of man. Perhaps it was from Bacon and Boyle that naturalists like Ray took their cue – but at least as early as the 1650s, Ray had shown an interest in mines and their production. Even his 1673 *Collection of English Words* had appended an *Account of preparing some of our English Metals and*

35 Robert Boyle, *Some Considerations Touching the Usefulness of Experimental Philosophy*, Part I, pp. 27–8.

36 Gunther, ed., *Further Correspondence*, pp. 218, 236–7. Ray to Lhwyd, May 27, 1691; Ray to Lhwyd, March 22, 1692.

37 Gunther, ed., *Further Correspondence*, pp. 219–24. Ray to Lhwyd, August 17, 1691; September 25, 1691; Raven, *John Ray Naturalist*, pp. 430–1, 469–75. On the links between Tillotson and Burnet, see Kubrin, "Providence and the Mechanical Philosophy," pp. 50–1, 145–6.

Minerals.[38] At first glance, his reports of alum mines at Whitby or wire-works at Tintern in Monmouthshire seem out of place. But they are no more so than his rival John Woodward's accounts of "suffocating" and "fulminating" damps in his 1695 *Essay toward a Natural History of the Earth*. This perplexing problem had a distinguished history in the Royal Society.[39] Decades ago, Robert Merton noticed how seventeenth-century travelers like Ray, or Newton's friend Francis Aston, were encouraged to report on mines and metallurgy during their continental travels. Such was the view of the naturalist Robert Plot in 1686 that "the knowledge of the means whereby they [mining-damps] seem to be occasion'd, may not perhaps be thought less profitable to the Philosopher, than the cure of them to the Collier."[40] The damps were to remain a chronic problem – but, significantly, Plot proposed philosophers should examine them. Here we have the signs of the maturation of the link between religious sentiment in the fragmented English community and a puzzling out of man's purpose in post-Noachian nature.

The sentiment in the seventeenth century for the overlap of the interests of philosophers and practitioners constituted an emergent ideology whose impact was very slow to be felt. One of the most widespread pronouncements of such a framework came in Ray's popular *Wisdom of God in the Works of Creation*, which appeared in 1691. This work was easily his most influential serving perhaps as the foundation for the even more admired Boyle Lectures of his disciple William Derham.[41] The real significance of the *Wisdom of God* lies not in its commonplace reassertion

[38] John Ray, *A Collection of English Words* (second edition, London, 1691); Raven, *John Ray Naturalist*, pp. 169–70. For the general background to this approach, see Richard S. Westfall, *Science and Religion in Seventeenth Century England* (New Haven, 1958), pp. 52, 91–2.

[39] John Woodward, *An Essay toward a Natural History of the Earth: And Terrestrial Bodies, Especially Minerals: As also of the Sea, Rivers, and Springs. With an Account of the Universal Deluge: And of the Effects that it had upon the Earth* (London, 1695), pp. 205–7; see also Lister, "An Extract of a Letter," July 28, 1675, *Philosophical Transactions*, X, no. 117 (August–September 1675), 391–5; B.L., Add. MSS. 25095. John Woodward, "Two Discourses on Metals," n.d.; and Oldenburg to Boyle, August 25, 1664, in Birch, ed., *Works of Boyle*, VI, pp. 150–1; and Kathleen H. Ochs, "The Failed Revolution in Applied Science: Studies of Industry by Members of the Royal Society of London, 1660–1688" (Ph.D. thesis, University of Toronto, 1981), pp. 121–8.

[40] Quoted in Robert K. Merton, *Science, Technology and Society in Seventeenth Century England* (second edition, New York, 1970), p. 149. Note the recommendation of "Experimental Observators" by the Puritan physician and mathematics teacher John Webster in *Metallographia: Or, an History of Metals* (London, 1671), pp. 35ff.; and "A Letter of Mr. [John] Beaumont containing his Observations of the Fiery Damps in Mines," in [Robert Hooke], *Philosophical Collections*, No. 1 (London, 1679), pp. 6–9.

[41] Raven, *John Ray Naturalist*, p. 452. Derham's lectures were frequently reprinted throughout the eighteenth century as *Physico-Theology*.

of design in nature, but in its confident portrayal of the uses of nature. This, indeed, was much of the intent of Ray's reflections. His biographer, Charles Raven, remarked that Ray "cannot escape the contemporary confidence that its [nature's] primary purpose is its utility to man. . . ."[42] The point needs to be made that such a confidence in the unity of providence and utility was widespread in the seventeenth century. Ray was exceedingly influential in its articulation.

Derived in part from the works of Cambridge Platonists like Henry More, Ray's *Wisdom* might easily be taken as yet another of those innumerable seventeenth-century confutations of atheism. But Ray was a transitional figure between the Cambridge Platonists and the eighteenth-century physicotheologians. His purpose was not confined to the acknowledgment of design; it was rather to broaden the teleology he had inherited. Ray could point to the ability ". . . to make large Additions to Natural History, to advance those other Sciences, and to benefit and enrich thy Country by increase of its Trade and Merchandize." From this he was able to instruct his reader to "Go thither . . . and bring home what may be useful and beneficial to thy Country in general, or to thy self in particular."[43] Providence had translated commerce into a commandment.

Ray recognized clearly that this emphasis on utility raised as many questions as it purported to answer; but it also heightened his enthusiasm for the challenge nature afforded. It may be objected that this advances Ray as an early apologist for the industrialization of the world. But the usefulness of things derives from his powerful expression of the unity of nature and Scripture.[44] Of course, Ray's definition of utility was much broader than that inherent in human intervention; he was still the heir of a nature revealed and contemplated and thus seen as a validation of the existence of God. Even so, the active and the reflective were not of separate worlds. They were, in effect, edges of a spectrum only then coming into focus. The text of Scripture proved no limit to the texture of nature. Ray argued carefully that there were purposes that we may not readily perceive:

[42] Raven, *John Ray Naturalist*, p. 455.

[43] John Ray, *The Wisdom of God Manifested in the Works of Creation*, seventh edition, corrected, London, 1717 (reprint edn., New York, 1977), pp. 160–4; Raven, *John Ray Naturalist*, p. 465; cf. William Coleman, "Providence, Capitalism, and Environmental Degradation," *Journal of the History of Ideas*, XXXVII (1976), 30–2.

[44] See the perceptive study of Ray's apologetics in Neal C. Gillespie, "Natural History, Natural Theology, and Social Order: John Ray and the 'Newtonian Ideology,'" *Journal of the History of Biology*, XX (Spring 1987), 1–49, esp. 41–4.

. . . I believe there are many Species in Nature, even in this sublunary World, which were never yet taken Notice of by Man, and consequently of no Use to him, which yet we are not to think were created in vain; but may be found out by, and of Use to those who shall live after us in future Ages. But though in this Sense it be not true, that all Things were made for Man; yet thus far it is, that all the Creatures in the World may be some way or other Useful to us, at least to exercise our Wits and Understandings, in considering and contemplating them, and so afford us Subject of admiring and glorifying their and our Maker. Seeing then, we do believe, and assert, that all Things were in some Sence made for us, we are thereby oblig'd to make use of them for those Purposes for which they serve us, else we frustrate this End of their Creation.[45]

Ray understood it was hard to see the benefit in the vast multitude of insects that were "noxious and pernicious to Man." To this he could only reply that their uses may not yet have been discovered because ". . . the World is too great for any man, or Generation of Men, by his, or their utmost Endeavours, to discover and find out all its Store and Furniture, all its Riches and Treasures."[46] Here was a faith of eternal delight; purpose was its promise. What was required was effort and patience, Ray reminded his readers; the treasures of nature were inexhaustible. He cited Psalms 111:2: "The Works of the Lord are great, sought out of all them that have pleasure therein" – which gave him the opportunity to decry the sorry state of the "real Experimental Philosophy in this [Cambridge] University; and that those ingenious Sciences of the Mathematicks are so much neglected by us. . . ."[47]

Nehemiah Grew and invention

The convergence of the utilitarian and the inspirational was confined neither to the lingering influence of Bacon nor to the reluctant John Ray. There were many whose interests in practical matters helped to produce a teleology much less sophisticated than the complex fabric of nature glorified by Ray. One such was the naturalist Nehemiah Grew, a nonconformist who became a successful London physician and who found time to make important, and enormously popular, contributions to botanical literature.

Nehemiah Grew's most important work, the *Cosmologia Sacra*, was also the one of least obvious scientific value. Written a decade after Ray's *Wisdom of God*, Grew's effort nonetheless was part of that continuing

[45] Ray, *Wisdom of God*, p. 177.
[46] Ray, *Wisdom of God*, pp. 368–70.
[47] Ray, *Wisdom of God*, p. 173.

assertion of the evidence of God in the works of creation. Creation, as Grew succinctly put it, "doth necessarily infer Providence."[48] It is, admittedly, a remarkable example of the depth of his faith that he was able to explain the geographical diversity of human populations and, as he put it, "the People sever'd one from another, with such Vast Seas? Not knowing, that it was a most Excellent Forecast, of all the Improvements of Navigation." The dependents in 1707 of the drowned sailors of the calamitous Admiral Sir Cloudesley Shovell would soon have good reason to sneer at Grew's optimism and it is hard not to join them. To do so, however, would hardly help us to grasp the mentality of Grew's world – for Grew, I suspect, was making a broader argument for the ability of men to solve those puzzles that only seemed insurmountable. What Grew was concerned to show, in the first instance, were the opportunities:

> How little is the Mischief, which the Air, Fire, or Water, sometimes doth; compared with the Innumerable Uses, unto which they daily serve? Besides Seas and Rivers, how many wholesome Springs are there, for one that is poisonous?. . . .
>
> Of that great Variety, we have of Minerals, Animals, and Plants; how few are Noxious, compared with those which are Friendly to us? And in every Species, of those which are the most Useful, we have the greatest Plenty. Of all Metalls, what so useful and necessary, in any degree, as Iron? Nor is there any other of so universal Growth.[49]

We must be careful not to dismiss Grew's optimism as the simple musings of a contented country physician. Grew lived what he preached. He was a successful and seemingly aggressive London projector in his own right. The cosmopolitan and brutal world of improvers was one that he came to know very well. In a nice example of making abundance useful, it is very likely that he was involved with the patentees of an invention to make salt water drinkable. This joint-stock, which licensed engines from the patentees' office in Lothbury near the Exchange, had published its proposals along with reports of Grew's chemical analysis of the scheme's results. In effect, the Doctor was providing scientific sanction for the 1683 patent – with the apparent approval of the Royal Society and of the Royal College of Physicians.[50] The degree of Grew's

[48] Nehemiah Grew, *Cosmologia Sacra: OR A Discourse of the Universe As it is the Creature and Kingdom of God. Chiefly Written, To Demonstrate the Truth and Excellency of the Bible; which contains the Laws of his Kingdom in this Lower World* (London, 1701), p. 85.

[49] Grew, *Cosmologia Sacra*, pp. 98–9.

[50] Nehemiah Grew, *New Experiments and Useful Observations Concerning Sea-Water made Fresh, According to the Patentees Invention*, ninth edition, September 29 (London,

association with the Company is unknown, but the medical implications undoubtedly interested him.

Grew was also an inventor in his own right, although he was to learn even patents held no guarantees. He discovered a method of producing epsom salts that he demonstrated to the Royal Society in 1679. Grew attempted to manufacture the salts commercially, but by 1693, he soon found himself in competition with the chemists George and Francis Moult. Although he was still able to obtain a patent in 1698, undoubtedly with the aid of influential friends at Court, the Moults were not to be stopped – a fact that even the Royal Society recognized. Grew's claim was clearly acknowledged by both the Royal Society and the Royal College of Physicians, but ultimately Grew sold his patent to Josiah Peter. By 1700, the Moults were evidently producing 200 pounds of the salts each week at their works on Shooters Hill.[51]

In Grew, therefore, we find a naturalist who was active in the exploitation of the uses of nature. Grew had a passion for patents of medical value – but he was also familiar with financial innovations like joint stocks, which became the vehicle of numerous patentees. His knowledge of the world of finance was sophisticated, at least sufficiently so for him to take advantage of opportunities represented by the financial revolution. He was one of those involved in the Land Tax loan in 1693, and in 1707, Grew subscribed £5000 to the Government annuity. This was a very considerable sum and far more than that invested by his contemporaries

1684), pp. 1–2; and "The Opinion of several Eminent Physicians of the Colledge, and others, concerning Sea-Water made Fresh," in *A Collection of Medical Advertisements*, no. 157 [B.L., shelf mark, C.112. f. 7]. The patentees were Robert Fitzgerald, Theophilus Oglethorp, William Bridgman, Thomas Maule, and Patrick Trant. See W. R. Scott, *The Constitution and Finance of English, Scottish and Irish Joint-Stock Companies to 1720*, 3 vols. (Cambridge, 1912; reprint edn., Gloucester, Mass., 1968), I, p. 315; III, p. 474. Fitzgerald was a cousin of Robert Boyle. See E. S. de Beer, ed., *The Diary of John Evelyn*, IV (Oxford, 1935), pp. 388–9, n1; "A Letter of Mr Boyl's to the learned Dr. John Beal . . . concerning Fresh Water made out of Sea Water, printed at the desire of the patentees, in a tract intitled, Salt Water Sweetened, by R. Fitzgerald," in Birch, ed., *Works of Boyle*, IV, pp. 593–5; and "An Account of the Honourable Robert Boyle's Way of Examining Waters, as to Freshness and Saltness," in Birch, ed., *Works of Boyle*, V, pp. 744–50.

[51] This account is based on Christine MacLeod, "Patents for Invention and Technical Change in England, 1660–1753" (Ph.D. thesis, Cambridge University, 1982), pp. 98–100; MacLeod, *Inventing the Industrial Revolution*, pp. 69–70; and Alex Sakula, "Doctor Nehemiah Grew (1641–1712) and the Epsom Salts," *Clio Medica*, XIX (1984), 1–21. See Nehemiah Grew, *Tractatus de Salis Cathartici Amari in Aquis Ebeshamensibus* (London, 1695), published under the imprimatur of both the Royal Society and the Royal College of Physicians. On Grew and the College, see Hunter, *Establishing the New Science*, p. 276.

John Arbuthnot and Hans Sloane.[52] For Grew, practicality and the rewards of providence were never far apart.

One of the great difficulties in assessing the efforts of men such as Ray or Grew is that they were proponents of a view of the universe that seems utterly naive. At worst, they appear as apologists for an emerging natural philosophy that actually needed no excuse.[53] Two things at least must be obvious about such physicotheologians. They accepted the view, as Samuel Parker had remarked about natural philosophers, that they should "but take the Divine Wisdom into their Mechanicks and make their several ways of mechanism the effects of this contrivance. . . . " Revelation was thus central to their vision of nature. Secondly, from a recognition of providence, Parker and Cumberland, for example, were able to derive a morality and not incidentally defend natural philosophy in the process.[54] The generation of Ray and Grew was the heir of this moral imperative. And it was partly this link between nature and morality that accounts for the increasing popularity of natural philosophy.

There is a tendency, hard to avoid, of historians to judge the works of former philosophers by measures that have no relevance to the seventeenth and eighteenth centuries. It may well be that Nehemiah Grew exhibited a less complex appreciation of natural design than did Ray – but what meaning can we now take from that?[55] Is it not worthwhile considering that Grew may have represented a faith that nature was sufficiently constructed for man's use so as to make projects and improvements desirable? If this was the case, then it is Nehemiah Grew, whose academic or theological advancement was curtailed by his nonconformity, and so deprived of his Fellowship in the Royal College of Physicians in the reign of James, who takes us to the fringes of the multitude of improvers of post-Revolutionary England. Although there appears to be no evidence of a direct connection between Ray and Grew, it would have been remarkable if the two Fellows of the Royal Society had no association. Ray certainly was aware of Grew's work and both had entertained the Royal Society.[56]

52 In 1707, Hans Sloane and the projector Charles Povey appeared on the list of the Land Tax loan. See P. G. M. Dickson, *The Financial Revolution in England, 1688–1756* (London, 1967), pp. 267, 426–8.

53 Cf. Westfall, *Science and Religion*, pp. 60–9, 96–100, 131.

54 Samuel Parker, *A Demonstration of the Divine Authority of the Law of Nature, And of the Christian Religion* (London, 1681), pp. xvii–xviii, 84.

55 Westfall, *Science and Religion*, p. 97; and K. Collier, *Cosmologies of our Fathers*, pp. 143–8.

56 It does no real service to elevate Ray above Grew's "simple and too anthropomorphic

The issue arises of whether the natural theology of John Ray or Nehemiah Grew was particular only to them. We need not look far to find numerous examples of the ecstasy over design. There were enough sermons to reflect the rhapsody over nature – such as William Derham's later Boyle Lectures. Such sermons might frequently mention Ray or Grew, or likewise Cumberland, as the source for an argument or example. Grew's patient, the Presbyterian clergyman John Shower, was apt to refer to Ray, Grew, or more curiously to Burnet or even Hooke for inspiration. Obviously, the readers of scientific controversy were not confined to the Royal Society. The task at hand, Shower once announced, was "To reverence the Divine Power, and Providence" – and this was not mere sentiment for he meant in practice as well.[57]

In providence, as Grew's contemporaries understood it, there was an implication of directly engaging nature that the Doctor put into practice in his projects and investments. Did this, however, strike a tone peculiar to Grew? There is likewise, in the early Newtonians, evidence of their concern with using nature as fully as possible. It seems unlikely, therefore, that Grew was unique in this at all. That is not to say projectors or investors were unmotivated by the very material quest about which Richard Cumberland had worried. But in the works of Ray and Grew, and especially of those who understood the power of the Newtonian frame of the world, there were deeper motives – a nature employed by man was a nature glorified by him.

It was partly this providential vision that was behind Grew's commitment to the Protestant Cause and his wish to enhance the expansion of English power. His belief in such a course is detected in his investments in government funds during Anne's reign, ventures that helped finance war against Catholic France. At the height of the war, in 1707, he also presented to the Queen, probably through her favorite Abigail Masham or Masham's close friend Dr. John Arbuthnot, a plan for "a most Ample Encrease of the Wealth & Strength of England." Undoubtedly thinking of his own bitter experience, Grew argued that it was necessary to protect

teleology." Cf. Raven, *John Ray Naturalist*, pp. 200–1; 455. Gunther, ed., *Further Correspondence*, pp. 69, 193. Oldenburg to Lister, Dec. 19, 1674; and Ray to Lhwyd, August 11, 1689.

57 John Shower, *Practical Reflections On the late Earthquakes in Jamaica, England, Sicily, Malta, &c. Anno 1692. With a Particular, Historical Account of those, and divers other Earthquakes* (London, 1693), pp. 63, 154, 156; Shower, *Winter Meditations: Or, A Sermon Concerning Frost, and Snow, and Winds, &c. and the Wonders of God Herein*, third edition, corrected (London, 1710), p. 22; and Shower, *Enoch's Translation. A Funeral Sermon Upon the Sudden Death of Dr. Nehemiah Grew, Fellow of the College of Physicians. Who died March 25, 1712* (London, 1712), *passim*.

investors in England more effectively than had proved possible. He referred to the case of a German who had attempted to set up a brass manufactory in Surrey some fifty years earlier who was harried by interlopers and their lawsuits not only to the destruction of the scheme but also to the loss to the nation of such a potential benefit. There was more than a slight echo here of his own battles. Thus, he recommended to Queen Anne that inventions must be seriously protected:

> For promoting the Invention of any Manufacture, if proof first made of its being Beneficiall to the Publick, to encourage the Author, as in France, would be of great use. Especially, to put it out of the Power of Knaves, for their private Gaine, to lay any Remoras in the Investor's way, to the Damage of the Publick.[58]

It is impossible now to tell whether this had any influence at Court and we know nothing of his relations with the indefatigable Masham. It may be the case that Grew's efforts played some role in promoting the notion that the Royal Society should become directly involved in the assessment of applications for patents even if nothing seems to have come of it.[59] In any case, the French model of support for patents was an issue that Denis Papin was soon to raise with the Royal Society.[60]

As far as Grew was concerned, there was far more involved here than private self-interest. His entire proposal, following a lead established in Andrew Yarranton's 1677 *England's Improvement* (and by Sprat, John Beale, and John Houghton in the Royal Society), was an assessment of the means by which the power of England might be extended.[61] For this reason, he suggested that the improvement of manual arts and inventions could be aided by repositories in every county in which would be placed patterns, models, and engines used in various manufactures:

[58] Huntington Library, San Marino, MS. HM 1264. Nehemiah Grew, "The Means Of a most Ample Encrease of the Wealth & Strength of England In a few years. Humbly Represented to her Majestie In the 5th year of Her Reign," pp. 87–8. See also BL., Lansdowne MS. 691; and MacLeod, *Inventing the Industrial Revolution*, p. 195.

[59] Royal Society MSS., Council Minutes (C), Vol. II (1682–1727), p. 259. January 29, 1712/13.

[60] On the background to this problem, see Alice Stroup, "Royal Funding of the Parisian Academie Royale des Sciences During the 1690s," *Transactions of the American Philosophical Society*, LXXVII (1987), esp. Chapter VI; and MacLeod, *Inventing the Industrial Revolution*, p. 187.

[61] Stubbs, "John Beale, Philosophical Gardener of Herefordshire. Part II," pp. 349–53. For Beale's recommendation of Yarranton's work, see *Philosophical Transactions*, no. 131 (January, 29, 1676/7), 795–8. Note the example of Hartlib's "Office of Address." Mendyk, "*Speculum Brittaniae*," pp. 140–1; and see Sprat's notion of the Royal Society as a "general Banck" of natural science. Alan G. Gross, *The Rhetoric of Science* (Cambridge, Mass., and London, 1990), p. 167.

The viewing of wch, would often give occasion to Ingenious men, either to think of a New Invention, or Improve an Old one. And so, tho' some of those we have may be stol'n by Foreigners, yet we shall have others still to succeed them. For in that no Bounds can be sett, to the Varietys of Matter and Motion; not to Humane Will in the Manageing hereof; there will be Inventions New and Infinite to the end of the World.[62]

Grew's doctrine was a manifesto of enduring providence. Such, in effect, was a variant of the view already proposed by John Ray in his *Wisdom of God* that the uses of things had not been fully discovered.[63] By the time it was expressed by Ray, this was an old argument. It had been employed, for example, in 1668 in Joseph Glanvill's apologia for the Royal Society, which described the "inexhaustible variety of Treasure which Providence hath lodged in Things, that to the Worlds end will afford fresh Discoveries, and suffice to reward the ingenious Industry and Researches of those that look into the Works of God and go down to see his wonders in the deep."[64] If there was now a difference, Grew's bestiary contained those creations of human design.

The bounty of nature had become a conventional motif by the early eighteenth century. Throughout the period following the Restoration, defenders of the Royal Society were instrumental in asserting a positive role for natural philosophers in the exploitation of the natural world. It mattered little that their rhetoric far exceeded the reality of their achievements. The explorations of nature alone were sufficient to glorify the Creator and so the credibility of natural philosophy was to be established.

William Derham and the rage of physicotheology

The concurrent emergence of financial and commercial innovation in the 1690s created a new stage for philosophers. Philosophers could also claim to be the guardians of civic virtue; they were to become regarded seriously as the preservative against projectors. Therefore, natural philosophy was inherently linked to the justification of the exploitation of Creation while claiming to be able to prevent the excesses of unscrupulous projectors who preyed on the naive and the enthusiastic. In the absence of any formal policy or program of the Royal Society in the 1690s, this was a

[62] Hunt. Lib., MS. HM 1264, pp. 88–9.

[63] Ray, *Wisdom of God*, pp. 369–70.

[64] Joseph Glanvill, *Plus Ultra: Or, the Progress and Advancement of Knowledge Since the Days of Aristotle. In an Account of some of the most Remarkable Late Improvements of Practical, Useful Learning* (London, 1668), p. 7, quoted in Christine MacLeod, "Patents for Invention and Technical Change in England, 1660–1753," p. 329.

remarkable achievement.[65] Such was the effect of the pulpit and the text. The link between projectors and philosophers proved to be an effective foundation for raising the profile of natural philosophy in the public consciousness. The public awareness of natural philosophy evolved from the providential claims of the naturalists.

Probably no work so reveals the acceptability by the eighteenth century of the doctrine of providence than the many editions of William Derham's *Physico-Theology*. Derham was John Ray's literary executor and he was also demonstrably the heir to the tradition of Ray and Grew.[66] Published in 1713 as his "Survey of the Creation," the *Physico-Theology* was an enlargement of his Boyle lectures of 1711 and 1712.[67] Derham's delight in the framework of the living world equaled that of Grew and Ray; and Derham also found a way to assert the significance of invention and commerce in the Christian universe. It is difficult now to tell what the original text of Derham's sermons were really like, but it is clear why the printed versions were such a success. When the sermons were published, they contained not only detailed references to Scripture and to Ray, Grew, Hooke, and Halley, to name a few – but they were also based on a precise understanding of Newton's own contribution to the role of God in the world. By 1711, Derham had the advantage of an ascendent natural philosophy. In addition, unlike Richard Bentley, Derham did not get trapped in metaphysical tangles over the inherent properties of matter. It reveals some sense of the difficulties for Derham to ascribe the cause of gravity, following "the Principles of the Newtonian (the most rational of any) Philosophy," to "... that universal Law of Matter, imprinted on it at it's [sic] Creation by the infinite Creator, namely Attraction: Which is congenial with all the Matter in the Universe.... "[68] Metaphysics was not a web in which Derham would be caught.

65 Cf. Gillispie, "Natural History, Natural Theology, and Social Order," pp. 34–5.

66 A. D. Atkinson, "William Derham, F.R.S. (1657–1735)," *Annals of Science*, VIII (December 1952), 368–92.

67 Atkinson, "William Derham," p. 383. In 1713, he also edited for publication the third edition of Ray's *Three Physico-Theological Discourses*. William Derham, *Select Remains of the learned John Ray, M.A. & F.R.S. With his life...* (London, 1760), p. 65.

68 William Derham, *Physico-Theology: Or, A Demonstration of the Being and Attributes of God, from his Works of Creation. Being the Substance of XVI Sermons Preached in St. Mary le Bow Church, London at the Honble Mr Boyle's Lectures, in the Years 1711 and 1712*, second edition, with additions (London, 1714), pp. 31–2. His Doctorate in Divinity was bestowed in 1730 because of "the Service he had done to Religion by applying his Discoveries in Natural Philosophy to the Promotion of it." Bodleian Library, MS. Rawl. J. fol. 3, f. 25, n.d. See also Gillispie, "Natural History, Natural Theology, and Social Order," pp. 46–9.

Derham's Boyle Lectures can be taken to be one of the many vehicles that made Newton accessible – notably even before the *Principia*'s General Scholium appeared. But that is not its most significant effect. Derham was even more important for the manner in which he explicitly widened the scope of the natural theology of Richard Cumberland, Samuel Parker, Joseph Glanvill, and their heirs. For example, it was the materials of the vegetable world and their uses and not merely their structure he found worthy of praise.[69] Likewise, Derham had great interest in the provenance of mechanical contrivances; his curiosity about invention had led to the earlier publication of his *Artificial Clock-maker*. However, he was not simply concerned with the mechanism – his curiosity also ranged to the impact of human invention.

In a remarkable chapter in his *Physico-Theology*, Derham explored the ability of man to overcome what may first seem as limits imposed by Creation. As the Rector of Upminster, we might expect he should have focused on the effects on the spread of Christianity; "that the great Improvements made in the last, and present Age, in Arts and Sciences, in Navigation and Commerce, may be a means to transport our Religion as well as Name, throughout all the Nations of the Earth."[70] Had Derham's arguments merely foreseen the spread of a proper religion, he would have been yet another exponent of the Protestant Cause – but this would not have made him especially unique. What was new, and impressive, about Derham was the way in which he was able to articulate the commercial consequences of providence and innovation. There was, he argued, a duty that arose from the souls of men "to pursue the Ends, and answer all the Design of the Divine Providence in bestowing his Gifts and Graces upon him."[71] Therefore, man's intervention in and utilization of nature was elevated from crass opportunism to the rank of a duty of glorification – precisely as Richard Cumberland had argued for a morality arising out of the uses of nature. There were, Derham reminded his audience, scriptural commands that demanded man exercise his dominion over nature:

> . . . we must improve and employ our Gift to the Glory of the Giver; or in that Ministration, that Use of the World, for which he gave it. Our Stewardship, our Craft, our Calling, be it that of the Ambassadors of Heaven, committed to us,

[69] Derham, *Physico-Theology*, pp. 415–35.

[70] Derham, *Physico-Theology*, p. 289. For a discussion of Derham's use of invention, see William Coleman, "Providence, Capitalism, and Environmental Degradation. English Apologetics in an Era of Economic Revolution," *Journal of the History of Ideas*, XXXVII (1976), 32–6.

[71] Derham, *Physico-Theology*, pp. 290–1.

as 'twas to Timothy, by the laying on of Hands; or be it the more secular Business of the Gentleman, Tradesman, Mechanick, or only Servant; nay our good Genius, our Propensity to any Good, as suppose to History, Mathematicks, Botany, Natural Philosophy, Mechanicks, &c. I say all these Occupations, in which the Providence of God hath engaged Men, all these Inclinations to which his Spirit hath disposed them, ought to be discharged with that Diligence, that Care, and Fidelity, that our great Lord and Master may not say to us, as was said to the unfaithful Steward, Luk. 16:2. *Give an Account of thy Stewardship, for thou mayest be no longer Steward*: but that he may say, as 'tis in the Parable before cited, Matt. 25, 21. *Well done thou good and faithful Servant, thou hast been faithful over a few things, I will make thee Ruler over many things, enter thou into the Joy of thy Lord.*[72]

Natural theologians like William Derham did not invent mechanization or industrialism. But from the Newtonians among them came the case for the link between science and the uses of nature out of which a technological revolution might emerge. In Derham's mind, natural philosophy was merely the first step to fulfilling the command to make use of the creation. The great effort put to finding ways of ensuring water supply was to be sanctioned by such a creed. Derham, for example, explored the spring waters near his home in Upminster and went on to speculate on the cause of the rise of waters in the mountains by reference to Francis Hauksbee's recent experiments before the Royal Society on capillary action. From Hauksbee's efforts, Derham concluded that the "ascent of liquors" followed not from atmospheric pressure, but from "the Newtonian Attraction."[73] Science and Scripture were effectively united in Derham's approbation of improvement.

There was undoubtedly an amazing confidence that arose from the proposition that philosophers – especially the apostles of Newton – had the tools to make the discoveries and proper applications of the laws of nature. Such assertions were the result of a widespread sanction and a legitimacy bestowed by natural theology. So the *Spectator*, in the midst of Derham's sermons, repeated for the gentle reader the arguments for divine providence that were derived from natural history. In the finest

[72] Derham, *Physico-Theology*, pp. 291–2; see also Coleman, "Providence, Capitalism, and Ecological Degradation," pp. 35–6; cf. Margaret C. Jacob, *The Newtonians and the English Revolution, 1689–1720* (Ithaca, 1976), pp. 180–1.

[73] Derham, *Physico-Theology*, pp. 51–2. There is an extensive discussion of Derham and this issue in Stephen Switzer, *An Introduction To a General System of Hydrostaticks and Hydraulicks, Philosophical and Practical.... Containing in General A Physico-mechanical Enquiry into the Original and Rise of Springs, and of all the Hypotheses relating thereto; as also the Principles of Water-works,... Deduc'd from the Theory of Archimedes, Gallileo, Torricelli, Boyle, Wallis, Plot, Hook, Marriotte, Desaguliers, Derham, Hawksbee, and others* (London,1729), Book I, pp. 25–9, 45–6, 50–1, and notes.

Newtonian fashion, it compared the instincts of animals with the law of gravity, which could not simply be explained by properties inherent in material bodies.[74] By the end of the reign of Anne, a social consensus was being forged that included the Newtonian frame of world. In 1713, at the request of Henry Newman of the Society for Promoting Christian Knowledge, Steele's *Guardian* and the *Englishman* praised the publication of Derham's *Physico-Theology.*[75]

Providence and the religious philosophers

The genre soon found numerous promoters. The Reverend John Theophilus Desaguliers in 1718, in a letter to John Chamberlayne of the S.P.C.K. and translator of Bernard Niewentyt's manifesto of natural theology, *The Religious Philosopher*, expressed the opinion that such works brought back the religious sense of providence. *The Religious Philosopher* was placed within the same tradition of popular works like Derham's *Physico-Theology*, which used nature to "prove the Existence of a Deity and the Divine Authority of the Scriptures."[76] Likewise, in 1734, the lecturer Benjamin Martin cited the *Spectator* on creation and providence.[77] Natural philosophers were then accepting what popular papers had made a commonplace. The topology of providence was ultimately convincing. How much confidence and certainty was there by 1740 in the words of Charles Lamotte, Chaplain to the Princess of Wales, in his survey of nature?

> . . . what is more conducive to the Happiness and Advantage of Men, than the stately Trees of the Forest; which besides their Beauty and Ornament, not only shelter us by their Leaves from the scorching and sultry Heats, not only furnish Materials and Timber of our Houses, without which, we should live but very

[74] *The Spectator*, no. 120 (Wednesday, July 18, 1711).

[75] Leonard W. Cowie, *Henry Newman. An American in London, 1708–1743* (London, 1956), pp. 157–8; *The Englishman*, no. 175 (Thursday, October 1, 1713).

[76] Royal Society MSS., Journal Book (C), XI (1714–20), p. 178. May 2, 1717. Chamberlayne to Desaguliers; Bernard Niewentyt, M.D., *Het Regt Gebruik Der Werelt Beschouwingen* (Amsterdam, 1715); Niewentyt, *The Religious Philosopher: Or, the Right Use of Contemplating the Works of the Creator . . .*, trans. John Chamberlayne. To which is Prefix'd, A letter to the Translator, by the Reverend J. T. Desaguliers [dated February 2, 1717/8] (London, 1718–19). See Ben Vermeulen, "Theology and Science: The Case of Bernard Nieuwentijt's Theo-Logical Positivism," in Sergio Rossi, ed., *Science and Imagination in XVIIIth-Century British Culture* (Milan, 1987), pp. 379–90. On Chamberlayne, the first secretary of the S.P.C.K. and F.R.S., see W. K. Lowther Clarke, *A History of the S.P.C.K.* (London, 1959), *passim.*

[77] Benjamin Martin, *The Philosophical Grammar; Being a View of the Present State of Experimental Physiology, or Natural Philosophy* (London, 1735), pp. x–xi.

uncomfortable Lives, and be exposed to all the Injuries of the Air, and the Severities of the Seasons, but also by producing Materials for our Ships contribute to the Safety of our People, advance the Riches and Honour of our Nation, and convey the Wealth and Treasures of foreign Countries to our happy Shores. . . .[78]

Lamotte's wisdom was a reflection of what Derham had helped turn into a widespread motif. It was a convention that was ultimately to make industrialism desirable and legitimate.

For natural philosophers, their role had to be more clearly defined – and Desaguliers was to make their proper claim. It is not very surprising that Desaguliers should have been involved in the translation of the famous and much published physicotheology of the Dutch physician Bernard Niewentyt.[79] Physicotheological sentiments were widespread and the Newtonians gravitated to them quickly. Providence was the keystone, and proper philosophy the supporting pillars, of the entrance to the universe of Desaguliers and his associates. Significantly, in 1717, Desaguliers proposed both a note of caution and a reason for sound philosophy in a sermon before the King. He argued that even in prosperity, we ought to be warned by the punishments of others.[80] This was not gratuitous moralism. For those soon to face the South Sea Bubble, his words were prophetic. And for natural philosophers drawn to the projectors, this was an early formulation of an argument Desaguliers and others were to make for the benefit of true philosophy. Moreover, physicotheology of the kind found in Grew, Derham, or Niewentyt provided credence to the argument Desaguliers consistently made about the need for philosophers in the world of innovation. This was the next claim for natural philosophy – one not of contemplation but of direct action.

There can be little question that the physicotheologians were widely influential. The effect of William Derham, for example, can be seen later in the work of the engineer Martin Triewald, who was also one of Desaguliers' students. It was Triewald's opinion that inventions were effectively bestowed by God – as in the case of Newcomen in which "what happened was caused by a special act of providence."[81] This particular

[78] Charles Lamotte, D.D., *The Greatness of God's Works in the Vegetable World, and the Pleasure that attends the Searches and Enquiries after them* (Stamford, Lincs., 1740), p. 10; see also Benjamin Martin, *The Philosophical Grammar; Being a View of the Present State of Experimental Physiology, or Natural Philosophy* (London, 1735), p. vi.

[79] Wolfgang Philipp, "Physico theology in the age of the Enlightenment: appearance and history," *Studies on Voltaire and the Eighteenth Century*, LVII (1967), 1248–9.

[80] John Theophilus Desaguliers, A.M. and Chaplain to the Right Honourable James Earl of Carnarvon, *A Sermon Preach'd before the King, at Hampton-Court; on Sunday, Sept. 29, 1717* (London, 1717), p. 14.

[81] Quoted in Svante Lindqvist, *Technology on Trial. The Introduction of Steam Power Technology into Sweden, 1715–1736* (Upsalla, 1984), p. 306.

providence might have been going too far for many Anglicans. Invention and discovery were not so much the products of luck, but of a providence made available to those who took the trouble to become conversant with the laws of nature.[82] Providence, which had matured as an ideology in the seventeenth century, by the early eighteenth had become a doctrine of the assertion of man over nature with all of the virtue Church and Scripture could possibly allow. Such, for example, was the popular message of Daniel Defoe's *Robinson Crusoe* of 1720, when many suffered their own financial shipwreck. The unity of Christianity and economy helped create what Simon Schaffer has called "a ringing endorsement of a collective, polyglot, commercial and divinely sustained natural philosophy."[83] Some might have recourse to a particular providence to explain an invention or to justify a revolution; others like William Derham might more commonly refer to a general providence:

> Of these matters [inventions and improvements], no satisfactory Account can be given by any mechanical Hypothesis, or any other way, without taking in the Superintendence of the great Creator and Ruler of the World: who oftentimes doth manifest himself in some of the most remarkable Transactions of his Providence, or by some great Revolution or other happening in the World thereupon.[84]

The Newtonian apostles were singularly active in proving that natural philosophy made possible what clergymen were anxious to propound. In many cases, proponent and Newtonian activist were the same.

It was the physicotheologians whose battles had fractured the superficial calm of the Restoration and created an intellectual revolution. Newton's natural philosophy was part of that upheaval; but it also helped to fashion a change that was as much a political and social transformation as well. The difficulty historians have often faced is the hurdle of bringing together the philosophical with the practical. It is for this reason that it is inappropriate to single out the Royal Society as the monolithic, supreme authority over the scientific interests of Englishmen. We must attack the issue on two fronts, the social as well as the philosophical, in order to

[82] Martin Triewald, *The Art of Living Under Water*, quoted in Lindqvist, *Technology on Trial*, p. 304. Cf. Christine MacLeod, "Patents for Invention," pp. 329–30.

[83] Simon Schaffer, "Defoe's Natural Philosophy and the Worlds of Credit," in John Christie and Sally Shuttleworth, eds., *Nature Transfigured. Science and Literature, 1700–1900* (Manchester, 1989), p. 26; G. A. Starr, *Defoe and Spiritual Autobiography*, pp. 85–7, 185–97. On the background to this notion, see Charles and Katherine George, "Protestantism and Capitalism in Pre-Revolutionary England," in S. N. Eisenstadt, ed., *The Protestant Ethic and Modernisation* (New York, 1968), pp. 155–76.

[84] Derham, *Physico-Theology*, pp. 288–9; see also Coleman, "Providence, Capitalism, and Ecological Degradation," p. 34.

see the place of Newton's science at the interface of these two seemingly unique realms.

Providence was probably the single most widely adopted motif of English political culture at the end of the seventeenth century. It was a notion simple enough to satisfy the most absolute minds, and mysterious enough to allow of the most elaborate constructions for those unable to find a revelation in the complex fabric of both the political and the natural worlds. This proposition – expressed as economic providentialism – helped both attenuate and justify the ideology of private self-interest. Active and interventionist, the idea of providence was powerful because it permitted the numerous political, social, and intellectual threads of the fabric of revolution to be sewn in a convincing relation to one another.

The appropriation of providence by Newton's natural philosophy – significantly in the defense of Christian orthodoxy – helped to transform the notion of what was possible in the social as well as in the intellectual world. And what became possible, because of Newton and his minions, was the justification of the conquest of nature that has since come to characterize much of Western society. That, alone, would make the idea of providence worthwhile examining. But what the Newtonians managed to do was convince the eighteenth century that they had discovered the active powers operating in the universe and that, as natural philosophers, their claim to the key to the conquest of nature was not an idle one.

The involvement of the followers of Newton in the enterprises of the early eighteenth century linked providence and opportunity with a desire for social validity on the part of entrepreneurs and philosophers alike. Such sanction proved elusive. Legitimacy needed to be cloaked in an intellectual vestment fashioned from the view that the methods of philosophers might actually be worthwhile. It was to be located in the epistemology of experimentalism that the Royal Society had so assiduously cultivated from its foundation.[85]

[85] On the problems of an experimental epistemology, see Steven Shapin and Simon Schaffer, *Leviathan and the Air-Pump. Hobbes, Boyle, and the Experimental Life* (Princeton, 1985), esp. Chapter 2.

3

Whiston, Clarke, and the crisis of doctrine

The promotion of natural philosophy was hazardous. Thomas Sprat's faith in the potential of experimental philosophy had obviously served the interests of the Royal Society. Yet the sullen gaze of Stubbe and South had fallen upon Sprat, Glanvill, and all those who would promote the new science. The great unease about the practices and new discoveries of philosophers proved impossible to ignore. But there was a way out of this dilemma, the implications of which were extensive. Public science might overcome private speculation. Sprat's intuition was of undeniable import; he forcefully pointed out that the subtleties of private knowledge were vulnerable to "the loss of a Library, [to] the overthrowing of a Language, or [to] the death of some few Philosophers. . . . " In an age that lived in the wake of the Civil War, and in a book still widely known in the aftermath of the Glorious Revolution, Sprat had asked, "Why was Learning the first thing, that was constantly swept away, in all the destructions of Empire, and forein inundations?" His answer must have struck a few nerves. "It is, because Philosophy has been spun out, to so fine a thread, that it could be known but only to those, who would throw away all their whole Lives upon it. It was made too subtile, for the common, and gross conceptions of men of business." Such was not true of natural knowledge made common knowledge, more vigorous than that of philosophers "shut up in the shades of their walks."[1]

It is important to recognize that Sprat had linked the Royal Society to a doctrine of public knowledge. Of the ancient philosophy Sprat asked,

> What help did it ever bring to the vulgar? What visible benefit to any City, or Country in the World? Their Mechanicks, and Artificers (*for whom the True*

[1] Thomas Sprat, *History of the Royal Society* (London, 1667), pp. 118–19.

Natural Philosophy should be principally intended) were so far from being assisted by those Abstruse Doctrines; that perhaps scarce any one of those Professions, and Trades, has well understood Aristotles Principles of Bodies, from his own Time down to ours. Hence then we may conclude, that those first Times, wherein these Arts were made, had been nothing dammag'd; if, instead of raising so many Speculative Opinions, they had only minded the laying of a solid ground-work, for a vast Pile of Experiments, to be continually augmenting through all Ages.

And I will also add; that, if such a course had been at first set on foot, Philosophy would by this means have been kept close to material things; and so, in probability, would not have undergone so many Eclipses, as it has done ever since.[2]

Here lay the essential link between public knowledge and the experimentalism that was to form the vehicle of popular science. Sprat clearly understood that the activities of natural philosophers required broader sanction than their own curiosity could create. Why else write a *History*?

Providence and revolution politics

Natural philosophy could not exist independent of the religious and social order. This recognition was hardly less important for Newton than for his disciples. There were plenty who understood more was at risk here than a natural philosophy that might otherwise have been able to survive less troubled in the common rooms of Cambridge. Political allegiance and debates over the elasticity of Scripture by gospel-swilling clerics might still inform the disputes of the colleges – even in the relatively low-stakes poker of university dons and young turks.[3]

Newton's *Principia* was overshadowed by the urgency of political crisis. At the moment of publication, the universities were swamped by the political tides then building to engulf the intransigent James II.[4] It is little wonder that Newton's work remained virtually unknown to most Oxford and Cambridge undergraduates who undoubtedly found in the immediacy of the political moment much more to interest them. Cambridge, in particular, was in an uproar over the attempt by a Papist king to enforce

[2] Sprat, *History*, pp. 117–18. My italics.

[3] Cf. John Gascoigne, "The Universities and the Scientific Revolution: The Case of Newton and Restoration Cambridge," *History of Science*, XXIII (1985), 391–434; and John Gascoigne, *Cambridge in the Age of the Enlightenment. Science, Religion and Politics from the Restoration to the French Revolution* (Cambridge, 1989), Chapter 6.

[4] Even Huygens is supposed to have thought of Newton's mathematical work as extremely obscure. Edward G. Ruestow, *Physics at 17th and 18th-Century Leiden* (The Hague, 1973), p. 111. The authority cited is Jean LeClerc. See also I. B. Cohen, *Introduction to Newton's Principia* (Cambridge, Mass., 1971), pp. 145–61; and Richard S. Westfall, *Never at Rest: A Biography of Isaac Newton* (Cambridge, 1980), pp. 473ff.

the granting of a degree to a Benedictine monk by dispensing with the customary subscription to the articles of the Church of England. There was to be certain irony in this position given later Newtonian doubts about Anglican doctrine. A Catholic, however, was beyond the pale. Newton, then Lucasian Professor at Cambridge, was incensed by the continued efforts to undermine the Protestant Succession. He became a member of the University delegation to the Ecclesiastical Commission to lodge a formal objection. By 1689, Newton was the University's representative in Parliament, and hence saw from the center both the short, inglorious, reign of James II and the establishment of the House of Orange.[5] Cambridge was awash in revolution politics.

As religion was at the center of the revolution in Cambridge, so, too, would it be wherever the Protestant succession was in question. Throughout the next decade, when the revolution was still in danger and the succession still uncertain, the justification of William's reign was a matter difficult for the Church to ignore. But there were also deeper currents in the seas swirling about the new Calvinist King – and consequences that could not be avoided. Doctrine, too, seemed in danger. The Church was not unwary of the apparent growing immorality and increasingly bold pamphlets of alleged atheists and deists who undermined orthodox Anglicanism. Such an atmosphere spontaneously created conspiracy theories and a deep resentment of efforts to undermine the faith. Well into the reign of Queen Anne, the High Church remained on alert. Yet out of this apprehension of both High Church and Low, despite the chronic High-Church vituperation of the hostile Henry Sacheverell, grew campaigns of moral renovation and Societies for the Propagation of the Gospel and for the Reformation of Manners.[6]

[5] Frank E. Manuel, *A Portrait of Isaac Newton* (Cambridge, Mass., 1968), pp. 109–14; Millicent Barton Rex, *University Representation in England, 1604–1690* (London, 1954), pp. 302–9; and Gascoigne, *Cambridge in the Age of the Enlightenment*, pp. 77, 83, 95–7. It has been argued that the *Principia* was extensively an assault on the tenets of Catholic theology. See George Grinnell, "Newton's *Principia* as Whig Propaganda," in P. Fritz and D. Williams, eds., *City and Society in the 18th Century* (Toronto, 1973), pp. 181–92; and the comment in Steven Shapin, "Social Uses of Science," in G. S. Rousseau and Roy Porter, ed., *The Ferment of Knowledge. Studies in the Historiography of Eighteenth-Century Science* (Cambridge, 1980), pp. 97–8.

[6] Tina Isaacs, "The Anglican Hierarchy and the Reformation of Manners, 1688–1738," *Journal of Ecclesiastical History*, XXXIII (July 1982), 391–411; and T. C. Curtis and W. A Speck, "The Societies for the Reformation of Manners. A Case Study in the Theory and Practise of Moral Reform," *Literature and History*, III (1976), 45–64. Cf. Eamon Duffy, "Primitive Christianity Revived; Religious Renewal in Augustan England," in Derek Baker, ed., *Renaissance and Renewal in Christian History* (Oxford, 1977), pp. 287–300.

The early Boyle Lectures

The Low-Church clergy was just as anxious to avoid the suspicion aroused by its schemes for religious comprehension, and still more so by bold world makers at Court like Thomas Burnet. That is why part of the reaction to Burnet's adventures came from Robert Boyle in July 1691. At that time, Boyle's will established the lectureship that came to bear his name, specifically, "for proving the Christian Religion, against notorious Infidels, viz. Atheists, Theists, Pagans, Jews, and Mohametans, not descending lower to any Controversies, that are among Christians themselves. . . . " The early adherents of Newton soon seized the opportunity of this annual series of eight sermons. The Newtonian pulpit made its contribution to the war against impiety through attacks on Hobbes, against the spoor of Spinoza, and the swarm of deists about to overrun the Church. Or so it seemed. Beyond the requisite damnation of Hobbes, the purpose for some of the Newtonians was to affirm the usefulness of the new philosophy in defense of religion even if it ultimately meant, as it did for the Reverend Samuel Clarke, metaphysical hazards.[7]

As it happened, few of the Boyle lecturers like Samuel Clarke and William Derham were to explicitly employ Newton's natural philosophy. Margaret Jacob has ventured the opinion that the Boyle Lectures were one of the most important means by which Newtonian science became disseminated.[8] The significance of these sermons must be acknowledged, at least in conjunction with the multiplying number of public lecturers on science in Anne's reign. The effect of Boyle's legacy is quite another matter. Just as with the public lectures, it is difficult to gauge the result of any sermon. Can we not expect in the eighteenth century that parishioners dozed in the middle of yet another denunciation of deism? Unless, of course, the minister was Dr. Henry Sacheverell, and then his listeners were treated more to spectacle than to substance.

Even so, the printed versions of sermons like the Boyle Lectures made a most lasting impression – served up sometimes with a heavy dose of Newtonianism. This was obviously the case with the numerous editions of the Reverend William Derham's *Physico-Theology*, which began as Boyle Lectures in 1711 and 1712 and then became one of the most widely

[7] Margaret C. Jacob, *The Newtonians and the English Revolution, 1689–1720* (Ithaca, 1976), p. 144; and Jacob, "John Toland and the Newtonian Ideology," *Journal of the Warburg and Courtauld Institutes*, XXXII (1969), 308.

[8] Jacob, *Newtonians and the English Revolution*, pp. 172, 175–6. Cf. Michael Hunter, "The Crown, the Public and the New Science, 1689–1702," *Notes and Records of the Royal Society*, XLIII (1989), 110–11.

sold works of natural philosophy in the eighteenth century. Whether Derham's reputation or Newton's philosophy was the greater beneficiary is an interesting question. For the most part, however, many of the Boyle lectures had very little scientific content, often merely referring to the common argument from design in order to humble the freethinkers. The lectures of John Williams, the Bishop of Chichester and Chaplain to William and Mary, were a case in point. Williams actually declared in his third sermon in 1696 that it mattered little if "Philosophers and Naturalists write Volumes, and wrangle eternally" as long as his interest was not affected – unless, of course, matters of revelation were disputed by what could never really be "learned by the Book of Nature or Reason."[9] Even with a scathing glance at Thomas Burnet, Williams' general aversion to speculative philosophy was more typical of the majority of the lecturers. The Newtonian lecturers really remained an impressive minority.

In a world apparently rife with irreligion, the choice of the Boyle lecturers could be of great significance. The decision was to be in the hands of the trustees of the lectureship, the most important of whom were Thomas Tenison, from 1692 Bishop of Lincoln, and John Evelyn. Boyle had originally assigned the rent of his house in Crooked Lane for the support of the sermons. This proved uncertain and it was the generosity of Tenison that allowed Boyle's intention to be carried out.[10] Their first selection in 1692 was Richard Bentley, chaplain to Edward Stillingfleet, the Latitudinarian Bishop of Worcester. This was an interesting choice and the circumstantial evidence at least suggests that Newton may have had some role in it.[11] Whatever the provenance of this decision, Bentley was deeply interested in the theological consequences of the new philosophy. He had recently examined Edmund Halley on this issue when the Savilian Professorship of Astronomy had become vacant in 1691. Suspicions of atheism had scuttled Halley's hopes.[12]

Bentley adopted for his sermons "The Folly and Unreasonableness of Atheism," a topic of common clerical wringing of hands. Bentley was clearly aroused by the antagonism in which the debates over natural theology were then being conducted and probably by his experience with

[9] Jacob, *Newtonians and the English Revolution*, pp. 163–5; and John Williams, *The Characters of Divine Revelation. A Sermon Preached at St. Martins in the Fields, March 4, 1694/5. Being the Third of the Lecture For the Ensuing Year, Founded by the Honourable Robert Boyle, Esquire*, second edition (London, 1697), pp. 3–4.

[10] Edward Carpenter, *Thomas Tenison, Archbishop of Canterbury* (London, 1948), p. 34n1.

[11] Jacob, *Newtonians and the English Revolution*, pp. 153–6.

[12] Kubrin, "Providence and the Mechanical Philosophy," pp. 246–7. See also Simon Schaffer, "Halley's Atheism and the End of the World," *Notes and Records of the Royal Society of London*, XXXII (July 1977), 17–40.

the unfortunate Halley. He could safely follow the lead set by Tenison, who several times during 1691 had spoken out against the rising tide of atheism, ever a safe topic for an ambitious clergyman. In February, Tenison had preached on "The Folly of Atheism" before the Queen at Whitehall, and in November "Against the present Atheisme of this Wicked Age." In the existence of evil, there was infinite occupation for the clergy. According to John Evelyn, who witnessed Tenison's November sermon, the Bishop argued that the organization of the universe, "the regularity of every thing subservient to some end, for the good & support of the whole; are evident Miracles of the being & providence of a Superior & Infinite Wise Director."[13] This was also a posture that Thomas Sprat had adopted a generation earlier.[14] Tenison had summed up the fundamental axiom of the golden age of physicotheology. And it was in this vein that the Boyle lectures were to proceed.

The Newtonian philosophy was one more weapon in the campaign against irreligion. But by 1692, Bentley's objective was more specific. As an adherent to the Church of Tillotson, he clearly understood the implications of the wars of the world makers. In 1692, Thomas Burnet had also published his *Archaeolgiae Philosophicae*, which extended the attack upon Moses. Of course, Burnet was to be the object of the abuse of nervous divines, including the followers of Newton. Thus, Bentley used the forum of the Boyle Lectures to descend directly into the controversies of the world makers on the grounds that Burnet's propositions tended to the furtherance of atheism. It will be recalled that Burnet had supposed the original state of the earth to have been a perfect sphere, the crust of which was destroyed by natural forces. Bentley, in his final lecture of December 1692, identified this specific hypothesis with atheism because the design of the earth in his view, even the supposed imperfections of mountains and valleys, was evidence of a beneficent creator.[15] Bentley's sermons fit comfortably into the mold of John Ray's popular *Wisdom of God* of the previous year. Boyle's legacy had immediately enlarged the arsenal of physicotheology.

It was critical for the advocates of Newtonianism that they not attract

[13] E. S. de Beer, ed., *The Diary of John Evelyn* (Oxford, 1955), V, pp. 43, 75–7; and Margaret C. Jacob, "The Church and the Boyle Lectures: The Social Context of Newtonian Natural Philosophy" (Ph.D. thesis, Cornell University, 1969), p. 90.

[14] Sprat, *History*, pp. 131–2.

[15] H. V. S. Ogden, "Thomas Burnet's *Telluris Theoria Sacra* and Mountain Scenery," *ELH, A Journal of English Literary History*, XIV (1947), 148; David Charles Kubrin, "Providence and the Mechanical Philosophy: The Creation and Dissolution of the World in Newtonian Thought. A Study of the Relations of Science and Religion in Seventeenth Century England" (Ph.D. thesis, Cornell University, 1969), pp. 219–20.

the hostility generated toward Burnet; hence, it was imperative they distance themselves from him. It was precisely this strategy that made the efforts of physicotheologians like John Ray so important. Bentley's sermon on "Revelation and the Messias," before the commencement exercises at Cambridge in July 1696, showed the difficulty of defending reason while attacking the deists who glorified it. For Bentley, the denial of a divine mystery underscored the inability of pure reason to grasp "the notions of the Messias, his incarnation, mediation, satisfaction."[16] But reasonableness in religion – that creed of the Latitudinarians – established the Anglican departure "from the errors of Popery, and that we know where to stop; neither running into the extravagancies of Fanaticism, nor sliding into the indifference of Libertinism."[17] Here was the precarious balance for religion and for natural philosophy: between the warring factions of atheists and deists on the one side, and the orthodox champions of revelation on the other including High-Church Tories, Non-Jurors, and even Jacobites. It was here that the balance had to be struck nonetheless, in a place where reason had a positive role in religion "for there is certainly a wide difference between what is contrary to Reason, and what is superior to it, and out of its reach."[18]

The comets of William Whiston

During the 1690s, the crisis over the creation and dissolution of the world absorbed numerous enthusiastic combatants – and a number of increasingly alarmed observers.[19] In 1695, John Woodward, F.R.S., Professor of Medicine at Gresham College, and close friend of the Newtonian divine John Harris, used his skill in geology and his knowledge of fossils to argue that Burnet was probably correct in his assertion about the destruction of the original frame of the earth. But he did not agree that the dissolution could have been brought about solely from natural causes, but required the "Assistance of a Super-natural Power."[20] This was an

[16] Richard Bentley, *Eight Sermons Preach'd at the Honourable Robert Boyle's Lecture*... The Fifth edition. *To which is added a Sermon Preach'd at the Publick-Commencement at Cambridge July V; MDCXCI. when he proceeded Doctor of Divinity* (Cambridge, 1724), p. 348.

[17] Bentley, *Eight Sermons*, p. 351.

[18] Bentley, *Eight Sermons*, p. 352.

[19] The best general account is Roy Porter, *The Making of Geology. Earth Science in Britain, 1660–1815* (Cambridge, 1977), Chapters II–III.

[20] John Woodward, *An Essay Towards a Natural History of the Earth* (London, 1702), p. 164, cited in Macklem, *Anatomy of the World; Relations between Natural and Moral Philosophy from Donne to Pope* (Minneapolis, 1958), p. 31; John C. Greene, *The Death of Adam: Evolution and Its Impact on Western Thought* (New York, 1961), pp. 61–3.

astute rejoinder. Not only was it superficially orthodox, but it also indicated that mechanical explanations need not necessarily undermine Scripture. After having read Woodward's *Essay* in 1695, Bishop Moore's chaplain William Whiston corresponded with Woodward on the problem of biblical chronology.[21] By this point, as that great defender of modernity William Wotton observed, the deluge was "the Subject of most of the Philosophical Conversations of the Virtuosi about the Town."[22]

In the midst of all of this uncertainty, Whiston's *New Theory of the Earth* took up the challenge of dealing with the thorny issue of the Mosaic account of creation.[23] Essentially, Whiston agreed that the waters from rainfall alone could not create a general flood and that subterranean waters were necessary to provide some explanation for the rainfall in the Mosaic account. To this end, we find the introduction of a comet, the force and tail of which could explain mechanically the events surrounding the life and world of Noah. The approach of a comet, which Whiston later identified with Halley's comet of 1680, was used to account for the flood waters in two ways: first, the force of the tail would create sufficient pressure to cause cracking and straining of the crust until the release of subterranean waters, and, second, the vapors of the tail added to the atmosphere would induce the rains associated with the biblical flood. Moreover, the near approach of the comet would cause the axis of the earth to tilt toward the ecliptic, thus producing the seasons that follow the fall of Adam, as well as creating the diurnal rotation of the earth at the same time.[24]

The intercession of a comet was not Whiston's special revelation. It was derived from Newton's own speculations in the *Principia* on the tails of comets as well as from suggestions of Edmund Halley on the causes of the diurnal rotation of the earth. In 1694, Halley had presented to the Royal Society the notion of the approach of a comet, although he did not give any consideration to the effect on the sea. His paper was not published for thirty years "lest by some unguarded Expression he might

See also Joseph M. Levine, *Dr. Woodward's Shield. History, Science, and Satire in Augustan England* (Berkeley and Los Angeles, 1977), Chap. II.

21 Maureen Farrell, "The Life and Work of William Whiston" (Ph.D. thesis, University of Manchester, 1973), p. 2.34.

22 Quoted in Porter, *The Making of Geology*, p. 84.

23 Whiston, *A New Theory of the Earth, From its Original, to the Consummation of all Things. Wherein the Creation of the World in Six Days, the Universal Deluge, and the General Conflagration, As laid down in the Holy Scriptures, Are shewn to be perfectly agreeable to Reason and Philosophy* (London, 1696). For a general account, see Greene, *Death of Adam*, pp. 27–34.

24 Farrell, "Life and Work of Whiston," pp. 2.37–2.40.

incur the Censure of the Sacred Order."[25] Halley had very good reason to know the cost of such censure. In 1691, he had applied for consideration for the Savilian Professorship at Oxford, but he soon found that he had been accused of defending the eternity of the world, which would have been undoubted heresy.[26]

Whiston was hardly the last word on the matter – although he was certainly one of the loudest. Responses came from John Harris, who later reproduced much of Woodward in his *Lexicon Technicum*, and from the Oxford professor John Keill.[27] Keill had very little sympathy with Whiggish elements in the Church and, although he agreed that Whiston's approach was an improvement upon Burnet, he argued that mechanical causes could never be established. The magnitude of the deluge could only have been the work of God's direct providence.[28] Like Burnet, Whiston was faced with the accusation that he had imposed a limitation upon providence. Such was the quicksand of natural theology. As late as 1717, we find, in his *Cause of the Deluge Demonstrated*, Whiston's assertion that it was "physically demonstrable, that a Comet did pass by at that time [2349 B.C..], and was, under the Conduct of the Divine Providence, and as his Instrument in punishing a wicked World, the Cause of that Deluge."[29]

Whiston's purpose was not to prove that reason might explicate revelation. That philosophical proposition he took as given, despite the obstinacy of paranoid churchmen. He wished, however, to steal their thunder. His real aim was to elaborate the reality and meaning of Scripture prophecy. This meant finding such evidence in nature and in history that indeed proved Scripture correct. It is clear that some Newtonians knew very well what Whiston intended as a defense against Burnet. Bentley's

[25] Farrell, "Life and Work of Whiston," pp. 2.36–2.37; Greene, *Death of Adam*, p. 28; and Marjorie Nicolson and Nora M. Mohler, "The Scientific Background to Swift's *Voyage to Laputa*," in *Science and Imagination* (Ithaca, 1956), pp. 131–2.

[26] Kubrin, "Providence and the Mechanical Philosophy," pp. 2.46–2.47. For this problem, see Simon Schaffer, "Halley's Atheism and the End of the World," *Notes and Records of the Royal Society of London*, XXXII (July 1977), 17–40; and W. R. Albury, "Halley's Ode on the *Principia* of Newton and the Epicurean Revival in England," *Journal of the History of Ideas*, XXXIX (Jan.–Mar. 1978), 36–7

[27] John Redwood, *Reason, Ridicule and Religion: The Age of Enlightenment in England, 1660–1750* (London, 1976), p. 128; and Porter, *Making of Geology*, pp. 86–7. A useful if outdated survey of the debate is Katherine B. Collier, *Cosmogonies of our Fathers* (New York, 1934).

[28] John Keill, *An Examination of Dr. Burnet's Theory of the Earth, together with some remarks on Mr. Whiston's New Theory of the Earth* (Oxford, 1698), pp. 177–8; and Redwood, *Reason, Ridicule and Religion*, p. 123.

[29] William Whiston, *The Cause of the Deluge Demonstrated*, bound with his *Astronomical Principles of Religion, Natural and Reveal'd* (London, 1717), pp. 1–2.

Trinity disciple Roger Cotes wrote, in 1703, that Whiston's anticipated "exposition of ye Revelations" was acceptable news – and this might relate closely to the Rev. John Smith's own reasons why he "could not wholly acquiesce in Dr Burnet's account of ye Millenium."[30] As Margaret Jacob has shown, Whiston's Boyle Lectures in 1707 on prophecy were laced with millenarianism; he found the beast of the Old Testament in the empire of Louis XIV.[31] Natural theology would therefore champion both Scripture and the Protestant Cause.

Whiston's method was probably best revealed in a little known correspondence with Richard Allin of Sidney Sussex College, Cambridge. Allin seems later to have established a minor reputation as a botanist. But in the days when natural theology was the rage, he was concerned with perfecting Whiston's general hypothesis with evidence as various as it was suggestive. Allin took whatever he could find from the Scriptures and biblical chronology, from geographical and botanical evidence provided by travelers in the sinews of trade, and indeed from pure speculation. Anticipating continental drift, he suggested that before the deluge the continent of America was one with that of Africa, and that "the largeness of ye Northern part of America above ye Southern, & as of ye Western part of this Continent above ye Eastern, may in great part arise from ye Sediment of ye Deluge which mostly fell on this side of ye Globe."[32]

Whiston and Allin tried to work out the precise time of the flood and its immediate aftermath. Their correspondence is full of the concordance between astronomical phenomena and biblical chronology. Whiston constantly maintained astronomical incidents to be linked to the presence of a providential God – but the speculations of the correspondents were continually under revision. In 1707, during the summer hiatus of Whiston's Boyle lectures of prophecy, Allin attempted to improve the accuracy of Whiston's 1702 *Short View of the Chronology of the Old Testament*. Allin's efforts reveal that a serious student of the heavens found much to illuminate the Bible:

I cannot but propose a bold conjecture to you. You know ye Perihelion, from ye most accurate observations of its motion, ought to be in ye 12th degree of

30 Trinity College Library, Cambridge, MS. 4.42, no.1. Roger Cotes to Rev. John Smith, Rector of Gate Burton in Lincolnshire, December 11, 1703. Smith was the father of the Newtonian Robert Smith.

31 Jacob, *Newtonianism and the English Revolution*, pp. 130–3.

32 Leicestershire Record Office, Field-Barker MSS., f. 90(2). Ri. Allin to Whiston, November 29, 1699. See James E. Force, *William Whiston. Honest Newtonian* (Cambridge, 1985), p. 16.

Taurus at ye deluge; (for as to what you say, that Sr Isaac Newtons Calculation corrected, does not allow 56, but 53 1/3 degrees to its motion since ye deluge; his Theory of ye Moon evinces ye contrary, where he gives, with Mr Flamsteed, 50″ to ye Annual motion:) I therefore fix upon that day for ye deluge, when ye Earth was in ye 12th degree of Taurus: viz. Nov. 21st, at ye Meridean of Pekin, & suppose that ye Comet passed by abt noon. The Moon then wanted about 43 degrees of its conjunction with ye Sun, & consequently if ye Orbit of ye Comet were placed ye contrary way of what you suppose it . . . ye situation of ye Moon & ye Earth will be most accurate according to ye full moon. But I answer; first, ye place of ye Perihelion is a *Natural* phaenomenon, arising from ye laws of motion, which as you elsewhere observe, are rigid things, & not at all compliable with our fancies; . . . [33]

The preoccupation of the seventeenth- and eighteenth-century divines, and indeed of early Newtonians like Whiston, with such issues had a great deal to do with the public perception of science. It was clearly difficult, and undesirable, for Newton's apostles to conceive of separating Scripture from natural philosophy. This unification often left them open to charges of heterodoxy. But it also enabled them to claim, with justification, that their natural philosophy was founded on the existence of a providential creator.[34] When Whiston looked at the astronomical evidence for comets, he concluded that the one examined by Halley in 1680 was "none other than that very comet which came by the Earth at the Beginning of Noah's Deluge, and which was the Cause of the same."[35] By 1736, when he had already paid a heavy price for his unfettered conjectures on Scripture, Whiston still speculated on an apparent increase in the incidence of meteors and northern lights in the period since the Hanoverian Succession. To him this was deep with meaning. Mechanical powers, he argued, were ". . . deriv'd all from the constant Influence of the Supreme Being himself; as is now, upon full Evidence, forc'd to be allow'd by the very best Philosophers." But there were times when the "Mechanical Causes here fail us; and we must therefore have recourse to such as are Immechanical, as we have in Cases everywhere."[36]

[33] LRO, Field-Barker MSS., f. 94. Allin to Whiston, August 29, 1707.

[34] Roy Porter, "Creation and Credence: The Career of Theories of the Earth in Britain, 1660–1820," in Barry Barnes and Steven Shapin, eds., *Natural Order: Historical Studies in Scientific Culture* (London and Beverley Hills, 1979), pp. 99–101; and Whiston, *New Theory of the Earth*, pp. 357–61.

[35] Whiston, *Cause of the Deluge Demonstrated*, pp. 2–3.

[36] William Whiston, *An Account of a Surprising Meteor, Seen in the Air, March the 6th, 1715/16 at Night*, second edition (London, 1716), pp. 70–1. Whiston's reflections on the next two decades are in his hand in the margins on page 67 of this edition. British Library, Shelf mark 873.l.2 (1); and in Whiston, *An Account of a Surprising Meteor, Seen in the Air March 19. 1718/19 at Night*, second edition (London, 1719), pp. 33–34. British

So long as the relationship between natural philosophy and revelation initiated heated disputes, it was also inevitable that there would be those who would query where Newton stood. It was not at all the case that Newton had set out to provoke religious debate about the consequences of his natural philosophy. Far from it. But the very fact that followers of Newton like William Whiston paused merely momentarily to enter the battles of Burnet only reinforced prevalent suspicions. The defense of the Christian religion intended by Boyle's will explains why the Lectures could be a useful vehicle for Newton's disciples. But as far as Newton was concerned, the issue was a great deal more complicated than the artful lunacy of the debates over the Mosaic account of creation. During the last three decades of the century, Newton was anxiously exploring the questionable provenance of scriptural notions of the Trinity.[37] And in Newton's physical and metaphysical principles, there lurked profound and complex dangers. At stake remained the relation between natural philosophy and Church doctrine.

Long before Leibniz took delight in drawing attention to Newton's speculations concerning God, Newton's supporters like Richard Bentley or Roger Cotes had trouble coping with the properties of matter and space.[38] Properly understood, Newton's principles were formidable weapons. For his part, Dr. John Harris was content to accept the notion that the passivity of matter could provide a defense of the immateriality of God. In his Boyle Lectures of 1698, Harris argued that the existence of phenomena that had yet to be philosophically explained, such as gravity, light, sound, magnetism and electricity, could not lead us to deny their existence. So it was with the existence of God.[39] But what might

Library, Shelf mark 873.l.2 (2). See also G. S. Rousseau, "'Wicked Whiston' and the Scriblerians: Another Ancients-Modern Controversy," *Studies in Eighteenth-Century Culture*, XVII (1987), 17–44.

[37] William Andrews Clarke Memorial Library, Los Angeles. Newton, MS. "Out of Cudworth," ca. 1680s, from Cudworth, *The True Intellectual System of the Universe* (1678); and MS. "Pradoxical Questions concerning ye morals & actions of Athanasius & his followers," ca. 1690s. See Robert Charles Iliffe, " 'The Idols of the Temple': Isaac Newton and the Private Life of Anti-Idolatry" (Ph.D. thesis, University of Cambridge, 1989), pp. 62–63, 126; and Richard H. Popkin, "The Crisis of Polytheism and the Answers of Vossius, Cudworth, and Newton," in James E. Force and Richard H. Popkin, eds., *Essays on the Context, Nature, and Influence of Isaac Newton's Theology* (Dordrecht, Boston, and London, 1990), pp. 9–25.

[38] Newton to Bentley, January 17, 1692/3. H. S. Thayer, ed., *Newton's Philosophy of Nature* (New York, 1953), p. 53; and Gideon Freudenthal, *Atom and Individual in the Age of Newton. On the Genesis of the Mechanistic World View* (Dordrecht and Boston, 1986), esp. pp. 47–59.

[39] John Harris, "The Atheist's Objections against the Immaterial Nature of God, etc. Refuted," in S. Letsome and J. Nicoll, eds., *A Defense of Natural and Revealed Religion* (London, 1739), I, pp. 371, 397.

be explored experimentally, such as the interaction of material bodies, might also lead to notions about matter and the powers in the universe. Such was the route to considerations about the attributes of God. Ignorance about the natural world was among the causes of atheism. Harris's lectures followed the well-worn path of Cudworth, Bentley, and Ray; he quoted Romans, I:20 that "The invisible Things of God may be understood by the Things that are made, and his eternal Power and Godhead discovered by this Means."[40]

One might easily get the impression from Bentley and Harris that the Newtonian philosophy had become a major weapon against the alleged torrent of deism and atheism that haunted the post-Revolutionary regime. Yet the design and frame of the world was not an issue that Newton was immediately hailed for having resolved. Still less was this the case with his disciples, who managed to complicate matters by venturing publicly into scriptural battle grounds. The structure of the world was not solely the preserve of the Party of Newton.

The attributes of God

The pulpits frequently rang with dire warnings about those who would undermine revelation. This was especially true after the revolution, when there existed great tension within the Church regarding loyalty to the new regime, a conflict hardly eased by troubled consciences over oaths of allegiance and the doctrine of passive obedience. To this there was added alarms of a faith under seige by atheists, Unitarians, Socinians, and Deists thus completing the bestiary of religious heterodoxy. The paranoid fantasies of the orthodox fed on this overwhelming confederacy of ill-principled wretches who would unseat a King and emasculate a Church.

Among the Boyle Lecturers, the champions of orthodoxy were not at all silent. By 1708, some were clearly fed up with the speculations of some of Newton's followers even if Newton could not be held directly responsible. A more blunt approach was taken by John Turner when he declared that men simply were no judges of what is unnecessary, absurd, or impossible in a well-attested Revelation. Turner, who was the Vicar of Greenwich, gave his lectures immediately following those of William Whiston, who was at that point beginning to engage in a great deal of noisy speculation about the validity of the Athanasian Doctrine. By the time Turner reached the pulpit, he was certain that natural philosophers

[40] Harris, "The Atheist's Objections," p. 380.

were to blame for much of the rise of deism and atheism. Here was a chronic refrain that thundered across the pews of more than one parish church. In a criticism of the deist Matthew Tindal, published while Whiston's own lectures were being delivered, Turner pilloried the physicotheologians and their followers:

> The Endeavours of the learned philosophers of this age, to rob their maker of his honour, &c. by tying down all the phenomena of the heavenly bodies to their natural causes, and thereby denying the agency of Providence in all those things, is now grown so scandalous, that they are justly contemn'd by good men, everyway as knowing in those things, as themselves.[41]

Turner had deliberately linked the troubles of the Church to the speculations of natural philosophers. Although he had not mentioned the Newtonians by name, the implication was clear enough. Turner proved prophetic. The deviation from orthodoxy during Anne's reign by such leading Newtonians as Samuel Clarke and William Whiston would inevitably be viewed as an assault on the foundation of the Church of England.

The conflict over the interpretation of Scripture was a struggle over the proper place of natural knowledge in the postrevolutionary world. The epistemological foundations of a social revolution were to be forged in this contest. Thus, in an age where Thomas Burnet's *Sacred Theory* appeared in most libraries, the effect of the manifold presentations of Newton's science could be far- reaching.[42] Yet, since the first appearance of Newton's *Principia* in 1687 and before the publication of the first edition of the *Opticks* in 1704, only once had a knowledgeable follower of Newton been chosen to give the Boyle Lectures. Notwithstanding Bentley, that had been in 1698, when the ambitious John Harris had given the sermons, and at that point, there was little to distinguish his philosophical persuasions. Only recently, in 1696, had Harris been made a Fellow of the Royal Society, where, it became increasingly obvious, there were opportunities in the Party of Newton. Of course, by this stage, natural philosophers were still doing battle over the carcass of Burnet. In the midst of the battle, Harris would come to the aid of Dr. John Woodward and, it is likely, that Harris's selection as a Boyle Lec-

[41] John Turner, *A Vindication of the rights and privileges of the Christian Church* (London, 1707), quoted in Redwood, *Reason, Ridicule and Religion*, p. 25. On the breadth of Turner's campaign, see Gascoigne, *Cambridge in the Age of the Enlightenment*, p. 46.

[42] G.S. Rousseau, "Science Books and their Readers in the Eighteenth Century," in Isabel Rivers, ed., *Books and their Readers in Eighteenth-Century England* (Leicester, 1982), pp. 236ff.

turer was owed to Woodward's intervention.[43] But the general defense of natural philosophy obviously attracted Harris. This was the foundation of the Newtonian interest in the Boyle Lectures.

The sermons were initially just another vehicle in the war against infidels of all sorts, and the blundering of Burnet in particular. It was not until 1704 that the Newtonians seriously exploited the wider polemical possibilities in Boyle's legacy. During 1704 and 1705, the sermons were given by Samuel Clarke, at that point Newton's closest confederate, then in 1707 by William Whiston, followed in 1711 and 1712 by William Derham. This does not make the Boyle Lectures a Newtonian vehicle, but there is little doubt that the early Newtonians were anxious to accept access to a polite audience afforded by the Lectures.[44]

It would be fairer to describe the Boyle Lectures as an extension of the designs of the natural theologians. This had been true of Richard Bentley and it was, to a great extent, also the case with Samuel Clarke. There is no doubt that Clarke was provoked by the Deist John Toland's distortions of Newton's philosophy in Anne's reign. Clarke was also well on the way to becoming Newton's bulldog, snapping at the heels of those who dared challenge the new and popular Newton of the *Opticks*[45] But Clarke's own sermons reveal a foundation that had been constructed by natural theologians in the seventeenth century. This was the tradition to which the Boyle Lectures truly belonged. In his sermons in 1705, Clarke stated clearly the proposition that was at the root of all their physicotheology: "A constant and sincere observance of all the Laws of Reason and Obligations of Natural Religion, will unavoidably lead a Man to Christianity. . . ."[46] And the basis of this reason was to be in the observations of nature and the creation. This he had sought to demonstrate in his sermons of the previous year through reference to, among others, Boyle, Ray, and Newton.[47]

[43] Jacob, *Newtonians and the English Revolution*, p. 179; Avelina Bardwell, "Arts and Science in English Encyclopedias: The Lexicon Technicum of John Harris and the Cyclopedia of Ephraim Chambers" (M.A. thesis, University of Toronto, 1976), pp. 16–17; and Joseph M. Levine, *Dr. Woodward's Shield*, pp. 39–40.

[44] Cf. Jacob, *Newtonians and the English Revolution*, pp. 145–6.

[45] Jacob, "John Toland and the Newtonian Ideology," pp. 321–4; and Robert E. Sullivan, *John Toland and the Deist Controversy: A Study in Adaptation* (Cambridge, Mass., 1982), pp. 178–9.

[46] Samuel Clarke, *A Discourse Concerning the Unchangeable Obligations of Natural Religion, and the Truth and Certainty of the Christian Revelation. Being Eight Sermons Preach'd at the Cathedral Church of St. Paul, in the Year 1705 at the Lecture founded by the Honourable Robert Boyle, Esq.* third edition, corrected (London, 1711), pp. 14–15.

[47] Clarke, *A Discourse Concerning the Unchangeable Obligations*, p. 10; and Clarke, *A*

As a proponent of Newton, Samuel Clarke had an understanding of natural philosophy that was more sophisticated and more abstruse than many of his brethren. Nevertheless, Clarke's theology gave every appearance of being as grounded in nature as had the views of John Ray or Richard Cumberland a generation earlier. To many of his readers, however, it must have seemed that Clarke's forays into the recesses of metaphysics took them far beyond the comfortable realm of abundant nature to which they had been made accustomed.

Samuel Clarke was a more complex and subtle thinker than normally graced a parish pulpit. He was also one not easily dismissed. Clarke proved sufficiently indebted to the physicotheologians to build his metaphysics on the commonly accepted argument for the existence of God – "from the excellent Variety, Order, Beauty, and Wonderful Contrivance and Fitness of all things in the World, to their proper and respective Ends."[48] The World as we know it, he argued, could be described in terms of matter and the laws of motion. But, under no conditions, could matter and motion originate any form of life. From this, he concluded, the Epicurean hypothesis of a creative accident must fail. And thus he added, anticipating Newton's arguments of the "General Scholium," "we may by the by observe the Usefulness of Natural and Experimental Philosophy, sometimes even in Matters of Religion."[49] What began as an objection to the Epicureans and materialism in general became, in the tradition from Sprat to Newton, a polemic for natural philosophy as the preservative of religion. By 1705, Clarke extended this creed in a form that lent assurance to the idea that physical creation was indeed providential for those who would make use of it. From the attributes of God, Clarke maintained, we could know His intentions:

> All the same Reasonings and Arguments, which discover to Men the natural Fitness or Unfitness of Things, and the necessary Perfections or Attributes of God, proving equally at the same time, That which is truly the Law of Nature, or the Reason of Things, is in like manner the Will of God.[50]

For Clarke, physicotheology could not rest there. He was now prepared to press the argument from design into more remote metaphysical realms.

Demonstration of the Being and Attributes of God: More Particularly in Answer to Mr. Hobbs, Spinoza, And their Followers... Being the Substance of Eight Sermons Preach'd at the Cathedral-Church of St. Paul, in the Year 1704. at the Lecture Founded by the Honourable Robert Boyle (London, 1705), pp. 118–19, 141–2, 230–1.

[48] Clarke, *A Demonstration of the Being and Attributes of God*, pp. 118, 227–8.

[49] Clarke, *A Demonstration of the Being and Attributes of God* p. 120.

[50] Clarke, *A Discourse Concerning the Unchangeable Obligations*, third edition (1711), p. 118.

Clarke had reasserted what natural theology had long professed. Nature and nurture were the same in God's intention. But to accept that purpose also created some unease among the literal minded. One such was William Whiston, who was soon to suffer the gaze of the High Church quite as much as Clarke. Whiston later recalled that about 1706, Clarke brought his printed Boyle Lectures to Cambridge as a gift, but Whiston was much troubled by the book's subtleties. Showing Clarke a nettle in his garden, Whiston remarked, "That Weed contained better Arguments for the Being and Attributes of God than all his Metaphysicks." Clarke, reportedly, admitted as much, but suggested it was useful to employ the very weapons of Hobbes and Spinoza in defense of religion.[51] Whiston could clearly sense this led down old, and dangerous, pathways from which natural philosophy had tried to escape.[52] But it was not metaphysics that made Whiston a particular victim of High-Church fervor when he was ejected from Cambridge in 1710. It was his fervent righteousness.

Wicked William Whiston

The nature of the relationship between Clarke and Whiston is ambiguous, obscured in part by Whiston's churlish biography of his colleague. Nor was Whiston the most reliable of witnesses. His memoirs of his tumultuous life and friendships, colored by time and self-justification, were once described by a contemporary as "the very pinbasket of all his Rhapsodys."[53] It is certain that Whiston and Clarke had known each other since William's reign when they had come under the patronage of John Moore, the latitudinarian Bishop of Norwich.[54] They maintained contact

[51] William Whiston, *Historical Memoirs of the Life of Dr. Samuel Clarke. Being a Supplement to Dr. Sykes's and Bishop Hoadley's Accounts* (London, 1730), p. 11. This was a position to which Clarke would adhere throughout his life, arguing that "The Eternity of God, can no otherwise be proved, than by considering *a priori* the Nature of a Necessary or Self- Existent Cause. The Temporary phenomena of nature, prove indeed demonstrably *a posteriori*. . . . " See Clarke, *A Letter Concerning the Argument a priori In the Proof of the Being of God*... [added to] *A Discourse Concerning the Connexion of the Prophecies in the Old Testament, and the Application of them to Christ. Being an Extract from the Sixth Edition of a Demonstration of the Being and Attributes of God*, second edition (London, 1725), p. 57.

[52] See William Whiston, *Astronomical Principles of Religion, Natural and Reveal'd*... *Together with a Preface, Of the Temper of Mind necessary for the Discovery of Divine Truth; and of the Degree of Evidence that ought to be expected in Divine Matters* (London, 1717), p. 109.

[53] Geoffrey F. Nuttall, ed., *Calendar of the Correspondence of Philip Doddridge D.D. (1702–1751)*, Historical Manuscripts Commission, XXVI (London, 1979), p. 316n 1550. John Barker to Philip Doddridge, November 11, 1749.

[54] Whiston, *Historical Memoirs of the Life of Dr. Samuel Clarke*, pp. 7–8.

at least until the days of Whiston's Boyle Lectures in 1707 on the *Accomplishment of Scripture Prophecies*. By 1705 or 1706, it is also clear, both had severe doubts about the validity of the Athanasian doctrine, choosing to regard it as an embellishment of the later Church fathers and not of the primitive Church. Which of the two had the greater influence in reaching this conclusion is impossible to tell. We do know, however, that Whiston's enquiries into the Apostolical Constitutions also enlisted the aid of other friends like Richard Allin with whom he had long speculated on the nature of biblical chronology.[55] Whiston's irascible personality would not permit his doubts about Athanasius to remain hidden for long.

By the summer of 1708, Whiston chose to break his silence. He wrote both to the Archbishop of Canterbury and to the Archbishop of York requesting a thorough examination of the doctrine of the Trinity. Certainly, the Low Church did not wish to be embroiled in doctrinal controversies, and Tenison was rather curt in reply. Even a High Churchman like Sharp of York tried to advise further counsel with friends before Whiston rushed into print. The besieged Lord-Treasurer Godolphin, anxious to avoid further trouble with Convocation's high flyers, tried to pressure Whiston into laying aside his discussions.[56] It was an exercise in futility.

Whiston's old friend Bishop Lloyd of Worcester, who had ordained him and refused the same privilege to Sacheverell, seized the initiative and wrote to Whiston in July. Lloyd was alarmed by rumors that Whiston was "running into Socinianism," which if true, despite their friendship, Lloyd would be forced to regard Whiston as a "Subverter of Souls." Lloyd's warnings reflected a nervousness, when the Church was far from unified, that Whiston was capable of great damage, "carried away as you have been, by an extravagant fancy greedily set upon hunting after Novelties into things not only untrue but also hurtful in every booke that you have published of late years."[57] Whiston seemed genuinely surprised by Lloyd's tone and took a month to reply. His defense was that he was simply gathering those parts of Scripture that referred to the Trinity and Incarnation, that he intended to publish them ". . . and wtht any Hypothesis at all of my own have exactly followed those antient Testi-

[55] *Ibid.*, pp. 12, 31. Cf. Eamon Duffy, " 'Whiston's Affair': The trials of a Primitive Christian 1709–1714," *Journal of Ecclesiastical History*, XXVI (April 1976), p. 134.

[56] B.L., Lansdowne MSS. 1024, f. 189. White Kennett's Journal, July, 1708; Whiston, *Memoirs of the Life and Writings of Mr. William Whiston* (London, 1749), pp. 139–43; and Duffy, "Whiston's Affair," p. 135.

[57] B.L., Add. MSS. 24197, f. 2, Lloyd to Whiston, July 30, 1708.

monies."[58] By that point Dr. Lany, Vice-Chancellor at Cambridge, had refused Whiston a license to publish his *Essay upon the Apostolic Constitutions*. But Whiston was just warming to the battle. He was evidently angered by the suppression of the questions he had raised. "Wt is Popery & Priest Craft if this be not so?" he asked Lloyd.[59] Lloyd was having none of it, probably alarmed at the consequences such a conflict would create in Convocation:

> . . . you began with Philosophical matters. From thence you went on to Scripture History, and from thence to ye prophecies of Scripture, Now at last you are come to matters of faith; Wherein I doubt not to find you are as grossly mistaken as I am sure you are, and I will prove it to you, in every one of ye books you have published hitherto. I except only ye first book because in that as I understand Mr Keil has taken you to task and I am told he handle's you somewhat severely wth what Justice he doe's it I leave to ye Judgement of ye Virtuosi.[60]

Whiston made few friends with his scriptural speculations. His preaching of the Arian doctrine during his lectureship at a parish church in Cambridge got him into more trouble. Evidently, John Moore, Bishop of Ely, to whom Whiston had once been chaplain, was forced to dismiss him from his lectureship because of the growing criticism.[61] Even those who favored Whiston's excursions had to be extremely cautious about their relations with him. Joseph Hallett, who kept a Dissenting Academy at Exeter, corresponded with Whiston in the spring of 1710, but begged him not to reveal the fact lest Hallett be ruined.[62] Dissenters had enough to be concerned with, but by 1709, the High-Church rant of Henry Sacheverell sent a cold shiver through those dependent on the moderation of the revolution settlement.

The apostles of Newton thus became the focus of an intense campaign against the very heterodoxy the High Church had long anticipated from new philosophy. Reason proved no respite from the perils of religion, least of all from the railing of High-Church hysteria. It did not help that William Whiston was the victim of his own enthusiasm. Whiggish White Kennett, the Dean of Peterborough, remarked of him that he "is sunk at last into what may be modestly called Madness."[63] Whiston had a

[58] B.L., Add. MSS. 24197, f. 9, Whiston to Lloyd, August 26, 1708.
[59] B.L., Add. MSS. 24197, f. 12, Whiston to Lloyd, September 18, 1708; Duffy, "Whiston's Affair," pp. 135–6; and Farrell, "Life and Work of Whiston," p. 5.23.
[60] B.L., Add. MSS. 24197, f. 14, Lloyd to Whiston, n.d.
[61] Duffy, "Whiston's Affair," p. 136; and Thomas Hearne, *Remarks and Collections*, vol. II, ed. C. E. Doble (Oxford, 1886), p. 309.
[62] Whiston, *Memoirs of the Life and Writings of Mr William Whiston*, p. 146.
[63] Bodleian, MS. Ballard 7, f. 122. White Kennett to Arthur Charlett, January 29, [1709?].

knack for making powerful enemies and, it must be admitted, he made himself an easy target. Yet, as the Whig Bishops understood only too well, the power of Latitude depended on constraining the beast of enthusiasm. Since the revolution, the spectre of sectarianism was barely suppressed. Sects like the French Prophets had recently surfaced in London. And, periodically, divines like Thomas Burnet or Whiston threw away caution for the elaboration and purification of gospel. If ever Whiston had seriously hoped for an examination of Athanasius, those hopes evaporated in the blazing righteousness of Dr. Henry Sacheverell, chaplain of St. Saviour's, Southwark, and crier of the High Church.

High Church and Henry Sacheverell

On August 15, 1709, Henry Sacheverell preached before the Derby assizes on "The Communication of Sin" and managed to connect the rebellion against Charles I to both Puritans and Jesuits. By November 5, invited to preach before the Lord Mayor, Sacheverell's sensational canvas had widened. The master of fulmination elevated the sense of crisis to a dizzying height:

> . . . whosever presumes to Recede the least Tittle from the express Word of God, or to explain the Great Credenda of Our Faith in New-fangl'd Terms of Modern Philosophy, must publish a New Gospel, Un-God his Saviour, and Destroy his Revelation; . . .[64]

Faith may have been at risk, but Sacheverell's motives were more mundane. His immediate quarry was the Whig and Low-Church influence under Queen Anne; his sermon marked a sudden turn in a political crisis that restored the Tories to influence.

St. Paul's Cathedral echoed with the furious damnations of dissent, Latitude, and the Godolphin ministry. Sacheverell soon silenced the scoffing of those accustomed to dismissing his paranoid rages. There was little new in "The Perils of False Brethren both in Church and State," but, as Gordon Rupp describes it, "this renovated diatribe breathed fire and

[64] Henry Sacheverell, *The Perils of False Brethren, both in Church and State* (London, 1709), p. 9. Apart from the assault on the ministry, there was nothing essentially new in Sacheverell's attack. See Sacheverell, *The Natural Mischief of Prejudice and Partiality, Stated in a Sermon Preach'd at St. Mary's in Oxford, At the Assizes Held There, March 9 1703/4*, second edition (London, 1708), esp. pp. 11, 23. Echoed here are the complaints of the nonjuring George Hickes just after the Revolution about "the modern Latitude, wch hath brought all things to confusion, and will in time bring the church to ruine." Bodleian Library, MS. Ballard 12, f. 74, Hickes to Arthur Charlett, November 14, 1691.

Figure 1. "To Preach Up Truth": The handsome Henry Sacheverell (with a portrait of Charles I) in the aftermath of the 1710 sermon. By permission of The Huntington Library, San Marino, California.

slaughter from the beginning to the end."[65] The political assault on Godolphin gave Sacheverell's usual demagoguery new depths and he blasted the ministry for permitting the Dissenting academies to exist where

All the Hellish Principles of Fanaticism, Regicide, and Anarchy, are openly Profess'd, and Taught, to Corrupt and Debauch the Youth of the Nation, in all Parts of it, down to Posterity, to the Present Reproach, the Future Extirpation of our Laws, and Religion. Certainly the Toleration was never intended to Indulge, and Cherish such Monsters, and Vipours in our Bosom, that scatter their Pestilence at Noon-day, and will Rend, Distract, and Confound, the firmest and Best-settl'd Constitution in the World.[66]

Sacheverell had created grand theater, but not all were amused. Quite apart from the reaction of the Whigs, it was the fiery cleric's High-Church sympathizers who were also dismayed. There was much fear that Sacheverell's railing would prove counterproductive and obviously would hardly convince those already disaffected from the true cause. Even nonjuring George Hickes gave an scathing portrayal of Sacheverell as "a man of so little or no Learning so he is remark'd for several Blunders and odd mistakes in this Sermon, and particularly for his making Comets burst their Orbs, which is reckon'd as absurd as what he once observ'd in another Sermon about Parallel Lines meeting."[67] As a cleric, Sacheverell had been taken seriously by neither High Church nor Low. But as a performer, he had an ability to stir the mob, and that made him extremely dangerous. William Whiston later described Sacheverell as more of a player than a clergyman. Even Whiston's jaundiced view had a lot of truth in it.[68]

[65] Gordon Rupp, *Religion in England, 1688–1791* (Oxford, 1986), p. 66; Geoffrey Holmes, *The Trial of Doctor Sacheverell* (London, 1973), pp. 51, 66–7; and [White Kennett], *The Wisdom of Looking Backward, To Judge the Better of One Side and T'Other By the Speeches, Writings, Actions And other Matters of Fact on Both Sides, For the Four Years Last Past* (London, 1715), pp. 1–2. See also J. P. Kenyon, *Revolution Principles. The Politics of Party, 1689–1720* (London and New York, 1977), Chapter 6.

[66] Henry Sacheverell, *The Perils of False Brethren, both in Church and State: Set forth in a Sermon Preach'd Before the Right Honourable The Lord-Mayor, Alderman, and the Citizens of London, at the Cathedral-Church of St. Paul, on the 5th of November, 1709* (London, 1709), p. 15.

[67] Hearne, *Remarks and Collections*, II, pp. 320–321. See also Richard Parkinson, ed., *The Private Journal and Literary Remains of John Byrom*, Vol. I, Part I. printed for the Chetham Society, XXXII (1854), pp. 9–10. John Byrom to John Stansfield, December 17, 1709.

[68] This confirmed the view of John Hough, the president of Magdalene College, Oxford, during Sacheverell's student days: "He is really Proud and thinks too well of himself wch I have often told him in plain terms and wth as much severity as I could tho to little purpose. – I am sorry to see the follies of the Boy grow up with the Man." B.L., Lansdowne MSS. 1024, f. 113. John Hough to William Lloyd, June 26, 1697; and

Sacheverell could not go unchallenged. By November 20, White Kennett, Dean of Peterborough, unleashed his indignation in his *True Answer to Dr. Sacheverell's Sermon*. Two days later Richard Chapman, Prebendary of Chichester, called for the civil authorities to take some action. And the Whig Junto really had little choice. The Sacheverell sermon with its assault on the "insidiousness of such wiley Volpones" was correctly regarded as an attack on the ministry.[69] Even the Whig Bishop of Carlisle, William Nicolson, thought Sacheverell's main offense lay not in the sermon so much as in the printing of it.[70] Inducing popular disaffection against the Government could not be tolerated. And in December, the House of Commons ordered his impeachment.

When his trial opened before the Bar of the House of Lords on February 27, 1710, the scene was as promising, and entrance as much sought after, as the most popular play in Covent Garden. On that very day, the Secretary of the Royal Society, John Harris, delivered a blast against the Doctor entitled *The Evil Mischief of a Fiery Spirit*. In it, Harris accused Sacheverell of avowing "Persecution for the Sake of Religion." Harris's return volley implied a connection between the High Church and "the open Profession of no other Church but that of Rome."[71] But before the Lords, Sacheverell was the model of restraint, unlike the London mob that set to the sport of burning Dissenting Meeting houses. His was a

Geoffrey Holmes, *The Trial of Doctor Sacheverell* (London, 1973), pp. 7–9. On the attitude to the High-Church of Hough, later Bishop of Lichfield, see Kenyon, *Revolution Principles*, p. 99. In the aftermath of the Sacheverell trial, Whiston published a confirmation of Bishop Hough's opinion from other sources: "That he [Sacheverell] was fitter to make a Player than a Clergyman: That in particular, he was dangerous in a Family, since he would, among the very Servants, jest upon the Torments of Hell." In William Bissett, *The Modern Fanatick. Part II. Containing What is Necessary to Clear all the Matters of Fact in the First Part; and To Confute what has been Printed in the Pretended Vindication of Dr. Sacheverell, relating to myself...* (London, 1710), p. 20. Whiston to Henry Clements, January 17, 1710/11.

[69] Holmes, *Trial of Sacheverell*, pp. 77–89; Rupp, *Religion in England*, p. 67; G.V. Bennett, *White Kennett, 1660–1728. Bishop of Peterborough* (London, 1957), p. 104; White Kennett, *The Wisdom of Looking Backward*, pp. 2–4; and Richard Chapman, *Publick Peace Ascertain'd; with some Cursory Reflections upon Dr. Sacheverel's Two Late Sermons. In a Sermon Preach'd On Tuesday, Nov. 22. 1709* (London, 1709), "To the Reader," n.p. During January, as debates in the House progressed, White Kennett, out of personal interest and with a view perhaps to pamphleteering, sat in the gallery. Kennett had been a particular target of the High Church. A poem intended to satirize the deist Tindal managed to include Kennett as well, thus reinforcing the connection of deist and latitudinarian Sacheverell had denounced. Bennett, *White Kennett*, p. 107.

[70] Holmes, *Trial of Sacheverell*, p. 75; and B.L., Add. MSS. 6116, f. 18, Nicolson to Wake, January 5, 1709[/10].

[71] John Harris, D.D., *The Evil and Mischief of a Fiery Spirit. A Sermon Preach'd in the Parish Churches of St. Mildred Bread-Street and St. Matthew Fryday-Street, On February 27, 1709/10* (London, 1710), p. 9.

trial not about theology, but about politics and the direction of the Church in Anne's reign. The third article of Sacheverell's impeachment had accused the Doctor of sedition for suggesting that under the Godolphin ministry the Church was in danger. His defense was to name numerous contemporary authors in the tide of "Theists, Deists, Tritheists, Socinians, Revilers of Holy Scripture." He pointed to the works of Matthew Tindal, Thomas Firmin, Charles Blount, Thomas Burnet, the Earl of Shaftesbury, John Toland, and notably William Whiston's *Sermons and Essays*, which had only recently appeared.[72] For Sacheverell, they were the spectrum of discord facing the defenders of the faith.

Convicted before the end of March, Sacheverell's sentence to have his sermon burned and restraint placed for three years on his advancement was so light as to constitute grounds for High-Church celebrations. While Sacheverell and his entourage toured the countryside in the spring and summer of 1710, Bishop Nicolson nervously watched the "Spirit of Mobbism," which he compared to a school of herrings "most Numerous on the Coasts of Scotland. . . . But all agree that the Herrings travel round the Island in their proper Seasons."[73] Jacobitism always lurked in the shadows of Low-Church minds. The sermon of Sacheverell was a blunt reminder of the violence that the frustrated champions of the High Church could direct at unorthodox thinkers.

The spectacle of Sacheverell's triumph delighted the Tory countryside and spelled doom for the Whigs in the election of 1710. It was a sight that was to haunt Whiston. His *Sermons and Essays* had been named in the trial on March 6, thus making Whiston a prominent target that High-Church Tories could hardly ignore. Shortly after Sacheverell's celebration in Oxford, Whiston wrote in reply to the nonjuring Robert Nelson that he would not give up his cause even at "the hazzard of all my hopes & preferment, of my family, nay of my life it self in ys world."[74] Whiston

[72] *The Tryal of Dr. Henry Sacheverell, Before the House of Peers, for High Crimes and Misdemeanours; upon an Impeachment By the Knights, Citizens and Burgesses Assembled, in the Name of Themselves, and of all the Commons of Great-Britain: Begun in Westminster-Hall the 27th Day of February 1709/10, and from March following*, Published by Order of the House of Peers (London, 1710), pp. 276–7, 295–308. Whiston's work was *Sermons and Essays upon several subjects. . . . To which is added, Incerti acutoris De regula veritatis, sive fidei; vulgo, Novatiani De Trinitate liber* (London, 1709). This apparently appeared between Sacheverell's Derby and London sermons, probably in August 1709. See Hearne, *Remarks and Collections*, II, p. 252; and Eamon Duffy, " 'Whiston's Affair,' " 136.

[73] BL., Add. MSS. 6116, f. 22, Nicolson to Wake, May 11, 1710. See Holmes, *Trial of Sacheverell*, Chapter 10.

[74] Leicestershire Record Office, Field-Barker MSS., f. 97, Whiston to Robert Nelson, July 31, 1710.

remained defiant in the face of possible persecution. Martyrdom was in his makeup, but he may be forgiven for assuming he might have company at the gallows.

Samuel Clarke, science and scripture

The attractions of the Newtonians for religious controversy did not make them unique. They were, to some degree, victims of an atmosphere of suspicion about natural philosophy in general, a condition obvious to anyone familiar with the rage evoked by Thomas Burnet. But many were also victims of their own enthusiasm; Burnet, Bentley, Whiston, and Clarke were all churchmen with a taste for natural theology. By the end of the reign of Anne, Samuel Clarke had a firm reputation as a Low Churchman. In his Boyle Lectures in 1704 and 1705, Clarke had enunciated the fundamental proposition that Newtonians owed to the natural theologians:

> That the same God who created all things by the Word of his Power, and upholds and preserves them by his continual Concourse, does also by his All-wise Providence perpetually govern and direct the issues and events of things . . . [75]

Yet, it was also Clarke who broke the bounds of the conventional natural theology of the Boyle Lectures. This followed from his profound understanding of Newton. However appealing Richard Bentley's sermons, the performance was not surpassed until Clarke's comprehension of Newton's active realm in his sermons of 1704 and 1705. About that time, Clarke was also translating Newton's *Opticks* into Latin at Newton's request.[76] Clarke's Boyle Lectures in these years quickly slipped into what Whiston soon regretted as "the pernicious Consequence [of] Metaphysical Subtleties."[77] Apart from Whiston's complaint, Clarke's sermons strike a particular note to the reader of natural theology. They reach a much greater degree of sophistication than one expects from the sermons of the enthusiasts of design. And they are no less surprising for

75 Clarke, *A Discourse Concerning the Unchangeable Obligations of Natural Religion, And the Truth and Certainty of the Christian Revelation. Being Eight Sermons Preach'd at the Cathedral Church of St. Paul, in the Year 1705, at the Lecture Founded by the Honourable Robert Boyle, Esq.*, third edition, corrected. (London,1711), p. 253.

76 Whiston, *Historical Memoirs of Clarke*, pp. 12–13.

77 *Ibid.*, p. 131.

Figure 2. William Whiston, nemesis of Sacheverell and the High Church, ca. 1720. By permission of The Huntington Library, San Marino, California.

that. It is impossible to gauge the reaction of many contemporaries to Clarke's complexities, but there was no lack of reprinting of the lectures. By 1725, they had already gone through six editions with corrections and additions – interrupted and reinforced by the publication of his important letters against Leibniz in 1717 in which Newton had a hand.[78]

In 1704, there had seemed a special urgency. As Clarke then explained patiently to Whiston, he intended that the weapons of the Hobbesians – and, by implication, of deists and freethinkers – should be turned against them. Clarke's early lectures had the particular objective of an answer to the freethinking John Toland. This author of *Christianity Not Mysterious* and defender of Thomas Burnet was no convert to the Newtonian ideal of providence and active powers.[79] In 1704, Toland's *Letters to Serena* specifically cited Newton while distorting Newton's fundamental principle of the passivity of matter. Toland had translated the final portion of Newton's Definition II on inertia in a manner that emphasized the relative – and therefore vulgar – impressions of motion and of rest. It followed, he argued, that the idea of "defining Matter only by extension, of making it naturally inactive, and of thinking it divided into real Parts every way independent of one another" necessarily resulted in the "erroneous consequence" of a void space.[80] For Toland, matter became infinite and dynamic and he reinterpreted Newton's laws to include gravity as an inherent quality of body.

Clarke's sermons marked the firm establishment of the Party of Newton.[81] A sect they may have become, but it was necessary that Toland be answered. Newton was so alarmed by Toland that he began to draft a response between 1704 and 1706 of what was to become the twenty-third query to the Latin *Optice*.[82] The result was essentially the position

[78] *Ibid.*, pp. 131–2; Alexandre Koyre and I. Bernard Cohen, "Newton and the Leibniz-Clarke Correspondence," *Archives internationale d'histoire des science*, XV (1962), 63–126; and Richard S. Westfall, *Never at Rest*, pp. 778–9. In his biography of Clarke, Benjamin Hoadly, no stranger to controversy himself, claimed that Clarke's Boyle Lectures had established the unification of natural and revealed religion. Such hyperbole might be excused, but Hoadly was always ready to support an ally. See John Clarke, ed., *Sermons... Published from the Author's Manuscript*. Vol. I. *With a Preface, giving some Account of the Life, Writings, and Character, of the Author: By Benjamin, Lord Bishop of Salisbury*, second edition (London, 1730), pp. xv, xxvi.

[79] Jacob, *Newtonians and the English Revolution*, pp. 238–41; Jacob, "John Toland and the Newtonian Ideology," pp. 320–4; Robert E. Sullivan, *John Toland and the Deist Controversy*, pp. 175ff. On Toland's political agenda, see Simon Schaffer, "Newtonian Cosmology and the Steady State" (Ph.D. thesis, University of Cambridge, 1980), pp. 165–71.

[80] John Toland, *Letters to Serena* (London, 1704; reprint ed., Stuttgart, 1964), pp. 172–3, 231.

[81] Clarke, *A Discourse Concerning the Unchangeable Obligations*, pp. 13–14.

[82] Jacob, "John Toland and the Newtonian Ideology," pp. 323–4. See also J. E. McGuire

Figure 3. Samuel Clarke, metaphysician and anti-Trinitarian. Courtesy of Thomas Fisher Rare Book Library, University of Toronto.

Samuel Clarke was then to take in his Boyle Lectures. Although his rebuttal proved every bit as momentous as the assault upon Burnet a decade earlier, Clarke was far from alone in meeting the freethinking challenge. Humphry Ditton, a nonconformist teacher of mathematics who owed his position at Christ's Hospital Mathematical School to the influence of Newton, took on the challenge of Toland. He replied to Toland with sarcasm in the preface to his Newtonian *General Laws of Matter and Motion*. Moreover, William Wotton, chaplain to the family of Heneage Finch, and who apparently had some role in the selection of the Boyle Lecturers, quarreled with Toland's citation of Newton's law of gravitation and his refusal to admit a void space. "The truth is," said Wotton, "a Man that can comprehend the Grounds upon which Mr Newton builds that noble Discovery will hardly ever afterwards talk with Gravity against a Void."[83] Likewise, Samuel Clarke had Toland specifically in mind in his Boyle Lectures when he ascribed to atheists the idea of motion as essential to matter.[84]

The immediate intentions of Samuel Clarke's metaphysical excursions were clear enough. We should not presume, however, that such was the sole impetus that carried Clarke through perpetual quarrels with numberless advocates of orthodoxy ultimately including Newton's Hanoverian nemesis Gottfried Wilhelm Leibniz. More than obscure metaphysics occasioned these battles; at risk were important theological and religious issues. Clarke's Boyle Lectures had shifted subtly away from conventional physicotheology. He moved through the dark recesses of his metaphysics into scriptural assertions reached secretly by Newton and dramatically by more adventuresome spirits like William Whiston.

The greatest crisis of Newton's disciples would come over the fundamental doctrine of the Athanasian Creed. Whiston learned of the doubts of Clarke and Newton on the Trinity in about 1705 and 1706, that is, roughly at the time of Clarke's Boyle Lectures and his translation of the

and P. M. Ratansi, "Newton and the 'Pipes of Pan,'" *Notes and Records of the Royal Society*, XXI (1966), 118–119. As Helene Metzger so clearly pointed out, the concept of attraction that the Newtonian disciples proved so determined to defend was essentially a metaphysical notion. The relation between matter and motion was a function of the immediate presence of God in the universe reflected from time to time in providence and in prophecy. Helene Metzger, *Attraction universelle et religion naturelle chez quelques commentateurs anglais de Newton* (Paris, 1938), pp. 38–43.

83 [Wotton], *A Letter to Eusabia: Occasioned by Mr. Toland's Letters to Serena* (London, 1704), p. 66.

84 See Wotton, *Letter to Eusebia*, p. 71; and Clarke, *A Demonstration of the Being and Attributes of God* (London, 1705), pp. 45–59.

Opticks.[85] Indeed, in 1704, there were some hints of Clarke's concern over the principle of the unity of God. At that point, he was willing, if troubled, to reassert that belief ought to follow what was "declared and made known to us by clear Revelation."[86] What Clarke's congregation might have absorbed from the narrow arguments of his sermons is anyone's guess. But an attentive reader in 1705 might have noticed that Clarke meant to address the Trinity in referring to the "Diversity of Persons."[87]

By 1712, Clarke took a fateful step. He declared openly in his *Scripture Doctrine of the Trinity* what his natural theology had already led him to conclude. This was Clarke's most audacious work as it challenged directly the scriptural evidence for Anglican doctrine. Perhaps Clarke was sincerely attempting to meet the reservations of Whiston, who had complained of Clarke's "nice metaphysick Definitions."[88] But Clarke's scriptural exegesis still did not remove Whiston's aversion to the "Subtilty of a great Metaphysician."[89] Clarke's agitation over Athanasius proved quite enough for him to be dismissed in 1713 as one of Queen Anne's Chaplains in Ordinary, and in 1714, loudly denounced in Convocation. The footsteps of Thomas Burnet still echoed in Kensington Palace.

Religion was the singular fact that made all the accumulated resentment of decades rise to the surface. This was even more deeply felt by the summer of 1714 when the excesses of the alliance of Bishop Atterbury and Viscount Bolingbroke, and the subsequent divisions amongst the Tories, strongly united the Whig Bench in opposition. At risk in the heat of the late summer of 1714, when Anne slipped finally away from the life to which she had tenaciously clung, was the eroding and inept influence of High-Church zealots like Francis Atterbury. During that fateful summer, as accusations of Jacobitism drove Tories to the breaking point, Whig forces gathered confidence in the bonfires celebrating the new Protestant King. Bolingbroke's greatest fears took shape in the flames and by spring would drive him to his fate with the Pretender.

Church doctrine proved central to succession politics. In this atmosphere, the Boyle Lectures proved a dismal failure as a vehicle of Newtonian science. Not only were they to enter into the doctrinal conflicts

[85] Whiston, *Historical Memoirs of Clarke*, p. 12; James E. Force, *William Whiston*, pp. 15–18, 105–9; and Maureen Farrell, "Life and Work of Whiston," pp. 5.16–5.24.

[86] Clarke, *A Demonstration of the Being and Attributes of God*, pp. 95–6.

[87] Clarke made this absolutely explicit by 1711 in the third edition of his lectures by adding a reference to "the ever-blessed Trinity" and its attributes, but "always excepting Self-Origination, Self-Existence, or absolute Independency." Clarke, *A Discourse Concerning the Being and Attributes of God*, third edition, corrected (London, 1711), p. 51.

[88] Whiston, *Historical Memoirs of Clarke*, pp. 41–67, esp. 55–6, and pp. 70ff.

[89] *Ibid.*, p. 71.

among Christians, the Newtonians owed some responsibility for adopting Boyle's legacy to that exact end. Samuel Clarke, cognizant of the difficulties in Newton's philosophy, retreated into obscure debates in the maze of metaphysical maze of his Boyle Lectures. Although frequently reprinted and amended, they remained the passion of a privileged few. The rhetorical limits of the textual tradition, which might appeal to an intellectual elite, were not overcome by the Boyle Lectures. And Boyle's fears of textual and rhetorical controversies amongst Christians themselves were neatly ignored.

The rarified metaphysics of the Trinity ultimately justified the appeal of the public world. Whiston's hostility to metaphysics was readily reinforced by Samuel Clarke's adventures. The politics of the Trinity proved hard slogging for them both, but at least Whiston could avoid the intensity of the Newtonian crisis with Leibniz. This is not to say the Newtonian Royal Society refused to acknowledge the deeper philosophical problems of the mechanical philosophy – as Clarke and Desaguliers later revealed in their interest in the *vis viva* controversy.[90] But it was William Whiston who understood, at least, that public vindication would follow best from what was most easily demonstrable. Seeing was believing.

Whiston's wars

Whiston's star had once risen as Newton's successor to the Lucasian professorship at Cambridge. But because of his unorthodox teachings on the nature of the Trinity, he had come under censure not only by the University, but also by Convocation. In the rising Tory storm of Anne's last years he had become anathema, especially in High-Church circles, and was expelled from Cambridge in 1710 for heresy.[91] In the circumstances, it is surprising that any one agreed to come to his assistance. But appear they did in Henry Newman, the American Secretary for the Society for Promoting Christian Knowledge, and Richard Steele, the Whig

[90] Royal Society MSS., Letter Book (C), XIX, 192. "A Letter from the Revd. Samuel Clarke to Mr. Benjamin Hoadly, occasioned by the present Controversy among Mathematicians concerning the Proportion of Velocities and Force in Bodies in Motion". See *Philosophical Transactions*, XXXV, no. 401 (Jan.–March, 1728), 381–8; Register Book (C), XI, pp. 221–30. Desaguliers, "An Account of some Experiments . . .," November 8, 1722

[91] On the background to this crisis, see Eamon Duffy, "Primitive Christianity Revived; Religious Renewal in Augustan England," in Derek Baker, ed., *Renaissance and Renewal in Christian History* (Oxford, 1977), pp. 298–9; and Duffy, " 'Whiston's Affair': The Trials of a Primitive Christian 1709–1714," *Journal of Ecclesiastical History*, XXVII (April 1976), 137–9.

author, pamphleteer, and, in 1713, candidate for Parliament. In the summer of 1713, Whiston had written a letter that was published in Steele's *Guardian* concerning a scheme to discover the longitude while at sea. His project was the result of discussions with Humphry Ditton. At the same time, Steele and his partner, Joseph Addison, the Whig essayist and M.P. for Malmesbury, arranged for Whiston to deliver a series of lectures on mathematics at Button's Coffee House. Although recommended by Newman, Steele was warned to "conjure him silence upon all topics foreign to mathematicks in his conversation or lectures at the coffee house."[92] Steele was not easily intimidated and would suffer his own expulsion from the House of Commons for his inability to restrain his pen.[93] Obviously, Newman was referring to the pugnacious Whiston's theological tendencies, which were barely suppressed, if at all. Newman had good reason to sound a caution; that year was no time for expressing publicly anything that suggested heterodoxy. High-Church leaders like Bishop Atterbury, with the aid of Viscount Bolingbroke, were then gathering their forces for an assault on the Dissenters. Steele was no friend of the high flyers. And Whiston, who smacked of apostasy, was far too prominent a target to hope to escape attention.

Holy Scripture and high politics made a corrosive combination. William Whiston had a point that there was "a great deal of Policy and Cunning" in the proceedings against him. His assertion that the tactics of Sacheverell were now ascendent in Convocation was not mere paranoia. Whiston took particular exception to an apparent reference to his *New Theory of the Earth*, which, complained Atterbury's anti-Whig *Representation of the Present State of Religion*, had

> advanced nice Explications of Mysteries which are inexplicable; by the Misapplication of Mathematical Reasonings to Matters of meer Revelation; by the Attempts made to shew, how all the Steps taken in the great Work of Omnipotence, the Creation of the World, were to be accounted for by the known Laws

92 Calhoun Winton, *Captain Steele: The Early Career of Richard Steele* (Baltimore, 1964), pp. 157–8, n2. Newman likely knew of Whiston through his Boyle lectures of 1707, although Whiston was an early member of the S.P.C.K. from which he resigned in 1710 to avoid "contests and disputes." Newman later recommended another Boyle lecturer, William Derham. Leonard W. Cowie, *Henry Newman: An American in London, 1708–1743* (London, 1956), pp. 156–8. Whiston only remained at Button's for a few months, moving to larger adjoining quarters in the Piazza, Covent-Garden. *Daily Courant* (February 8, 1714)

93 Winton, *Captain Steele*, pp. 181–2, 187–8. During his expulsion, Steele like Whiston, proved a formidable adversary of High-Church policy, publishing a pointed attack on the motives of those behind the Schism Bill in 1714. See Steele, *A Letter to a Member of Parliament Concerning the Bill for Preventing the Growth of Schism* (Edinburgh, 1714).

of Motion; and that the Destruction of it, by the Deluge, might in like manner be explain'd.[94]

On the contrary, like other physicotheologians and Boyle Lecturers before him, Whiston claimed that it was his intention "To shew that Mathematicks, Reason and Philosophy do all attest to Revelation . . . [and] accord with, and necessarily suppose an Almighty God, and an All-wise Providence in the Government of the World."[95] One thing, at least, was obvious. Such a conflict over revelation and a providential nature was not going to be resolved by the disputes of scholars or churchmen.

Throughout 1712 and 1713, various attempts were made to revive the proceedings. By this point, the Queen retreated into vacillation, perhaps worn down by the acrimony surrounding Atterbury, perhaps aware of the reflections on her own reign in the constant reminders of religious infidelity. It seems the first representations from Convocation to the Queen concerning the Whiston affair were mislaid.[96] By 1712, Samuel Clarke's own reviews of the scriptural proofs of the Trinity added his name to Atterbury's list of blasphemers.[97] The sheer frustration of the Tory Rector of St. Anne's, Westminster, Dr. John Pelling at this spreading infection led to his laying charges of heresy against Whiston, probably at the instigation of Henry Compton, Bishop of London.[98] It was the intention to drag Whiston before the ecclesiastical courts because Convocation had proved a dismal failure.

Whiston's adventures in the last year of Anne's reign reveal how thoroughly partisan the whole theological furor had become. He was only

[94] *A Representation, of the Present State of Religion, with regard to the late excessive Growth of Infidelity, Heresy, and Profaneness: Unanimously agreed upon by a Joint Committee of Both Houses of Convocation, of the Province of Canterbury, and Afterwards rejected by the Upper House, but Passed in the Lower House* (London, 1711), p. 9.

[95] William Whiston, *Account of Convocation's Proceedings in Relation to Mr. Whiston. With a Postscript, Containing a Reply to the Considerations on the Historical Preface, and the Premonition to the Reader. To which is added, A Supplement to the foregoing Account of the Convocation's Proceedings* (London, 1711), p. 60. Dated July 25, 1711. Whiston constantly maintained this principle. See Whiston, *Astronomical Principles of Religion, Natural and Reveal'd* (London, 1717), p. 257.

[96] Lambeth Palace, Misc. MSS. 1029, f. 114, [John Moore, Bishop of] Ely to Tenison, February 10, 1711[/12]. Gibson MSS., Vol. XIII, no. 36, Tenison to Lord Dartmouth, June 20, 1713. At this time, Whiston also wrote to Sacheverell complaining of the references made to his works during the impeachment and in the Queen's presence. See Whiston, *Several Papers Relating to Mr. Whiston's Cause Before the Court of Delegates . . .* (London, 1715), pp. 148–151. "Mr Whiston's Letter to the Reverend Dr. Henry Sacheverell, Rector of St. Andrew's, Holborn," April 23, 1713.

[97] Duffy, "Whiston's Affair," p. 145.

[98] Historical Manuscripts Commission, *Portland MSS.*, VII (London, 1901), p. 172. Dr. William Stratford to Edward Harley, November 19, 1713; Force, *William Whiston*, pp. 18–19; and Duffy, "Whiston's Affair," pp. 146–8.

just beginning to make his mark as a public lecturer in natural philosophy and the Whigs, Steele and Addison, had initiated Whiston's lectures at Button's Coffee House. Evidently, Steele's patron, the Duke of Newcastle, helped defray some of Whiston's legal expenses before the Court of Delegates. Sir Peter King and Nicholas Lechmere, who had both served the prosecution in the Sacheverell impeachment, offered their services to Whiston without fee.[99] Whiston could not have made their job an easy one. His pen was never still in his own defense. He happily identified himself with Galileo, and bragged that the prosecutions made it easier for him to seek subscriptions to his works. Late in 1711, Bishop Compton complained that Whiston was promising to conceal the names of his subscribers lest they fear censure from Convocation.[100] Whiston's ultimate answer to Pelling's charges was further calculated to infuriate the high flyers – he compared his case to that of Sacheverell, even quoted Sacheverell in his own defense. Persecution merely enhanced a reputation. It had worked for Sacheverell whom the Queen had presented to the rich living of St. Andrew's, Holborn – William Whiston's own parish! Here the war began another round with Sacheverell determined to deny Whiston communion.[101]

William Whiston and the Whigs

The new science ultimately came to benefit from the patronage of the Court and of the Whigs. This was obviously the case with the lectures of William Whiston, who by January 1713 was performing with the elder Francis Hauksbee in Hind Court as well as privately at his home in Cross-Street, Hatton Garden.[102] Shortly thereafter, he encountered the Whig Richard Steele, who in 1712 had rented premises with the intention of promoting science as entertainment. Steele hoped to employ scientific lecturers interspersed with musical and dramatic engagements. In his "Censorium," it was hoped that "All the works of Invention, All the

[99] Duffy, "Whiston's Affair," p. 148; and Force, *William Whiston*, pp. 98–9. For Lechmere's involvement with Samuel Clarke, see Whiston, *Historical Memoirs of Samuel Clarke*, pp. 99–100.

[100] HMC., *Portland MSS*, V (Norwich, 1899), p. 76, Compton to Robert Harley, Earl of Oxford, August 27, 1711.

[101] [Whiston], *Several Papers Relating to Mr. Whiston's Cause*, pp. 12, 15, 21–4, 150; Rupp, *Religion in England*, p. 69. Whiston's letter to Sacheverell followed the Doctor's presentation to St. Andrew's by just ten days. B.L., Lansdowne MSS. 1024, f. 409, White Kennett's journal, April 13, 1713.

[102] *Daily Courant* (January 21, 1713).

Sciences, as well as mechanick Arts will have their turn. . . ."[103] Steele was ready to turn his entrepreneurial talents to a philosophical venture.

During the reign of Anne, there was a dramatic increase in the number of experimental lecturers. These soon included Whiston, James Hodgson, Humphry Ditton, Desaguliers, the elder Hauksbee as well as Hauksbee's competitive nephew. Often two of these would unite in a joint venture, like Whiston and Ditton, who lectured alternately with the elder Hauksbee, although by 1712, Ditton was engaged with the nephew.[104] By 1713, Ditton's place among the lecturers had been virtually taken over by William Whiston with whom he became fast friends. In the spring of 1713, Whiston was busy lecturing not only with the younger Hauksbee in Crane Court, but also on mathematics at Douglas's Coffee House in St. Martin's Lane near Covent Garden and at the Marine Coffee House in Birchin Lane near the Royal Exchange. It was at the Marine where Harris, Hodgson, and Ditton had lectured successively.[105]

After his Cambridge expulsion in 1710, Whiston rapidly rose to prominence in the galaxy of London mathematical practitioners. While designing instruments to be sold by John Senex, he gave lessons at this home in Hatton-Garden.[106] Apparently Whiston's lectures were very popular even with the fear concerning his "wrong notions in Religion," which brought him the label of heretic.[107] Whiston revelled in public scrutiny; a private martyrdom was not his goal and Tory John Gay later obliged with a parody on "Socianism, Arianism, and Whistonism."[108] Ever since the campaigns of Glanvill and Sprat, High-Church circles had regarded the new science as prejudicial to religion; even so, Whiston's lectures in 1713 were attended by Alexander Pope, who was increasingly

[103] John Loftis, "Richard Steele's Censorium," *Huntington Library Quarterly*, XIV (1950), 43, 53; Marjorie Nicolson and G. S. Rousseau, *"This Long Disease, My Life," Alexander Pope and the Sciences* (Princeton, 1968), p. 143.

[104] E. G. R. Taylor, *The Mathematical Practitioners of Tudor and Stuart England*, (Cambridge, 1954), pp. 296–7, 302; *Daily Courant* (January 10, 1712).

[105] *Daily Courant* (March 18, 1712). Mathematics was also being taught at the same time at the nearby Union Coffee House by Robert Baker. E. G. R. Taylor, *Tudor and Stuart England*, p. 439.

[106] E. G. R. Taylor, *Tudor and Stuart England*, p. 285; E. G. R. Taylor, *The Mathematical Practitioners of Hanoverian England* (Cambridge, 1966), pp. 7, 9, 143, 146–8; Nicholas Hans, *New Trends in Education in the Eighteenth Century* (London, 1966), pp. 142–3.

[107] Cited in Nicolson and Rousseau, *Alexander Pope*, p. 155. See also Howard Erskine-Hill, *The Social Milieu of Alexander Pope* (New Haven and London, 1975), p. 154. William, Lord Digby, to Whiston, December 15, 1712.

[108] [John Gay], "God's Revenge against Punning" (1716), in J. Swift and A. Pope, *Miscellanies* (London, 1732), III, p. 53. On authorship see G. S. Rousseau, "'Wicked Whiston' and the Scriblerians," pp. 33–4.

inclined to the Tories. Pope, however, was in rapture over the limitless boundaries of space he discovered in Whiston's company.[109] And it has been suggested that much of the public sense of science in the early eighteenth century can be traced to the influence of his poetry. Whiston may, indeed, be the source.[110]

Throughout his long career as a public lecturer, Whiston never gave quarter. In 1742, he combined a short course on a controversial method for discovering longitude with lectures at the Horn Tavern on "Models and Schemes of the Tabernacle of Moses, and of Solomon's, Zerobabel's, Herod's, and Ezekial's Temples at Jerusalem." Both courses included an account of a recent meteor.[111] Moses and meteors may have been glorious amusement, but to the novelist Samuel Richardson at Tunbridge Wells about 1750, the aging Whiston merely struck dumb the gaping rabble: by "... showing eclipses, and explaining other phenomena of the stars, and preaching the millenium, and anabaptism (for he is now, it seems, of that persuasion) to gay people, who, if they have White teeth, hear him with open mouths, though perhaps with shut hearts; and after his lecture is over not a bit wiser, run from him... to flutter among the loud-laughing, young fellows upon the walks like boys and girls at a breaking- up."[112] By this point, there were many who distrusted philosophers pandering to the masses.

Profound philosophical debates that put the Trinity on the rack would not solve what the simplicity of primitive Scriptures solely could assert.

[109] On Pope's politics, see Bertrand A. Golgar, *The Curse of Party: Swift's Relations with Addison and Steele* (Lincoln, 1961), pp. 134–5; G. S. Rousseau, "Science," in Pat Rogers, ed., *The Eighteenth Century* (London, 1978), p. 162. Pope's contact with Whiston has occasioned interest as a letter of August 14, 1713, refers to "dialogues with WHISTON." This has been assumed to mean Whiston's lectures at Button's Coffee House. Whiston, however, did not move from Douglas's to Button's, Covent Garden, until August 24, 1713. *Daily Courant* (August 24, 1713). There can be no doubt, at least, that Whiston was lecturing before the date of the letter. Cf. James E. Force, *William Whiston*, pp. 26, 166n77. In a later satire on Whiston, Pope's friend John Gay described Whiston's small audience as composed of tradesmen and apprentices who believed his prophecies of the end of the world. [Gay], "A True and Faithful Narrative of What pass'd in London during the general Consternation of all Ranks and Degrees of Mankind..." in Swift and Pope, *Miscellanies* (London, 1732), III, pp. 255–76. See G. S. Rousseau, " 'Wicked Whiston' and the Scriblerians," p. 27.

[110] Marjorie Nicolson and G. S. Rousseau, "*This Long Disease, My Life*," *Alexander Pope and the Sciences* (Princeton, 1968), pp. 25–6, 88–9. On this effect, see John Harris, *Astronomical Dialogues Between a Gentleman and a Lady*, second edition (London, 1729), p. 136.

[111] B.L., Add. MSS. 47012B, f. 92. Egmont Papers, from *London Daily Post, and General Advertiser* (Saturday, January 2, 1741–2).

[112] S. Richardson to Miss Westcomb, n.d., ca. 1750, in Anna Laetitia Barbauld, ed., *The Correspondence of Samuel Richardson*, III (London, 1804), pp. 318–19.

Public demonstration, accessible to all, alone might establish the fundamental acceptance of the Newtonian philosophy. Whiston was driven by his own theological passions; but for his contemporaries amongst Newton's apostles, metaphysics was traded for the manipulation of the natural world. This was a process that began with the moral economy and united the ideology of experiment with the universe of the materialist and the merchant. Experimental natural philosophy would become a commodity to be traded in the marketplace.

The experiment of the Boyle Lectures had seduced some natural philosophers beyond the interpretation of the codes of nature. On the one hand, Newton's apostles may merely have magnified the vogue of natural religion. Yet the intent of Boyle's will was also denied. The wounds of the Church were salted with this irony. Metaphysical and theological skirmishes explored the chronic feuds amongst Christians. Clergymen came to repeat old arguments and explore new, and increasingly obscure, eddies in the search for scriptural justification. On the other hand, the lectures had also magnified the providential fashion of natural religion. Nature was more malleable than not; invention meant improvement and had the providence of God to give it sanction.

PART II

The rise of public science

Superior Beings, when of late they saw
A mortal Man unfold all Nature's Law,
Admir'd such Wisdom in an earthly Shape,
And show'd a NEWTON as we show an *Ape*.

– Alexander Pope, *An Essay on Man. In Four Epistles to H. St. John, Lord Bolingbroke* (London, 1733–4), Epistle II, ll:31–34.

. . . said Adams, "I never write anything but Sermons, and I assure you I am no Enemy to Trade, whilst it is consistent with Honesty; nay I have always looked upon the Tradesman, as a very valuable Member of Society, and perhaps inferior to none but the Man of Learning." "No, I believe he is not, nor to him neither", answered the Host. "Of what use would Learning be in a Country without Trade? What would all you Parsons do to clothe your Backs and feed our Bellies? Who fetches you your Silks and your Linens, and your Wines, and all the other Necessaries of Life? I speak chiefly with regard to the Sailors."

– Henry Fielding, *The History of the Adventures of Joseph Andrews, And his Friend Mr. Abraham Adams* (London, 1742), Bk. ii, Ch. xvii.

4

Entrepreneurs of science

In the midst of political danger and the ardent passions of religion, the reception of Newton's natural philosophy was, at best, uncertain. The rages of philosophers and clerics who traded grim denunciations failed to settle the distemper of "Church in Danger" or the torments over the Trinity. Even the brilliance of the *Principia* had been eclipsed by events. It would prove difficult to assert the widespread intelligibility of Newton's principles in such circumstances.

There were other currents that would come to require a public presence for natural philosophy. Newton's reluctance to accept the concordance of philosophical validity with vulgar knowledge was as true for him of science as it was of Scripture. In its very construction, the *Principia* was a reluctant participant with a wider audience than its mathematics would allow. Only after Halley's urgent sollictations could Newton be convinced not to suppress that part which became "The System of the World" – for "those that will call themselves philosophers without Mathematicks, which are by much the greater number."[1] Newton's attitude was to prove increasingly anachronistic. The translation of private knowledge into public certainty had long been at the heart of competing visions of the Royal Society. Thomas Sprat had recognized this in the early days, and his rhetoric affirmed it. No new Atlantis was possible unless " . . . Mechanick Laborers shall have Philosophical heads; or the Philosophers shall have Mechanical hands."[2] Thus, the principle of accessibility to the arcana of philosophers, the move from private laboratories to public spaces, would

[1] *Correspondence of Isaac Newton*, ed. H. W. Turnbull (Cambridge, 1960), II, p. 443. Halley to Newton, June 29, 1686.

[2] Stephen Pumfrey, "Ideas Above his Station: A Social Study of Hooke's Curatorship of Experiments," *History of Science*, XXIX (March 1991), 17.

be the means to achieve the social legitimation that some of the early Fellows had once sought.[3]

Epistemology and public knowledge

Newton's work was not well understood. The revolutionary appearance in 1687 of Newton's *Principia Mathematica*, a difficult work of mathematical physics, occasioned scant hosanahs of revelation. William Petty's dismay that "I have not met one Man that puts an extraordinary value upon his Book" was one early reaction to the *Principia*.[4] The reason may be that few were equipped to deal with it, much less understand it. Petty's disappointment proved premature. But, even in the universities, those who had the knowledge to realize the importance of Newton's achievement were in the distinct minority. This was true even among those capable of comprehending the niceties of the calculus. Yet, it must be recognized that partly because of the mathematical analysis, it was deemed a considerable feat in the eighteenth century to understand Newton's system. The *Principia* presented a formidable edifice. In 1691, William Wotton inquired on behalf of a friend how best to tackle its mathematics. The mathematician John Craig, sufficiently cautious of his own capacities, told Wotton that he should tell his friend that "nothing less than a thorough knowledge of all that is yet known in most curious parts of the Mathematicks can make him capable to read Mr. newton's book. . . ."[5]

The mathematics were generally regarded an imposing obstacle to understanding Newton. Locke was forced to consult with Christian Huygens on the validity of the mathematical propositions. Locke's confusion was not much different than that of accomplished mathematicians such as David Gregory or John Craig, although the blind professor, Nicholas Saunderson, seems to have made better progress in making Newton intelligible to Cambridge undergraduates.[6] Newton was often misunder-

[3] Robert Charles Iliffe, "'The Idols of the Temple': Isaac Newton and the Private Life of Anti-Idolatry" (Ph.D. thesis, University of Cambridge, 1989), pp. 226, 236–7; Steven Shapin, "Who was Robert Hooke?" in Michael Hunter and Simon Schaffer, eds., *Robert Hooke. New Studies* (Woodbridge, 1989), esp. pp. 275–82; Shapin, "The House of Experiment in 17th-Century England," *Isis*, LXXIX (September 1988), 373–404; and Jan V. Golinski, "A Noble Spectacle. Phosphorous and the Public Culture of Science in the Early Royal Society," *Isis*, LXXX (March 1989), esp. 34–8.

[4] Quoted in James L. Axtell, "Locke, Newton and the Two Cultures," in John W. Yolton, ed., *John Locke: Problems and Perspectives* (Cambridge, 1969), p. 170. See also Iliffe, " 'The Idols of the Temple,' " p. 264.

[5] Trinity College Library, MS. R.4.7, no. 9, Jo: Craig to William Wotton, June 24, 1691.

[6] Dr. Thomas Nettleton, Rev. Mr. John Boldero, et al., "Memoirs of the Life and Character of Dr. Nicolas Saunderson," in Saunderson, *The Elements of Algebra* (Cambridge, 1740),

stood by his contemporaries, and in some cases even by his associates. Two decades after the publication of the *Principia*, the Reverend Thomas Byrdall, who had followed Newton's work closely, complained to Halley that "tis almost as difficult, fully to comprehend the Author as sufficiently to admire him."[7] Nevertheless, by that time, Newton was the beneficiary of a campaign that set out to subdue the limits imposed by his calculus. His work became known and praised, and the powers of attraction upon which his system was based became a fashionable mode of expression – from Richard Bentley's Boyle Lectures in 1692,[8] to William Derham's immensely popular *Physico-Theology* in 1713,[9] to Voltaire's approval of attraction as "the great Spring by which all Nature is mov'd."[10]

As the ripples of Newtonianism spread ever outward from Newton's own little-understood work, so then did the phenomenon of popularization become more evident. This was to be founded upon demonstrations of the principles of matter and motion that were revealed, in the Newtonian system, by way of experiment. At least, so it was believed. Within two years of Newton's death, John Conduitt, who had married Newton's niece Catherine Barton, was concerned that the philosopher's epitaph should make proper mention of the essential significance of experiment.[11] It was precisely this belief brought to fruition in the public lectures on natural philosophy that helped to secure the influence of Newton.

The public lectures represented the mediation between the experimentalist and his audience, which might grant assent and authority. John Conduitt, especially, idealized the doctrine of experiment, and Henry

pp. v–vi; G. A. J. Rogers, "Locke's *Essay* and Newton's *Principia*," *Journal of the History of Ideas*, XXXIX (1978), 224–7; Axtell, "Locke, Newton and the Two Cultures," pp. 176–8; and Richard S. Westfall, *Never at Rest: A Biography of Isaac Newton* (Cambridge, 1980), pp. 468, 470–1.

7 Royal Astronomical Society, MSS. Add. 4, f. 12, Byrdall to Halley, December 10, 1710.

8 Richard Bentley, *The Folly and Unreasonableness of Atheism Demonstrated from the Advantage and Pleasure of a Religious Life, The Faculties of Humane Souls, the Structure of Animate Bodies, & The Origin and Frame of the World: In Eight Sermons Preached at the Lecture Founded by the Honourable Robert Boyle, Esquire; in the First Year, MDCXCII*, fourth edition (London, 1699), p. 226

9 William Derham, *Physico-Theology: Or, A Demonstration of the Being and Attributes of God, from his Works of Creation. Being the Substance of XVI Sermons Preached in St. Mary le Bow Church, London, at the Honble Mr Boyle's Lectures, in the Years 1711 and 1712*, second edition (London, 1714), p. 52.

10 Voltaire, *Letters Concerning the English Nation* (London, 1733), p. 136.

11 B.L., Add. MSS. 6209, f. 261, John Conduitt to J. Ward, Nov. 4, 1730. Ward was Professor of Rhetoric at Gresham College. As Simon Schaffer has pointed out, this credo of the perceptibility of the effects of nature's laws laid claim to significant epistemological and theological positions. Schaffer, "Natural Philosophy and Public Spectacle in the Eighteenth Century," *History of Science*, XXI (1983), 1–2.

Pemberton linked the acceptance of Newton's ideas to confirmation through experience. But, the refraction of Newton's sunbeams revealed that written reports alone could not guarantee assent. It was not necessarily the experiment that was crucial of itself but the audience at the event or at its replication.[12]

Conduitt's appreciation was stirred by the generation of experimental lecturers who were anxious to move into waters that Newton recognized but did not chart. The refusal to feign hypotheses and the emphasis on experimentalism were reinforced in Conduitt's propaganda. Public lectures, of course, did not preclude private experiments. The chemical lectures of Peter Shaw, Francis Hauksbee, the younger, and Henry Pemberton were establishing new territory where they were followed by the electrical researches of Hauksbee and Stephen Gray. By the end of the 1730s, electricity had become an exciting frontier that provided both dramatic entertainment and a paying audience. By the 1740s, at the end of our period, the drama of electricity was all the rage although some like the engineer John Smeaton reacted with ambivalence:

> I don't take yt showing ye wonders of electricity for money is much more considerable than ye shewing any other strange sight for ye same end; however if £200 could be got by a worthy employment in yt way I don't see where is ye harm as there is no fraud or dishonesty in it.[13]

Smeaton's reservation about science in the public markets was a reaction against showmen who gave experimentalists a bad name. It was also a reflection of pressures neither the Royal Society nor natural philosophers found possible to ignore. The evolution of a system of public lectures, that advent of a popular Newtonianism, had within it a confirmation of practicality – an element of the structure of public science that has seldom been acknowledged.[14]

The tempests over the theologians were often also rooted in an epistemology founded in the problematic schemes of the experimentalists.[15]

[12] Cf. Simon Schaffer, "Glass Works: Newton's Prisms and the Uses of Experiment," in David Gooding, Trevor Pinch, and Simon Schaffer, eds., *The Uses of Experiment. Studies in the Natural Sciences* (Cambridge, 1989), Chapter II, esp. pp. 92–8.

[13] Schaffer, "Natural Philosophy and Public Spectacle," pp. 7–9; A. E. Musson and Eric Robinson, *Science and Technology in the Industrial Revolution* (Toronto, 1969), pp. 52–3. On Stephen Gray, see Lesley Murdin, *Under Newton's Shadow. Astronomical Practices in the Seventeenth Century* (Bristol and Boston, 1985), *passim*; and Michael Ben-Chaim, "Social Mobility and Scientific Change: Stephen Gray's Contribution to Electrical Research," *British Journal for the History of Science*, XXIII (March 1990), 3–24.

[14] Musson and Robinson, *Science and Technology*, pp. 27–9; and Margaret C. Jacob, *The Cultural Meaning of the Scientific Revolution* (Philadelphia, 1988), esp. Chapter V.

[15] Steven Shapin and Simon Schaffer, *Leviathan and the Air-Pump. Hobbes, Boyle, and the Experimental Life* (Princeton, 1985), esp. pp. 92–9, 310–16.

As Benjamin Martin later recalled, Burnet's and "... Mr. Whiston's, and all the other World-mongers Systems and Theories, dissolve into a philosophical Nothing, which want actual and repeated Experiments to support them."[16] The establishment of a new set of attitudes did not arise out of what was self-evident or readily apparent by demonstration. Rather, the very fact that experiments could be repeated publicly meant that acceptable knowledge was increasingly dependent on general consensus. Thus, Samuel Clarke's assertions about the self-evident contradiction of the Trinity or even the evidence for a vacuum were not merely claims of logic. They were demands for acknowledgment of the *evidence* upon which his conclusions were based. Thus, it was crucial to its public image that the Newtonian attack on the corruption of Trinitarianism was linked to the assertion of the power of evidence – whether scriptural or experimental.[17] This implication lay squarely behind Roger North's scepticism of the force with which the Newtonians glorified the results of experiment: despite Robert Boyle's continued "vexations of compound Bodys ... he could Never discover what any one of those products consisted off [sic]."[18] Scriptural evidence and experimental results were supposed to be evasions of the argument *a priori*. Such was the real power of the new physics and its experimental foundation. And the new epistemology emerged as public knowledge and consensus.

This had been the problem of Thomas Sprat's *History* – the tension brought on by competing forces of orthodoxy and revision. Dismissing the darkness of the alchemists, Sprat had claimed that "the true Philosophy is like to receive the noblest Improvements."[19] By such proclamations on behalf of experimentalists, Sprat was prepared to fashion a social credibility as well – "And seeing we cast Enthusiasm out of Divinity it self, we shall hardly sure be perswaded, to admit it into Philosophy." In a religious world, disputes could be settled by Church Councils or by force. But in a less deliberate social realm, where the pressures of transformation were ubiquitous, the assertion of legitimacy was to follow from the improvement of society. Sprat understood this to be a question of virtue: "... if they [the chemists] could be brought to content themselves

[16] Benjamin Martin, *The Philosophical Grammar*, third edition (London, 1748), p. 18. See also Roy Porter, "Creation and Credence: The Career of Theories of the Earth in Britain, 1660–1820," in Barry Barnes and Steven Shapin, eds., *Natural Order: Historical Studies in Scientific Culture* (Beverley Hills, 1979), p. 102.

[17] See Iliffe, "'Idols of the Temple,'" pp. 126, 148, 153.

[18] B.L., Add. MSS. 32548, f. 55; see Simon Schaffer, "Natural Philosophy and Public Spectacle," pp. 1–43.

[19] Thomas Sprat, *The History of the Royal Society of London, For the Improving of Natural Knowledge* (London, 1667), p. 37.

with moderate things, to grow rich by degrees, and not to imagine, they shall gain the Indies out of every Crucible: there might be wonderful Things expected of them."[20] With such claims, science would reach the public world with promises, if not proof, of widespread benefit when men "... injoy'd the pleasure of the true Speculative Philosophy, and the profit of the Practical."[21] In the end, this would produce a revolution in epistemology and social authority. No longer would the *text* be sufficient. Although abstruse metaphysics must prove impossible to resolve, true philosophy would emerge in the social world as that knowledge that made mines workable and engines more reliable. Such was public knowledge – no longer the possible illusions of metaphysical subtleties, but, as Sprat proclaimed, "they must pardon us, if we still prefer the joynt force of many men."[22] Consensus was all.

It would take generations between the Civil War and the Industrial Revolution for the crisis of evidence to be resolved. The authority ultimately accorded Newton's science demanded not only an audience, but a clear articulation by his closest apostles. By the early eighteenth century, much would depend on the existence of a group of Newtonian disciples that gave the appearance of a "sect" both within and beyond the Royal Society, whose purpose it was to propagate the gospel according to Newton. But can we truly speak in terms of a close-knit cabal of Newtonians who responded in lectures, sermons, and publications to a growing public demand for natural philosophy? Given the difficulties with Leibniz in the early eighteenth century, there can be no doubt that some like the brothers Keill, Dr. John Freind, or the Reverend Dr. Samuel Clarke understood themselves to be in a philosophical war. The famous remark made by John Keill within days of the death of Queen Anne and the impending Succession that "I hope Mr Leibnits ... will not have the impudence to show his face in England. If he does I am persuaded that he will find but few friends" distinctly suggests the Newtonians were prepared to deal with the Hanoverian.[23] By then Richard Mead, Humphry Ditton, William Whiston, John Harris, Francis Hauksbee, and Desaguliers had all become exponents of a new and seemingly more convincing natural philosophy than that of their continental rivals.

Newton's acolytes recognized themselves as "an enduring group of

[20] *Ibid.*, p. 38.
[21] *Ibid.*, p. 39.
[22] *Ibid.*, p. 39.
[23] Cambridge University Library, Add. MSS. 3985, f. 13, John Keill to Newton, August 6, 1714. See also A. Rupert Hall, *Philosophers at War: the quarrel between Newton and Leibniz* (Cambridge, 1980).

adherents" – to use Thomas Kuhn's phrase. Ephraim Chambers included the Newtonians in his definition of "SECT," which was "a collective term, comprehending all such as follow the doctrines or opinions of some famous divine, or philosopher, etc." This made them distinct from the disciples of Descartes' "philosophical Romance" or from those pedants who shunned mathematics. It was, Desaguliers later argued, "to Sir Isaac Newton's Application of Geometry to Philosophy, that we owe the routing of this Army of Goths and Vandals in the philosophical World."[24] So at the Succession, "Mr. Leibnits party," in John Keill's words, had been set up against Bernoulli's view of the "Idolatorum Newtonianorum."[25] On this basis, not only did Newtonianism assume an intellectual definition in relation to interparticulate forces, universal gravitation, or rules or reasoning, or whatever variations, it was also transformed into a social force through the propaganda and experimental practice of Newton's disciples. Newtonianism existed not in the text as natural philosophy, but in the text transmuted into gospel. Even the essential meaning of the scientific texts themselves is a function of the community interpreting them.[26]

Almost a century after Sprat, the problem of experimentalism was posed by George Horne, follower of the anti-Newtonian John Hutchinson. Wary of the overwhelming influence of the apostles of Newton, Horne complained of the performers of experiments "where no farther end is proposed by it, and it produces nothing but a stupid admiration, is a very low and servile employment for a man of genius." It was the wheelwright and the husbandman who deserved more admiration than "he who spends his time in measuring the tail of a comet, only to surprise and terrify mankind with a formidable range of cyphers." Indeed, as Horne claimed, many public scientific lectures may well have been "little better

[24] Desaguliers, *A Course of Experimental Philosophy*, second edition, corrected (London, 1745), I, pp. v–vi.

[25] Thomas Kuhn, *The Structure of Scientific Revolutions* (Chicago, 1962), p. 10; E. Chambers, *Cyclopaedia: Or, an Universal Dictionary of Arts and Sciences. . .*, fourth edition, corrected (London, 1741), II, s.v. "Sect"; *Correspondence of Isaac Newton*, VI, pp. 21–2. Leibniz to J. Bernoulli, August 8, 1713; p. 44, J. Bernoulli to Leibniz, November 25, 1713; p. 128, John Keill to Newton, May 17, 1714; p. 123, Leibniz to Bernoulli, March 29, 1715. In 1717, *The Newsletter* referred to Newton's "physical and theological disciples"; and in 1719, the anti-Newtonian George Gordon complained of Newton's admirers as "a party, produced and subsisting upon the same motives as other parties, and altogether the same nature with them." Quoted in Simon Schaffer, "Newtonian Cosmology and the Steady State" (Ph.D. thesis, University of Cambridge, 1980), p. 141.

[26] Robert E. Schofield, *Mechanism and Materialism: British Natural Philosophy in the Age of Reason* (Princeton, 1970), *passim*. Schaffer and Shapin, "Making Newton: On Interpreting Scientific Texts," unpublished.

than a raree-show. . . . " In Horne's mind, public lectures were designed to dupe the unwary by generating false evidence.[27] But those lectures had also brought natural philosophy beyond the limits of the text. It was to be through Newton's disciples and their demonstrations that science proved able to conquer the public culture of eighteenth-century Britain. The issue was public knowledge – what was public was not simply that confined to the country homes of the aristocracy or to those whom the Court chose to patronize.

John Harris and the Protestant cause

The notion that Newtonian philosophers as well as the Protestant Succession had much to fear from the vengeance of High-Church Tories, and perhaps of Jacobites, was not a fear unique to Whiston or to Clarke. This was dramatically demonstrated in the somewhat obscure history of John Harris, who was one of the first to turn his entrepreneurial talents to lectures on Newtonian natural philosophy. As a veteran of the Burnet wars and as the author of the *Lexicon Technicum*, Harris was one of most active exponents of Newton's philosophy. More than many of his contemporaries, Harris displayed a hectoring ambition with an ideological commitment as well to the Low-Church cause.[28]

[27] George Horne, *A Fair, Candid, and Impartial State of the Case Between Sir Isaac Newton and Mr. Hutchinson* (Oxford, 1753), pp. 55–6. Horne's tale in 1751 of Newton's apparent alarm over an instrument maker who could provide experimental evidence against the vacuum was really intended to erode the claims to certainty of public experiments. George Horne, *The Theology and Philosophy in Cicero's Somnium Scipionis Explained* (London, 1751), pp. 27–8. See also John R. Millburn, *Benjamin Martin. Author, Instrument-Maker and "Country Showman"* (Leyden, 1976), pp. 45–7.

[28] John Harris's personal history has always proved elusive. Some evidence suggests that he may have been the son of Edward Fitzharris, an Irish baronet who schemed with Catholics during the Popish Plot and who was executed for treason in 1681. Perhaps his enthusiastic Whiggism was the exorcism of a Tory devil. More certainly, the events of 1688 alerted a young scholar to the possibilities that followed from proper connection. After leaving Trinity College, Oxford, he was apparently the beneficiary of a family connection with Daniel Finch, Earl of Winchelsea, second Earl of Nottingham, and brother of Heneage Finch, who was concerned in the management of the Boyle Lectures. His career was seemingly also promoted by Simon Patrick, the Bishop of Chichester and later of Ely, one of the defenders of the Latitudinarians who came to power in the Church following the rise of Tillotson after the Revolution. See Grant Carr-Harris, "The Ancestry of John Harris (1667–1719) – Father of the Encyclopaedia," *Notes and Queries* (February 1978), 19–23; *D.N.B.*, IX, 13–14; Margaret C. Jacob, *The Newtonians and the English Revolution, 1689–1720* (Ithaca, 1976), pp. 43–8, 149–50. The Finch family was also patrons of John Sharp, Archbishop of York, who was probably related to Abraham Sharp, Flamsteed's assistant. See William Cudworth, *Life and Correspondence of Abraham Sharp* (London, 1889), pp. 267–9; and Henry Horwitz, *Parliament, Policy and Politics in the reign of William III* (Manchester, 1977), p. 68.

Harris's ascendancy was intimately tied to Church politics. In Whig preferment, he sought his salvation. In one of the few views that we have of him in the early years of the century, William Wake, then Bishop of Lincoln, described Harris as "a troublesome seeker" of preferment.[29] During the reign of Anne, Harris secured the support of Sir William Cowper, the Lord Chancellor, a moderate Whig and a Fellow of the Royal Society. Harris became chaplain to Cowper, who then presented him to a prebend in the Cathedral at Rochester. Significantly, Cowper was later one of those hot in pursuit of the high-flying Henry Sacheverell in 1710. Harris clearly began to flaunt his enthusiastic Whig political and theological sympathies after Cowper refused to accommodate the Tory ministry of Robert Harley in the summer of 1710.[30] Harris was certainly no friend of the High Church, which he regarded as the vanguard of Jacobitism.

Like other Newtonian clergymen, Harris proved a lusty advocate of the Protestant Succession. His sermons in the early days of Hanoverian rule displayed his passionate search for Jacobites.[31] As Rector of St. Mildred's, Bread Street, Harris stood guard against anti-Hanoverian agitation. One chronic irritation was Charles Humphreys, an old friend of Sacheverell and a lecturer of the parish.[32] In the heat of Humphreys' prayers and sermons there was a dangerous disaffection to the Government. And the Rising of the "Fifteen" proved the Jacobites were no phantoms. In October 1715, shortly after the House of Commons agreed to the arrest of Members of Parliament suspected of aiding the Pretender,

[29] Lambeth Palace Library, MS. 1770. Wake Diary, fols. 16, 21. May 4, 1706 and August 13, 1706.

[30] Geoffrey Holmes, *British Politics in the Age of Anne* (London, 1967), pp. 112, 242; *D.N.B.*, IX, p. 13. The *D.N.B.* article is based extensively on John Nichols, *Literary Anecdotes of the Eighteenth Century*, IX (London, 1815), pp. 773–4. Both must be used with caution. For a general account of Harris, see D. McKie, "John Harris and his *Lexicon Technicum*," *Endeavour*, IV (1945), 53–7. Simon Schaffer has suggested that Harris adapted his political associations to "country" interests in these years and consequently was excluded from the Wake group of Court Whigs. The evidence for a shift of allegiance from the Junto seems scant at best. If anything, it would seem Harris's political views were consistently hostile to the Jacobites. He may have alienated Wake and others, but that may simply have followed from the fact that he was so relentless in his pursuit of place. Cf. Schaffer, "Newtonian Cosmology and the Steady State," pp. 198–203; and Jacob, *Newtonians and the English Revolution*, p. 179n49.

[31] For example, see Harris, *The Wickedness of the Pretence of Treason & Rebellion for God's Sake. A Sermon Preach'd on November 5th. 1715. In the Parish-Church of St. Dunstan in the East, in the City of London* (London, 1715); and Harris, *A Sermon preach'd in the Parish-Church of Stroud, near Rochester in Kent, on August the 1st, 1715. Being the Day of Thanksgiving appointed for King George's Happy Accession to the Crown* (London, 1715).

[32] B.L., Lansdowne MSS. 1024, f. 222, White Kennett's journal, September 21, 1710.

Harris heard complaints that Humphreys had included in his preaching favorable comments on those deprived of their rights who were hiding in foreign countries.[33] To the nervous Harris, this was an open incitement to rebellion. When he complained to the Bishop of London – an implacable enemy of Samuel Clarke – concerning Humphreys' activities and got no satisfaction, this merely confirmed his suspicions. Harris felt compelled to dismiss Humphreys from the lectureship. It was no great surprise that Humphreys was to find a home with Henry Sacheverell in St. Andrew's Holborn, where they both met their match in that other Newtonian propagandist, William Whiston. Whiston lived in the parish and loudly demanded his right to worship even if it was under the nose of Sacheverell. As Whiston delicately suggested, it was God's church and not Sacheverell's.[34] Succession politics defined the field for bitter skirmishes.

John Harris's fastidious defense of the Church may have sold well in some quarters. To others, he was an unfortunate zealot. Even the Whig Archbishop of Dublin, William King, saw fit to warn Archbishop Wake in 1717 that Harris's search for preferment in Ireland in the entourage of the Lord Lieutenant was met with apprehension as "he brings no agreeable reputation with him."[35] Harris was one of that swarm of clergy forced to make a career of proper connection. The constant refrain that "Dr. Harris expects Preferment" ultimately soured an ally like Whiston.[36] Like many, Harris was unsuccessful in entirely mastering the political winds. But when he died in September 1719, the *Lexicon Technicum* was his lasting contribution to the victory of Newtonianism. On the other hand, Jacobites like the Oxford antiquary Thomas Hearne reviled him, complaining in 1713 that "The Books Harris had already printed are very

[33] [John Harris], *The Picture of a Highflying Clergyman: or, A True State of the Case Between John Harris, D.D. Rector of St. Mildred's Bread Street London; and Charles Humphreys, LL.D. late lecturer of that Parish* (London, 1716), pp. 7–8. The *D.N.B.* mistakenly ascribes this to Humphreys. B.L., Lansdowne MSS. 1024, fols. 483–4, White Kennett's journal. Some of the London clergy were suspected of Jacobitism and feelings ran so high in 1715 that there was instituted the practice of Whig "notetakers" in attendance at High-Church sermons on the lookout for sedition. J. Doran, *London in Jacobite Times* (London, 1877), I, p. 278

[34] *The Weekly Journal, OR, British Gazetteer* (Saturday, January 31, 1719), p. 1269; and (Saturday, February 7, 1719), p. 1276; and, William Whiston, *Mr. Whiston's Account of Dr. Sacheverell's Proceedings In order to Exclude him from St. Andrew's Church in Holborn* (London, 1719), *passim*.

[35] B.L., Add. MSS. 6117, fols. 66–7. William King to William Wake, August 28, 1717. King is not to be confused with the satirist of the same name who attacked Hans Sloane.

[36] Whiston, *Historical Memoirs of Clarke*, p. 155; see also James E. Force, *William Whiston. Honest Newtonian* (Cambridge and New York, 1985), p. 161n41.

silly. In his Lexicon Techn. things are repeated over and over again."[37] Harris's demise did nothing to blunt Hearne's savage sarcasm when he announced that "Last Week dyed Dr. John Harris (commonly call'd technical Harris), wch is no great Loss, he being a most rank Whigg, & a sad, vile, loose Wretch."[38] Harris had a desperation for preferment that antagonized even those who might have been helpful. His self-interest was an extreme instance of the Whig and Newtonian causes represented by Samuel Clarke and Clarke's disciple John Jackson.[39]

Harris and the growth of public science

For John Harris, practicality provided a vocation when preferment, as often as not, failed him. In his *Astronomical Dialogues Between a Gentleman and a Lady*, which he dedicated to Lady Cairnes, his student was made to suggest that the Newtonian philosophy might be taught in an "easy and delightful way."[40] But Harris had more in mind than drawing-room delights. He was intent on access to the concerns and fast pace of commerce as well as to the leisure of polite society. Not long after the Revolution, he taught pupils at his home in Amen Corner, near Paternoster Row, in the shadow of St. Paul's, which at that time was a hive of booksellers. More important, however, was his contribution to a series of public mathematical lectures that he delivered in London from 1698 to 1707.[41] Although the tradition of the scientific lecturer had an earlier start, the lectures for which Harris is known were founded in 1698 by Charles Cox, a Southwark brewer, who later, in 1701, was a Member of Parliament for the borough.[42]

[37] D. W. Rannie, ed., *Remarks and Collections of Thomas Hearne* (Oxford, 1898), IV, p. 164. April 23, 1713.

[38] *Remarks and Collections of Thomas Hearne* (Oxford, 1906), VII, p. 46. September 14, 1719. Harris was followed within two months by the death of his old nemesis Humphreys. *Weekly Journal or British Gazetteer* (November 28, 1719), p. 1455. Hearne's expressions about rank Whigs were at least consistent, for he held the same view about John Laughton, a friend of Newton at Trinity. See Gascoigne, "Politics, Patronage and Newtonianism," p. 11.

[39] Larry Stewart, "Samuel Clarke, Newtonianism, and the Factions of Post-Revolutionary England," *Journal of the History of Ideas*, XLII (Jan.–Mar. 1981), 58–61.

[40] John Harris, *Astronomical Dialogues Between a Gentleman and a Lady*, second edition (London, 1729), pp. 146–7.

[41] Harris advertised by way of a trade card, a method increasingly utilized by instrument makers. See H. R. Calvert, *Scientific Trade Cards in the Science Museum Collection* (London, 1971), p. 24, no. 173; E. G. R. Taylor, *The Mathematical Practitioners of Tudor and Stuart England* (Cambridge, 1954), p. 284; Taylor, *The Mathematical Practitioners of Hanoverian England* (Cambridge, 1966), pp. 124–5.

[42] Both Cox and the other Southwark M.P. and brewer John Cholmley subscribed to

Figure 4. John Harris, ca. 1719. By permission of The Huntington Library, San Marino, California.

John Harris's public lectures were instituted expressly "for the public good." Read first without fee in Southwark and then in the City at the Marine Coffee House, they soon catered primarily to the shipping interests and investors of Birchin Lane. Harris founded a career eagerly selling natural philosophy to such audiences. Of his lectures, it was said, "All thinking men will at last be able to distinguish between nice and useless speculation, and solid substantial, *profitable* Learning." As Thomas Sprat had shown, what was useful was not necessarily narrowly defined. By 1705, the lectures on geometry were linked directly to "the principles of true Mechanick Philosophy."[43]

At the beginning of Anne's reign, there were those like Dr. John Arbuthnot who attempted to unite mathematics and the mechanical philosophy with improvement and the ideology of public benefit. Arbuthnot had praised Charles II's establishment of the Mathematics School at Christ's Hospital and the institution of mathematical lectures during King William's wars "to breed up Engineers and Officers."[44] It was to one of these Army lectureships that Newton recommended Halley in 1697.[45] Popular natural philosophy soon mimicked public policy. The ideology of the public good and Harris's relationship with the shipping community undoubtedly explains why his initial lectures were devoted to the solution of navigational problems. It was likely through such contacts that Harris was induced to compile his *Navigantium atque Itineratium Bibliotheca; or, A Complete Collection of Voyages and Travels*, which included an account by Harris of navigational technique.[46] Meanwhile, amidst the bustle of Birchin Lane Harris made the acquaintance of Daniel Midwinter,

Harris's *Lexicon Technicum*. Taylor, *Tudor and Stuart England*, p. 284; *D.N.B.*, IX, 13; Samuel Wilks and G. T. Bettany, *A Biographical History of Guy's Hospital* (London, 1892); Robert Walcott, *English Politics in the Early Eighteenth Century* (Cambridge, Mass., 1956), p. 164. Little is known of Cox, although he seems to have had an abiding interest in natural philosophy as he also subscribed to Pemberton's *View* in 1728. He had entered the Brewer's Company in 1686. Guildhall Library, MS. 5449a, Brewer's Company, Freemen and Livery List, 1665–1754. Both Harris and Cox also appear in the list of petitioners who obtained a patent for an "Amicable Society for a Perpetual Assurance Office." Cecil T. Carr, ed., *Select Charters of Trading Companies, A.D. 1530–1707. Selden Society*, XXVIII (1913), pp. 249–50.

43 Taylor, *Tudor and Stuart England*, pp. 284, 294; Aytoun Ellis, *The Penny Universities: A History of the Coffee-Houses* (London, 1956), pp. 126–7; Bryant Lillywhite, *London Coffee Houses* (London, 1963), p. 360. My italics. The lectures in Southwark are referred to in the dedication to Cox in F. Ignat Pardies, *Short, but yet Plain Elements of Geometry and plain Trigonometry*, trans. John Harris, second edition (London, 1702).

44 [Arbuthnot,] *An Essay on the Usefulness of Mathematical Learning, in A Letter from a Gentleman in the City to his Friend at Oxford*, second edition (Oxford, 1721), p. 31.

45 *Correspondence of Isaac Newton*, IV, p. 229. Newton to Halley, February 11, 1696/7.

46 Harris, *A Collection of Voyages and Travels, Some now Printed from Original Manu-*

who became publisher of his lecture texts. It was Midwinter who was to engage with a group of seven booksellers to undertake the famous *Lexicon Technicum*.[47]

Within the bounds of the mathematical lectures, Harris and his followers fostered an interest in mechanical philosophy. He was drawn into the rage for experimental lectures exploited soon after the Revolution through the reign of Queen Anne by the chemists George Wilson and Edward Bright, and that blossomed in the rivalry between the instrument makers Francis Hauksbee, senior and nephew, ultimately absorbing university lecturers like William Whiston and J. T. Desaguliers.[48] The interest in experimental demonstrations continued when Harris was succeeded in 1707 by James Hodgson, who gave a course on mechanics. By then Hodgson, was a veteran of the public lectures having left the employ of the astronomer Flamsteed in October 1704 to join with the elder Francis Hauksbee in lectures on experimental philosophy. This had caused Abraham Sharp some unease as he wrote to Flamsteed in 1704 that Hodgson's venture "is a great undertaking for one, especially so young a man, to teach all parts of experimental philosophy, and the apparatus or instruments in order thereto which he intimates is now making for him. It will, in my opinion, be too great a charge for a single purse."[49] Little did Sharp know what an industry the lectures were to become. In January 1705, Hodgson advertised classes in natural philosophy and astronomy to "lay the best and surest Foundation for all useful knowledge" at the Hand and Pen in St. Paul's Churchyard. This was formerly Ayers' Writing School.[50] By 1706, Hodgson was lecturing on experimental philosophy

scripts. Vol. I (London, 1704). The market for such books was as competitive as that of the public lecturers. It may be that Harris intended a rival to the *Collection of Voyages and Travels* published by the noted Whigs John and Awnsham Churchill who were also responsible for Flamsteed's *Historia Coelestis Britannica*. Nichols, *Literary Anecdotes*, IX, p. 771; *D.N.B.*, IX, 13; and J. F. Scott, ed., *The Correspondence of Isaac Newton*, IV (Cambridge, 1967), p. 438.

47 Taylor, *Tudor and Stuart England*, p. 284; *The History of the Works of the Learned*, IV (July 1702), p. 417.

48 On the lectures of the chemists, see Musson and Robinson, *Science and Technology*, pp. 25–6, 37. In the late part of Anne's reign, Edward Bright lectured at his "Laboratory" in White-Fryars, off Fleet Street; by 1714, his lectures on chemistry were expanded to include *materia medica* "very applicable to the Practice of Physick." *Daily Courant*, no. 3237 (February 28, 1712), no. 3815 (January 15, 1714)

49 Margaret E. Rowbottom, "The Teaching of Experimental Philosophy in England, 1700–1730," *XIe congres international d'histoire des sciences (1965)*. Actes IV (1968), 48–9; Cudworth, *Life and Correspondence of Abraham Sharp*, p. 81.

50 Alan Morton, "Public Science in Mid-18th Century London: The King George III

bustle of Birchin Lane Harris made the acquaintance of Daniel Midwinter, until 1709 or 1710, having obtained more certain appointment at Christ's Hospital Mathematical School at the end of 1708. The lectures at the Marine were then adopted by Humphry Ditton. Not surprisingly, Hodgson's interests during this period also tended toward problems faced by mariners. Just prior to his engagement at the Marine Coffee House, he had published two works, one on the longitude and the other on general problems of navigation.[51]

It is notable that both Ditton and his predecessor were teachers at Christ's Hospital.[52] The origins of Christ's Hospital Mathematical School in 1673 rested partly in Samuel Pepys's fear of the inadequacy of navigational practice in naval officers. Despite some difficult beginnings, Christ's Hospital was ultimately in competition with other schools in the teaching of navigation and mathematics in order to fit boys to become apprentices to the captains of ships.[53] The fact that Ditton and Hodgson were active in the public lectures suggests that a community of interest existed well beyond the more circumscribed intentions of Christ's Hospital. The connection between it and the Marine Coffee House suggests that education, whether institutional or merely popular, was increasingly characterized by a deliberate utility.

While James Hodgson initially was engaged with the elder Hauksbee by 1706, his earliest courses were advertised to be given at Ayers' Writing School in St. Paul's Churchyard. This may have followed from Hodgson's friendship with Thomas Ayers, a writing master and teacher of navigation, whom Hodgson nominated to fellowship in the Royal Society in 1707. Ayers' practical interests were indicated in his early activities in the Society when he discussed the numerous inventions and chronic complaints of

Collection at the Science Museum" (unpublished paper, 1990), p. 5. I thank Alan Morton for permitting me to examine his essay.

[51] He was made F.R.S. in 1703, just before he joined Hauksbee in his lectures. *D.N.B.*., IX, 961; cf. Taylor, *Tudor and Stuart England*, pp. 288–9, 419, 422; Lillywhite, *London Coffee Houses*, p. 360; Rowbottom, "Teaching of Experimental Philosophy," p. 49; and Morton, "Public Science in Mid-18th Century London," p.11. Until 1710, Hauksbee's shop was located in Giltspur Street near Christ's Hospital. At that point, Hauksbee moved to Wine Office Court, Fleet Street, near Senex and the Royal Society. *Dictionary of Scientific Biography*, VI (New York, 1972), p. 169.

[52] Taylor, *Tudor and Stuart England*, pp. 288–9, 293; and William Cudworth, *Life and Correspondence of Abraham Sharp* (London, 1889), p. 96. Flamsteed to Abr. Sharp, March 24, 1709.

[53] Nicholas Hans, *New Trends in Education in the Eighteenth Century* (London, 1966), pp. 213–14; Ralph Davis, *The Rise of the English Shipping Industry* (Newton Abbot, 1972), pp. 124–6; and Geoffrey Howson, *A History of Mathematics Education in England* (Cambridge, 1982), pp. 35–8.

Denis Papin. Furthermore, Ayers was deeply concerned in the agreement signed with Sir James Lowther in 1715 to exploit the Newcomen engine, a matter with which many Newtonians were soon to be involved.[54]

St. Paul's Churchyard, haven of booksellers, increasingly became a focus for the activities of the teachers and lecturers. Probably before 1706, when Humphry Ditton was exploring the calculus, he was living near Child's Coffee House in London House Yard, which was just a stone's throw from Harris's residence in Amen Corner, where mathematics was being taught. The transition into the increasingly public world of philosophers and lecturers imposed new forces that proved difficult to manage. When he threw over the dissenting ministry for London mathematics, Ditton was not prepared when commerce overwhelmed contemplation.

Ditton soon had a horribly jaundiced view of the world of London's practitioners. It was probably early in Anne's reign that Ditton petitioned Cotes at Cambridge to enter into a correspondence with him in mathematics. One senses great bitterness in his finely drawn portrait of the price demanded by the competition that intruded on private contemplation: "Wee people in London, are so humbly immersed in slavish business, & taken up with providing for a wretched Carkasse; yt there is nothing allmost, but what is grosse & sensuall to be gotten from us." Ditton, who had trained for the ministry, was overwhelmed by the pressures of the metropolis. In the midst of London's practitioners, he complained to Cotes of his isolation: "Our Mathematicians at London, are either Ignorant & Honest, or Ignorant & vicious, or Skilful & vicious; upon which account I cannot cordially correspond wth any of ym."[55] Nonetheless, John Harris, who knew these pressures intimately, gave Ditton an opportunity to publish material on mathematics in the *Lexicon Technicum*.[56] Ditton was then to follow Harris and Hodgson at the Marine Coffee House. By 1712, he was also assisting the younger Hauksbee

[54] Taylor, *Tudor and Stuart England*, pp. 141–2, 276; Royal Society MSS., Council Minutes, II (1682–1727), p. 192, July 2, 1707. See also Alan Smith, "Steam and the City: The Committee of Proprietors of the Invention for Raising Water by Fire, 1715–1735," *Transactions of the Newcomen Society*, XLIX (1977–8), 8–9. Ayers died in 1715.

[55] Trinity College Library, MSS. R.4.42, no. 4, Ditton to Cotes, n.d.

[56] Taylor, *Tudor and Stuart England*, p. 140; B.L., Sloane MSS. 4058, f. 245, Ditton to Sloane, n.d. On relations between Child's and the philosophers, see Lillywhite, *London Coffee Houses*, p. 157. Harris described the mathematical entries in the *Lexicon* as having been "communicated to me by Mr. Humphry Ditton, a Person very skilful in these Matters." *Lexicon Technicum*, I (1704), s.v., "Maximis" and "Minimis"; "Fluxions." Harris seems to have encouraged the work of mathematicians like Brian Robinson and William Hawney.

in a "Course of Hydrostatical, Pneumatical, Magnetical and Optical Experiments" in Crane Court.[57]

The utilitarian lectures at the Marine, involving Harris, Hodgson, and Ditton, were not unique in Anne's reign. The incalculable difficulties of the longitude at sea preoccupied numerous mathematical practitioners. This may explain why John Harris, the first of these lecturers, spent so much space in the *Lexicon Technicum* of 1704 discussing problems of navigation. Practicality was the key. Harris's works on algebra, on the use of globes and the pocket quadrant, and on plain trigonometry, were presumably of some interest to his mercantile contacts. During Anne's reign, John Harris was busily involved with the commercial networks of London. For example, both Harris and his patron Charles Cox were among the petitioners who obtained a patent for an "Amicable Society for a Perpetual Assurance Office." They were later among the Directors of the Union, or Double Hand in Hand Fire-Office for Goods in 1716. It is likely that Cox and Harris met through their insurance interests.[58] The links between the popular science of the Newtonians and the mercantile community were to be far more deliberate than has been generally assumed.

The development of the enterprise of public lectures in the early eighteenth century parallels the growing recognition of Newtonianism. John Harris's part in the elucidation of Newton's work was an important one and not to be easily separated from the transition of Newtonianism into its public form in the lectures in the coffeehouses, where popular concerns were also often practical ones. I do not mean to argue that popular Newtonianism was confined to practical concerns. The antagonism between the legions of Leibniz and the disciples of Newton in 1709 and 1710 was clearly crucial. But it was in 1710, when John Harris was completing the second volume of his *Lexicon*, and when the *Acta Eruditorum* began its pursuit of the brothers Keill, that the elder Francis Hauksbee pronounced an early manifesto of the public lecturers and agents of the Newtonian defense:

[57] *Daily Courant* (January 10, 1712). Interestingly, Ditton's friend William Whiston had replaced him with the Hauksbees early in 1713, at which point Whiston was also lecturing on mathematics at the Marine. *Daily Courant* (January 21, 1713; March 18, 1713).

[58] Cecil T. Carr, ed., *Select Charters of Trading Companies, A.D. 1530–1707*. Selden Society, XXVIII (1913), 249–50; *Daily Courant*, no. 4678 (Wednesday, October 17, 1716). It is possible, though unlikely, that he was also John Harris, author of the *London Merchant's Mirror or the Tradesman's Guide, being... tables composed for the ready casting up bills of exchange, etc.* (London, 1705). See R. V. Wallis and Peter Wallis, *Biobibliography of British Mathematics and its Applications*, Part II, 1701–1760 (Newcastle, 1987), p. 20.

The learned World is now almost generally convinc'd that instead of amusing themselves with Vain Hypotheses, which seem to differ little from Romances, there's no other way to Improving Natural Philosophy but by Demonstrations and Conclusions founded upon Experiments judiciously and accurately made.[59]

As John Conduitt later understood, the relationship between the demonstrations of the lecturers and the experiments of the natural philosophers was an important one. It became fundamental to the propaganda of Newtonianism. The demonstrations were repetitions of the experiments, simplified and clarified, to illustrate the basic principles of natural philosophy.[60] The public courses, thus, transformed the rhetoric of Newtonian epistemology in fundamental distinction from the supposedly hypothetical musings of Cartesians or obdurate Leibnizians. Through experimental lectures were revealed the "general Laws of Attraction and Repulse, common to all Matter," and so on to "establishing the true System of Nature, and explaining the Great Motions of the World."[61] Public lectures meant public authority for natural philosophy. And this was to be an authority ultimately reinforced by claims to improve the world through knowledge of the principles of nature.

Here was fine theater. Obviously, there might be the atmosphere of a "raree show" for bored gentlemen – as with the display of the skeleton of a whale marooned near Greenwich in 1713, of the museum of monsters in Don Saltero's Coffee House, or of demonstrations of astounding strength in the taverns or the public markets.[62] Amusement had been the basis of the mechanical wonders of Winstanley's water theater until its

[59] F. Hauksbee, F.R.S., *Physico-Mechanical Experiments on Various Subjects. Containing An Account of Several Surprizing Phenomena Touching Light and Electricity, Producible on the Attrition of Bodies. With many other Remarkable Appearances, not before observ'd. Together with the Explanations of all the Machines* (London, 1709), "Preface." This was dedicated to John, First Lord Somers, then President of the Privy Council, and a leading member of the Whig Junto. See Robert M. Adams, "In Search of Baron Somers," in Perez Zagorin, ed., *Culture and Politics From Puritanism to the Enlightenment* (Berkeley, 1980), pp. 165–202. Similarly, John Harris's *Lexicon Technicum* in 1710 was dedicated to the Junto Whig William Cowper, Lord Chancellor, and a key figure in the prosecution of Sacheverell. Geoffrey Holmes, *British Politics in the Age of Anne* (London, 1967), pp. 239, 242. Cf. Schaffer, "Newtonian Cosmology and the Steady State," pp. 197–204.

[60] Simon Schaffer, "Natural Philosophy and Public Spectacle in the Eighteenth Century," *History of Science*, XXI (1983), 1–43.

[61] Hauksbee, *Physico-Mechanical Experiments*, "Preface." On Hauksbee's experimental Newtonianism, see Arnold Thackray, *Atoms and Powers* (Cambridge, Mass., 1970), pp. 74–80; and *D.S.B.*, VI (New York, 1972), 170–3.

[62] *Daily Courant* (May 25, 1713); Richard D. Altick, *The Shows of London* (Cambridge, Mass., and London, 1978), pp. 17–19. See Desaguliers' analysis of the amazing feats of Thomas Topham, of London; in Desaguliers, *Course of Experimental Philosophy* (1734), I, pp. 265, 289ff.

demise in August 1720 – perhaps as a result of the competition from the dramatic lectures of the natural philosophers.[63] It was easy to denigrate public display as pandering to vulgar superstition or baffling the ignorant. Yet a popular scientific culture had a momentum increasingly difficult to ignore. If the lectures could provide moral entertainment, a kind of play of self-improvement, they were also deliberately self-evident and determined to engage the audience in claims to a greater certainty. They required, in other words, the suspension of belief to take part in the new, experimental, improving, universe. Public entertainment might also further the aims of natural philosophy by investigating unexplained phenomena, such as puzzling electrical attractions, capillary movement of liquids, or even the involuntary motions of animals. The lecturers felt they could make their own contribution and some like Hauksbee or Desaguliers even had an important role in the Royal Society. But useful experiments might also come from those like Humphry Ditton, William Whiston, or Benjamin Martin, who never achieved election. The public lectures emerged as the popular counterpart of Newton's queries to the *Opticks*. Through the multiplication of sites, and by breaking the constraints of a Society of Fellows, they were part of the extension of experimental Newtonianism.

The rise of John Theophilus Desaguliers

The victory of Newtonian epistemology, through the claims to certainty that rested on mathematics and on experiments, was ideally suited to a coffeehouse universe. The translation from Newton's esoteric *Principia* to the popular expositions in the commercial world of London made natural philosophy into an industry. This industry prospered precisely because of public acknowledgment of that confident certainty in a Newtonian method that could be readily displayed in the instrument makers' shops and by the lecturers who came down from the Universities in Anne's reign. But this was more than a university extension; many of the lecturers and authors had little connection with the colleges.

Public science and its claims to certainty flourished amid the difficulties of Anne's reign. The Newtonian lecturers soon came to depend upon the demonstrable confidence of experiment. While the experimental lectures were beginning in London, John Keill was expounding on the Newtonian philosophy at Oxford. Keill began a course of lectures at Hart Hall that

[63] Richard D. Altick, *The Shows of London* (Cambridge, Mass., and London, 1978), pp. 78–9.

was to be attended by John Theophilus Desaguliers, soon the most active of the Newtonian lecturers. Desaguliers briefly succeeded Keill at Oxford in 1709, but he was drawn by the attractions of London in 1712. By January 1713, he was apparently lecturing in London, and in the following May, he had inaugurated lectures on behalf of the widow of the elder Hauksbee.[64] This was the beginning of an eventful career that defined the fashions of natural philosophy in the first half of the century. By 1734, he was engaged in his one hundred and twenty-first course. He had withstood increasing competition, especially from the younger Francis Hauksbee, who gave demonstrations while lectures were read by William Whiston, recently sent down from Cambridge and who had once been employed by the elder Hauksbee.[65]

Desaguliers quickly experienced success in the community of experimentalists and his timing was impeccable. The elder Francis Hauksbee died in April 1713, leaving the Royal Society without a curator of experiments – a position that a man of Desaguliers' talents could obviously fill and for which his own public lectures could recommend him more readily than the apostate Whiston. Possibly as a result of the inflaming of the controversy with Leibniz, Newton was anxious to have some critical experiments performed and he looked to Desaguliers for assistance. This was a logical choice for Newton and it may have been John Keill, then active as Newton's champion, who brought his former student to the President's attention.

Although Desaguliers was first proposed as a Fellow by Hans Sloane in July 1713, curiously nothing had come of it. After virtually a year of providing numerous experiments to the Society, it was not until July 8, 1714, that Desaguliers' name was put forward again – this time by Newton, who seemingly had come to appreciate how valuable an agent Desaguliers might turn out to be. This coincided not only with Desaguliers' resurrection of the *experimentum crucis* on Newton's light and colors, but also with the impending crisis of a Hanoverian succession and the

[64] Little is known of Desaguliers' motives for leaving Oxford and none of his biographers suggests anything more than economic necessity because he had married in October 1712. *D.S.B.*, IV (New York, 1971), 43–5; Margaret E. Rowbottom, "John Theophilus Desaguliers (1683–1744)," *Proceedings of the Huguenot Society of London*, XXI (1968), 196–218; Rowbottom, "Teaching of Experimental Philosophy," 47–51; Nicholas Hans, *New Trends in Education*, pp. 138–41; *Daily Courant* (May 4, 1713). On Keill's Oxford Lectures, see Anita Guerrini, "Newtonian Matter Theory, Chemistry, and Medicine, 1690–1713" (Ph.D. thesis, Indiana University, 1983), pp. 158–65.

[65] Musson and Robinson, *Science and Technology*, p. 40. Whiston was teaching mathematics privately at his home in Cross-street, Hatton-Garden. *Daily Courant* (January 21, 1713). Apparently, the younger Hauksbee had established a rivalry with his uncle by 1712. See *Daily Courant* (January 10, 1712, and January 14, 1712).

possible descent of Leibniz on London. Three days before the death of Queen Anne, Desaguliers was elected a fellow to reaffirm the Newtonian hegemony. Within a week, Keill reminded Newton that a battle was still possible should Leibniz be impudent enough to follow George from Hanover.[66] In July 1714, Desaguliers did repeat for the Society Newton's crucial prism experiments, which he was evidently already incorporating in his public lectures by the end of 1713.[67] Desaguliers had tied his public performances to the needs of the Royal Society. What proved critical for public science was the confidence of method.

Desaguliers' public lectures were reflected in his duties for the Royal Society. This was the case even since his appearance among the London lecturers in the spring of 1713. It was not long before Newton noticed Desaguliers, who was then in the employ of Hauksbee's widow. Soon after the Succession, the diplomatic world beat a path to Desaguliers' door. Before the summer of 1715, he reproduced some of Newton's crucial optical experiments for a French delegation in London to view the solar eclipse. In March, he had also entertained the members of the Dutch embassy, including the secretary Wilhem 'sGravesande. By the following February, he repeated several of his experiments for the ambassadors of Spain, Sicily, and Venice, soon to be followed in the queue by the Russian Envoy, Count Volkra.[68] Thus, Desaguliers became instrumental in the spread of Newton's philosophy to the Continent long before he undertook his lecture tours to the Low Countries. In a short

[66] Royal Society MSS., Journal Book (C), vol. X (1703–14), p. 501 (July 2, 1713); p. 553 (March 18, 1713/14); p. 574 (May 27, 1714); p. 580 (June 10, 1714); vol. XI (1714–28), p. 5 (June 24, 1714); pp. 11–13 (June 8, 1714); p. 16 (June 22, 1714); p. 17 (July 29, 1714). That Newton had such a program in mind is suggested by the proposal for a committee on experiments. See Council Minutes, II, 261–2 (March 30, 1713). I owe this reference to Simon Schaffer. See also Cambridge University Library, Add. MSS. 3985, f. 13. John Keill to Newton, August 6, 1714; Desaguliers, *Physico-Mechanical Lectures. OR, An Account of What is Explain'd and Demonstrated in the Course of Mechanical and Experimental Philosophy... Wherin the Principles of Mechanics, Hydrostatics and Opticks, are demonstrated and explain'd by a great Number of Experiments. Design'd for the Use of all such as have seen, or may see Courses of Experimental Philosophy* (London, 1717), pp. 73–4; and Schaffer, "Glass Works: Newton's Prisms and the Uses of Experiment," pp. 94–6.

[67] *D.S.B.*, IV (New York, 1971), 43–4; Rupert Hall, "Newton in France: A New View," *History of Science*, XIII (1975), 243–4; Royal Society MSS., Council Minutes, II (1682–1727), f. 280 (July 15, 1714); *A Catalogue of the Experiments in Mr. Desaguliers's Course*. (This course was to begin on December 14, 1713.) See *A Collection of Advertisements* (B.L., Shelf Mark C112. f.7), no. 181, Lecture 21.

[68] Rupert Hall, "Newton in France," p. 243; and Schaffer, "Glass Works: Newton's Prisms and the Uses of Experiment," pp. 92–8. See Royal Society MSS., Journal Book (C), XI (1714–20), p. 57 (March 24, 1714/15); p. 58 (April 7, 1715); pp. 104–5 (February 23, 1715/16); p. 110 (March 8, 1715/16).

time, he agreed with the refugee Pierre Coste, a former Huguenot minister in Amsterdam and devotee of Locke, to translate Newton's *Opticks* into French. This was completed with the assistance of Desaguliers and was published in Amsterdam in 1720 and followed two years later by an improved edition in Paris.[69]

Newton's *Opticks* was thus the source of much that attracted the experimental lecturers. His speculations gave them the opportunity to operate in the vanguard of natural philosophy. The dynamic corpuscularity, the nature of attractive and repulsive forces, seems to have been the major object of attention. This can readily be seen in the work of Desaguliers. In 1717, he published his *Physico-Mechanical Lectures* in which he attempted to explain most of the fundamental tenets of the Newtonian natural philosophy. These lecture notes are probably similar to the material presented that year. He sought to illustrate the attractive forces in which he listed gravity as a universal principle of matter, and cohesion, which he called "the Electrical Attraction unexcited," which force was so to captivate his contemporaries.[70] Desaguliers continued his activities, designing optical experiments that could be done by candlelight as it was usual for the courses to begin at 6 in the evening.[71] It is particularly interesting that by 1725 Desaguliers was still careful to note that experiments could overcome a lack of skill in mathematics in understanding the phenomena of nature. Obviously, by taking this tack, Desaguliers

[69] Westfall, *Never at Rest*, p. 795; Thackray, *Atoms and Powers*, p. 93; and *D.N.B.*, XII, 275.

[70] J. T. Desaguliers, M.A., F.R.S., *Physico-Mechanical Lectures*, p. 2. This was sold by Richard Bridger and William Vream. Both had been employed by the elder Hauksbee and later sold apparatus to Watts' Academy. Bridger, who had been apprentice to the elder Hauksbee, got into a dispute with the nephew over the design of air pumps that Francis, jun. claimed to have improved. *Daily Courant* (January 9, 1714, and January 19, 1714). See Nicolas Gauger, *Fires Improv'd: Being a Method of Building Chimneys, So as to prevent their Smoaking . . .*, trans. J. T. Desaguliers (London, 1715), "Appendix"; *Daily Courant* (May 3, 1716); Taylor, *Hanoverian England*, pp. 1, 113, 145; and Rowbottom, "John Theophilus Desaguliers," p. 206.

[71] Desaguliers, *Physico-Mechanical Lectures* (1717), pp. 72–3; Desaguliers, *Lectures of Experimental Philosophy. Wherein The Principles of Mechanicks, Hydrostaticks, and Opticks, are Demonstrated and Explained at large, by a great Number of curious Experiments: With a Description of the Air-Pump, and the several Experiments thereon: of the Condensing-Engine; as also of the different Species of Barometers, Thermometers, and Hygrometers; with several Experiments to prove and explain Sir Isaac Newton's Theory of Light and Colours, as performed in a Course of Mechanical and Experimental Philosophy. To which is added, A Description of Mr. Rowley's Machine called the Orrery, which represents the Motion of the Moon about the Earth, Venus, and Mercury about the Sun, according to the Copernican System: All carefully Examined and Corrected by Mr. Desaguliers*, second edition (London, 1719), pp.187–91; Desaguliers, *A Course of Mechanical and Experimental Philosophy* (London, 1725), p. 2; for example, see *Daily Courant* (January 9, 1719).

followed the lead of Harris and Ditton to vastly increase the appeal of natural philosophy.

Desaguliers had found his role to play. By 1719, judging from the frequent advertisements that appeared in the London newspapers, he was easily the most active and probably one of the most successful of the Newtonian lecturers. In his prospectus for his *Course of Mechanical and Experimental Philosophy*, he described his method as mathematical, but it was so only analogically, as "a chain of Propositions proving each other; but instead of Definitions, Axioms, and Postulata, purely Geometrical, the Experiments made at the First Lecture, prove the precepts of the Second, and so on:. . . ."[72]

This strategy was fundamental to capturing a philosophical audience. On the one hand, entertainment had to be assured and the auditors involved in witnessing experimental phenomena that were present and actual – much as in the best plays where all happened on stage. Intimacy was not confined to Drury Lane or to the new popular novels. On the other hand, the lecture courses had also to be sold as uniquely advantageous – whose logic and simplicity made natural philosophy accessible and, hence, secured as broad an appeal as possible.

The mathematical analogy of Desaguliers was essential to the marketing of philosophy. In his course in 1725 at his home in Channel Row, he claimed, "Things which otherwise would be merely speculative, being by this means rendered Objects of the Senses, and better understood in a Month or six Weeks, than in a Year's close Application to books only." This understandable promotion to avoid the limits of the text was mimicked a generation later by William Griffis' *Short Account of a Course of Mechanical and Experimental Philosophy and Astronomy* in his avowal to overcome "*several Years* Application to Books only."[73]

For Desaguiliers and his disciples, there was a close relation between the mathematical and mechanical in experimental courses. Hence, "a great many Persons get a considerable Knowledge of Natural Philosophy by

[72] Bodleian Library, MS. Rawl. D. 871, *A Course of Mechanical and Experimental Philosophy* (n.d., post-1719).

[73] Spalding Gentlemen's Society, Spalding, Lincolnshire. Maurice Johnson MSS., Drawer 1, no. 2. John Theophilus Desaguliers, *A Course of Mechanical and Experimental Philosophy, Whereby any one, although unskill'd in Mathematical Sciences, may be able to understand, all those Phaenomena of Nature, which have been discovered by Geometrical Principles, or accounted for by Experiments; and Mathematicians may be diverted in seeing those Machines used, and Physical Operations perform'd, concerning which they have read* (London, 1725), p. 2; Will. Griffis, *A Short Account of a Course of Mechanical and Experimental Philosophy and Astronomy. In which will be Exhibited the Experiments Necessary for the Explanation of Natural Philosophy in general, Mechanics, Hydrostatics, Hydraulics, Pneumatics, Optics, and Astronomy* [1748], p. 2.

Figure 5. John Theophilus Desaguliers, at the height of experimental glory, ca. 1725. By permission of The Huntington Library, San Marino, California.

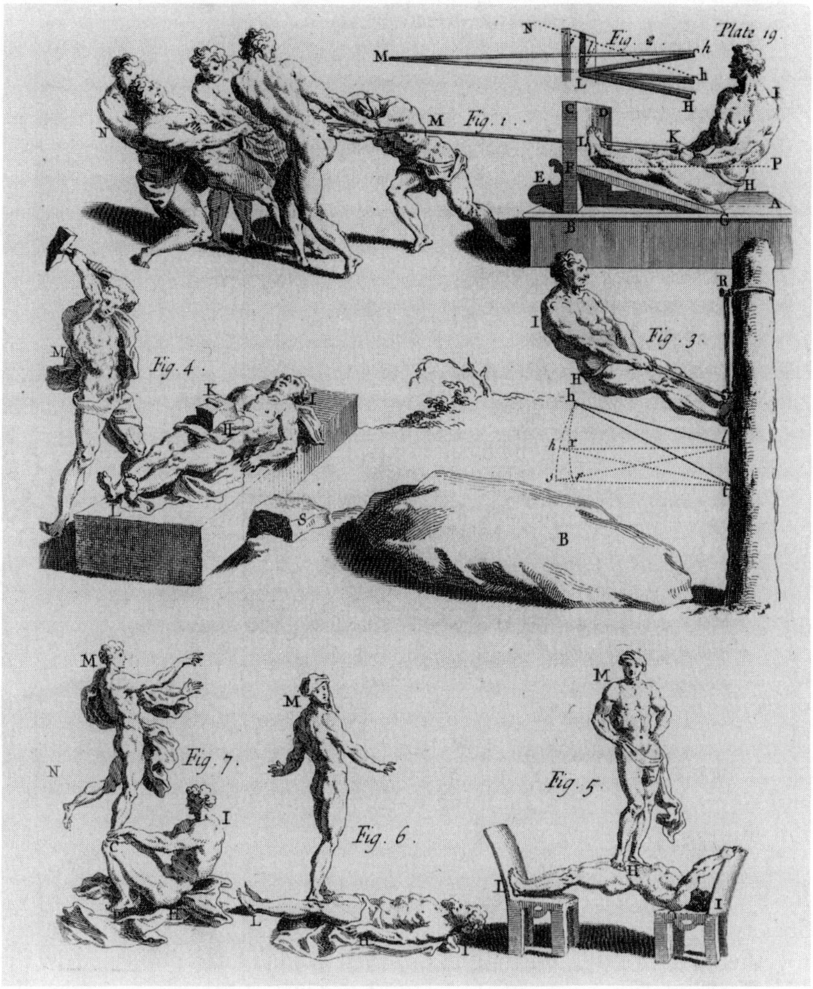

Figure 6. "Pretended Sampsons" according to Desaguliers. From *A Course of Experimental Philosophy*. By courtesy of Thomas Fisher Rare Book Library, University of Toronto.

way of Amusement; and some are so well pleas'd with what they learn that way, as to be induc'd to study Mathematicks, by which they at last become eminent Philosophers."[74] Even allowing for hyperbole, entertainment was part of a mission to reveal true mechanical principles so that it was much more difficult for the public to be duped by showmen or projectors. It was partly this that lay behind Desaguliers' interest in the strongmen of the taverns and the marketplaces. This led to several experiments before the Royal Society and in his public courses for which he designed an apparatus so that great feats of strength might more easily be repeated. "I don't hear that any o these Sampsons have attempted since to impose upon People in the same manner in or near London," he later bragged.[75]

The demonstration of the basic principles of mechanics became a staple in Desaguliers' lectures. He had taken particular notice of the show of strength of Thomas Topham in bending kitchen pokers and pulling against horses.[76] In one of his lectures in London in 1740, Desaguliers employed Topham as his operator of experiments before an audience that included the engineer John Grundy, junior, and the scientific lecturer Peter Shaw. Topham's strength, which so astonished his audiences in the provinces and in the London markets, could then be used by Desaguliers to demonstrate the advantages of mechanical apparatus. At Shaw's request, Topham performed naked so that the functions of the muscles could be better observed – at least this way no trickery was up his sleeve. Sweat and sinew brought new meaning to public philosophy.[77] In the market for philosophy, seeing was believing.

Desaguliers' design of new experiments – and he had claimed over 300 for his courses in 1714 – was precisely what he continued for the Royal Society.[78] In the summer of 1718, he and John Harris experimented together with a large burning glass, their results later published in the *Philosophical Transactions*.[79] In December 1721, at Newton's request, he

[74] J. T. Desaguliers, *A Course of Experimental Philosophy*, second edition, corrected, I (London, 1745), pp. viii–ix.

[75] *Ibid.*, p. 265.

[76] *Ibid.*, p. 288–90.

[77] Spalding Gentlemen's Society, Maurice Johnson MSS., Drawer 2, no. 84, John Grundy, junior to Maurice Johnson, May 3, 1740.

[78] *Daily Courant* (April 5, 1714).

[79] "An Account of some Experiments tried with Mons. Villette's Burning Concave," *Philosophical Transactions*, no. 360 (1719), 976–7. For one discussion of the significance of these experiments, see John Laurence, *A New System of Agriculture. Being a Complete Body of Husbandry and Gardening ... in Five Books. Containing All the best and latest, as well as many new Improvements, useful to the Husbandman, Grazier, Planter, Gardener and Florist* (London, 1726), pp. 18–19.

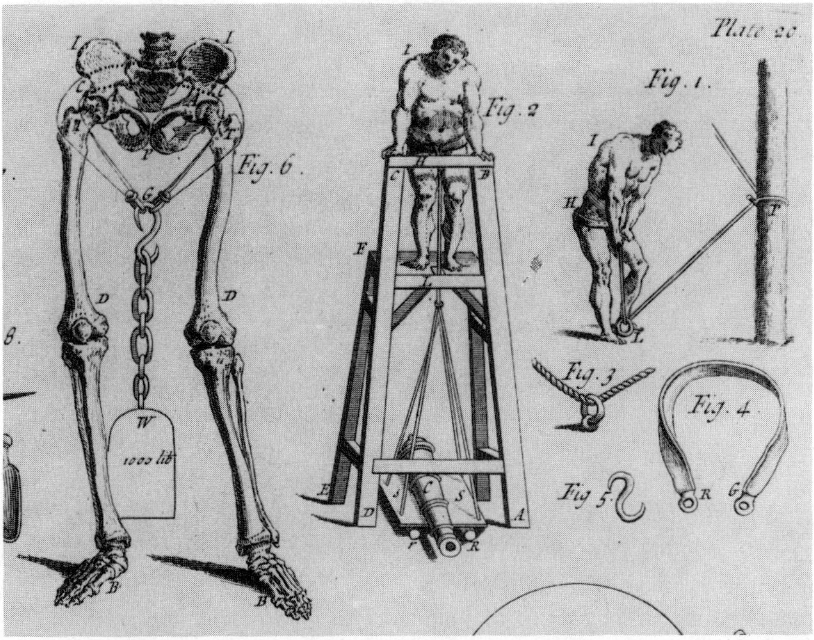

Figure 7. Feats of Strength, intended to show the naked Thomas Topham. From Desaguliers, *A Course of Experimental Philosophy*. By courtesy of Thomas Fisher Rare Book Library, University of Toronto.

provided the Society with an experiment "to shew that when a heterogeneous body descends in a Fluid with a uniform Motion, the whole weight of the body is communicated to the Fluid" precipitated by "Mr Leibnitz's new principle for explaining the Causes of variation of the length of the Mercury in the Barometer."[80] In the previous month, Desaguliers engaged in discussions relating to attempts made to produce perpetual motion. Indeed, not only was Desaguliers constantly fending off the enemies of the Newtonians, he was also often called to evaluate a proposal that appeared to violate the very rules of nature that Newton had helped establish. Even after Newton's death, Desaguliers was still attempting to defend him in a controversy that had arisen in 1727 as a result of attacks by Giovanni Rizzetti on the optical experiments. De-

[80] Royal Society MSS., Register Book (C), XI, pp. 168–70. Desaguliers later noted Newton's order when the issue arose in 1731. On Desaguliers' efforts in the *vis viva* controversy, see Carolyn Iltis, "The Leibnizian–Newtonian Debates: Natural Philosophy and Social Psychology," *British Journal for the History of Science*, VI (1973), 367ff.

saguliers set up apparatus to repeat the experiments yet again in his house in Channel Row, Westminster, and invited representatives of the Royal Society to witness the results.[81]

The experimentalists were essentially entrepreneurs cashing in on the fashions of philosophy. From Desaguliers to the popular lecturers of midcentury, there was an evolution in demonstrations and considerable investment in increasingly elaborate apparatus. By 1729, James Stirling had laid out £250 for instruments which by 1730, included an orrery. The lecturers Benjamin Martin and William Griffis soon faced the same necessity of investment in entertainment.[82] But all of this recognized a concordance of self-interest with public service.

Clearly Stirling's lectures at the Academy in Little Tower Street and Desaguliers' experiments for the Society might also provide a source of income. Stirling claimed to take in about £200 a year at the Academy.[83] To that end in 1729, Desaguliers suggested that the Copley experiment for that year should be directed against a recent translator of Vitruvius who had claimed an invention that worked without friction. It was Desaguliers' belief that those of the public who might have been induced into the expense of building such a machine might be best served by mathematical and experimental evidence of the "mechanical mistakes."[84] Desaguliers repeatedly faced such chronic problems. He had come to have great suspicion of the swarm of projectors who tried to draw others into building mechanical contrivances contrary to the laws of nature.

81 British Library, Sloane MSS. 4049, f. 215, Desaguliers to Sloane, August 8, 1728; Royal Society MSS., Register Book (C), XIII, 179–213. "Optical Experiments made in the Beginning of August 1728, before the President and several Members of the Royal Society and other Gentlemen of Several Nations upon the occasion of Sign. Rizetti's Opticks . . . "; see *Philosophical Transactions*, XXXV (December 1728), 596–629; Westfall, *Never at Rest*, p. 796; Schaffer, "Glass Works: Newton's Prisms and the Uses of Experiment," pp. 97–8.

82 Private Collection, Stirling of Garden MSS. James Stirling to ?, April 18, 1728; July 22, 1729; Spalding Gentlemen's Society, Maurice Johnson MSS., Drawer 1, no. 2. Will. Griffis, *A Short Account of a Course of Mechanical and Experimental Philosophy and Astronomy. In which will be Exhibited the Experiments Necessary for the Explanation of Natural Philosophy in general, Mechanics, Hydrostatics, Hydraulics, Pneumatics, Optics, and Astronomy* [1748], p. 8; Millburn, *Benjamin Martin*, pp. 47, 61; and Wallis and Wallis, *Biobibliography*, p. 310.

83 Private Collection, Stirling of Garden MSS., James Stirling to his brother, March 21, 1730.

84 British Library, Sloane MSS. 4050, f. 244. Desaguliers to Sloane, December 12, 1729. Desaguliers performed such experiments on friction in January 1730. Royal Society MSS., Register Book (C), XV, pp. 192–7, 211–13. A survey of these perpetual issues is Alan Gabbey, "The Mechanical Philosophy and Its Problems: Mechanical Explanations, Impenetrability, and Perpetual Motion," in Joseph C. Pitt, ed., *Change and Progress in Modern Science* (Dordrecht, 1985), esp. pp. 50ff.

Such errors and frequent deceptions, of course, were not confined to mechanics. But they were to have an immediate bearing on the careers of philosophers.

Desaguliers was often called to adjudicate philosophical claims. Early in 1731, Hans Sloane received a letter from a John Brown in Paris that threw Sloane head first into the Newtonian ether. Brown claimed to have proof that an elastic medium about which Newton had speculated must act on all bodies according to their mass. Sloane was utterly puzzled and passed it on to Desaguliers, who immediately recognized the implications. Desaguliers acknowledged the existence of a subtle medium that acts upon light – as those who had read Newton would recognize. However, the notion that such a medium might be capable of mechanical effects so as to cause gravity he suggested was "only propos'd by Sir Isaac in his last English Edition of the Optics by way of Query." If such a medium could be demonstrated to cause not only gravity but also elasticity in bodies, then such a discovery would be astonishing. But Desaguliers, who had become increasingly drawn to electrical attractions, was suitably skeptical. His reply reflected the impact of Newtonian experimentalism on the generation weaned on the *Opticks*:

> . . . if the same Cause did extend to Electricity and Magnetism; it is very probable that this wou'd be a Discovery of the first or second Causes. Whereas now our Principles are Four or Five at least, whose Cause we do not know, nor all the Laws of some of them, viz. Gravity, Attraction of Cohesion, Electricity, Magnetism, and Elasticity: we only know that there are such powers in Nature, & that they produce Effects which are the Subjects of our Contemplation.[85]

The understatement masked a great deal of thought on the ether. This likely had been in conjunction with Desaguliers' old friend, Stephen Gray, who had spent a quarter century wrestling with the secrets of electrical attractions and repulsions. As Heilbron has shown, at the very moment when Desaguliers replied to Sloane he was giving electricity increasing attention, clearly aware that Newton's ether queries had left the matter in great ambiguity. It is not easy to perceive this from the publication of Desaguliers' *Course of Experimental Philosophy* in 1734, although his long association with Gray culminated in his introduction of Gray's electrical research program to the Royal Society in 1729.[86] By 1744, however,

[85] British Library, Sloane MSS. 4052, fols. 69–71, John Brown to Sloane, February 21, 1731; MSS. 4051, fols. 200–1, Desaguliers to Sloane, March 14, 1730/1. Royal Society MSS., Register Book (C), XV, pp. 255–8.

[86] Cf. J. L. Heilbron, *Electricity in the 17th and 18th Centuries* (Berkeley and Los Angeles, 1979), pp. 55, 290–4. On the background, see Gad Freudenthal, "Early electricity be-

the second volume of his *Course* not only acknowledged the awkward queries, but declared that they "contain an Excellent Body of Philosophy, and upon Examination appear to be true. . . ."[87] By then, Desaguliers had been Newton's exponent for thirty years. In taking seriously the queries to the *Opticks*, he had also succeeded in transforming the meaning of Newtonianism. This was the ultimate victory of Newton's epistemology. The experimentalism of the middle of the century moved aggressively into territory upon which Newton had merely speculated.

The ways of patronage and the Royal Society

Desaguliers' success amongst the London experimentalists had been won more suddenly than he had expected. Like John Harris, he had hedged his bets. In 1710, he had taken Deacon's orders from Henry Compton, Bishop of London, but had stayed in Oxford for another two years. It was not until he had established himself in London that another door opened. He was admitted chaplain in 1716 to James Brydges, then Earl of Carnarvon, F.R.S., and later First Duke of Chandos. This was the most significant connection of his career. No observer would have understood the weight of such a preferment at the time, especially as Brydges had no political position from which to launch his associates to lucrative ventures. This association turned out to be critical to the encouragement of his scientific activities. In 1716, however, Desaguliers was still dependent on income from his lectures and his small clerical living.

Desaguliers was a man on the rise. In 1717, he preached before the King and gave lectures on experimental philosophy to George I at Hampton Court and to the Prince of Wales and Princess Caroline, who had been, during the previous two years, the conduit between Leibniz and Samuel Clarke. Desaguliers was close to the center of the storm. By the end of 1717, he had taken priest's orders and was presented to the living of Bridgham in Norfolk by William, Lord Cowper. Cowper had been John Harris's Junto patron in Anne's reign and had a revival under the

tween chemistry and physics: The simultaneous itineraries of Francis Hauksbee, Samuel Wall, and Pierre Poliniere," *Historical Studies in the Physical Sciences*, XI (1982), 203–29; and Michael Ben-Chaim, "Social Mobility and Scientific Change: Stephen Gray's Contribution to Electrical Research," *British Journal for the History of Science*, XXIII (March 1990), 15–16. The co-operation in experimentation between Desaguliers and Gray appears to have begun early in 1716. See Royal Society MSS., Journal Book (C), XI (1714–20), p. 106 (March 1, 1715/16); p. 125 (May 17, 1716).

[87] Desaguliers, *A Course of Experimental Philosophy*, II (London, 1744), p. 403; and R. W. Home, "Newton on Electricity and the Aether," in Zed Bechler, ed., *Contemporary Newtonian Research* (Dordrecht and Boston, 1982), pp. 191–213.

Hanoverians, being installed Lord Chancellor after the Succession. Desaguliers had taken pains to establish powerful friends and his talents as a lecturer were in constant demand, frequently among the astonished Royal family. Much later, in 1731, Desaguliers commemorated in horrendous verse an incident at Bath when Amanda Smith, governess to the young Prince William, had rescued some fish from Desaguliers' air pump and flung them into the river rather than see "each convulsive Strife, When rolling Eyes fortell departing Life. . . . " The day wasn't a complete loss, however:

> In darkned Room we view a little Ray,
> And all its Colours with our Glass display:
> Surprizing Charms, unknown before, we find.
> In that bright Orb, whose Lustre Unconfin'd
> Had barr'd our Search, and Struck th'Observer blind.[88]

Desaguliers was more successful in the marketing of natural philosophy than as a poet. He demonstrated simple propositions confirmed by experiments – a method that he had learned from John Keill. But this was not mere entertainment; poems and experiments were intended to secure public approval. A comment scribbled in a copy of a syllabus of one of Desaguliers' lectures indicates his technique was not lost: "The man who combines usefulness with pleasure has covered every point," wrote a listener.[89] One temptation is to see the public lectures as completely separated from Desaguliers' activities as a demonstrator for the Royal Society. Such a divorce may suggest degrees of sophistication that sometimes may not have existed at all. Utility in the public lectures did not always mean excessive simplification.

Desaguliers' efforts for the Royal Society were frequently controversial. He was caught by the inability of the Society to resolve the salary of its demonstrators. He sought a significant income from the experiments he performed, but frequently had difficulty collecting what he felt was his due. The Royal Society was often ambivalent about such payment – as Denis Papin had discovered while seeking its patronage in 1709. About a decade earlier, the Society had established a policy to support those who proposed and performed experiments at the direction of the Soci-

[88] B.L., Add. MSS. 32556, fols. 191–4, Desaguliers to Rev. Cox Macro, April 14, 1731. Macro was a celebrated antiquary who became chaplain to George II.

[89] Geoffrey Cantor, *Optics after Newton: Theories of light in Britain and Ireland, 1704–1840* (Manchester, 1983), pp. 43–6, 220n15. The comment was in *Physico-Mechanical Lectures* (London, 1717). I wish to thank Dr. Cantor for pointing this out to me.

ety.[90] But this issue continued to fester while the public was increasingly well served by lectures in natural philosophy. Papin's various proposals to the Royal Society evidently had a pension in mind, although in December 1708, just before they agreed to reimburse him for experiments already performed, Papin had requested the assistance of Hans Sloane in seeking a place to teach "either experimental philosophy or some parts of mathematicks."[91] For some, public demonstrations and experiments before the Society were a natural association. This was the view of Hauksbee and of Desaguliers, his successor as curator of experiments – and that role prevented neither of them from lecturing publicly. One cannot but wonder whether the failure of the Society to establish permanent pensions may have played a major part in the foundation of the coffeehouse lectures.

From 1715 until 1743, Desaguliers received an irregular, but significant, income from the Royal Society. At first, he was able to do so because of a benefaction of Sir Godfrey Copley, sometimes receiving extra amounts for the performance of the annual Copley experiment and after 1731 winning the Copley medal three times. These sums were not large nor were they inconsiderable, ranging from £30 in 1715, to £40 in 1720, to 10 Guineas in 1743.[92] The amounts appeared to vary in relation to Desaguliers' appearances before the Society, but even then matters were not quite so simple.

Despite Newton's patronage, and the evident usefulness of Desaguliers, there were occasions when the Society felt he had not lived up to expectations. In 1725, for example, Desaguliers wrote to Newton in complete frustration at complaints he had heard about the infrequency of his demonstrations. This would prove to be an especially trying year for Desaguliers; his hydraulic concerns in water supply had gotten him into very difficult waters at York Buildings on the Strand. He was, nevertheless, willing to oblige the gentlemen of the Council, but, he complained, it was not clear what was expected of him. He did not wish to offend by making experiments that were not new. Yet, as he told Newton slyly, "I cou'd easily do something every Thursday; having by me a very large Apparatus, which I us'd at my Courses of Philosophy." Desaguliers'

[90] Royal Society MSS., Council Minutes, II (1682–1727), f. 143 (January 25, 1698); f. 206 (January 12, 1708/9); f. 208 (February 23, 1708/9).

[91] British Library, Sloane MSS. 4041, f. 257, Papin to Sloane, December 18, 1708.

[92] *D.S.B.*, IV (New York, 1971), 43–4; Royal Society MSS., Council Minutes, II (1682–1727), III (1727–47), *passim*. Many of Desaguliers' experiments and communications, some of which appeared in the *Philosophical Transactions*, are listed in Royal Society MS., Register Book (C), XI–XXI.

exasperation demanded instructions "in writing, if I am wrong in my Notion of what is required of me."[93] How the Society felt about being compared to the audience at his public lectures would be interesting to know. Desaguliers was also being a little disingenuous; it is likely his lack of attention to the Society was the result of the large number of other activities in which he was then engaged, from lectures to involvement in various schemes requiring some technical expertise. He always had the air of a projector. But perhaps it was inevitable that he should discover private witnesses evolved into public, and paying, auditors.

Similar difficulties arose in the 1730s about Desaguliers' contribution to the Society. Possibly in order to avoid the antagonism of Hans Sloane, the secretary of the Society, Cromwell Mortimer requested an annual accounting by Desaguliers of his demonstrations. These became a common feature of the reports to Council before Desaguliers could be paid. His letters to Mortimer are quite revealing not only because they provide some detail of the expenses Desaguliers incurred to produce the necessary apparatus; they also give us some appreciation of the way in which his chronic attacks of the gout often interrupted his efforts. His gout, however, was also compounded by bad weather and by travel. During 1731 and 1732, his absence from the Society and the reduction of his salary in those years can be accounted for by two extended lecture tours of Holland, his return often delayed because of contrary winds and the painful swelling of his legs.[94] It is clear that this affliction was a serious and recurrent problem magnified perhaps by the rigors of the North Sea in winter. But the problem was still with him in 1743 near the end of his life when he was "confin'd at Norwich by Philosophy and the Gout."[95]

Thomas Watts and the entrepreneurs

The M. P. Richard Steele's promotion of public lectures is further evidence of wide demand for experimental philosophy. Although his intended "Censorium" seemingly did not to operate until 1715, after the Hanoverian Succession, this may be accounted for by the intensity of

[93] Cambridge University Library, Add. MSS. 4007(B), f. 669, Desaguliers to Newton, April 29, 1725.

[94] Royal Society MSS., English Letters, D.2, fols. 71–3, Desaguliers to Mortimer[?], October 29, 1733; November 8, 1735. There also exists an undated, unaddressed letter begging for the intervention of someone with Sloane to ensure payment of Desaguliers' salary. B.L., Sloane MSS. 4050, f. 40. On the connections with Holland, see Johan A. Van Reijn, "John Theophilus Desaguliers, 1683–1983," *Thoth*, V (1983), 165–204.

[95] Royal Society MSS., Letters and Papers, Decade I, volume V, no. 228, Desaguliers to Sloane?, Nov. 17, 1743.

Steele's political activity in these years. The demand remained, however, and its extent suggests one reason why the lectures became less mathematical in content, emphasizing instead the principles of philosophy through wonderful demonstrations. In the spring of 1716, Whiston was still engaged by Steele, this time commenting on astronomical and meteorological phenomena around London, which had afforded endless speculation for the superstitious and for those prone to political prophecy. Whiston's lectures were probably similar to the descriptions of various astronomical systems and Newton's improvements that he was still expounding upon in 1724 with Francis Hauksbee. In any case, by 1719, Desaguliers had become involved with the Censorium when he began a course of experiments, this time in French, possibly for the French refugees in the capital.[96] The courses at the Censorium were also expanded by a series introduced in the evenings by Benjamin Worster and Thomas Watts.[97] The involvement of Worster and Watts in the Censorium is an important indication of the widening circle of connection among the lecturers and their patrons. By 1720, Worster was employed by Watts at his Academy in Little Tower Street to teach experimental natural philosophy.

Although the Academy is well known, it was only one of several such establishments. Thomas Weston, formerly an assistant to Flamsteed, opened a school at Greenwich in 1706 where courses in experimental philosophy were given. The most famous in the eighteenth century was probably the Soho Academy founded about 1718 by Martin Clare, later a Fellow of the Royal Society. In 1735, Clare published the *Motion of Fluids* with the assistance of Desaguliers and Clare's works became increasingly reliant upon Newtonian principles.[98] For its part, Watts' in-

[96] This may also have been an attempt to respond to a more cosmopolitan aristocratic audience. Since 1714, Desaguliers had been advertising that his lectures would be given "in Latin, French, or English, as the Auditors shall require." *Daily Courant* (April 5, 1714); likewise, Bodleian Library MS., Rawl. D. 871, n.d.

[97] John Loftis, "Richard Steele's Censorium," *Huntington Library Quarterly*, XIV (1950), 53–62; *An Experimental Course of Astronomy; Proposed by Mr Whiston and Mr Hauksbee* (definitely 1724 or later). Cf. Hans, *New Trends in Education*, p. 141. Desaguliers' lectures were published by one of his students, Paul Dawson, who was enabled to attend through the encouragement of Steele. Dawson dedicated the volume to the politician. On the introduction of courses by Watts and Worster, see *Daily Courant* (January 12, 1719; February 25, 1719; September 28, 1719; November 11, 1719). With Richard Steele, see *Daily Post*, no. 26 (November 2, 1719).

[98] On Weston, see Wallis, *Biobibliography*, pp. 158–9; and *Daily Courant* (December 23, 1725). On Clare, see Margaret C. Jacob, "Scientific Culture in the Early English Enlightenment: Mechanisms, Industry, and Gentlemanly Facts," in Alan Charles Kors and Paul J. Korshin, eds., *Anticipations of the Enlightenment in England, France and Germany* (Philadelphia, 1987), pp. 139–140. In 1719, Clare published an immensely suc-

stitution seems to have predated the Soho Academy by about three years, being first located in Abchurch Lane near Exchange Alley in Cornhill. In 1719, Watts moved further to the east, to Little Tower Street, closer perhaps to the instrument makers' shops near the Tower and the river.[99] Originally called the Accomptant's Office, it catered primarily to the education of clerks and accountants, but the rage for natural philosophy resulted in new additions to the courses.[100]

In 1722, Benjamin Worster published a text intended for the use of his students, entitled *A Compendious and Methodical Account of the Principles of Natural Philosophy*. Worster included here a detailed discussion of attraction and repulsion, explaining that they were not to be considered occult qualities. His discussion of this issue clearly indicates that his students were soon immersed in the contentious issues surrounding the Newtonian philosophy. Worster obviously kept abreast of the latest investigations of the experimentalists, following Desaguliers' views on electrical attraction suggesting that it is "nothing else but the Attraction of Cohesion, excited by a strong Attrition to act with less Force in a larger Sphere."[101] By 1727, Worster's successor as joint Principal of the Academy, William Watts, delivered "A Course of Mechanical and Experimental Philosophy" with the mathematician James Stirling, Peter Brown, and William Vream, the instrument maker, given both at the Academy and at the new haunt of the Whigs in the Bedford Coffee House in Covent Garden. All but Vream, who made the instruments, were members of the Academy.[102] This course followed the usual trend, including me-

cessful *Youth's Introduction to Trade and Business*. Clare was elected F.R.S. on the recommendation of Desaguliers, Ephraim Chambers, and Alexander Stuart, among others. Royal Society MSS., Certificates, I (1732–50), no. 85 (March 27, 1735).

99 Musson and Robinson, *Science and Technology in the Industrial Revolution*, pp. 41, 119; Geoffrey Holmes, *Augustan England: Professions, State and Society, 1680–1730* (London, 1982), p. 51; and Mary M. Robischon, "Scientific Instrument Makers in London During the Seventeenth and Eighteenth Centuries" (Ph.D. thesis, University of Michigan, 1983), p. 166

100 Hans, *New Trends in Education*, p. 141; Taylor, *Hanoverian England*, pp. 7, 146. The poet James Thomson, known for his *Poem Sacred to the Memory of Sir Isaac Newton*, was briefly a member of the Academy in 1726. See J. L. Heilbron, "Robert Symmer and the Two Electricities," *Isis*, LXVII (March 1976), 11.

101 Benjamin Worster, A.M., *A Compendious and Methodical Account of the Principles of Natural Philosophy: As they are Explain'd and Illustrated in the Course of Experiments, perform'd at the Academy in little Tower-Street* (London, 1722), pp. 9–10, 29. On Worster, see E. G. R. Taylor, *Hanoverian England*, p. 151.

102 Taylor, *Hanoverian England*, pp. 113, 134, 144–5, 166–7. Stirling evidently owed his post in the Academy to Newton. He left by 1735 for employment with the Scots Mining Company at Leadhills in Lanarkshire. Stirling likely was engaged earlier with this group, probably through the connections of Watts. Musson and Robinson, *Science and Technology in the Industrial Revolution*, pp. 41–2; Huntington Library, Stowe MSS. 57,

chanics and numerous experiments relating to inventions, but also demonstrating "The Galilean and Newtonian Philosophy" and explaining optics "according to Sir Isaac Newton's Theory."[103]

The role of Thomas Watts and the relationships amongst the figures involved in the Academy is an important clue to the expansion of natural philosophy into the world of practical enterprise in the early eighteenth century. Watts, who is known mainly for a rather popular *Essay on the Proper Method for Forming the Man of Business* published in 1716, had an extensive network of contacts through which he was able to establish access to the political and financial establishment of Newtonian England. There is much about the links between the commercial world and the mathematics teachers we do not know.[104] But the efforts of Clare and Watts to fill the need for practical approaches to mathematics was one measure of a much wider movement in popular education.

The same issue of connection arises with the lecturers on natural philosophy. Samuel Clarke's translation of Rohault's *Traite de physique* provided one of the vehicles through which Newtonianism gained access to the Universities. Rohault was a particular favorite amongst educators, and in 1716, Watts published a translation of Rohault's *Treatise of Mechanicks*. This may indicate a design for a work to compliment a series of lectures planned by the public lecturers whose ranks Watts was soon to join. More significant, however, is the genesis of this project, which Watts dedicated to Dr. Richard Mead. The original proposal had come from Humphry Ditton, but his death in 1715 interrupted the translation. It may well have been Ditton's friend William Whiston who undertook the work, which by 1716 was well advanced. Whiston was responsible for putting a draft into the hands of Thomas Watts who, in May 1716, had decided "that some competent Understanding of Mechanicks was not altogether foreign to the Design I was then forming, and have lately enter'd upon. . . ."[105] This project confirms Watts' close links with the

LXV, f. 135. James Brydges, Duke of Chandos to Ferguson (his steward), October 28, 1724. On Stirling, see Charles Tweedie, *James Stirling: A Sketch of His Life and Works along with his Scientific Correspondence* (Oxford, 1922); Wallis and Wallis, *Biobibliography*, p. 84; and *Dictionary of Scientific Biography*, XIII (New York, 1976), pp. 67–70. Desaguliers had a long association with Vream, whom he later described as his "Operator for Philosophical Machines." *A Course of Experimental Philosophy*, II (London, 1744), p. 377. See also Thomas Watts, *An Essay on the Proper Method for Forming the Man of Business*, ed. Arthur H. Cole (Boston, 1946).

103 *A Course of Mechanical and Experimental Philosophy, Consisting of Seven Parts* (London, 1727).

104 Ditton, for example, wrote a preface to John Hill, *Arithmetick both in Theory and Practice*. Such connections need to be more fully explored. Taylor, *Tudor and Stuart England*, p. 304.

105 [J. Rohault,] *A Treatise of Mechanicks: Or, The Science of the Effects of Powers of*

Newtonian lecturers Whiston and Ditton, along with those he established with Worster and Desaguliers' assistant William Vream. It is also possible Thomas Watts may have been one of the merchants whom Whiston and Ditton consulted before petitioning Parliament in 1714 to establish a reward for discovering longitude while at sea. The existence of this network has an even wider significance.

The activities of many of these lecturers constantly reveals contact with the commercial world. Over fifty years ago, Sir Lewis Namier noted the existence in Parliament of a Thomas Watts, Whig M.P. for St. Michael and later Tregony. This Thomas Watts was once also the mathematics teacher and founder of the Academy. Before his death in 1742, Watts was active in a broad range of financial and commercial activities, some of which sought to exploit the learning of the natural philosophers. He was likely related to John Watts who, in 1707, was among the trustees of the insurance scheme of the projector Charles Povey. This ultimately became known as the Sun Fire Office in the formation of which John Harris was also concerned.

This raises some complex and intriguing relationships. It is known that John Harris played an active role in the new insurance companies that took shape at the turn of the century. He was probably the John Harris, clerk, who appears among the list of clergymen who were trustees in the Society of Assurance for Widows and Orphans in 1699.[106] More certainly, Harris was among the petitioners for the Amicable Society in 1706. Here he was in important company with William Lloyd, John Stillingfleet, White Kennett, as well as Captain Thomas Savery, a Joseph Watts, and Charles Cox, the brewer who originally employed Harris in lectures at the Marine Coffee House. The Amicable Society was formed shortly before Charles Povey's project, which soon led to the Sun Fire Office and Harris's involvement there in 1710.[107]

The connection between the insurance projects and the Newtonian

Moving Forces, as apply'd to Machines, demonstrated from its first Principles. Done out of French. In this Translation are inserted Particular References to the several Propositions in Euclid, on which Demonstrations are built: With considerable Additions, whereby the Whole is more Compleat and Universal, and particularly the Rising and Falling of the Quicksilver in the Weather-Glass explain'd and accounted for. Useful for all Artificers, as well as Natural Philosophers and Mathematicians (London, 1716).

106 B.L., [shelf mark] 816.m.10(128). *Assurance for Widows and Orphans.*

107 On these insurance societies, see Cecil T. Carr, ed., *Select Charters of Trading Companies, A.D. 1530–1707*. Selden Society, XXVIII (1913), 249–50; W. R. Scott, *The Constitution and Finance of English, Scottish and Irish Joint-Stock Companies to 1720*, 3 vols. (Cambridge, 1912; reprint edn., Gloucester, Mass., 1968), vol. I, p. 383; vol. III, pp. 368, 389–90; and P. G. M. Dickson, *The Sun Insurance Office (1710–1960)* (London, 1960), pp. 21, 268.

lecturers was quickly established. It is not clear in Harris's case whether involvement arose from Christian altruism, the assertion of the public good, or merely followed from his mathematical skill. An early link to the Watts family is suggested by the appearance of Joseph Watts in this group in 1706. The following year, the link with John Watts seems to confirm a family interest. John Watts was a printer with whom Benjamin Franklin worked while in London. His relationship with the Sun, of course, would not have ensured Thomas Watts' advancement, although by 1720, he was a member of the Court of Assistants of the Sun Fire.[108]

It is significant that Watts was deeply entangled in the activities of Desaguliers' patron, James Brydges, the first Duke of Chandos. An examination of the share transfers of the Sun indicates that for some years before his death, Watts was acting as a broker, on behalf of Chandos and his family, who were large shareholders, dealing not only in Sun shares, but in the shares of many other companies in which the Duke was also interested. Likewise, the lecturer Benjamin Worster and Thomas's brother William, who took over the Academy, were also engaged in these transactions. Frequently, as Desaguliers realized, much hinged on mechanical devices. For example, the York Buildings Company depended upon the success of a variety of mechanical ventures from mining, to salt pans, to water supply.[109] There was often much urgency in these matters. As a result, Watts and Chandos were habitually instrumental in shaping Courts of Directors to do their bidding.[110] Thomas Watts appears to have been quite successful; at the very least, he had his hands in so many of the important financial pies of the early eighteenth century that his residence in Threadneedle Street in 1740 close by the Bank of England was more than symbolic.[111] Here he could keep a watch on the major currents in the tricky waters of eighteenth-century finance. His expertise in these

[108] Sir Lewis Namier, "Brice Fisher, M.P.: A Mid-Eighteenth- Century Merchant and his Connexions," *English Historical Review*, XLII (October 1927), 520; Dickson, *The Sun Insurance Office*, p. 268; Guildhall, London, MS. 11931, Vols. I–III. Sun Insurance, General Committee Minutes; R. V. Wallis and P. J. Wallis, *Biobiliography*, p. 78; P. J. Wallis, "Thomas Watts. Academy Master, Freemason, Insurance Pioneer, MP," *History of Education Society Bulletin*, XXXII (Autumn, 1983), 51–3. On Watts' political career, see Romney Sedgwick, ed., *The House of Commons, 1715–1754* (New York, 1970), II, p. 525.

[109] Hunt. Lib., Stowe MSS. 57, XXVII, fols. 234–5. Chandos to Col. Samuel Horsey, January 24, 1725/6.

[110] Hunt. Lib., Stowe MSS. 57, XXVII, f. 240, Chandos to Thomas Watts, January 28, 1725/6; XXVII, fols. 306–7, Chandos to Watts, February 25, 1725/6.

[111] Guildhall MS. 12019C. Share ledger, 1720–43; MS. 12023. Share transfers, 1720–5. In 1730, Chandos placed a young student in Watts' Academy to be taught merchant's accounts in order to employ him in the African Company. Hunt. Lib., Stowe MSS. 57, XXXVI, fols. 2–3, Chandos to Thomas Watts, October 4, 1730.

matters made him indispensable to Chandos. It is not inconceivable that Watts' later ascent into the world of politics had much to do with these associations initially cultivated in his Academy and with the public lecturers like Desaguliers. The promotion of philosophy and the politics of connection had gone hand in hand. The popular victory of Newtonianism could hardly have done Watts' reputation any harm. He had been instrumental in it.

By the time of Newton's death in 1727, the system of public lectures rested on a strong foundation of public and private patronage. The market for lectures continued without pause. The younger Hauksbee delivered lectures with the physician Peter Shaw, who, with Ephraim Chambers, had translated the chemical lectures of Boerhaave. By 1733, Shaw had relocated to Scarborough and was delivering lectures on chemistry while practicing medicine in the increasingly fashionable spa. These lectures he published in a work dedicated to Henry Lowther, Viscount Lonsdale, and Lord Privy Seal. Shaw was later appointed physician-in-ordinary to the King.[112]

Not all of the lecturers could count on such notable support, however. The social hierarchy was reflected in groups other than the Royal Society. At one end of the social scale was the provincial Spalding Gentlemen's Society with a large and prominent membership. In London, at the opposite end of the scale, was the Spitalfields Mathematical Society, composed originally of Huguenot weavers who met in the White Horse, Wheeler Street in Spitalfields, and whose activities, judging from their library, ranged from merchants accounts, to longitude, through the breadth of natural philosophy.[113] Spitalfields was only one of several such groups in the early eighteenth century. In 1708, there was a mathematical society meeting at Jones's Coffee House in Finch Lane behind the Royal Exchange. From about 1723, a mathematical society was meeting in which discussion took place about navigational improvements. This might be

[112] Peter Shaw, *Chemical Lectures, Publickly Read at London, In the Years 1731, and 1732; And since at Scarborough, in 1733; For the Improvement of Arts, Trades and Natural Philosophy* (London, [1734]). See F. W. Gibbs, "Peter Shaw and the Revival of Chemistry," *Annals of Science*, VII (September 28, 1951), 211–37; and Jan V. Golinski, "Peter Shaw: Chemistry and Communication in Augustan England," *Ambix*, XXX (March 1983), esp. 21–2.

[113] H. J. J. Winter, "Scientific Associations of the Spalding Gentlemen's Society during the period 1710–1750," *Archives internationale d'histoire des sciences*, III (1950), 77–88; John Nichols, *Literary Anecdotes*, VI (London, 1812), 69–122; Margaret C. Jacob, "Scientific Culture in the Early English Enlightenment," pp. 148–9; and Henry H. Cawthorne, "The Spitalfields Mathematical Society (1717–1845)," *Journal of Adult Education*, III (1929), 155–66. Much of the Spitalfields' Library exists today in the Royal Astronomical Society, London.

associated with the Spitalfield's society. The members included John Milton, Peter Pooley, the mathematician William Heath, the writing master William Hodgkin, a former mathematics teacher William Pardon, who was an accountant to the Company "for smelting down Lead with Pit Coal," and Stephen Plank who sought an award from the Board of Longitude.[114]

These societies could have a significant impact on the livelihood of natural philosophers. For the botanical lecturer John Martyn, Fellowship in the Royal Society was to prove no security, nor was his position as Professor of Botany at Cambridge. As early as 1721, he tried to emulate the experimentalists by giving a course of lectures to the Botanical Society, which met in the Rainbow Coffee House in Watling Street.[115] With the encouragement of Hans Sloane, Martyn tried to bridge the gap between the London public and the colleges of Cambridge. To overcome the vicissitudes of the market, he published his *Historia planatarum rariorum* over a period of ten years from 1728 by dedicating sumptuous plates to individual subscribers with their titles and coats of arms engraved. But his career remained one of desperate survival.

Public lecturers evidently did not always draw much distinction between their audiences in the coffeehouses and the students they might face in formal institutions like Watts' Academy or Christ's Hospital. Newton's editor, Henry Pemberton, gave a course of public lectures on chemistry at Gresham College in 1730 and 1731 to rival those of Peter Shaw. Pemberton had been associated with Gresham since 1728, at which time he was chosen physics professor to succeed the combative John

[114] Peter J. Wallis, "Book Subscription Lists," unpublished lecture to the Bibliographical Society (April 1973), p. 12; Royal Society MSS., Classified Papers, VIII (2), fols. 151–6, February 19–29, 1731/2; Register Book (C), XVI, 399–403 (March 2, 1731/2); XVII, 19–20 (March 19, 1731/2). For Pardon, see B.L., Add. MSS. 47009A, f. 38, Egmont Papers, William Pardon to Lord Percivale, July 17, 1739. Cf. E. G. R. Taylor, *Hanoverian England*, p. 139.

[115] Blanche Henrey, *British Botanical and Horticultural Literature before 1800*, II (London, 1975), pp. 52–3. There were many such coffeehouses of the same name. This appears to have been at the end of Watling Street farthest from St. Paul's Church Yard, by St. Antholm's Church. See Lillywhite, *London Coffee Houses*, p. 475; and Colin Russell, *Science and Social Change, 1700–1900* (London, 1983), pp. 83–4. In 1724, Martyn became secretary to the Botanical Society and was elected F.R.S. The next year, he issued proposals for a public course of botany at London. This, in turn, was followed by a similar proposal for Cambridge. Henrey, *Botanical and Horticultural Literature*, pp. 53ff. Martyn probably began his lectures in 1721. His proposal for a course, dated March 1, 1725/26, suggests that there had been at least one former course. B.L., [Shelf Mark] 443.d.27(2); Raymond Williamson, "John Martyn and the Grub-Street Journal, with Particular Reference to his Attacks on Richard Bentley, Richard Bradley and William Cheselden," *Medical History*, V (1961), 362–3.

Woodward who died that year.[116] But Pemberton may have had cause to reflect on the limitations of his listeners. There had long been a tension in the broadening base of public knowledge of natural philosophy and the degree of sophistication a lecturer might impart. This was the reason for the reservation underlying his remark in 1728 that "... now, since Sir Isaac Newton's doctrine has been fully established by the universal approbation of all, who are qualified to understand the same; it is without doubt to be wished, that the whole of his improvements in philosophy might be universally known."[117] This would be the aim of the public philosophers. They would assume Sprat's doctrine of making the secrets of nature accessible to artisans, to unite a Baconian practicality and philosophical inductivism by more than rhetoric. And this would provide a greater sanction for natural philosophy than ever the Royal Society could create.

[116] John Ward, *The Lives of the Professors of Gresham College* (London, 1740), p. 301.

[117] Henry Pemberton, *A View of Sir Isaac Newton's Philosophy* (London, 1728), p. 2; and Richard Yeo, "Genius, Method, and Morality: Images of Newton in Britain, 1760–1860," *Science in Context*, II (Autumn 1988), 263.

5

The Newtonians and the English transformation

Any pursuit of the lecturers of natural philosophy inevitably leads to the coffeehouses like the Marine in Birchin Lane or Button's, Covent Garden. Such associations seem to have existed virtually since the founding of the Royal Society in the seventeenth century. But there was something slightly unsavory about coffeehouses that smacked of the rabble and sedition. It was not solely Roger North's dyspeptic toryism that asked, "Who can apply close to a subject with his Head full of the Din of a Coffee-House?" Certainly Glanvill had his own reservations about the levelers of the coffeehouses.[1] This, however, was far from the view promoted by a convivial few in the Royal Society. North's contemporary Robert Hooke continually attended meetings of natural philosophers either at Man's, near Charing Cross, or at Garraway's, in Exchange Alley. Even the Royal Society watering hole at the Crown Tavern in Threadneedle Street witnessed experiments.[2] And the link between the Royal Society and the Grecian, in Devereux Court in the Strand just beyond

[1] Roger North, *The Lives of the Right Hon. Francis North, Baron Guilford; The Hon. Sir Dudley North; and the Hon. and Rev. Dr. John North*, ed. Augustus Jessop, Vol. III (London, 1890), p. 292. On the dangers of coffeehouses, see Roger North, *Examen: Or, An Enquiry into the Credit and Veracity of A Pretended Complete History* (London, 1740), pp. 320–1, 370, 597; and James R. Jacob, *Henry Stubbe, radical Protestantism and the Early Enlightenment* (Cambridge, 1983), p. 84. Some hints at the social issues involved are found in Peter Albrecht, "Coffee-drinking as a Symbol of Social Change in Continental Europe in the Seventeenth and Eighteenth Centuries," *Studies in Eighteenth-Century Culture*, XVIII (1988), 91–103.

[2] Henry W. Robinson and Walter Adams, eds., *The Diary of Robert Hooke, M.A., M.D., F.R.S., 1672–1680* (London, 1935), *passim*, esp. pp. 276–7, 279, 431; Bryant Lillywhite, *London Coffee Houses* (London, 1963), pp. 180–4, 216–20, 351–6.

Temple Bar, may well have confirmed some High-Church suspicions of the dark alleys where philosophers might lead.[3]

As early as the reign of William, the coffeehouses had become a convenient gathering point in London for those interested in Newton's achievements. The conglomeration of interests made the coffeehouses an appropriate setting for lectures on natural philosophy. Significantly, the coffeehouses provided a focus for the London intelligentsia where clergymen, translators, and index-makers sold their talents to wealthy traders and to the aristocracy. To the coffeehouses, gentlemen could repair for an afternoon or an evening, as the astute Addison often did, to pick up useful information in the din of debate, as many papers must have been written as were read in such places. The promoter John Houghton reported on a conversation with a gentleman who held the universities and the coffeehouses in equal esteem as they had both improved useful knowledge, suggesting "that he who has been well-educated in the schools, is the fittest man to make good use of the coffeehouses. . . ."[4] It was the lecturers who helped to erode the social distinctions the Royal Society maintained. The bustling opportunism of the nearby markets and the stock exchange was a riot of social fluidity, of the rise and fall of fortunes, of hopes dashed or futures made.[5]

Coffeehouse philosophers

After the Restoration, coffeehouses like the Marine proved to be more than merchants' clearinghouses. They were also transformed into a movable feast of natural philosophy, accessible to those capable of the fee for a course of lectures.[6] Their dynamism was fed by the arteries of commerce, by transactions over a bowl of coffee, reinforced by lectures on mathematics and natural philosophy, or by rumors of projects or of patents. In 1712, the *Spectator* reported,

[3] Aytoun Ellis, *The Penny Universities: A History of the Coffee-Houses* (London, 1956), p. 82; Lillywhite, *London Coffee Houses*, pp. 243–4.

[4] Quoted in Ellis, *Penny Universities*, pp. 28–9.

[5] Ellis, *Penny Universities*, p. xv; W. A. Speck, *Stability and Strife. England (1714–1760)* (London, 1977), pp. 74–6. On mobility, see Roy Porter, *English Society in the Eighteenth Century* (Harmondsworth, 1982), pp. 64–7, 92–3, 245; Gary Stuart de Krey, *A Fractured Society. The Politics of London in the First Age of Party, 1688–1715* (Oxford, 1985), p. 221; and Lawrence Stone and Jeanne Stone, *An Open Elite? England 1540–1880* (Oxford, 1984), Chapter VII.

[6] A. E. Musson and Eric Robinson, *Science and Technology in the Industrial Revolution* (Toronto, 1969), pp. 57–8; Michael Foss, *Age of Patronage: The Arts in England, 1660–1750* (Ithaca, 1971), pp. 82–3.

. . . There was a fellow in town some years ago, who used to divert himself by telling a lye at Charing-Cross in the morning at eight of the clock, and then following it through all parts of the town until eight at night; at which time he came to a Club of his friends, and diverted them with an account what censure it had at Will's in Covent Garden, how dangerous it was believed at Child's and what inference they drew from it with relation to stocks at Jonathan's.[7]

Here was the world in which Newton's philosophy prospered.

In these surroundings, the virtuosi of the Royal Society met and discussed science and stocks, new voyages, and the problems of the longitude. Jaded observers of the Royal Society relished the irony. Public science had presented problems for the restricted vision of the Royal Society. William King, in one of the satires that so enraged Hans Sloane, noted that it was the coffeehouses that had reportedly increased trade and perhaps even improved useful knowledge.[8] The Royal Society had long frequented the Grecian Coffee House in Devereux Court in the Strand. Samuel Pepys, Newton, Halley, and Sloane are known to have been part of the company at the Grecian. An alternative was provided by Child's in St. Paul's Churchyard, which was also a favorite of the clergy. Humphry Ditton had lived conveniently next to Child's. It was here in 1720 that an abortive scheme was hatched by Halley and Sloane to obtain membership in the Royal Society for William Whiston.[9] Besides being frequented by the Royal Society and the clergy, Child's was also regarded as the physicians' coffeehouse. The origins of the Royal Society dining club may also have come out of nearby gatherings at the Sun in St. Paul's Churchyard.[10]

The philosophical brotherhood of the coffeehouses sought out the company of prominent politicians. William Whiston initiated his public lectures at Button's Coffee House in 1713 after coming under the patronage of Richard Steele. Whigs like Steele stood between Whiston and the High Church. Through the influence of Steele and Addison, Button's became the favorite of the Whigs. It was to Button's that Whiston and Ditton

[7] *The Spectator* (Tuesday, October 28, 1712).

[8] [William King], *The Transactioneer With some of his Philosophical Fancies: In two Dialogues* (London, 1700), p. 86. King was a close friend of Jonathan Swift and Henry Sacheverell. See *The Original Works of William King*, I (London, 1776), pp. xix–xxiv; and [King], *A Satire on Learning, With Reference to the Royal Society* (1709), in Frederick George Stephens, *Catalogue of Political and Personal Satires* (London, 1978), II, 296–301.

[9] Ellis, *Penny Universities*, p. 105; Lillywhite, *London Coffee Houses*, p. 157; James Force, *William Whiston*, pp. 23–4.

[10] T. E. Allibone, "Edmond Halley and the Club of the Royal Society," *Notes and Records of the Royal Society of London*, XXVIII (April, 1974), 195–205.

addressed their letter published in Steele's *Guardian* promising a method to discover the longitude at sea.[11] Political context did not hinder the philosophical activities of those who lectured at Button's. Political factors could not always be ignored, however. Even though Richard Steele originally conceived of his Censorium as nonpartisan in design, it was likely his involvement in the party rivalry after the middle of 1713 that was the cause of the delay in his plans.[12] The coffeehouse connections merely intensified these concerns.

Popular lectures and cheap pamphlets smashed the boundaries of public education. And Whig coffeehouses formed the matrix of Newtonian persuasion. Those who later taught at Thomas Watts' Academy also sometimes performed at the Bedford Coffee House in Covent Garden. This particular choice is of interest because it reflects the evolution of natural philosophy as entertainment, as in Steele's Censorium, as well as the confirmation of Whiggish connections. The Bedford, near Covent Garden Theatre, gained its popularity with the decline of Button's after the death of Addison in 1719. It was the Bedford, as well as the Marine Coffee House, that became the site of William Whiston's lectures in 1719. From Button's to the Bedford, natural philosophy trailed in the wake of London's Whig politicians, poets, and pamphleteers, like Steele, Richard Savage, or Martin Folkes. Moreover, after the destruction of Channel Row to make way for the construction of the new Westminster Bridge, the Bedford became both residence and lecture theater for Desaguliers until his death in 1744.[13]

The transition from the private patronage of a prominent merchant like Sir Charles Cox to the public of the coffeehouses was a venture into the marketplace for the entrepreneurs of science. The breadth of consumption for science meant an increasing demand for philosophers. At times of remarkable events, such as the solar eclipse of May 11, 1724, the public interest could spawn numerous imitators. For example, at the Cock and Greyhound in Hoxton, Archibald Mitchell intended to set up his instruments to assuage the terror of the townspeople. While Desaguliers and the Royal Society made their observations from Bath, in London, the other lecturers had the field to themselves. It was rumored that one of the law societies was to set up a series of platforms 50 feet high to accommodate all of the astronomers and their auditors. This impending

[11] *The Guardian*, no. 107 (July 14, 1713), 91–3.

[12] John Loftis, "Richard Steele's Censorium," *Huntington Library Quarterly*, XIV (1950), 53.

[13] Lillywhite, *London Coffee Houses*, p. 114; Ellis, *Penny Universities*, p. 154; *Daily Courant* (March 27, 1719), *D.S.B.*, IV (New York, 1971), p. 45.

pyramid of philosophers did not intimidate Whiston, however. Striking while the market was hot, Whiston and Senex published an account of the transit and sold glasses properly darkened for viewing. At the Marine and at Button's, Whiston gave special lectures on the solar system, and on the day of the eclipse, in a large room prepared at the Horse Guards, he explained the event with an orrery and charged no less than one-half guinea to those who came solely to view the sun.[14] The public demand that followed the path of a comet or an eclipse was part of a long tradition dependent on the dramatic.

This wider audience was no longer solely royalty and aristocracy. The addition of merchants and mechanics as well might also partly explain the inclusion in the lectures of matters of technical interest. This was endorsed by Desaguliers, who, in 1719, advertised his Course of Experimental Philosophy to include an improvement on the Savery engine "of the greatest Use for draining Mines, supplying Towns with Water, and Gentlemens Houses. . . ."[15] Ultimately, practicality may have overcome many lingering political or religious reservations in subscription to such courses.

The lectures became an industry extending far beyond London. By the 1730s, Desaguliers had lectured in The Netherlands, Shaw had amused the aristocracy at Scarborough Spa (presumably not in the nude) and William Whiston combined scientific subjects with providence and the millennium to entertain his provincial listeners. Neither the provinces nor the continent was as remote as may first seem. By 1712, James Jurin, master of the Newcastle Grammar School, gave courses on natural philosophy that he advertised in the *Newcastle Courant* as designed for "Gentlemen concerned in Collieries and Lead-Mines."[16] Jurin's connections in the north meant more than his rise as Secretary of the Royal Society. The coal magnate William Cotesworth, who was anxious to adopt the newest technology from London, was possibly related to Jurin. In any event, the difficulties of flooded mines ensured sufficient interest in the possibilities frequently claimed for natural philosophy.[17]

[14] Royal Astronomical Society, Add. MSS. 88, *Astronomical Scrapbook*, fols. 41, 43, 112.

[15] *The Weekly-Journal or Saturday's-Post* (January 10, 1719).

[16] *Newcastle Courant* (January 24–6, 1712). Quoted in Alan Morton, "Public Science in Mid–18th Century London: The King George III Collection at the Science Museum" (unpublished paper, 1990), p. 33.

[17] Musson and Robinson, *Science and Technology*, pp. 159–60; F. J. G. Robinson, "A Philosophic War: An Episode in Eighteenth Century Scientific Lecturing in North-East England," *Transactions of the Architectural and Archaeological Society of Durham and Northumberland*, II (1970), 102; and Joyce M. Ellis, *A Study of the Business Fortunes of William Cotesworth c. 1688–1726* (New York, 1981), *passim.*

The promises of industrialization provided the foundation for the northern lecturers. Jurin was succeeded in the north by John Horsley, a Presbyterian minister, who lectured at Morpeth and Newcastle on natural philosophy in the early 1720s. Although he was later in Bath and London, and elected Fellow of the Royal Society in 1730, he continued to lecture at Morpeth until just before his death in 1732.[18] He appears to have had plenty of competition for an audience amongst the mine owners. Isaac Thompson lectured in Durham in 1730 and in Newcastle in the following year.

Increasingly, the itinerant lecturers sought the guarantee of a minimum number of auditors before they would consent to begin. In a course at Spalding in Lincolnshire in 1740, Thompson's rival, John Booth of Glasgow, required at least twenty subscribers and got forty.[19] Perhaps this minimum was the result of expensive apparatus necessary to provide experimental demonstrations, compounded by the demand to show the effects of machines. This is suggested more directly in the proposals of Isaac Thompson at Newcastle for an annual series of lectures once a guarantee of thirty pounds per annum had been subscribed. His course of mechanic, hydrostatic, and pneumatic experiments was designed explicitly "For the Service and Improvement of the Coal-Works near the River Tyne." Thompson intended, upon viewing the collieries, that "Models may be contrived and made answering the Structure of those Machines employed in the said Work; so that the whole may be fully illustrated and explained, and the Theory of Natural Philosophy not only laid down, and demonstrated from Physical Reasons, but the Practice of it may be clearly inculcated and applied to the Affairs of Life."[20] In response to the widening demand, and thus the broadening of the base of patronage, scientific lectures could be tailor-made to fit an audience.

Even Desaguliers may have considered capitalizing on the chronic difficulties of mines – like the explosive mine damps and the temperamental Newcomen engines. In 1741, the *Newcastle Courant* reported that Desaguliers intended to perform a course on experimental philosophy and that "He will be particularly serviceable to Gentlemen concerned in Col-

[18] The *D.N.B.* links Horsley with the York Buildings Company estate at Widdrington. See below, Chapter 12.

[19] Spalding Gentlemen's Society, Maurice Johnson MSS., Drawer 1, no. 2 [John Booth], *Course of Experimental Phylosophy* [sic], *Consisting of Five Parts* . . . [1740]; R. V. Wallis and Peter Wallis, *Biobibliography of British Mathematics and its Applications*, Part II, 1701–60 (Newcastle, 1988), p. 292.

[20] Carlisle Record Office, Lonsdale MSS., D/Lons./W7/1/324. Isaac Thompson, *Proposals For Establishing A Course of Mechanic, Hydrostatic, and Pneumatic Experiments, In the Town of Newcastle* (n.d.); Wallis and Wallis, *Biobibliography*, p. 302.

lieries &, as knowing an infallible Method to clear Coal Pits of Damp."[21] The lecturers need not have worried; Desaguliers did not appear with his extensive apparatus in tow. Practical problems in the mines and hydraulics provided a ready audience of natural philosophers. In 1736, the engineer and mathematics teacher John Grundy, senior, whose works Desaguliers recommended, offered to publish by subscription *A new method, grounded upon Philosophical Experiments, and Mathematical Calculations, for Draining all sorts of Low-Lands.*[22] Even the Royal Society soon acknowledged those who challenged the projectors' monopoly on practical ventures with the award of the Copley Medal.[23]

The works of the lecturers were scattered into the provinces beyond both the constraint and the focus that the Royal Society imposed. This is not to say that connection with London's philosophers was undesirable. Benjamin Martin recognized that being a Fellow of the Royal Society could provide more credibility than a country showman might obtain; he made a career of attempting to receive the honor, and in 1741, he enlisted Hans Sloane and Charles Lennox, the Duke of Richmond and F.R.S. Martin complained that it was to his disadvantage during his tours of the country towns to be unable to claim he was a Fellow.[24]

Martin was another of those who drifted between the worlds of teacher, public lecturer, and that of the instrument makers. It has been supposed that Martin may have been an assistant to Desaguliers although the evidence for this is suspect.[25] It is certain, however, that he kept a boarding school at Chichester during the 1730s and wrote works on mathematics and natural philosophy mainly for those seeking to educate themselves.

[21] F. J. G. Robinson, "A Philosophic War," pp. 103–4. On Desaguliers' interest in the problem, see Royal Society MSS., Register Book (C), XIX, 193–4; Desaguliers to Cromwell Mortimer, *Philosophical Transactions*, XXXIX (April-June 1735), 47. On ventilators, see D. G. C. Allen and R. E. Schofield, *Stephen Hales: Scientist and philanthropist* (London, 1980), pp. 81ff.

[22] Spalding Gentlemen's Society, Bound Pamphlets, vol. II, no. 7. John Grundy, *Philosophical and Mathematical Reasons, Humbly offer'd to the Consideration of the Publick; To prove that the Present Works, executed at Chester, to recover and preserve the Navigation of the River Dee Must intirely Destroy the same*... (London, 1736); Wallis and Wallis, *Biobiliography*, p. 205.

[23] Margaret C. Jacob, "Scientific Culture in the Early English Enlightenment: Mechanisms, Industry, and Gentlemanly Facts," in Alan Charles Kors and Paul J. Korshin, eds., *Anticipations of the Enlightenment in England, France and Germany* (Philadelphia, 1987), p. 147.

[24] John R. Millburn, "Benjamin Martin and the Royal Society," *Notes and Records of the Royal Society of London*, XXVIII (June 1973), 15–23; and Millburn, *Benjamin Martin: Author, Instrument-Maker and "Country Showman"* (Leyden, 1976), pp. 28–9, 35–8.

[25] Millburn, *Benjamin Martin*, pp. 45–7; cf. Augustus de Morgan, *A Budget of Paradoxes* (London, 1872), p. 91.

His first book, *The Philosophical Grammar*, which appeared in 1735, was enormously successful, being reprinted and translated into several languages by the end of the century. It is an interesting work because it directly criticized the very books that were instrumental in the spread of Newtonianism such as the efforts of Pemberton, Derham, and George Cheyne – as well as Newton's *Principia* because it was unintelligible.[26] Although a dedicated Newtonian, according to Martin, many of the books of Newton's disciples were either incoherent or too expensive. No wonder Fellowship in the Royal Society proved elusive.[27]

It was a natural progression that when Martin ultimately left Chichester in 1742, he should enter the cast of itinerant public lecturers. As early as 1741, and possibly sooner as his letters to Sloane reveal, Martin was attempting to make natural and experimental philosophy palatable in the provincial towns. Here his experiences indicated that such country lectures were not always an overwhelming success. He complained that some were so ignorant as to take him for a magician so that he was safe only in the company of the clergy or gentry who took his courses, a situation that he reported was not foreign even to the illustrious Desaguliers. Martin seems to have continued to lecture throughout the 1740s, probably largely in Bath, and possibly until 1756. By then, he was in Fleet Street selling instruments in a shop at the entrance to Crane Court, near the Royal Society to which he was refused access.[28]

By the middle of the century, we find ourselves back among the instrument makers of Fleet Street, where Desaguliers, Whiston, the Hauksbees, and Ditton had lectured and where, until the end of 1755, John Senex's widow continued to sell instruments in the shop facing St. Dunstan's Church. By then, the lecturers had produced a remarkable transformation in the world of natural philosophy; no longer merely entertainment and performance, practicality was part of the vogue – as in the lectures of the itinerant Isaac Thompson, who advertised his lectures to appeal to the coal owners of Tyneside.[29]

[26] Millburn, *Benjamin Martin*, pp. 15–16.

[27] Benjamin Martin, *The Philosophical Grammar*, third edition (London, 1748), p. vi; and Martin, *Bibliotheca Technologica: Or, A Philosophical Library of Literary Arts and Sciences* (London, 1737), pp. 640–1.

[28] Millburn, *Benjamin Martin*, pp. 35, 41–49, 84ff. See also E. G. R. Taylor, *The Mathematical Pratitioners of Hanoverian England* (Cambridge, 1966), pp. 184–5. For Martin's trade card, see H. R. Calvert, *Scientific Trade Cards in the Science Museum Collection* (London, 1971), no. 34.

[29] F. J. G. Robinson, "A Philosophic War," pp. 101–2. On developments in public lectures after midcentury, see also A. Q. Morton, "Lectures on Natural Philosophy in London, 1750–1765: S. C. T. Demainbray (1710–1782) and the 'Inattention' of His Countrymen," *British Journal for the History of Science*, XXIII (December, 1990), 411–34; Henry C.

By midcentury, the philosophical entrepreneurs had sufficiently established their credentials. Their influence brought acceptance of the value of the philosophers' practical interests as a preservative against false projects. It was Desaguliers, the philosopher Thomas Reid later remarked, who was owed some credit for stopping the disease of projects by "... shewing Men clearly upon principles of Science the utmost Effects that the Mechanical Powers can produce."[30] This was an echo of an early refrain that linked public benefit with the true philosophy of the Newtonian lecturers. The private patronage of natural philosophers by improving landlords had evolved into a public fascination. When what was once private becomes public, it is transformed into a commercial relation. In this sense, science was to welcome the economic transformation of the eighteenth century.[31]

Subscribing to philosophy

The development of publication by subscription in the late seventeenth century likewise allows us to bridge the false gap between philosophy and commerce, between nature and exploitation. Subscription, by which the supporters of an author would advance funds for the completion of a work, became a popular and important method by which an author or printer might defray some of the costs in the production of elaborate books. It was also a form of patronage, although one in which the middling sort might play a part. Although subscription publication became more frequent during the seventeenth century, it also represents part of the movement from a client economy to a mass market in the eighteenth century.[32] The flexibility of subscription met the fluidity of entrepreneu-

King, *Geared to the Stars. The Evolution of Planetariums, Orreries and Astronomical Clocks* (Toronto, 1978), esp. Chapters 10–12; John R. Millburn, *Benjamin Martin: Supplement* (London, 1986); and Millburn with Henry King, *Wheelwright of the Heavens. The Life and Work of James Ferguson, FRS* (London, 1988).

30 D. O. Thomas and Bernard Peach, eds., *The Correspondence of Richard Price*, I (Durham, N.C., 1983), p. 154. Thomas Reid to Price, ca. 1772/3.

31 On the impact of the lecturers, see F. W. Gibbs, "Itinerant Lecturers in Natural Philosophy," *Ambix*, VI (1960), 111–17; John R. Millburn, "The London Evening Courses of Benjamin Martin and James Ferguson, Eighteenth-Century Lecturers on Experimental Philosophy," *Annals of Science*, XL (1983), 437–55; Simon Schaffer, "Natural Philosophy and Public Spectacle in the Eighteenth Century," *History of Science*, XXI (1983), 1–43; Neil McKendrick, John Brewer, and J. H. Plumb, *The Birth of a Consumer Society. The Commercialization of Eighteenth-Century England* (Bloomington, 1982), *passim*; and Margaret C. Jacob, *The Cultural Meaning of the Scientific Revolution* (Philadelphia, 1987), Chapter 5.

32 McKendrick, Brewer, and Plumb, *The Birth of a Consumer Society*, pp. 224–5; On one such method of production of botanical works, see Blanche Henrey, *British Botanical*

rial sentiments. Again, the coffeehouses held an important key. In the journal of John Byrom, the inventor of shorthand, we find frequent mention of newly published scientific works and the circulation of their proposals for subscription, such as that for Chambers' *Cyclopaedia*, which was discussed at in the Sun in St. Paul's Churchyard the year before it was published.[33]

Thanks largely to the work of Ruth and Peter Wallis, we must now appreciate the importance of subscription publication in the sciences. The mandate of the Royal Society for the promotion of natural knowledge was always problematic when its finances were so uncertain. Efforts, for example, to encourage the printing of works in mathematics ran contrary to the heavy losses of the booksellers. As John Collins had repeatedly complained, "Mathematical learning will not . . . go off without a dowry. . . ."[34] It was these pressures on booksellers and philosophers alike that drove the gathering of subscribers.

From the early decades of the Royal Society, subscription publication became an effective means to overcome the limitations of financial resources. Second only to the theologians, scientists were quick to grasp the possibilities of such a method, especially in the production of large and lavishly illustrated texts. As early as 1672, the Royal Society had lured Dr. Nehemiah Grew from his medical practice in Coventry with the promise of support of his botanical research. However, the proposal to finance his research by a subscription among the Fellows was more laudable than real. As a result, Grew required subscription in order to complete the order of the Royal Society for a catalogue of its rarities. The *Musaeum Regalis Societatis* is an interesting work because it is one of the transition from private to public patronage. Not only did it have a large number of subscribers, but it also benefitted from the private benefaction of the merchant Daniel Colwall, who underwrote the expensive engravings.[35] Botanical works, because of their need for expensive illustration, figure prominently in this evolution. Subscription was

and Horticultural Literature before 1800, 3 vols. (London and New York, 1975), II, 51–2.

[33] Richard Parkinson, ed., *The Private Journal and Literary Remains of John Byrom*. For the Chetham Society, XXXII (1954), 197.

[34] *Correspondence of Isaac Newton*, II, 192. John Collins to Thomas Baker, February 10, 1676/7; 241–2. Collins to John Wallis, 1677/8.

[35] Michael Hunter, "Early Problems in Professionalizing Scientific Research: Nehemiah Grew (1641–1712) and the Royal Society, With an Unpublished Letter to Henry Oldenburg," *Notes and Records of the Royal Society of London*, XXXVI (February 1982), 201–3; Sarah L. Clapp, "Subscription Publishers Prior to Jacob Tonson," *The Library*, XIII (1933), 165–6. In 1680, Grew circulated a subscription proposal together with the names of 145 who had already subscribed. I owe this reference to Michael Hunter.

quickly regarded by printers as an effective means of publication. One such project was the publication in 1680 of the *English Atlas* by Moses Pitt, which had over eight hundred subscribers while being overseen by Fellows of the Royal Society who stood behind the subscription. Likewise, Joseph Moxon employed subscription in the collected edition of his *Mechanick Exercises.*[36]

The success of such projects depended not only on advertising the proposal, but on the drive of influential persons who might help generate enthusiasm amongst a literate public. The assistance of the Royal Society might be of great benefit even though it was often unable to provide financial support. In 1685, however, the printer of *A Treatise of Algebra* by John Wallis was able to secure the agreement of the Society to take the first sixty copies.[37] One of the surest methods of securing the subscription was to use political and social connections with wealthy merchants, and with the gentry and aristocracy who found it necessary to be in London while Parliament was in session. This was a common stratagem. For example, early in the century, the naturalist John Ray, then at Black Notley, requested the assistance of Hans Sloane and Henry Compton, Bishop of London, whose botanic garden at Fulham was the site of much scientific interest.[38] Ray's *Historia Plantarum* required expensive plates that he could not afford from his own resources. In 1702, Compton urged him "that what is done would best be done while the Town was full of Nobility & Gentry sitting [in] the Parliament" at the beginning of Anne's reign.[39]

By the reign of George I, this process had been worked to perfection. The public success of John Harris, Thomas Watts, or Desaguliers may have been the result of contacts with powerful patrons, sometimes with the Royal family, as well as with many Whigs whose fortunes were linked to the Hanoverian star. Of this Whig connection a great deal can be made in the history of Newtonianism.[40] If we leave aside for the moment the

[36] Clapp, "Subscription Publishers," pp. 169–70, 179–80; and David Woodward, "English Cartography, 1650–1750," in Norman J. W. Thrower, ed., *The Compleat Plattmaker* (Berkeley, 1978), p. 186.

[37] Clapp, "Subscription Publishers," p. 178.

[38] Edward Carpenter, *The Protestant Bishop* (London, 1956), pp. 365–7; Blanche Henrey, *British Botanical and Horticultural Literature*, II, 318, 335, 646.

[39] B. L., Sloane MSS. 4039, f. 49, Ray to Sloane, December 16, 1702.

[40] James R. Jacob and Margaret C. Jacob, "The Anglican Origins of Modern Science: The Metaphysical Foundations of the Whig Constitution," *Isis*, LXXI (June 1980), 251–67; Margaret C. Jacob, *Newtonians and the English Revolution, passim*; John Gascoigne, *Cambridge in the Age of the Enlightenment. Science, Religion and Politics from the Restoration to the French Revolution* (Cambridge, 1989), p. 147. Among the skeptics are Geoffrey Holmes, "Science, Reason and Religion in the Age of Newton," *British*

supposed metaphysical and ideological connections, can we see the results of any apparent patronage beyond Harris and Desaguliers? After all, was the world of early eighteenth-century England so divided into antagonistic camps of Whig and Tory to the point where we continually find ourselves face to face with the importance of the Whig adherence to prominent Newtonians? Or was this merely a function of Whig dominance and Tory drought? Certainly, the existence of early Tory Newtonians like John Keill, John Arbuthnot, or even Jacobites like Archibald Pitcairne should not be ignored.

It would be naive to wrap either patronage or politics in philosophical principle. Ideological committment may indeed have been a crucial component of alliances at the Revolution or in Anne's reign, and more so perhaps in religious connection than in City finance. But by the time of the machiavellian moment of the Robinocracy, or of the reign of George II, there were more instances where self-interest overwhelmed ideology in the end. What, after all, did Desaguliers' peculiar poem on the *Newtonian System of the World* represent at the end of the reign of George I more than an attempt to curry favor with a new regime? Patronage might mask more than it revealed.

Patronage, politics, and subscription

Politics and connection were critical in the success of a subscription. By Anne's reign, science had become fashionable among the English establishment.[41] Yet, there is no question that Whigs were increasingly the most prominent among subscribers to the philosophers – but not to the exclusion of the Tories or even to the point of failing to assist the works of those suspected of Jacobitism like the mathematician James Stirling, who both Newton and Desaguliers supported. What is interesting about the list appended to Stirling's *Lineae tertii ordinis Newtonianae* of 1717 is that it exhibits strong Tory, and possibly Jacobite, political links, including John and James Keill, and the physician John Freind, who was to be imprisoned in the Tower during the Atterbury plot. The most

Journal for the History of Science, XI (July 1978), 164–171; Michael Hunter, *Science and Society in Restoration England* (Cambridge, 1981), pp. 185–7; and Colin Russell, *Science and Social Change, 1700–1900* (London, 1983), pp. 57–9. On the problem of sorting metaphysical, philosophical, and political conflict in this issue, see Steven Shapin, "Of Gods and Kings: Natural Philosophy and Politics in the Leibniz-Clarke Disputes," *Isis*, LXXII (June 1981), 187–215; and James E. Force, *William Whiston: Honest Newtonian* (Cambridge, 1985), pp. 60–2, 91–5

[41] M. C. W. Hunter, "The Crown, the Public, and the New Science, 1689–1702," *Notes and Records of the Royal Society*, XLIII (1989), 109–11.

consistent subscriber of all those with a firm political identity was, not surprisingly, Newton himself. Besides Newton, among the prominent subscribers were the Tories Charles Boyle, Earl of Orrery and F.R.S., and Edward Harley. The Whigs, however, clearly dominate as they came to do politically, with Joseph Eyles, a Director of the East India Company, Richard Boyle, the Earl of Burlington and F.R.S., along with James Brydges, Martin Folkes, Thomas Herbert, Earl of Pembroke and a former President of the Royal Society, and Charles Townshend, F.R.S. and Secretary of State under Walpole.

Many of the Newtonian works of the early eighteenth century were published by subscription. Political persuasion was an important factor in the connections that supported subscription between the appearance of such prominent works as John Harris's *Lexicon Technicum* in 1704 and the final volume of J. T. Desaguliers' *A Course of Experimental Philosophy* in 1744. But politics was surely only one factor, and there are simply many subscribers who cannot be identified with any certainty.[42] It is impossible to draw any final conclusions as to any precise political motivation behind subscription to the Newtonian publications in general because some persons subscribed infrequently or possibly only once through the entire period. It is not surprising, for example, that James Brydges, later Duke of Chandos, should be among the most consistent subscribers – but what can one say about a man who once purchased books by the area they were to fill on the shelves of his estate except that subscription may be a different degree of commitment than mere purchase.[43]

In this respect, natural philosophy was no less a part of culture than the writing of poetry or of history. That is not to say politics alone would promote a subscription except perhaps if a work were of a particular polemical tone. But subscription can be viewed as a function of connection just in the same manner that connection formed the basis of faction in politics. There are numerous examples of the effect of such associations, as complex as some of them were in the early part of the century. The poet James Thomson had received support from Robert Walpole for the

[42] The best introduction to this issue is through the work of Peter Wallis. See F. J. G. Robinson and P. J. Wallis, *Book Subscription Lists: A Revised Guide* (Newcastle Upon Tyne, 1975); Wallis, "British Mathematical Bibliography," *Institute for Navigation Journal*, XX (April 1967), 200–4; Wallis, "Book Subscription Lists," unpublished lecture, Bibliographical Society (April 1973); Wallis and Robinson, "Scientists and Subscription Lists," *British Journal for the History of Science*, VI (1972), 227–8; and A. D. Burnett, "Essay Review," *Annals of Science*, XXXII (1975), 579–82.

[43] Hunt. Lib., Stowe MSS. 57, XII, Brydges to Anthony Hammond, his agent in Paris, September 26, 1715.

production of his poem "To the Memory of Sir Isaac Newton." An enthusiast of commerce, Thomson had an interest in natural philosophy and was suitably employed in Thomas Watts' Academy in 1726. But when Thomson published his famous work *The Seasons* in 1730, it was subscribed to by a large number of Whigs, including some who had split from Walpole's supporters. W. A. Speck has likewise shown that Government Whigs were prominent in the support of Henry Pemberton's *View of Sir Isaac Newton's Philosophy* in 1728. It is not possible, however, to reduce connection to ideology. James Stirling was similarly employed at the Academy by Thomas Watts, and his subscribers in 1717 were decidedly Tory. Christian Cole, who was probably involved with Watts' commercial ventures, had Whig subscribers to his political history in 1733. The motivation for subscription could be complex; politics was only one factor.[44]

Subscription was a means of financing a project and such projects might be of many kinds. In 1710, Richard Steele used his friendship with Charles Montague, first Earl of Halifax, to help promote a subscription to his edition of *The Lubrications of Isaac Bickerstaff*.[45] Likewise, in 1722, the Newcastle physician James Jurin, as Secretary of the Royal Society, wrote to the colonial judge James Dudley for a subscription among the Fellows to assist Thomas More, who had been sent to New England to study its natural history.[46] Subscription had become one theme of the chorus of

[44] W. A. Speck, "Politicians, Peers and Publication by Subscription, 1700–1750," in Isabel Rivers, ed., *Books and their Readers in Eighteenth-Century England* (Leicester, 1982), pp. 56–64. On James Thomson, see A. S. Collins, *Authorship in the Days of Johnson. Being a Study of the Relationship between Author, Patron, Publisher and Public, 1726–1780* (London, 1928), pp. 149ff.; J. L. Heilbron, "Robert Symmer and the Two Electricities," *Isis*, LXVII (March 1976), 11; Marjorie Hope Nicolson, *Newton Demands the Muse. Newton's Opticks and the Eighteenth Century Poets* (1946, repr. Westport, 1979); and John McVeagh, *Tradefull Mechants. The Portrayal of the Capitalist in Literature* (London, 1981), pp. 61–2, 74–5. Speck (p. 56) identifies as a Whig list that of Christian Cole, *Memoirs of Affairs of State* (London, 1733). As it happens, Cole, who was possibly resident at Venice in the embassy of the Duke of Manchester, was likely also a member of the management committee of the York Buildings Company, as well as involved in the Mines Royal and Mineral and Battery Works. Cole was probably used because of his valuable contacts at Court. Scottish Record Office, GD1/170, vol. I. Case Billingsley to Cole, January 6, 1718/9; Billingsley to Cole, February 21, 1718[19?]. Among the subscribers to Pemberton in 1728 were also Chandos, Charles De la Faye, Thomas Watts, John Hartopp, Sir John Eyles, and John Billingsley, Secretary of the York Buildings Company.

[45] British Library, Add. MSS. 7121, f. 69, Steele to Halifax, January 26, 1709/10. Halifax was formerly President of the Royal Society in 1695, and in 1696, appointed Newton as Warden of the Mint. On Newton and Halifax, see Richard S. Westfall, *Never at Rest: A Biography of Isaac Newton* (Cambridge, 1980), *passim*.

[46] Clements Library, University of Michigan, Ann Arbor, Miscellaneous Collection, James

patronage. Just a few days before Jurin's plea, the botanist Richard Bradley had written a desperate letter to Hans Sloane describing measures to restore his fortunes following losses he had sustained at the hands of booksellers and since the failure of his garden in Kensington. He had the promise of assistance from the botanist William Sherard and from the ubiquitous merchant Charles Du Bois. Bradley was inclined to go abroad "Even into the Most Dangerous country" despite the fears of friends at Court, reasoning that "to live upon Expectations at home is as bad as it can be to venture ones Life among Savages abroad. . . ."[47]

Bradley's vision of the anguish of the client was not overdrawn. Life could be very hard indeed, and his threat to go abroad may not have been an idle one. Earlier in 1722 in a similar situation, Sloane had received a request from the Duke of Chandos to recommend a botanist to the African Company. In April, the surgeon, James Houstoun, in dire need of employment, began a dismaying set of adventures when sent to the slave coast in Company service.[48] It was increasingly apparent that commercial contacts were one avenue to salvation. This was, of course, nothing new, but the degree of such associations reinforced the venue of the coffeehouse lectures and the flowering of subscription publication. New wealth meant new patrons and a chance to escape the clutches of the booksellers.

To those natural philosophers who depended upon private patronage or public subscription, their future was an uncertain proposition at best. One often senses an air of desperation. Although we really know little about the difficulties of securing subscriptions, a fascinating glimpse is given in the career of Richard Bradley. As early as 1710, he had considered subscription as a means of financing the cost of copperplates for "A Treatise of succulent plants." Over the next two decades, he was able to secure assistance from Charles Du Bois and the Duke of Chandos to

Jurin to Paul Dudley, July 2, 1722. On Thomas More, see Blanche Henrey, *Botanical and Horticultural Literature*, II, 348.

[47] B. L., Sloane MSS. 4046, f. 251, Richard Bradley to Hans Sloane, June 23, 1722. Both Sherard and Du Bois had been early members of the Temple Coffee House Botany Club. See Henrey, *Botanical and Horticultural Literature*, II, pp. 78, 176, 335, 431–2, 447–78, 452, 639. Du Bois, who was Treasurer to the East India Company, was also a subscriber to both volumes of John Harris's *Lexicon Technicum*. Humphry Ditton dedicated his *General Laws of Matter and Motion* of 1705 to Du Bois and Desaguliers mentions him in his translation of Gauger's *Fires Improv'd* in 1715. Bradley also refers to Du Bois' design of walls for the better ripening of grapes in his translation of G. A. Agricola, *The Experimental Husbandman and Gardener. Containing a New Method of Improving Estates and Gardens . . .*, second edition (London, 1726), p. 308.

[48] L. Stewart, "The Edge of Utility: Slaves and Smallpox in the Early Eighteenth Century," *Medical History*, XXIX (1985), 60–5.

whom Bradley dedicated part of his *New improvements of planting and gardening*.[49] It is difficult to know how James Brydges, not yet Duke of Chandos, came into contact with Bradley although it was likely through Hans Sloane or Desaguliers, or simply by the gift of one of his books in 1717. Brydges was always suspicious, for all the good it did him in the end. He asked his personal physician, Henry Levett, to try to discover the character and circumstances of a Mr. Bradley, a pensioner of the Charterhouse. Within a few months, Bradley was employed on the estate at Cannons. By 1719, however, this arrangement had gone sour as Brydges raged over mismanagement and poor workmanship in the construction of a physic garden and hot house even though Desaguliers had approved of the accounts.[50] This was neither the first nor the last of his complaints of being badly used by those he patronized.

Richard Bradley faced financial difficulties most of his life. This was undoubtedly exacerbated by the expensive business of publishing works of natural history that required so many plates. Besieged by his creditors, Bradley was sometimes to correspond by using a pseudonym although it seems an elaborate ruse.[51] Nonetheless, he did try to find ways to overcome the exorbitant costs of publication, at one point enquiring of Hans Sloane about some engravings the naturalist James Petiver had once used for an edition of John Ray's *History of plants*.[52] In 1727, Bradley was in such a bad way that Sloane had to send him a guinea so that he could afford the stamps necessary for the printing of his *Weekly miscellany for the improvement of husbandry, trade, arts, and sciences*.[53] Like so many of Bradley's projects, this, too, was doomed to failure. Even efforts to republish the works of the popular agricultural writer John Houghton failed to help. Bradley died in poverty in 1732, leaving a wife and child with nothing but a hope in a plaintive petition that Sloane received a

[49] Henrey, *Botanical and Horticultural Literature*, II, pp. 428–50.

[50] Hunt. Lib., Stowe MSS. 57, XV, f. 260, Brydges to Dr. Levett, June 21, 1718; XV, f. 275, Brydges to Bradley, July 9, 1718; XV, f. 295, Brydges to Bradley, August 8, 1718; XVI, fols. 407–10, Brydges to Bradley, December 16, 1719. Henrey, *Botanical and Horticultural Literature*, II, pp. 435. The complaint was sent through Desaguliers; Stowe MSS. 57, XVI, f. 410. Brydges to Desaguliers, December 16, 1719. On Bradley's contribution to hot houses, see [Nicolas] Gauger, *Fires Improv'd: Being a Method of Building Chimneys, So as to prevent their Smoaking*, trans. J. T. Desaguliers (London, 1715), p. 2; and Bradley, *A General Treatise of Husbandry and Gardening; Containing a New System of Vegetation* (London, 1726), Part IV.

[51] Henrey, *Botanical and Horticultural Literature*, II, p. 425.

[52] B. L., Sloane MSS. 4058, fols. 56–7, Bradley to Sloane, n.d. On Petiver and Bradley, see Frank N. Egerton, "Richard Bradley's Illicit Excursion into Medical Practice in 1714," *Medical History*, XIV (1970), 53–62.

[53] B. L., Sloane MSS. 4049, f. 31, Bradley to Sloane, Sept. 22, 1727; Henrey, *Botanical and Horticultural Literature*, II, pp. 449–50.

month after Bradley's death. Upon her marriage, his widow had brought with her enough to pay off the debts Bradley brought with him; now she was destitute.[54]

The client economy did not answer the desperation of many authors. Bitterness had long been the client's lot. Even the Elizabethan author John Norden had complained that "relieving deeds are the substance, wordes but the shadow, performance is the fire, and promise but the smoake."[55] Bradley's history may seem one of extreme circumstances, but the evidence suggests that it was not unique. For example, in 1730, Sloane received a request for assistance from the Whig pamphleteer Thomas Cooke, who was imprisoned with his family for debt. Although Sloane had already subscribed with many others like Walpole and the Earl of Pembroke to a proposed work, Cooke required further aid as he was close to securing his release.[56]

A much more prominent case was that of Newton's friend, the Swiss mathematician Nicholas Fatio de Duillier, who had disgraced himself early in the century in the affair of the French prophets. Fatio found temporary refuge as a projector having invented a method to overcome friction in timepieces by using rubies as bearings although he was ultimately obstructed by the opposition of the Clockmaker's Company. In 1736, he wrote to Sloane with a claim of having lost thousands of pounds by his invention. He sought Sloane's sympathy for proposing a voluntary subscription of the Royal Society to publish his discovery of the sun's parallax and thereby to recover his fortunes.[57] Subscription remained a potential release from the literary debtor's prison.

But importuning patrons was frequently ineffective and tiresome. By 1729, Stephen Switzer gave up subscription as a means of publishing his lavishly illustrated introduction to hydrostatics and hydraulics and simply

[54] B. L., Sloane MSS. 4058, f. 50, "The Case and Humbl [sic] Petition of Mary Bradley widow of Dr. Richard Bradley late Professor of Bottany in the University of Cambridge," n.d.

[55] Quoted in Stan A. E. Mendyk, "*Speculum Britanniae*." *Regional Study, Antiquarianism, and Science in Britain to 1700* (Toronto, 1989), p. 64.

[56] B. L., Sloane MSS. 4051, fols. 111–12, Thomas Cooke to Sloane, September 26, 1730; *D.N. B.*, XII, 95.

[57] B. L., Sloane MSS. 4055, f. 27, N. Facio to Sloane, December 28, 1736; David S. Landes, *Revolution in Time: Clocks and the Making of the Modern World* (Cambridge, Mass., 1983), pp. 135–8. On Fatio de Duillier, see Charles Andrew Domson, "Nicolas Fatio de Duillier and the Prophets of London: An Essay in the Historical Interpretation of Natural Philosophy and Millennial Belief in the Age of Newton" (Ph. D. thesis, Yale University, 1972). The cause of Fatio's circumstance was losses in speculation of the stock of the Charitable Corporation. See *Gentleman's Magazine*, IV (May 1734), 235–6. The list of the sufferers also included William Watts.

offered the work at a better price to those who had already ventured money.[58] In 1735, there was an attempted response to the problem by organizing a society under the guidance of Charles Lennox, Duke of Richmond and patron of the lecturer Benjamin Martin. It was intended that by the subscription of its members, a thousand guineas would be raised to assist authors and thereby "to institute a Republick of Letters for the promoting of Arts and Science. . . . " The group was composed of the Earl of Stanhope, the Newtonian poet laureate James Thomson, as well as a significant number of physicians, including Richard Mead, and Alexander Stuart who was probably one of Chandos's advisors.[59]

Gentlemen might easily be dunned by those who never finished a project or never intended to. Moreover, the natural unwillingness of printers to take the risk of an expensive publication onto their own shoulders meant that they would seek means to minimize the gamble by obtaining funds in advance from an interested public. But this had implications that went far beyond the works themselves. Trade publishers, of course, would be reluctant to undertake esoteric ventures unless accompanied by subscription. In addition, the broadening of the base of patronage among the merchants suggested a social transformation with which not all were equally comfortable. Perhaps this awareness lay behind the alarm sounded by Arthur Charlett, master of University College, Oxford, who wrote to Hans Sloane in 1697 in the hope that the two universities and the Royal Society might cooperate in publishing works of learning, in order to lessen the growing dependence "on ye men of Trade."[60]

Corruption and ideology

We are faced with the problem of the broad, discursive links between science and society. Margaret Jacob has revealed an ideological transformation in the early eighteenth century that linked Newtonianism with

[58] Stephen Switzer, *An Introduction to a General System of Hydrostaticks and Hydraulicks*, II (London, 1729), "Advertisement."

[59] B.L., Sloane MSS. 3984, fols. 34–5, "Plan for Managing the designs of the intended Society for the encouragement of learning and Learned Men . . . ," August 1, 1735; cf. Arnold Zuckerman, "Dr. Richard Mead (1673–1754). A Biographical Study" (Ph. D. thesis, University of Illinois, 1965), p. 133. The effort to extend the range of patronage for authors did not prevent a bitterness that was endemic to the system in the eighteenth century. See Thomas Malton, *An Essay Concerning the Publication of Works, on Science and Literature, By Subscription, Inscribed to the Nobility and Gentry of England . . .* (London, 1777). I thank Jan Golinski for pointing this out to me.

[60] B.L., Sloane MSS. 4036, f. 301, Charlett to Sloane, April 12, 1697.

Latitudinarianism and political stability. On the other hand, such an ideological framework does not make it easy for us to recognize the place of the "crafty, ill-principled men" who threatened to destroy that tenuous stability. They, too, were also closely connected with, if not children of, the post-Revolutionary order of capitalist and industrial self-interest, when greed and growth became synonymous.[61] What jaundiced contemporary observers like Roger North, and numerous historians since, have recognized is that the world of the Newtonians was one of widespread transition. As Philip Jenkins nicely pointed out in his recent study of the Glamorgan gentry, "we lack essential components for an account of the background to the Industrial Revolution" and to the reasons for "a great gulf [that] separates the life and thought of the generation that was old in 1720 from that of the mid eighteenth century."[62] It seems pointless to ignore the place of Newtonianism in that transformation.

Economic or technological development rested not so much on the achievements of Newtonians as on the dreams they were prone to indulge. It could not have been easy to ignore the promises of natural philosophers like John Harris or J. T. Desaguliers. Many found it impossible to do so. When one explores the traces of projectors on the prowl and the great lottery in which inventors played, one quickly encounters many who fancied themselves part of the world of natural philosophy and a few who even made a legitimate contribution to it. But most were committed to quick profits. Only incidentally did the lasting accomplishment of natural philosophy or technological advance enter their minds. Examples of such attitudes abound. It is amidst such opportunism that we make contact with the social world of Newton and his disciples.

In the Financial Revolution that followed 1688, there were the seeds of both innovation and social resentment. Alienation from this new world was based on a sense of corruption in the new forms of property and on a fear of the sudden power that economic alchemy might have wrought. Invention and projects held all of the risk that landed property happily disguised. New ventures meant new puzzles. So Daniel Defoe echoed an

[61] Jacob, *Newtonians and the English Revolution*, pp. 160, 183–184; J. G. A. Pocock, *Virtue, Commerce, and History. Essays on Political Thought and History, Chiefly in the Eighteenth Century* (Cambridge, 1985), pp. 108–10; and Margaret C. Jacob, *The Cultural Meaning of the Scientific Revolution*, pp. 130–1, Chapter V.

[62] Philip Jenkins, *The Making of a Ruling Class. The Glamorgan Gentry, 1640–1790* (Cambridge, 1983), pp. xvii–xviii. Cf. Sidney Pollard, *The Genesis of Modern Management. A Study of the Industrial Revolution in Great Britain* (London, 1965), pp. 23, 115; J. C. D. Clark, *Revolution and Rebellion. State and Society in England in the Seventeenth and Eighteenth Centuries* (Cambridge, 1986), pp. 37–9; and J. C. D. Clark, *English Society 1688–1832* (Cambridge, 1985), pp. 65–70.

earlier unease in his remark on the companies forged from the imaginations of projectors:

> Mine Adventures raise Annuities, when Funds fall; lose their Vein of Oar in the Mine, and yet find it in the Shares; let no Man wonder at these Paradoxes, since such strange things are practised every Day among Us?[63]

It was, in other words, a society highly susceptible to claims of invention and discovery, and thus to the hopes of projectors and of the practitioners of natural philosophy.

By contrast, many Newtonians had an urgent enthusiasm for the commercial world that had been given more precise shape since the Revolution – in the form of the growth of joint-stock and chartered companies. Margaret Jacob has argued that Boyle lecturers like the naturalist William Derham and technical John Harris saw Christianity affirming "the mutual Support of Society and Commerce . . . "[64] Such sentiments give rise to ideological conclusions that link early capitalism with the kind of Erastian and Latitudinarian views seemingly in favor among the early eighteenth-century Whigs. A great deal might be made of such a structure, especially if it is taken as an ideological reflection of widespread attitudes to changing economic relationships. If some country squires chose to dislike what they could not understand, there were sound economic reasons for their suspicion of the innovations of the City. The funding of the National Debt at the end of the seventeenth century created a field of investment that drew off monies that might otherwise have been invested in land. Land still had its attraction in terms of social prestige, but as an investment, it had a difficult time matching the returns of the new public funds, of the Bank, or of the trading companies.[65]

All was not lost, however. Even the most diehard of Tory landowners showed that they had not been inoculated against Whiggish innovation, if indeed that is what it was. In her study of the Tory party, Linda Colley has sensibly cautioned us against Tory cant that relies on the likes of

[63] Quoted in J. G. A. Pocock, "Early Modern Capitalism – The Augustan Perception," in Eugene Kamenka and R. S. Neale, eds., *Feudalism, Capitalism and Beyond* (Canberra, 1975), p. 79. See also Pocock, *The Machiavellian Moment: Florentine Political Thought and the Atlantic Republican Tradition* (Princeton, 1975), pp. 452ff.; and Joan Thirsk, *Economic Policy and Projects. The Development of a Consumer Society in Early Modern England* (Oxford, 1978), pp. 98–105. On the feudal basis of capitalism, see Alan Macfarlane, *The Culture of Capitalism* (Oxford, 1987), pp. 158–62, Chapter VIII.

[64] Jacob, *The Newtonians and the English Revolution*, pp. 180–1.

[65] H. J. Habakkuk, "English Landownership, 1680–1740," *Economic History Review*, X (February 1940), 9–13; G. E. Mingay, *English Landed Society in the Eighteenth Century* (London, 1963), pp. 80–5; J. V. Beckett, *The Aristocracy in England, 1660–1914* (Oxford, 1986), pp. 44–5, 80–6; Pocock, *Virtue, Commerce, and History*, pp. 112–13.

Bolingbroke or Jonathan Swift to link the most unsavory aspects of eighteenth-century society with the Whigs.[66] Bolingbroke and Swift were hardly likely to be the most objective observers even when they had cause to complain against corruption – whether political or financial. Indeed, the South Sea Company had been the creation of a government of which Bolingbroke himself had been a minister and he had been one of its directors.[67] This strategy of public virtue often serves to lock away the skeletons of the past.

The picture of early eighteenth-century capitalism as an instrument of City corruption and Whig connection induces nostalgia for a lost world, to be sure. But it can easily be forgotten that there were many Tory landlords who failed to see their politics as an impediment to investment or to the improvement of their estates. For example, the widespread Leveson–Gower estates and their partnership with some of the early industrialists placed the family in the forefront of mineral development in the first half of the century. The second Baron Gower actively sought ways to encourage entrepreneurs to open new mines on his estates.[68] Tory politics was no obstacle in the copper and coal developments in the west country involving Sir Humphrey Mackworth in the 1690s and later the opposition M.P., Richard Lockwood.[69] Similarly, the iron manufacturing family of Ambrose Crowley II traded Quaker sympathies for Tory politics with some success.[70]

It is clear that opportunity had as much impact on the adoption of techniques of investment and exploitation as an Opposition or Country ideology may have had. Whether linked to Court or Country, whether Whig or Tory, many of the large landowning families were in the vanguard of the adoption of the new technologies in order to make the exploitation of their mineral resources viable. This was especially true as mines had to be sunk lower and flooding became more of a problem as time went on. Massive investment poured down deep shafts. The Earls of Lowther provide an excellent example of this kind of Country adaptation. Sir John Lowther was the first governor in 1693 of the "Company for making

[66] Linda Colley, *In Defiance of Oligarchy. The Tory Party 1714–1760* (Cambridge, 1982), pp. 146–7.

[67] Shiela Biddle, *Bolingbroke & Harley* (London, 1975), pp. 188–90, 280–1; John Carswell, *South Sea Bubble* (London, 1961), pp. 57–8, 67–8.

[68] J. R. Wordie, *Estate Management in Eighteenth-Century England. The Building of the Leveson-Gower Fortune* (London, 1982), pp. 77–8, 107.

[69] Philip Jenkins, "Tory Industrialism and Town Politics: Swansea in the Eighteenth Century," *Historical Journal*, XXVIII (1985), 118.

[70] Colley, *Defiance of Oligarchy*, p. 148; see also Arthur Raistrick, *Quakers in Science and Industry* (Newton Abbot, 1950), p. 110.

iron with pit coal," only one of many of such groups to deal with that difficult technique. It was his son, the Country Whig Sir James Lowther, who signed an agreement with Thomas Newcomen in 1715 for the erection of a steam engine to help drain the collieries at Whitehaven in Cumberland.[71] Likewise, the Earl of Dudley erected a Newcomen engine at Coneygree to drain the coal mines in sight of Dudley Castle possibly as early as 1712.[72]

It was nevertheless the case that many were suspicious of the new finance. Their unease, perhaps, was as much directed against the abuse of credit as against such familiar commerce.[73] This was a point that Daniel Defoe sought continually to reassert.[74] The problem was not trade, but the discredit of trade. Even Defoe was not above satirizing the variety of fanciful schemes that threatened to swamp legitimate trade. In 1720, he announced the erection of "several Royal Societies for the general Consumption of Fish," which were not likely to succeed without a general Brew-House created by a joint-stock of £10,000,000 in the first Call of which preference would be given to the "Virtuosi of Gresham, and the two Universities."[75]

The morality of trade, in Defoe's mind, imposed a choice between private interest and public benefit. Properly cultivated, "Commerce is a friend to Learning; Trade makes the People rich, and their Wealth puts them upon the improvement of Arts and Sciences."[76] If knowledge im-

[71] A. Raistrick and E. Allen, "The South Yorkshire Ironmasters (1690–1750)," *Economic History Review*, IV (May 1939), 176; J. S. Allen, "The 1715 and other Newcomen Engines at Whitehaven Cumberland," *Transactions of the Newcomen Society*, XLV (1972–3), 237–68. The best treatment of the Lowther family is J. V. Beckett, *Coal and Tobacco. The Lowthers and the Economic Development of West Cumberland, 1660–1760* (Cambridge, 1981). Sir James Lowther held strong Country views of the "rogueries" of some company directors. See J. V. Beckett, "A Back-Bench M.P. in the Eighteenth Century: Sir James Lowther of Whitehaven," *Parliamentary History*, I (1982), 86; and Beckett, *Aristocracy in England*, Chapter VI, *passim*.

[72] John S. Allen, "The 1712 and other Newcomen Engines of the Earls of Dudley," *Transactions of the Newcomen Society*, XXXVII (1964–5), 57–84; and L. T. C. Rolt and J. S. Allen, *The Steam Engines of Thomas Newcomen* (New York, 1977), pp. 46–9.

[73] Neil McKendrick, John Brewer, and J. H. Plumb, *The Birth of a Consumer Society*, p. 199.

[74] See Isaac Kramnick, *Bolingbroke and His Circle. The Politics of Nostalgia in the Age of Walpole* (Cambridge, Mass., 1968), pp. 39–40; Pocock, *The Machiavellian Moment*, pp. 454–5; and Daniel Defoe, *The Complete English Tradesman in Familiar Letters* (1727; repr. New York, 1969), I, pp. 336–48.

[75] William Lee, ed., *Daniel Defoe: His Life, and Recently Discovered Writings* (London, 1869), II, pp. 221–2; see also Simon Schaffer, "Defoe's Natural Philosophy and the Worlds of Credit," in John Christie and Sally Shuttleworth, eds., *Nature Transfigured. Science and Literature, 1700–1900* (Manchester and New York, 1989), esp. pp. 24–8.

[76] Daniel Defoe, *A General History of Discoveries and Improvements* (London, 1727), p. 92; Schaffer, "Defoe's Natural Philosophy and the Worlds of Credit," p. 26.

plied a conquest of the world, the symbiosis between philosopher and projector was still a struggle between virtue and vice.

So many new projects seemed to siphon money away from land. For example, the 1690s were to see a great increase in the number and diversity of insurance schemes, based in part on the actuarial work of the seventeenth-century mathematicians. In the Royal Society, not only was its President the Chancellor of the Exchequer who oversaw these developments, but Newton also was intimately involved. In addition, both Edmund Halley and Abraham de Moivre wrote works of use to the new insurance industry.[77] Such circumstances were obviously not without consequence to natural philosophers. In 1691, a divine of the Church of England, in a tract entitled *A Plain and Easie Way for the Speedy Raising of Money*, cast a prophetic and jaundiced glance at the profits of the relatively new marine insurance industry.[78] The issue of financial innovation was fundamentally a question of political and social transformation in which natural philosophers were bound to play an important role.

The inducement to invest in stocks and lotteries, compounded by a seemingly burdensome land tax, merely raised further rancor amongst those who linked land and pedigree.[79] Hatred of the Whigs and the monied interest was at the root of Jonathan Swift's savage attack on the coinage of William Wood in the *Drapier's Letters*. By the 1720s, Tory suspicions of all that the Whigs had wrought since the Revolution had seemed to come to pass. When Swift sought to explore the basis for Wood's project of Irish coinage he saw an old pattern: "Hath he discovered the Longitude or the Universal Medicine? No. But he hath found out the Philosopher's Stone after a new Manner, by Debasing of Copper, and resolving to force it upon us for Gold." Newton, at the Mint, approved the new coin – and Wood's closest links were with the Whigs.[80] Such evidence as existed was enough to fuel fears that had festered since the Revolution – notwithstanding the obvious fact that the landowners Tories claimed to represent were also seduced by the public stocks as well as by numerous innovations to improve their estates.

[77] P. G. M. Dickson, *The Financial Revolution in England, 1688–1756* (London and Melbourne, 1967), p. 41.

[78] Dickson, *Financial Revolution*, p. 24.

[79] Dickson, *Financial Revolution* , pp. 25–6; and G. E. Mingay, *English Landed Society in the Eighteenth Century* (London, 1963), pp. 38, 54.

[80] J. M. Treadwell, "Swift, William Wood, and the Factual Basis of Satire," *Journal of British Studies*, XV (Spring 1976), 84–90; Brian W. Hill, *Sir Robert Walpole "Sole and Prime Minister"* (London, 1989), pp. 124–5.

The world of projectors

When Isaac Newton had arrived in London in the spring of 1695 as warden of the Mint he formally entered a world of entrepreneurs and practical men. Although he had been asked the previous autumn to comment on the controversy surrounding the efforts at recoinage, Newton had not, of course, traded his infinite universe for the more circumscribed world of government finance. He had simply discovered the benefits of government patronage. Yet he soon found that the laws of human motivations were not so easily defined as those universal laws of the creator he had sought to reveal.

Newton's new world was something of a shock. He encountered clippers, counterfeiters, and his predecessor as Master of the Mint, Thomas Neale, was a man of a dubious reputation. By the time of Newton's arrival, Neale had amassed extensive experience in projecting various schemes to enrich himself as well as to resurrect the royal finances. In 1693, Neale ran a private lottery resulting in a neat personal profit, which he followed in 1694 with the Million Lottery to improve the revenue of the Crown.[81] To any government engaged in such a means to raise revenue, the matter of chance or risk was of obvious concern and it was in this context in November 1693 that Samuel Pepys consulted Newton on this problem behind Neale's proposed lottery.[82] By 1695, Neale was promoting "An Act for granting to his Majesty certaine rates and duties upon Marriages Births and Burials and upon Bachelors and Widowers for the Terme of Five Yeares for carrying on the Warr against France with Vigour."[83] It was not long before Newton saw through Neale as a "Gentleman who was in debt & of a prodigal temper & by irregular practices insinuated himself into ye Office [of master]. . . ."[84]

Even so, Neale was merely one of many bred by the demands of state expenditure and of the National Debt. In Newton's predecessor, there lurked a long lineage of political arithmeticians and projectors whose achievements seldom reached the heights of the establishment of the Bank

[81] In 1692, Neale was the first Governor of the Shadwell Water Works. H. W. Dickinson, *The Water Supply of Greater London* (London, 1954), pp. 49–50. On the range of Neale's activities, see Norman G. Brett-James, *The Growth of Stuart London* (London, 1935), p. 343; Sir John Craig, *The Mint. A History of the London Mint from A.D. 287 to 1948* (Cambridge, 1953), pp. 179–80, 196–7; Dickson, *Financial Revolution*, pp. 45, 53–4.

[82] Richard S. Westfall, *Never at Rest. A Biography of Isaac Newton* (Cambridge, 1980), p. 498; see also Colin Brooks, "Projecting, Political Arithmetic and the Act of 1695," *English Historical Review*, XCVII (January 1982), 35.

[83] Brooks, "Projecting," pp. 39, 45.

[84] Quoted in Westfall, *Never at Rest*, p. 564.

of England in 1694.[85] More often than not they were the authors of sinister plots that produced more fraud and deceit than national benefit.[86] But that did not stop the breed from multiplying. With such schemes lay the necessity for the calculation of risk and rates of mortality determined partly by the efforts of the Baconian William Petty, Edmund Halley, Dr. John Arbuthnot, or James Hodgson of Christ's Hospital.[87] But here, too, lay the often fatal attraction of investors for sudden fortunes, an attraction that should not be reduced to mere gullibility.

Some of the most sophisticated entrepreneurs of the age like James Brydges, the future Duke of Chandos and patron of Desaguliers, made and lost fortunes in that roulette of invention, innovation, and projecting. Brydges was one of the early proponents of Hugh Chamberlen's land bank scheme in 1698. This involved as well Newton's patron at the Mint, Lord Chancellor Montagu, and was the start of a career for Brydges that survived many such projects.[88] As Fellows of the Royal Society, they were not alone in subscribing to various loans to the Government. They were later joined by Hans Sloane, Nehemiah Grew, and Dr. John Arbuthnot.[89] Nor was the Royal Society immune to the possibilities offered by trading in the stock of the chartered companies like the East India Company or the African Company. Royal Society minutes in the 1690s establish transfers of stock into the management of fellows like Sir Joseph Williamson or other brokers for sale when prices rose.[90] It is especially

[85] Brooks, "Projecting," pp. 39, 45.

[86] Dickson, *Financial Revolution*, pp. 6–18.

[87] Peter Buck, "Seventeenth-Century Political Arithmetic: Civil Strife and Vital Statistics," *Isis*, LXVIII (1977), 67–84; Buck, "People Who Counted: Political Arithmetic in the Eighteenth Century," *Isis*, LXXIII (March 1982), 29–30. In 1711, John Chamberlayne tried to urge Sloane to publish the calculations of Clarke, Arbuthnot, and King on the parity of the sexes. B.L., Sloane MSS. 4042, f. 329, Chamberlayne to Sloane, July 25, 1711.

[88] C. H. Collins Baker and Muriel I. Baker, *The Life and Circumstances of James Brydges, First Duke of Chandos* (Oxford, 1949), p. 31; On Chandos's investments, see L. Stewart, "The Edge of Utility: Slaves and Smallpox in the Early Eighteenth Century," *Medical History*, XXIX (January 1985), 54–70; on land banks, see Dickson, *Financial Revolution*, pp. 24–5. Cf. John Cannon, *Aristocratic Century. The Peerage of Eighteenth-Century England* (Cambridge, 1984), p. 127.

[89] Geoffrey Holmes, *Augustan England: Professions, State and Society, 1680–1730* (London, 1982), pp. 220–1; Dickson, *Financial Revolution*, pp. 267, 426. Arbuthnot, Grew, Sloane, as well as Pepys, Halley, and Hugh Chamberlen were all among those who gave their approbation to Houghton's proposed *Collection for Improvement of Husbandry and Trade*, no. 1 (Wednesday, March 30, 1692). On Grew's own efforts as a projector, see above, Chapter 2.

[90] Royal Society MSS., Council Minutes, Vol. II (1682–1727), fols. 111, 118, 138, 154–5. Sir Joseph Williamson erected a school at Rochester in 1707 for teaching mathematics. Geoffrey Holmes, *Augustan England*, p. 50. See also R. K. Bluhm, "Remarks on the

interesting that while the Royal Society may have had an aversion to projectors from its early days – a suspicion that Thomas Sprat shared with many others in Augustan England – it did not stop the Society from taking advantage of the opportunities trading in stock might sometimes afford. The antagonism to the more disreputable of mechanic projectors was tempered throughout the Restoration by the claims Sprat also made in his *History* that trade was not to be despised, so that the world might "become more active, and industrious. . . . if our Gentlemen shall more condescend to engage in commerce, and to regard the Philosophy of Nature."[91]

Any assumption that the Society or its Fellows were uniformly antagonistic to trade or to commerce does not stand up to scrutiny. Newton may have looked down his nose at the vulgar sort. But as Master of the Mint, he reached to the very heart of commerce in the efforts of the Crown to engage in recoinage and to establish the proper value of specie. Likewise, when the naturalist Nehemiah Grew chose to invest in government loans, his money simply followed his principles. To many botanists, the natural world suggested a teleology that men might easily adopt. Grew was also something of a natural theologian who saw in the efforts of men at trade the means to the improvement of the world. In his little known tract *The Means of a most Ample Encrease of the Wealth & Strength of England In a few years*, which he wrote for Queen Anne about 1708, he displayed no patience with those who contributed little to the growth of the nation, especially those

> . . . Idle and Unprofitable People called Scholars: the Abilitys of the greater part of whome, are starvd for want of Money. And those who go not to the Plough, a Trade, or the University; become Clerks to the Mischievous Brood of Attourneys and Pettifoggers.[92]

Whether or not Boyle or Newton was interested in practical questions is not the issue here. More significant was the contact between the entrepreneur and the inventor, and between the entrepreneur and the natural philosopher.

Royal Society's Finances, 1660–1768," *Notes and Records of the Royal Society*, XIII (1958), 82–103.

[91] Thomas Sprat, *History of the Royal Society*, p. 407.

[92] Hunt. Lib., MS. 1264, p. 160. For a commentary on the British Library version of the MS see E. A. J. Johnson, *Predecessors of Adam Smith. The Growth of British Economic Thought* (New York, 1937), pp. 117–38. See also Richard Grassby, "Social Mobility and Business Enterprise in Seventeenth-Century England," in Donald Pennington and Keith Thomas, eds., *Puritans and Revolutionaries. Essays in Seventeenth-Century History presented to Christopher Hill* (Oxford, 1978), p. 363.

In this respect, the early Newtonians represented the adherence to a long tradition. Whatever might be assumed about the official attitude of the Royal Society, it is evident that there antedated its foundation a long standing tradition of practical interests among natural philosophers. This is obvious, for example, in the various experiments in smelting and furnace design in the ironworks of the Forest of Dean before the Restoration.[93] Others have proposed, for example, that William Petty's notion for a *gymnasium mechanicum* for the education of tradesmen was the legacy of the influence of Bacon's *New Atlantis* of 1624.[94] There are numerous examples of the early Fellows attempting to find solutions to practical problems such as in the draining of mines or in the smelting of ores. Petty had long promoted such notions. He was the author of the "Apparatus to the History of the Common Practices of Dying" appended to Sprat's *History* and was instrumental in efforts to establish a lectureship in the history of trades endowed by Sir John Cutler of the Grocers' Company.[95] Although the arrangements fell through, and degenerated into prolonged litigation, the intended beneficiary Robert Hooke in 1665 lauded the merchant Sir John Cutler for endowing a lecture "for the promotion of Mechanick Arts":

> ... This Gentleman [Cutler] has well observ'd, that the Arts of Life have been too long imprison'd in the dark shops of Mechanicks themselves, & there hindered from growth, either by ignorance, or self-interest: and he has bravely freed them from these inconveniences: He hath not only obliged Tradesmen, but Trade itself: he has done a work that is worthy of London, and has taught the chief City of Commerce in the world the right way how Commerce is to be improv'd.[96]

Rhetoric, at least, had established common ground for philosophers and merchants in the improvement of trade.

Merchants and philosophers

Naturalists like Hooke took delight in the improving spirit. Hooke's remarks about commerce reflected much optimism over the support of

[93] Charles Webster, *The Great Instauration. Science, Medicine and Reform, 1626–1660* (New York, 1976), pp. 395–402, 514–15.

[94] Walter E. Houghton, Jr., "The History of Trades: Its Relation to Seventeenth-Century Thought," *Journal of the History of Ideas*, II (1941), 43; Buck, "Seventeenth-Century Political Arithmetic," p. 77. On Petty's Baconianism, see Mendyk, "*Speculum Britanniae*," pp. 143–5.

[95] Michael Hunter, *Establishing the New Science. The Experience of the Early Royal Society* (Woodbridge, 1989), pp. 283ff.; Sprat, *History*, pp. 284–306.

[96] Robert Hooke, *Micrographia; or, Some physiological descriptions of minute bodies made by magnifying glasses. With observations and inquiries thereupon* (London, 1665), "Preface."

the merchant community of London for the sciences in the seventeenth century. It was the merchant Thomas Gresham who constructed the first Royal Exchange, which was later destroyed in the Great Fire. The returns from the Royal Exchange helped provide financial foundation for Gresham College, which has long been associated with Restoration sciences whatever its failings as an educational institution.[97]

But the interest of the merchant community is most clearly revealed in schemes to improve navigation. Robert Hooke's invention of the spiral balance spring offered some hope for improvement of timepieces in the determination of longitude. It was not only merchants, but the Navy as well that encouraged the activities of the Gresham professors. Lawrence Rooke, one of the original Fellows of the Royal Society, and from 1652 Professor of Astronomy at Gresham College, tried to improve the observational data upon which navigational tables might be constructed. To this end, while Professor of Geometry from 1657, Rooke was engaged in making accurate observations of the occlusions of the satellites of Jupiter.[98] As late as 1704, in *A Collection of Voyages and Travels* published by the Whigs John and Awnsham Churchill, a scheme by Rooke for the collection of scientific information was reprinted.[99] This related to compass declination, dipping needles, tides, coastal charts, and changes in the wind and weather.[100] Rooke's predecessor as Professor of Astronomy, Henry Gellibrand, published numerous works on navigational problems. These continued to appear long after his death in 1636, particularly in the early eighteenth century through the interest of the mathematics teacher James Atkinson.[101]

The effort of natural philosophers and astronomers to improve the

[97] John Ward, *The Lives of the Professors of Gresham College* (London, 1740), pp. 12, 18, 20; cf. H. F. Kearney, "Puritanism, Capitalism and the Scientific Revolution," in Charles Webster, ed., *The Intellectual Revolution of the Seventeenth Century* (London, 1974), pp. 231–3; Francis R. Johnson, "Gresham College: Precursor of the Royal Society," *Journal of the History of Ideas*, I (1940), 422–3; Mordechai Feingold, *The Mathematician's Apprenticeship. Science, Universities and Society in England, 1560–1640* (Cambridge, 1984), pp. 174ff.

[98] Ward, *Lives of the Professors*, pp. 90–2; and Johnson, "Gresham College," pp. 429–30.

[99] Awnsham Churchill was a Whig M.P. for Dorchester (1705–1710) as well as one of the Commissioners for the Land Bank. See Clyve Jones and Geoffrey Holmes, eds., *The London Diaries of William Nicolson Bishop of Carlisle 1702–1718* (Oxford, 1985), *passim.*

[100] *A Collection of Voyages and Travels, Some now first Printed from Original Manuscripts . . .*, Vol. I (London, 1704), pp. lxxiii – lxxiv. This had been printed in the *Philosophical Transactions*, no. 8 (January 8, 1665/6).

[101] Ward, *Lives of the Professors*, pp. 81–5; E. G. R. Taylor, *The Mathematical Practitioners of Tudor and Stuart England* (Cambridge, 1954), pp. 205–6, 254, 348, 349, 386; Johnson, "Gresham College," pp. 432–6; cf. Feingold, *Mathematician's Apprenticeship*, pp. 145, 179.

methods of navigation had been a key element in the defense of the Royal Society. And it was reflected in the practicality of some of the early Newtonians. The resurgence of interest in navigation in the reign of Anne did not mean that any dramatic advances had taken place – especially in relation to the imponderable longitude. But natural philosophers were going to play a significant role in trying to solve the puzzle. Such a notion had helped lay the foundation of the Royal Observatory in the reign of Charles II, but little had been achieved. In the spring of 1704, when James Hodgson showed round the Royal Society the "Estimate" given to him of Flamsteed's projected *Historia britannica coelestis*, this was a work recognized to hold considerable potential for navigation. Yet, antagonism between Newton and Flamsteed delayed its printing. It is notable that the referees who sought financial support from Prince George indicated that "it tends to the perfiction of Astronomy & Navigation: so if it should be Lost, the Los would Be irreparable. . . ."[102] Beyond such publications, however, lay the attempts to improve the level of knowledge of seamen and merchants that had been one of the aims of Gresham College. This was an issue that Flamsteed had explicitly acknowledged to Pepys in 1697 when he recounted the advantages that might result from "Seminaries of well educated Seamen."[103]

By the turn of the century, there was clearly a strongly utilitarian impetus barely submerged below the rhetoric of mathematical and experimental philosophy. By then the concerns of the philosophers with longitude, the draining of mines, or with the recoinage were no different from those of the Interregnum. Newton was ensconced at the Mint, and Halley owed his position as "Engineer" at the Chester Mint to the philosopher.[104] No wonder that John Wallis had come to the view that mathematical learning was more mechanical than academical as it came to involve more traders, merchants, surveyors, and the like.[105]

By promoting the practical advantages of mathematical and natural knowledge, philosophers could find employment either through the vogue of public lectures or they might be employed on the estates of enlightened noblemen in various capacities.[106] This last was briefly the route taken by the young mathematician Fatio de Duillier who published

[102] Westfall, *Never at Rest*, pp. 655–9.
[103] B.L., Lansdowne MSS. 1024, f. 112. Journal of White Kennett, April 1697.
[104] Cambridge University Library, Additional MSS. 4007(B), fols. 419–20. Halley to Newton, February 13, 1696/7.
[105] Sir George Clark, *Science and Social Welfare in the Age of Newton*, second edition (Oxford, 1970), p. 88n1.
[106] Pollard, *The Genesis of Modern Management*, p. 28.

in 1699 *Fruit-Walls improved By Inclining them to the Horizon; or a way to build walls for fruit-trees.* Fatio's calculations of the incidence of the sun's rays, based indirectly on such examinations by Halley and Wallis, tended to exaggerate the effect of a sloping wall in extending the growing season of plants. In any case, the Duke of Devonshire and the Duke of Rutland were sufficiently impressed to have such walls built on their estates.[107] Here Fatio merely appears as an early version of the lecturers like Desaguliers and Bradley, or of that sophisticated artisan Stephen Switzer, who were all employed in the improvement of estates.[108]

The frantic bustle of the Financial Revolution helped to produce a hectic market for the public lecturers, mathematical practitioners, and instrument makers who offered technical knowledge to the many who were anxious to exploit it. And there was no shortage of such philosophical entrepreneurs. By the end of the first decade after the Revolution, the new financial experiments had found their way into the coffeehouses when shortly after the Revolution the stockjobbers removed from the Royal Exchange and congregated in Jonathan's or Garraway's, both in Exchange Alley.[109] Garraway's, for example, was the site of the meetings of the undertakers of a marvelous, though likely abortive, scheme to establish by way of a lottery two "Royal Academies" near the Royal Exchange and near Covent Garden. Two thousand winners (out of forty thousand tickets) would be given the choice of a variety of subjects, including fencing, languages, and a wide range of mathematics to be taught by Richard Sault and de Moivre.[110]

Garraway's was a short step from the Marine Coffee House, where

[107] Peggy Kidwell, "Nicholas Fatio de Duillier and *Fruit-walls Improved*: Natural Philosophy, Solar Radiation, and Gardening in Late Seventeenth Century England," *Agricultural History*, LVII (1983), 413. Compare the later efforts of Bradley and Charles Du Bois in G. Agricola, *The Experimental Husbandman and Gardener*, ed. R. Bradley, second edition (London, 1726), p. 308

[108] See Stephen Switzer, *The Practical Fruit-Gardener. Being the best and newest Method of raising, Planting, and Pruning all Sorts of Fruit-Trees, agreeably to the Experience and Practice of the most eminent Gardeners and Nursery-Men*, revised and recommended by the Revd Mr. Laurence and Mr. Bradley (London, 1724), pp. 294–303. On Switzer, see Blanche Henrey, *British Botanical and Horitcultural Literature* (Oxford, 1975), II, pp. 325–9, 495–8.

[109] John Houghton, *A Collection For Improvement of Husbandry and Trade*, no. 2 (Wednesday, April 6, 1692); Bryant Lillywhite, *London Coffee Houses*, pp. 219, 360; Geoffrey Holmes, *British Politics*, p. 154.

[110] Houghton, *Collection for the Improvement of Husbandry and Trade*, no. 134 (Friday, February 22, 1694/5). Sault kept a school in Adam's Court near the Royal Exchange and was the author of *A Treatise of Algebra, According to the late Improvements...* (London, 1695). See *D.N.B.*, L, 313; and Gilbert D. McEwen, *The Oracle of the Coffee House. John Dunton's Athenian Mercury* (San Marino, 1972), Chapter VIII, esp. pp. 120–2.

the mathematical lectures were first presented in 1698. In these warrens about the Royal Exchange the public demand for natural philosophy made real what Gresham had only barely imagined. And it was here that Ephraim Chambers, whose popular *Cyclopaedia* provided a rival for Harris's *Lexicon*, had been apprenticed to the instrument maker John Senex. Senex's shop in 1706 was next door to the Fleece Tavern, by the Royal Exchange, merely two or three doors from the Marine Coffee House across Birchin Lane where Harris, Ditton, and Hodgson lectured.[111] The explanation of instruments of surveying and navigation in Chambers' *Cyclopedia* was merely a reflection of the practical influences of the age originally defined in Harris's *Lexicon*.[112]

"Profit hunters" and adventurers could not have been immune to the public entertainment of philosophy. The immediate purpose of the lectures and the publications of the group, which included Harris, Chambers, Desaguliers, Whiston, Ditton and many others, had been to introduce to a curious public some of the principles of the Newtonian philosophy. The new public culture of the early eighteenth century was no longer dictated by the tastes and temperament of the gentlemen and aristocracy of London. Even some of the gentry, whose reading included the *Tatler* and the *Spectator*, and whose daily rounds of the coffeehouses brought them into contact with the wits and profit mongers, would have encountered the new science in its more comprehensible guise of experimental demonstration. Such an effort to make natural philosophy intelligible ran through the lectures and was easily detected even in works like Harris's *Astronomical Dialogues Between a Gentleman and a Lady*, which speaks not to libraries but to public learning.[113]

We are dealing with a world of the centrifugal forces of new finance

[111] E. G. R. Taylor, *The Mathematical Pratitioners of Hanoverian England* (Cambridge, 1966), p. 114; Lillywhite, *London Coffee Houses*, pp. 694–5; H. R. Calvert, *Scientific Trade Cards*, p. 38. Senex, a member of the Stationers' Company, was made F.R.S. in 1728. Sarah Tyacke, *London Map-Sellers, 1660–1720* (Tring, 1978), p. 142; Tyacke, "Map-Sellers and the London Map Trade c. 1650–1710," in Helen Wallis and Sarah Tyacke, eds., *My Head is a Map. Essays and Memoirs in Honour of R. V. Tooley* (London, 1973), p. 67. Senex was evidently a friend of William Derham. See B.L., Sloane MSS. 4058, f. 231. Derham to Sloane, n.d.. In 1721, Senex, with the engraver John Harris and Henry Wilson, petitioned for a patent for a new method of projecting sea charts. Public Record Office, State Papers (Domestic), 35/75/4. June 5, 1721; Taylor, *Hanoverian England*, pp. 125, 149–50

[112] Bardwell, "Arts and Science in English Encyclopedias," pp. 17–18, 107.

[113] John Harris, *Astronomical Dialogues Between and Gentleman and a Lady; Wherein the Doctrine of the Sphere, Uses of the Globes, and the Elements of Astronomy and Geography are Explain'd; In a Pleasant, Easy, and Familiar Way. With a Description of the famous Instrument called the Orrery*, second edition by George Gordon (London, 1729), pp. 146–7.

and of the decline of the fortunes of the lesser gentry whose cause Jonathan Swift and Viscount Bolingbroke sought later to promote. Behind their rhetoric were the anxieties of independent Tory squires who were increasingly estranged from the great Whig lords, from grandees like the Duke of Chandos, and from the greater merchants and financiers with whom some of the Newtonians associated themselves. This gave the appearance – perhaps often exaggerated but real nonetheless – of an increasing disaffection between Country and Court; between a view of political legitimacy founded traditionally upon land as the ballast of the nation on the one hand and, on the other, upon political and economic power associated both with Westminster and the City.[114]

The anonymous *Battle of the Bubbles* in 1720 made great sport of the parade of abortive projects proposed since the Revolution. Such were a "Project to Insure Uniformity amongst Protestant Dissenters . . . , Another to Insure it amongst the Orthodox . . . ,A Design to build Castles in the Air, for the use of all Towns-Folks, who either never had any Country Seats, or who have lately sold them . . . ," for "An Hydrostatical Air-Pump, to draw all manner of wind and Vapours out Human Brains . . . ," and ultimately "A Subscription for raising Money, to make an Engine to remove the South Sea House, intire, as it stands now, in all its Parts, to Moor-fields; and then, and there, to Incorporate it with Bethelem-Hospital."[115] It was precisely notions such as these that gave to Swift the character of the Academy of Lagado in *Gulliver's Travels* – to be viewed as a variation on common attacks on the Royal Society.[116] Satire may often seem too oblique to sense all the allusions, but soon the adrenaline of the stock jobbers overwhelmed amusement. Many projects were not at all ethereal, such as a "General Insurance from Losses by Fire," or schemes "For Supplying London with Sea-Coal . . . , For Erecting Salt-Pans in Holy Island . . . , For Improving the Art of Making Soap . . . , For Drying Malt by Hot Air . . . ," and "For Extracting Silver from Lead." Several Newtonians actively promoted these. In the City, in the whirl of the metropolis, amongst the company of the coffeehouses of Birchin Lane and Exchange Alley, there was a market in natural philosophy, for information about light, air pumps, barometers, and the steam engine that the lecturers found they could exploit. Among the prospec-

[114] G. E. Mingay, *The Gentry* (London, 1976), pp. 188–9.

[115] [Anon.,] *The Battle of the Bubbles. Shewing their Several Constitutions, Alliances, Policies, and Wars; From their first Suddain Rise, to their late Speedy Decay. By a Stander-By* (London, 1720), pp. 18–27.

[116] Pat Rogers, "Gulliver and the Engineers," *Modern Language Review*, LXX (1975), 265.

tors, financeers, and engineers, there was an audience that would pay for power over nature.

Patents and the stocks

The place of the Royal Society in the swarm of inventions seen by Revolutionary England was frequently uncertain. The authority of the Society was sometimes sought over schemes with which both the Crown and numerous financiers were inundated. It was obvious that the imprimatur of the Royal Society, even if it could only be informally obtained, might be of great benefit to the many who sought patents from the Crown. So, for example, in 1700, the Reverend John Davis wrote to Hans Sloane of a pump maker of his acquaintance who by a method discovered by Davis had constructed a device "yt drawes water seventy nine foot, and yt it goes easily." Davis, as became common, offered to reveal all to the Royal Society "upon such considerations as you think to oblige. . . ."[117]

The efforts of inventors to obtain the approval of the Royal Society was hardly surprising – and evidently it became a major irritation. In 1691, Sir John Hoskins, formerly President of the Society, convinced it to adopt a notion from Robert Boyle that it should employ a person to assist inventors establish their claims as well as to expose any frauds.[118] What became of this is unclear, but the problem surely did not disappear. This was inevitable as long as there were people associated with the Royal Society who were engaged in mechanical invention. We have already seen how the inventor Denis Papin hoped to obtain financial support from the Society throughout the reign of Anne for contrivances like his "Machine to get good air for respiration and vegetation" or for his design of a new furnace.[119] From such creations, disputes easily arose. Papin's claim to have improved on the Savery engine was challenged by Savery himself in a letter to John Harris in 1709. As usual Papin had offered to build his improved engine but required the financial backing of the Society to get started.[120] Each new idea produced further badgering. This finally

[117] Royal Society MSS., Letter Book D1, no. 45, John Davys to Sloane, February 19, 1699/1700. See also Rev. John Davis, "Account of the Wirtemberg Engine," *Philosophical Transactions*, XV (January 1684/5), 846.

[118] Royal Society MSS., Council Minutes, II, f. 108, January 21, 1690/1. See also Christine MacLeod, *Inventing the Industrial Revolution. The English Patent System, 1660–1800* (Cambridge and New York, 1988), pp. 186–8.

[119] Royal Society MSS., Classified Papers, 1660–1740, III (2), no. 2, January 18, 1708/9; B.L., Sloane MSS. 4041, fols. 328–35, Papin to Sloane, June 4, 1709.

[120] Royal Society MSS., Classified Papers, VI, no. 61, Savery to Harris, read May 18, 1709; Class. Papers, XVIII, no. 80, Papin's "Proposals about the engine for raising water by the force of fire," n.d..

culminated in the significant suggestion in 1711 that they erect a "Company or Society for New Inventions, for which Subscriptions may be made for £1000 Stock."[121] Once again, the matter was allowed to rest whether out of irritation or lack of interest, but hardly because they were averse to joint-stock companies in which by then the Society had invested its funds. Perhaps they merely felt the risk was too great.

Whatever the reason for the Royal Society's lack of enthusiasm for schemes of invention, it is also clear that within two years of Papin's proposals, Queen Anne's government came to see possibilities for the Society. Before the end of January 1713, the Society had received from Dr. John Arbuthnot an order from the Queen that "what ever new Mechanicall Invention shall be brought before the Privy Council in order to obtain a Patent, the same shall be referred to the Examination of the Royal Society before the Patent be passed." The Society dutifully responded to Viscount Bolingbroke with thanks to the Queen "in taking their Approbation of Invention" as well as for requiring officers of the Crown going abroad to assist the Society in its enquiries in natural philosophy.[122] With the way perhaps prepared by Grew, the Crown obviously thought the views of the philosophers worthy of consideration in the granting of patents, although there seems to be no evidence that the Society was formally involved in such issues as a matter of policy until very much later. Nonetheless, the Royal Society frequently faced these queries precisely because of a general appreciation of a link between the natural philosophers and the inventors.

On occasion, the support of the Royal Society could prove decisive. This is what many a prospective inventor understood. Yet patents bred disputes and the Royal Society could easily become embroiled in the controversy. Such was the case in 1723 when Ambrose Godfrey, the father of a family of chemists who operated a laboratory in Southhampton Street, sought a patent for an invention to extinguish fires. His claims were challenged by Charles Povey, who had a long standing reputation as a projector and who had been the originator of the Sun Fire Office (in

[121] Royal Society MSS., Journal Book (C), X, 278–9, April 26, 1711; Classified Papers (1660–1740), XVIII, no. 68, read April 26, 1711. This bears a remarkable resemblance to the proposal made in 1712 by the horologist Abbe Jean de Hautefeuille of Orleans to the Academie des Sciences for a "Compagnie des Nouvelles Decouvertes." See Alan Gabbey, "The Mechanical Philosophy and Its Problems: Mechanical Explanations, Impenetrability, and Perpetual Motion" in Joseph C. Pitt, ed., *Change and Progress in Modern Science* (Dordrecht and Boston, 1985), p. 52.

[122] B.L., Sloane MSS. 4026, f. 240, n.d. The Royal Society minutes establish the context fairly closely. See Royal Society MSS., Council Minutes (C), II, p. 259, January 29, 1712/13.

which John Harris was also involved) when it began its operation out of Causey's Coffee House near St. Paul's in 1710.[123] The dispute that erupted was one of priority of invention. Although Godfrey admitted to having imported the idea from Germany, Povey claimed that the notion of extinguishing fires with a mixture of water and gunpowder belonged to him and that Godfrey's application for a patent ought to be denied.

Like Denis Papin, Povey sought the metamorphosis of invention into a joint-stock. Povey also faced the inertia of the Royal Society and a coterie that seemed to revolve around Sloane. When Godfrey's first demonstration proved a failure, Hans Sloane was among those who supported a second attempt. The antagonists carried their conflict into public experiments in houses in Belsize Park, near Hampstead, and later near the Mulberry Garden on the north edge of the City. Accounts of the demonstrations and the squabbles of the inventors provided a summer's entertainment in the London press.[124] In the end, perhaps the government was reluctant to allow Povey a charter as it had already been singed by the heat of the South Sea chaldron. In any case, the members of the Royal Society like Sloane, Martin Folkes, and the instrument maker George Graham carried the day and the patent was granted to Godfrey.[125] It was impossible for the Society or for individual natural philosophers to remain immune from these quarrels. As projectors began to form limited partnerships and joint-stock companies to exploit their claims, it was belatedly discovered that it was in the public interest to assure some proper technical means of evaluation. Mechanical knowledge was to become a prominent feature in the promotion of public lectures.

[123] On Povey see P. G. M. Dickson, *The Sun Insurance Office, 1710–1960* (London, 1960), pp. 17ff.; and MacLeod, *Inventing the Industrial Revolution*, p. 46. Povey may have been known to Thomas Savery, who was accomptant to the Commissioners for the Sick and Wounded after the death of the Treasurer, Richard Povey, in 1705. Bodleian Library, MS. Rigaud 39, f. 52. On Ambrose Godfrey, see R. E. W. Maddison, "Studies in the Life of Robert Boyle, F.R.S., Part V. Ambrose Godfrey Hanckwitz, F.R.S.," *Notes and Records of the Royal Society*, XI (1954), 159–88; Maddison, "Notes on Some Members of the Hanckwitz Family in England," *Annals of Science*, XI (March 1955), 64–72; Musson and Robinson, *Science and Technology*, pp. 56–7. Godfrey was regarded as a man in Sloane's interest in the issue of the election of a new President for the Royal Society in 1728. See B.L., Sloane MSS. 1968, fols. 20–1, Richard Bradley to Sloane, New Years Day, 1727/8.

[124] Pat Rogers, "Gulliver and the Engineers," pp. 267–8; see *St. James's Journal* (Saturday, April 6, 1723).

[125] B.L., Add. MSS. 22675, fols. 107–8; PRO., State Papers (Domestic), 35/75/36, fols. 177–82. This contains certificates signed by Fellows of the Royal Society. Armand B. DuBois, *The English Bubble Company after the Bubble Act, 1720–1800* (New York, 1938), p. 379n6. Engines to extinguish fires were among the most popular preoccupations of inventors. See PRO., State Papers (Domestic), 35/63/1. Joseph Brooke, patent, September 2, 1726. On George Graham, see Taylor, *Hanoverian England*, pp. 120–1.

Denis Papin's suggestions for a joint-stock are noteworthy because they represent an aspect of the search for innovation with which the Royal Society was never quite comfortable. Joint-stock companies were increasingly familiar to Fellows of the Royal Society like John Houghton. They were a mechanism about which, for example, both John Harris and Thomas Savery were to gain some familiarity.[126] Nor was such knowledge peculiar to them. The careers of the public lecturers suggest they sought to exploit these commercial contacts made in the coffeehouses of London. This is reflected in the practicality of works like John Harris's *Lexicon Technicum* and John Desaguliers' *Lectures on Experimental Philosophy*. We should not be tempted to reduce practical concerns to mere philosophical inclination.

John Harris, for example, was an investor in various projects born in Exchange Alley. During 1709 and 1710 when Harris was Secretary of the Royal Society, he was also a member of the Court of Assistants of the recently united Mines Royal and Mineral and Battery Works while Newton was a Deputy-Governor.[127] The charter of this company attracted the attention of petitioners for a marine insurance scheme known as the Royal Exchange Assurance led by Case Billingsley of the longitude hunt, one of the arch-projectors of the age.[128] Harris was also among the subscribers to the Royal Exchange Assurance as well as fostering the formation of insurance ventures like the Sun Fire Office, which insured the buildings and merchandise of traders rather than their ships.[129] His technical interests obviously brought him into contact with those concerned in commercial innovation and thus with questionable elements of Whig

[126] For a later report on the stock in the Savery patent, see B.L., Add. MSS. 22675, fols. 27–8, cases and opinions of Sir Tho. Pengelly, Chief Baron of the Exchequer, July 7, 1720.

[127] Sir George Clark, *Science and Social Welfare in the Age of Newton*, second edition (Oxford, 1970), pp. 66–7; B.L., Loan 16, Records of the Mines Royal and the Mineral and Battery Works, Vol. II, f. 233; Vol. III, f. 98.

[128] For some of Billingsley's activities, see Carswell, *South Sea Bubble*, pp.167–8, 171–2; and Pat Rogers, "Gulliver and the Engineers," pp.263–4; A. J. G. Cummings, "The York Buildings Company. A Case Study in Eighteenth Century Corporation Mismanagement" (Ph.D. thesis, University of Strathclyde, 1980), *passim*; and A. J. G. Cummings and Larry Stewart, "The Case of the Eighteenth-Century Projector: Entrepreneurs, Engineers, and Legitimacy at the Hanoverian Court in Britain," in Bruce T. Moran, ed., *Patronage and Institutions. Science, Technology and Medicine at the European Court, 1500–1750* (Rochester and Woodbridge, 1991), pp. 235–61. There were very close ties between the Mines Royal, the Royal Exchange Assurance and the York Buildings Company. Those who petitioned for a charter for the Royal Exchange Assurance were all members of the Mines Royal. Barry Supple, *The Royal Exchange Assurance. A History of British Insurance, 1720–1970* (Cambridge,1970), pp. 24–5.

[129] P. G. M. Dickson, *The Sun Insurance Office, 1710–1960* (London, 1960), p. 268.

finance. He seems, then, to have fit remarkably the political demonology of his age. This may be one reason why Harris's death in 1719 was not lamented by some who took him to be the ultimate opportunist, especially the Oxford Jacobite Thomas Hearne, who reviled him as "technical Harris."[130] But the link between Harris, the Sun Fire, and the Royal Exchange Assurance was only one example of the numerous connections between the companies and the lecturers on experimental philosophy.

The broad range of the interests of John Harris as a proponent of Newtonianism and practical commercial ventures was not unusual. Nor was it a contradiction. This becomes especially clear when we recognize that the institution of the Royal Exchange may have had as important an impact on the propagation of science as did the Royal Society. Of course, the Society was notoriously lax about, if not hostile to, the suggestion of practical commercial ventures. But an examination of Harris's connections might well demonstrate that he was typical of Newton's disciples, especially those like John Theophilus Desaguliers or Henry Beighton, the editor of the *Ladies Diary*, both of whom later earned a considerable livelihood in the rough and tumble of Exchange Alley. It was John Harris's misfortune that his death in 1719 occurred at a time when the opportunities he had sought so desperately were just then coming into abundance in the shadow of the Exchange. Others, like Swift, would not rejoice at the result.

Emphasizing the interest in practicality paints the issue in very broad strokes. Yet, the combination of the theoretical and the useful was obviously promoted by public lectures. Some of the proponents of Newton's philosophy, like John Harris, were centrally involved in projects for improvement. Broad, ill-defined notions of the "public good" may have been critical to the appreciation of natural philosophy by traders and merchants. One might recall that Harris's patron Sir Charles Cox, along with John Cholmley, was M.P. for Southwark in 1701. Both were brewers and both subscribed to the *Lexicon*.[131] For Harris, the separation of application and theory in relation to the sciences was not something to which he would readily have submitted. This was clearly demonstrated in the *Lexicon* in Harris's description of techniques to establish specific

130 *Remarks and Collections of Thomas Hearne*, ed. Oxford Historical Society, VII (Oxford, 1906), p. 46.

131 Robert Walcott, *English Politics in the Early Eighteenth Century* (Cambridge, 1956), p. 164. Harris's translation of Pardies' *Short, but yet plain elements of geometry and plain trigonometry* in 1701 was dedicated to Cox. Besides his other technical and devotional works, Harris may also have been the author of *The London Merchant's Mirror or the Tradesman's Guide . . .*, an obviously practical work that suggests one of his many forays into the world of trade. Cf. Wallis and Wallis, *Biobibliography*, pp. 2–4, 20.

gravity to "most accurately discover, whether you are imposed upon by the Merchant, Vintner, or, Distiller, &c. in Quantities of Wine or Spirits . . . ," or by the inclusion of James Hodgson's astronomical tables for purposes of navigation.[132]

Newtonians like John Harris or Humphry Ditton moved easily between the financiers and projectors on the one hand and the world of the instrument makers and mathematical teachers on the other. Increasingly, natural philosophers had come to ignore any apparent social boundaries between them and the instrument makers who sold the apparatus of philosophers, merchants and navigators.[133] Instrument makers were likewise honored not merely by Fellowships in the Royal Society as was Senex, but also ultimately by burial in Westminster Abbey in the case of Thomas Tompion and his nephew George Graham, a recognition not restricted to the greatest of philosophers like Newton.[134] Similarly, the boundaries between philosophers and the world of commerce were not insurmountable. Hence, we find Ditton in 1705 dedicating his explanation of parts of Newton's *Principia*, in *The General Laws of Nature and Motion*, to Charles Du Bois, Esq., of the City of London. Du Bois was treasurer of the East India Company, patron of Richard Bradley, and he became a Fellow of the Royal Society in 1714.[135] Ditton was also in the habit of publishing advertisements in his works for the benefit of several of the London instrument makers and mathematical teachers without whom the merchant adventurers might have been a great deal less successful. In 1706 in his *Institution of Fluxions*, he had noted the teaching of mathematics of Ralph Snow with "their Application to Natural Philosophy." In 1714, Ditton advertised the air pumps and engines of the younger Hauksbee, as well as the experimental lectures delivered at Hauksbee's shop in Crane Court in Fleet Street.[136] By this point, Ditton's place in public lectures with Hauksbee had been taken over by his friend William Whiston.

[132] Harris, *Lexicon Technicum*, I (1704), s.v. "Specific Gravity."

[133] This was not merely an English phenomenon as is indicated by the van Musschenbroek family. See W. D. Hackmann, "The Growth of Science in the Netherlands in the Seventeenth and Eighteenth Centuries," in Maurice Crosland, ed., *The Emergence of Science in Western Europe* (New York, 1976), p. 98. A good introduction to this broad community is M. A. Crawford, "Evidence from Trade Cards for the Scientific Instrument Community," *Annals of Science*, XLII (1985), 453–554.

[134] Robsichon, "Scientific Instrument Makers," p. 176.

[135] *D.N.B.*, VI, 77. He was also a subscriber to both volumes of Harris's *Lexicon Technicum* and was later the Governor of the London Lead Company. See William Rees, *Industry Before the Industrial Revolution*, II (Cardiff, 1968), p. 517; and Ray Desmond, *Dictionary of British and Irish Botanists and Horticulturalists* (London, 1977), p. 198.

[136] Ditton, *The New Law of Fluids* (London, 1714), "Advertisement."

Mathematical learning had truly become mechanical. This was largely a response to the traders and merchants who formed the new element in the patronage of the natural philosophy. This was the basis for the success of Thomas Watts' Academy in Little-Tower street and is reflected to some degree in the geographic proximity of the public scientific lectures to the center of the financial activity of the City.[137] The broadening nature of patronage, in relation to the sciences, may be a consequence of the concerns of the merchants with the practical advantages they felt could be achieved through natural philosophy. Thus, we find the demonstration of steam principles in Desaguliers' lectures and its counterpart in the explanations of Chambers' *Cyclopaedia*. This response to the rise of merchant patronage was in turn indicated by the development of subscription publication, where traders and artisans basked in the prestige that had once been the preserve of the titled.[138]

Subscription suggests a social transformation as well as a cultural one. Appearing on a list with the Prince of Wales, the nobility, and a few bishops did one's social position little harm. Property and patronage had often been linked, but subscription was also a reflection of the search for status and credibility by the wealthy.[139] What lay behind the emergence of subscription and the lectures in rooms rented in the coffeehouses was not merely a fascination with the mysterious, such as the northern lights, which the philosophers strained to explain. A strange bluish light that may be induced by the fall of mercury in a barometer might provoke a passing curiosity, but it is unlikely that mere amusement would have contributed a vital momentum to natural philosophy that carried well into the last half of the eighteenth century. Those who supported the calls for subscription, who paid handsomely to hear Whiston or Desaguliers, or who endured the tortured sermons of Newtonian divines may have sensed that they were witnessing a transformation. This fascination with science was not based solely on a feeling that they were seeing the emergence of a dynamic universe. Subscription may have been one place

[137] Sir George Clark, *Science and Social Welfare*, second edition (Oxford, 1970), p. 88n1. This can be seen in the argument advanced in Thomas Watts, *An Essay on the Proper Method for Forming the Man of Business*, ed. Arthur H. Cole (Boston, 1946), pp. 24–5.

[138] Sarah L. C. Clapp, "Subscription Publishers Prior to Jacob Tonson," *The Library*, XIII (1933), *passim*; P. J. Wallis, "British Mathematical Bibliography," *Journal of the Institute of Navigation*, XX (April 1967), 200–5; Wallis and Wallis, *Biobibliography*, *passim*; Paul J. Korshin, "Types of Eighteenth-Century Literary Patronage," *Eighteenth-Century Studies*, VII (Summer 1974), 463–4; and Brewer, Plumb, and McKendrick, eds., *Birth of a Consumer Society*, pp. 224–7.

[139] Harold Perkin, *The Origins of Modern English Society 1780–1880*, second edition (Toronto, 1972), pp. 38, 45–56.

where public assent and experimental philosophy came to interact. The lecturers and instrument makers were the point of contact between the technology of early industrialization and the propagation of the new philosophy.

6

The Longitudinarians

The manifest concern with improvement and mechanics extended far beyond the policies of the Royal Society. The residual Baconianism of the early eighteenth century required a wider foundation in the social and intellectual world than that confined to the Fellows.[1] There was a remarkable concordance between the ideal of useful public knowledge and the concept of contract that was fundamental to the Whig polity. Although this democratization of interest may have been largely a fiction, popular concerns, such as with the improvement of navigation, provided an avenue of participation in the world of natural philosophers. The practical explanations in the works of Harris and Chambers or the principles of steam power demonstrated by Desaguliers were the legacy of a long provenance. It was a legacy that would prove philosophers were drawn by the allure of projects – not simply by claims of mechanical achievement, but by their capacity to distinguish between vice and virtue, between vile promoters and the common good.

Longitude and the public interest

The improvement of navigation had been central to the efforts of several of the Gresham professors in the seventeenth century. By the early eighteenth century, it was the longitude at sea that remained the crucial

[1] G. S. Rousseau, "Science Books and their Readers in the Eighteenth Century," in Isabel Rivers, ed., *Books and their Readers in Eighteenth-Century England* (Leicester, 1982), p. 200. On the background to this issue, see Charles Webster, *The Great Instauration. Science, Medicine and Reform, 1626–1660* (New York, 1975), pp. 495–6; and Michael Hunter, *Science and Society in Restoration England* (Cambridge and New York, 1981), *passim*.

obstacle to accurate navigation. Even *A Collection of Voyages and Travels* published in 1704 repeated the suggestions for navigators that Laurence Rooke had listed in the *Philosophical Transactions* forty years earlier – including instructions to record the declination of the compass, the inclination of the dipping needle, along with specific observations of meteorological phenomena.[2] Little had been achieved in the interim. Efforts to set longitudinal position required immense skill, and frequently proved impossible, especially at sea. Toward the end of the reign of Anne, according to Newton, the observation of lunar position was not sufficiently exact to prove helpful. And the attempt to set longitude by the eclipses of Jupiter's satellites was impractical because of the length of the telescope required and the hurdle of attempting such observations while on a moving ship. Despite the efforts of Edmund Halley in the last decade of the century, the phenomenon of compass variation was of little help.[3] But the age of projects was not to be overwhelmed by such a chronic, and frequently tragic, problem.

During 1707, part of the British fleet under Sir Cloudesley Shovell had been operating in the Mediterranean against the French. At the end of September, it had set out from Gibraltar bound for home. The weather was poor for most of the passage and on October 22, convinced they were in the mouth of the Channel, they ran eastward before a gale. At night, they found themselves among rocks and part of the fleet ran upon the Scilly Isles, resulting in the loss of nearly two thousand men including the Admiral, who, legend had it, was murdered for his jewels after having reached shore. Despite the common belief, this is not to say the problem of the longitude was the cause of Shovell's disaster.[4] But the magnitude of the catastrophe served to remind the London scientific community that, whatever their efforts at education, the hazards of navigation had not been overcome.

[2] *A Collection of Voyages and Travels, Some now first Printed from Original Manuscripts*, Vol. I (London, 1704), "An Introductory Discourse," pp. lxxiii–lxxiv.

[3] *The Correspondence of Isaac Newton*, IV (1694–1709), ed. J. F. Scott (Cambridge, 1967), pp. 193–4, 194n1. Newton to Halley, March 14, 1695/6; *The Works of the Learned*, I (1742), 53–5; Maureen Farrell, "The Life and Work of William Whiston" (Ph.D. thesis, University of Manchester, 1973), Chapter 3, esp. p. 3.14.

[4] Commander W. E. May, "The Last Voyage of Sir Clowdisley Shovell," *Journal of the Institute of Navigation*, XIII (196), 324–32; J. G. Pickwell, "Improbable Legends Surrounding the Shipwreck of Sir Clowdisley Shovell," *Mariner's Mirror*, LIX (May 1973), 221–3; Historical Manuscripts Commission, *Report of the MSS of the Marquess of Downshire*, Vol. I, Part II (London, 1924), p. 855. John Bridges (the younger) to Sir William Trumbull, November 14, 1707. In 1699, Newton apparently met with Shovell and Sir Richard Haddock, Comptroller of the Navy, to examine a proposal for finding the longitude. *Correspondence of Isaac Newton*, IV, 314. Newton to Haddock, August 12, 1699.

The Channel had always posed problems for navigators and even the charts available for the coast of England were unreliable. *Colson's Seaman's New Kalendar* of 1697, most likely to be found aboard ship, was terribly inadequate. The difficult currents of Channel navigation demanded more accurate charts. In 1701, a member of the Royal Society warned of the variation in compass headings in the vicinity of the Scilly Isles.[5] By 1702, Halley had published a new Channel chart that gave the variation and tides in the area of these lethal islands and in 1728 reprinted Nathaniel Cutler's *General Coasting Pilot* in his handsomely produced *Atlas Maritimus & Commercialis* to which over four hundred persons subscribed.[6] Navigational disasters were hardly new, but by 1707, the dramatic shipwreck of the Mediterranean fleet served to focus urgent attention upon the inadequacies of navigation, more lethal even than Queen Anne's wars. Finding the longitude, especially, proved to be a veritable industry.

At the time of the Shovell tragedy, the Whig Joseph Addison had been Under Secretary of State. Addison, therefore, had probably discussed the uncertainty of navigation even in the confines of the Channel. This may have contributed to his support of natural philosophers like William Whiston. Addison certainly assisted Whiston in his efforts to solve the longitude problem. Science was all the rage in the summer of 1713. In early July, Newton was about to publish the second edition of his *Principia*. And Addison and his associate Richard Steele, who was then engaged in an electoral war with the Tories, had arranged for Whiston to deliver a series of lectures at Button's Coffee House. This could not have been without some risk to Steele's chances as Whiston was a highly publicized heretic. Steele had received a recommendation for Whiston from Henry Newman, Secretary to the Society for Promoting Christian Knowledge in which Whiston was once instrumental.[7] But Whiston was already busy with his own plans. By August 24 he had placed an adver-

[5] B.L., Add. MSS. 9332, f. 88, Collections Relating to the Navy. "An Advertisement, Necessary to be Observed in the Navigation Up and Down the Channel of England."

[6] [Edmund Halley], *Atlas Maritimus & Commercialis; Or, A General View of the Worlds so far as relates to TRADE and NAVIGATION: Describing all the Coasts, Ports, Harbours, and Noted Rivers, according to the Latest Discoveries and Most Exact Observation. Together with a large Account of the COMMERCE Carried on by Sea between the several Countries of the World*... (London, 1728); E. G. R. Taylor, "A Note" to W. E. May, "Last Voyage," p. 332; Norman J. W. Thrower, "Edmund Halley and Thematic Geo-Cartography," in Thrower, ed., *The Compleat Plattmaker. Essays on Chart, Map, and Globe Making in England in the Seventeenth and Eighteenth Centuries* (Berkeley, 1978), pp. 221–2.

[7] Leonard W. Cowie, *Henry Newman. An American in London, 1708–1743* (London, 1956), pp. 156–7.

tisement in the newspapers announcing the revival of his mathematical lecture at the Marine Coffee House, as successor to Ditton and Hodgson, as well as the initiation of his lectures at Button's on August 25.[8] At the same time Whiston and Humphrey Ditton entered into their partnership to discover the longitude. In a letter dated July 11, 1713, addressed to Button's, Whiston and Ditton proposed that Parliament establish a reward for the discovery of the longitude while at sea.[9]

The Longitude Act

The inspiration of Whiston and Ditton was apparently fired by the celebration on July 7 for the signing of the Peace of Utrecht. Or it might even have been the memory of reports of the French guns off Beachy Head a quarter century earlier.[10] Whiston knew a dramatic event when he saw one. The Utrecht fireworks probably gave him the idea of combining the reports of mortars with their flashes, in order to determine the distance from a previously recognized point of origin. This proposal undoubtedly came out of experiments by William Derham to determine the velocity of sound.[11] Like Whiston, Derham had been recently drafted for the Boyle Lectures, and in 1713, he also had received the approbation of the S.P.C.K. and the recommendation of the *Guardian* for the published lectures in his *Physico-Theology*.[12] By December, Whiston and Ditton intimated in Richard Steele's new journal *The Englishman* that they had discovered a method that would prove practical while at sea. However, with their eye to a possible reward they published little of the details of their project.[13]

By spring, the lack of response seemed to indicate nothing would come

[8] Calhoun Winton, *Captain Steele. The Early Career of Richard Steele* (Baltimore, 1964), p. 157; *Daily Courant* (Monday, August 24, 1713); Marjorie Nicolson and G. S. Rousseau, "*This Long Disease, My Life,*" *Alexander Pope and the Sciences* (Princeton, 1968), pp. 142–3.

[9] *Guardian*, no. 107 (Tuesday, July 14, 1713); Calhoun Winton, *Captain Steele. The Early Career of Richard Steele* (Baltimore, 1964), pp. 170–6.

[10] William Whiston and Humphry Ditton, *A New Method for Discovering the Longitude Both at Sea and Land, Humbly Proposed to the Consideration of the Publick* (London, 1714), pp. 18–19.

[11] Whiston and Ditton, *New Method for Discovering the Longitude*, pp. 28–9. Derham had first presented his experimental results in the *Philosophical Transactions* in 1708. William Cudworth, *Life and Correspondence of Abraham Sharp* (London, 1889), p. 112, Flamsteed to Abr. Sharp, August 31, 1714.

[12] *The Guardian*, no. 175 (Thursday, October 1, 1713); Winton, *Captain Steele*, pp. 157–8n2.

[13] *The Englishman*, no. 29 (December 10, 1713).

of their scheme.[14] But Whiston and Ditton were not without allies. On April 19, 1714, they brought to the House of Commons a petition for a bill to provide for a reward for whatever person might discover the longitude while at sea. In their "Petition About Longitude," care was taken to link the discovery to the improvement of trade and the naval supremacy of Great Britain.[15] These issues could hardly be ignored in the immediate aftermath of Utrecht and Queen Anne's wars. Before this proposal gathered dust, the House of Commons received another petition on May 25 from naval officers, merchants of London, and the commanders of merchantmen arguing

> . . . That the Discovery of the Longitude is of such Consequence to Great Britain, for the Safety of the Navy, and Merchant Ships, as well as the Improvement of Trade, that for want thereof, many Ships have been retarded in their Voyages, and many lost; but if due Encouragement were proposed by the Publick, for such as shall discover the same, some Persons would offer themselves to prove the same, before the most proper Judges, in order to their entire Satisfaction, for the Safety of Mens Lives, her Majesty's Navy, the Increase of Trade, and the Shipping of these Islands, and the Lasting Honour of the British Nation. . . .[16]

These were familiar sentiments. Yet the argument and the language of this petition are so similar to that of April 29 that it is possible Whiston and Ditton were behind this as well.

Whiston and Ditton had important contacts in the community of mathematical practitioners and lecturers. Whiston's own works in natural philosophy and astronomy were published by John Senex and his partner John Maxwell from their new premises in Salisbury Court, Fleet Street. Senex we know as one of the more significant of the instrument and map makers.[17] Maxwell, who seems to have later taken religious orders, is far more elusive. He was the author of several short works on natural theology, including a brief foray into Samuel Clarke's biblical battles as *A*

[14] Maureen Farrell, "The Life and Work of William Whiston," p. 3.22. Abraham Sharp to Flamsteed, April 3, 1714.

[15] British Library, Shelf Mark 516.m.18, "Law Cases." I wish to thank John Parry for pointing this out to me. Sister Maureen Farrell has located another copy in the Houghton Library, Harvard University.

[16] *Journals of the House of Commons* (1803), XVII (December 7, 1711, to August 1, 1714), 641–2.

[17] Sarah Tyacke, *London Map-Sellers, 1660–1720* (Tring, 1978), p. 142; E. G. R. Taylor, *The Mathematical Practitioners of Hanoverian England* (Cambridge, 1966), p. 143; H. R. Calvert, *Scientific Trade Cards in the Science Museum Collection* (London, 1971), p. 38; and Sarah Tyacke, "Map-Sellers and the London Map Trade c. 1650–1710," in Helen Wallis and Sarah Tyacke, eds., *My Head is a Map. Essays and Memoirs in honour of R. V. Tooley* (London, 1973), pp. 79–80.

Discourse concerning God; with a Short Account of the Cape of Good Hope, and a version of *The English Atlas*. There exists one letter to John Harris relating to a manuscript on compass variation and the Cape of Good Hope that Maxwell requested Harris leave for him at Child's Coffee House near where Harris and Humphrey Ditton were then living. In 1713, Maxwell had presented to the Royal Society a copy of a "Scheme of the Solar System" done by Whiston from Halley's table of comets. Maxwell's interest in the problems of navigators was further established in 1721 when the Royal Society considered his observations of compass headings in the Atlantic dating from 1706 while on board a man of war.[18]

Whiston had even more important links with the universe of merchants. Shortly after their proposal for a reward had gone forward, Whiston revealed in 1715 that the original idea to calculate the time of the travel of sound had arisen from Ditton during a conversation with "some other common friends."[19] Who those friends might have been was not revealed by Whiston, but we can suggest that Whiston may have been referring to his contacts with Thomas Watts. Humphry Ditton died early in 1715, leaving Whiston with the task of nurturing his original notion. He also left a translation of Jacques Rohault's *Treatise of Mechanicks*, which Ditton had used with his students at Christ's Hospital. Upon Ditton's death, it seems Whiston undertook to complete it and ultimately put it into the hands of Thomas Watts. Watts was, at that time, looking for a work on mechanics for the use of students who were entering his Academy and ultimately revised and published the translation.[20] The links between Whiston and Ditton and the commercial world of Thomas Watts likely formed part of the support for their proposals on the longitude. Watts' known appreciation of the commercial risks of the period would have encouraged the petitions for a reward.

By June 10, 1714, Whiston and Ditton, possibly for the benefit of the Members of the House then considering their petition, had published in a handbill "Reasons for a Bill, Proposing a Reward for the Discovery of

[18] B.L., Sloane MSS. 3399, fols. 48–53, Maxwell to Harris, [1706]; cf. Sloane MSS. 4058, Ditton to Sloane, n.d.; Royal Society MSS., Journal Book (C), X (1702–1714), pp. 445–6 (January 29, 1712/13); Register Book (C), XI, pp. 166–7.

[19] *A New Method for Discovering the Longitude both at Sea and Land*, second edition (London, 1715), p. 18; quoted in Farrell, "Life and Work of William Whiston," p. 3.24.

[20] J. Rohault, *A Treatise of Mechanicks: or, The science of the Effects of Powers of Moving Forces, as apply'd to Machines, demonstrated from its first Principles . . . Useful for all Artificers, as well as Natural Philosophers and Mathematicians*, ed. Thomas Watts (London, 1716), "Foreword" by Whiston, May 17, 1716; dedicated by Watts to Dr. Mead, May 23, 1716. Cf. P. J. Wallis, "Thomas Watts. Academy Master, Freemason, Insurance Pioneer, MP," *History of Education Society Bulletin*, XXXII (Autumn 1983), 53.

the Longitude." For the first time, they publicly justified their proposal "Because it will prevent the Loss of abundance of Ships and Lives of Men; as it would certainly have sav'd all Sir Cloudsley Shovel's Fleet, had it been then put in Practice."[21] This was part of the campaign building in Parliament. On the next day, the committee was ready to make its report. This consisted largely of Newton's views of the failure of possible methods utilizing timepieces, the observation of the eclipses of Jupiter's satellites, and of the impracticability of the use of lunar observations. Instead, it was recognized that Ditton and Whiston's method might prove useful amidst the shallows, tides, and hidden shoals of coastal waters. With the support of Samuel Clarke, Whiston suggested that the incentive of a reward might hasten the solution of the problem. They were aided by three Whig members of the Commons' committee, Whiston's own patron Addison, General James Stanhope, and Robert Walpole – the last two having been advocates of Richard Steele when he ran into difficulties with the Tory majority and was expelled from the House.[22]

Even with the Tory majority, it was the Whig members who piloted the Longitude Bill through committee. They were merely helping their own. Accordingly, Whiston and Clarke, with the urging and assistance of Stanhope and Walpole, drafted the Bill that Stanhope presented to the House on June 16. It passed into law on July 3 as "An Act for providing a publick Reward for such Person or Persons as shall discover the Longitude at Sea." This provided a prize of £10,000 for the discovery within one degree, or 60 miles, up to £20,000 if it were accurate within 30 miles. The bounty was placed in the hands of Commissioners of the Longitude, who were to determine the worth of the proposals placed before it.[23] Immediately, the Commissioners were overwhelmed by a swarm of schemes more easily characterized by the lust of their inventors for a quick reward than by their utility.

The pertinacious Whiston was slightly more successful than most navigational projectors. The death of Humphrey Ditton shortly after their scheme had been issued did not deter Whiston from continuing to promote the original idea. In January 1715, Whiston issued a set of subscription proposals in which he suggested that a survey might be made of England and Wales based on their method of rockets fired into the air. Particular care was to be taken to determine the exact position of the points on the sea coast. Whiston pointed out that recent trials of a mortar

[21] "Law Cases," n.p..
[22] Winton, *Captain Steele*, pp. 200–3.
[23] *Journals of the House of Commons*, XVII, 642, 677–8, 686–7, 711, 715–16, 721.

fired at Blackheath had been confirmed by the observations of William Derham, who had already written on the speed of sound in the *Philosophical Transactions* in 1708. By this point, Derham was concerned about the variation in the performance of clocks produced by the weather and other shipboard conditions.[24] Derham was to be one of the receivers of Whiston's subscriptions, along with several booksellers, Ditton's widow at Christ's Hospital, instrument makers like Senex and Hauksbee, as well as his friends Cotes, Richard Laughton, and Richard Allin in Cambridge. In a covering letter to a copy of the proposal that he sent to Edmund Gibson, chaplain to Archbishop Tenison and later Whig Bishop of London, Whiston explained that part of the proceeds was to be used to provide for Ditton's widow and family.[25]

By the fall of 1714, Whiston was to expand the limits of public discourse. He announced a series of experiments in which all those in view of Hampstead heath could take part. Public lectures required attendance, but these experiments could include the entire population of the City and Westminster. Indeed, anyone within sixty miles was invited to send him a note on details of the observation of the flash of the fire ball from Whiston's mortar. For the next two years, Whiston's bombs echoed over Hampstead, and later at Gravesend, having been announced in the press so that a large population might be encouraged to send in reports. Even when mortars coincided with the eclipse in May 1715, Whiston, it seems, was not overwhelmed by public curiosity. Although the scheme was ultimately abandoned, it is certain observations were recorded at least at Greenwich. The surveyor Edward Laurence, brother of Whiston's friend John Laurence, proposed to survey counties or large tracts of land by Whiston's method. Others, however, were not as enamored of its practicality.[26]

Whiston's committment to this scheme was, above all, an experiment in public science. But he was too early. When Whiston fired his mortars over Hampstead Heath, at least a mile high he hoped, he was doomed

[24] Royal Society MSS., Letter Book (C), XV, f. 142, Derham to Mr. Watson, August 4, 1714.

[25] Lambeth Palace Library, Edmund Gibson MSS. 1742, fols. 74–5, Whiston to Gibson, January 6, 1714/15. On Laughton and Allin, see James E. Force, *William Whiston. Honest Newtonian* (Cambridge and New York, 1985), pp. 16–18. This, along with the sale of Ditton's library, was an attempt to support his survivors. *Daily Courant* (March 9, 1715).

[26] Maureen Farrell, "Life and Work of William Whiston," pp. 3.23–3.27; *Daily Courant* (July 15, 1715; May 13, 1717; June 5, 1717); Edward Laurence, *The Young Surveyor's Guide: Or, a New Introduction to the Whole Art of Surveying Land, Both by the Chain, and all Instruments now in Use. Now first Publish'd from an Original M.S.*, second edition (London, 1717), pp. 210–11.

to see his public science fail. In a remarkable exchange of letters with Roger Cotes, two different visions collided. Cotes warned Whiston that even the "least unfit" of the market towns would prove unreliable – and anyway they had no instruments or, without proper instruction, any knowledge of "clearly how to draw a Meridean, how to take ye bearing of Your light from that Meridean, & how to oberve the interval he found." He was encouraged by Whiston's consultation with Halley, but Cotes still saw no light in Cambridge.[27] What Whiston saw clearly, and what Cotes sixty miles away in darkest Cambridge could not, was that the things that were obvious, the phenomena we might all agree upon, would make science public and, thus, establish as great a claim to legitimacy as any form of human knowledge. If such vulgar rumblings from London reached Cambridge, a broader vision did not. For Whiston, those college gates that had once slammed shut behind him would soon be irrelevant in a public world.

Whiston soon shifted to another avenue of assault on the longitude. By 1721, he had obtained virtually £500 by subscription for a method of discovering the longitude at sea by the dipping needle. Forgetting the multitudes, this time among his subscribers, were King George, the Prince and Princess of Wales upon whom Whiston had come to rely, as well as James Brydges, by then first Duke of Chandos, Martin Folkes of the Royal Society, Thomas Parker, the Lord Chancellor, and his relation Thomas Anson, brother of the Admiral, and possibly the dissenting merchant Jeremiah Sambrooke and his brother-in-law William Hedges.[28] Whiston's notion of a coastal survey was not revived until 1739. By then, it was based on patient astronomical observations rather than mortars whose momentary burst was dispersed by wind and weather. To this end, in 1741, he was able to convince the Commissioners of the Longitude to grant him £500 for the purchase of instruments.

Whiston's vision of overcoming the constraints imposed by a wrong-headed Church hierarchy coincided with his committment to public philosophy. This strategy Whiston never relinquished. By the fall of 1740, he was still requesting all observers of the eclipses of Jupiter's satellites to send him their reports, while captains of ships should assist him to correct Halley's sea chart with their observations of compass variations

[27] Trinity College Library, MS. R.4.42., no. 8, Whiston to Cotes, November 26, 1714; no. 9. Cotes to Whiston, December 2, 1714; no. 11. Whiston to Cotes, April 7, 1715.

[28] *Biographia Britannica*, VI, pt. II (1766), p. 4210. Whiston reported on his method before the Royal Society in 1722. Royal Society MSS., Journal Book (C), XII, p. 219, March 22, 1722. On Hedges and Sambrooke, see Gary Stuart De Krey, *A Fractured Society. The Politics of London in the First Age of Party 1688–1715* (Oxford, 1985), pp. 76, 199.

over the forty years since its publication.[29] Whiston had refused to surrender to Cotes's pessimissm. Early in 1742, he advertised two new courses of lectures, one at the Temple Exchange Coffee House on latitude and longitude, to alternate with a course on religion at the Horn Tavern. During the nights of the longitude lectures, his mortars would fire from a demiculverin on Shooter's-Hill and Whiston, across the river at the Temple Exchange, would demonstrate the advantages to navigation.[30] Whiston's drama must have shattered the calm of nearby Greenwich, and public response was sufficient for a subscription to assist him further. He finally produced a large chart that contained insets of more dangerous obstacles such as the tragic Scilly Isles.[31]

Hunting the longitude

The most immediate result of the Longitude Act was a profusion of satire. On July 7, 1714, Whiston and Ditton had published a description of their method of rocket ships stationed near the coasts, firing mortars like providential warnings of imminent danger. Had their scheme proved practical, their timing would have been perfect as the Act had just passed. However, Doctor John Arbuthnot, a Fellow of the Royal Society and a friend of Jonathan Swift, had an entirely different opinion. On July 17, Arbuthnot wrote to Swift:

> Whetstone [Whiston] has at, last publish'd his project of the longitude; the most ridiculous thing that ever was thought on, but a pox on him he has spoild one of my papers of Scriblerus', which was a proposal for the longitude not very unlike his to this purpose, that since ther was no pole for East & west that all the princes of Europe shold joyn & build two prodigious poles upon high mountains with a vast Light house to serve for a pole Star. I was thinking of a calculation of the time charges & dimensions. Now you must understand his project is by light houses & explosions of bombs, at a certain hour.[32]

[29] Royal Astronomical Society, Add. MSS. 88, *Astronomical Scrapbook*, no. 112 (September 8, 1740).

[30] B.L., Add. MSS. 47012B, f. 92, from *The London Daily Post, and General Advertiser* (Saturday, January 2, 1741/2).

[31] Farrell, "Life and Work of Whiston," pp. 3.54–3.58.

[32] Harold Williams, ed., *The Correspondence of Jonathan Swift*, II (Oxford, 1963), p. 70. Arbuthnot to Swift, July 17, 1714. See also [Arbuthnot], *The Longitude Examined* in *Miscellaneous Works of the Late Dr. Arbuthnot. With an Account of the Author's Life* (London, 1770), II, pp. 295–310. Whiston was a continual object of satire. A quarter of a century later, one writer reported of an ode composed to Whiston and Ditton in reference to chamber pots supposedly adorned with their picture:

Let Whiston and Ditton
Be p--st on, and sh-t on.

See Raymond Williamson, "John Martyn and the Grub-Street Journal," *Medical History*, V (1961), 372.

To another projector who was foolish enough to solicit the assistance of the jaundiced Swift, the Dean merely replied, "One of my Projectors cut his Throat, and the other was found an Imposter. This is all I can say; but am confident you would deceive others, or are deceived yourself."[33] The image of Whiston's stir-crazy sentinels, isolated at strategic points in the North Sea and "reduced to streights for want of Firing," was great fodder for satirists like Arbuthnot.[34] Such a reaction was partly a reflection of deeply rooted Tory frustrations.

If Swift's or Arbuthnot's remarks appear unfair, there were reasons enough to sneer. John Keill complained that many who had demanded the reward scarcely understood the longitude, never mind proved capable of finding it.[35] John Flamsteed, Astronomer Royal, and, by virtue of his position, a member of the Board of Longitude, constantly faced the most lunatic of schemes. Only slightly more than a month after the reward had been offered, Flamsteed wrote to Abraham Sharp:

> About the middle of this month a couple of young Nonconformist ministers from Worksop in Derbyshire came hither to have my approbation of some method they had to propose for finding the longitude at sea, and I shall tell you because it will make you laugh abundantly. They proposed to have two or more strong vessels made of earth or glass, with a vent and stop-cock. The vent was to be filled with an agate, or some very hard stone or metal that might admit a very small hole to be drilled in it, and then, this hole being closed, all the air was to be pumped out of the vessel by the help of an air-pump, and then afterwards they were to open the small hole in the agate, and by a good pendulum clock to see how long time the vessels should be filled again with air, for this being known they calculated that as fast as the vessel was discharged it would fill again in the same time precisely, and they had a contrivance as proper as this to know when the vessel was full. They also proposed to find it by the times of the high waters in the open sea, which they would find by a barometer and a timekeeper with a thermometer annexed to it, with a table for correcting of the clock. I received them very civilly, but refused to give them my hand to testify that I had seen their proposals, and advised them not to print them. They told my servant Joseph that they had come 150 miles purposely for this business, but I believe they took my advice, for I heard no more of them.[36]

[33] *Correspondence of Jonathan Swift*, III (Oxford, 1963), p. 240. Swift to John Wheldon, September 27, 1727.

[34] [John Arbuthnot], *To the Right Honourable the Mayor and Alderman of the City of London: The Humble Petition Of the Colliers, Cooks, Cook-Maids, Blacksmiths, Jack-makers, Brasiers, and Others* [1716], in [Jonathan Swift] *Miscellanies*, Vol. III (London, 1732), p. 78.

[35] John Keill, *An Introduction to the True Astronomy* (London, 1721), p. 189.

[36] Cudworth, *Life and Correspondence of Sharp*, pp. 111–12. Flamsteed to Abr. Sharp, August 31, 1714.

By comparison, Flamsteed thought Whiston's and Ditton's notions merely impractical rather than ludicrous.

Finding the longitude at sea floundered in the mania for the reward. Yet the elusive grail of the longitude provoked an endless stream of obscure crusaders who claimed to have solved the riddle. There were, of course, the works of Fellows of the Royal Society like James Hodgson, who published in 1706 *The Theory of Navigation Demonstrated*. Even before it had occurred to Whiston and Ditton to secure a reward, projectors frequently laid claim to a solution to forestall others. In 1710, an anonymous author printed a tract for "the Universal Benefit of Mankind" after having heard of a presentation before the Royal Society by a Mr. Keith. Notably, the tract claimed *A New and Easy Method To find the Longitude at Land or Sea. From the Observation of the Altitudes of the Coelestial Bodies only; without having any Recourse to Eclipses, the Satellites of Jupiter, the Doctrine of Appulses, or the Magnet; and without the use of Clock-work, Dials of any sort, Sand-Glasses or Pendulums; or any other Instrument, but a Sextant, Quadrant, or Astrolabe*. This would have been a remarkable performance.[37] But the measure of time was widely understood to be the crux of the matter. It was the solution of James Clarke, a Wiltshire physician, to construct an elaborate clock based on the flow of some fluid that would be unaffected "by the different Gravity, or other Changes of the Atmosphere. . . ."[38]

It was the lure of a lucrative prize that really drove the London mathematical community into paroxysms of fantasies. Within a week of the reward being announced, Flamsteed received a printed proposal from William Hobbs, a watchmaker, which suggested that a device be provided to offset all of the motions of a ship so as not to affect a pendulum clock.[39] Hobbs had no further revelations. At the same time, Case Billingsley, one of the originators of British maritime insurance companies and possessor of a truly remarkable imagination for various plots to produce

[37] The British Library names the author as Charles Hayes, the Deputy Governor of the African Company, who published *A Treatise on Fluxions* in 1704. See R. V. Wallis and Peter Wallis, *Biobibliography of British Mathematics and its Applications*, Part II, 1701–1760 (Newcastle, 1988), p. 14.

[38] James Clarke, *An Essay, Wherein a Method is Humbly Propos'd for Measuring equal Time with the utmost Exactness; Without the Necessity of being confin'd to Clocks; Watches, or any other Horological Movements; in order to discover the Longitude at Sea* (London, 1714), pp. 7–9.

[39] Cudworth, *Life and Correspondence of Sharp*, p. 112; William Hobbs, Philo. Mathem., *A New Discovery for Finding the Longitude. Humbly Submitted to the Right Honourable of the Lords Spiritual and Temporal . . .* (London, 1714). For various approaches, see Isaac Hawkins, *An Essay for the Discovery of the Longitude at Sea, By several Methods fully and Particularly laid before the Publick* (London, 1714).

immense wealth, published a tract obviously directed at his competitors entitled, *The Longitude at Sea, Not to be found by Firing Guns, nor by the Most Curious Saving-Clocks or Watches.* Typical of Billingsley, a frontal assault was thought to achieve the best results. On August 23, Billingsley printed an open letter to the Commissioners of the Longitude in which his contrivance was described. He suggested a timekeeper that was to be enclosed in an airtight vessel, along with a thermometer, and another instrument by which variation in gravity might be accounted for. By all of this, and through references to tables, Billingsley hoped to eliminate the variations that would affect a pendulum clock that, in turn, was attached to a so-called Balance-Wheel to compensate for the motion of the ship. This was merely a more elaborate version of the proposal made by William Hobbs. In addition, Billingsley proposed to equip the ship with a large telescope planted firmly upon a platform that did not move with the roll of the vessel, so as to enable the observation of the eclipse of heavenly bodies. Flamsteed managed to persuade him that the shock and motion of the ship would nevertheless be transferred to the pendulum so that, as Flamsteed reported, "he held his peace, which is as much as I can expect from persons that have swelled themselves with the hopes of getting £20,000."[40] Flamsteed had unwittingly captured the chronic agitation of Billingsley's mind. Almost twenty years later, Billingsley was still making claims that his scheme was preferable to the work of Halley having been "seen, examin'd & approv'd by the great Sir Isaac Newton & the Lords of the Admiralty" and therefore worthy of a patent even if the Board of Longitude was uninterested.[41] Billingsley's testimony was always a little suspect. In fact, Newton's review of years of proposals for the Commissioners of the Admiralty revealed that Billingsley's notions for a new clock would be as affected as any other by the variations of gravity in varying latitudes.[42] Billingsley's suggestion,

[40] Cudworth, *Life and Correspondence of Sharp*, p. 112; Wallis and Wallis, *Biobibliography*, p. 62; Case Billingsley, *A Letter to the Commissioners Appointed by the Parliament of Great Britain, for Discovering the Longitude at Sea, (from Case Billingsley) Further to Explain his Late Proposal for that Discovery, by the Sun, Moon, and Stars; and a True Time- keeper* (London, 1714). This was printed by, among others, the map seller Richard Mount and a John Billingsley, under the Royal Exchange. Mount was formerly a partner of John Senex. Tyacke, "Map-Sellers and the London Map Trade," p. 76. This John Billingsley had been apprenticed to his father Benjamin, of the Stationers' Company, in 1690. D. F. McKenzie, ed., *Stationers' Company Apprentices, 1641–1700* (Oxford, 1974), p. 14; Henry R. Plome et al., eds., *A Dictionary of Printers and Booksellers who were at work in England, Scotland and Ireland from 1668 to 1725* (Oxford, 1922), p. 35.

[41] University of London Library, Autograph Letters 5, Billingsley to the Earl of Westmorland, April 17, 1732.

[42] Cambridge University Library, Add. MSS. 3972, f. 30, "Sir Isaac Newton's Record upon the Petition of Mr Jacob Rowe" (August 26, 1725).

and also his trials, were rather typical of the way in which the treasure hunters lusted after the elusive longitude.

One of the most interesting efforts to secure the reward was that of Jeremy Thacker of Beverley in Yorkshire. Although he proposed that a chronometer was the only practical solution to the problem – and thereby making the first printed reference to such a device – his tract was more revealing in its criticisms of his competitors. Despite his evident removal from the center of the London philosophical community, Thacker displayed considerable awareness in making a connection between the longitude interest of Whiston and Ditton and the Latitudinarian theology that had led to Whiston's exile from Cambridge:

> If I should follow the Example of the greatest Part of the Longitudinarians, that wrote after the first Attempt, I should take Mr. W—n and Mr. D—n in Hand, shipwreck their Hulls, drown their Sounds in a Tempest, lift up the Waves to intercept the Sight of their Fires, and break their General Peace; but Gratitude forbids: They sprung the Twenty Thousand Pounds, and as I hope to get it, I ought to be civil to them. Then let Mr. D—n rest assur'd, that I will neither meddle with his Longitude, nor his *New Law of Fluids*, and poor Mr. W—n has been so often handled as a Longitudinarian, and a Latitudinarian, so amaz'd at the bursting of his Mortar, and one of his Shells, and so frighten'd with the Fall of another unlighted one, that it would be as barbarous as ungrateful for me to insult over him.[43]

Thacker was well informed of events in London. He seems to have understood that natural philosophy was not foreign to the pressures of projects and politics in these years. But Thacker's barbs were not only directed against Whiston and Ditton. Any potential competitor felt the sharpness of his wit. He began by attacking Hobbs and then he offered to help Case Billingsley, and assure his success, providing that he would first prove Newton's first law of motion to be false.[44] Thacker had a good grasp of the problems involved, suggesting that only by improvement of either astronomical observation or by the perfection of time pieces would the longitude be determined while at sea. But he then uttered the most

[43] Jeremy Thacker, *The Longitudes Examin'd. Beginning with a short Epistle to the Longitudinarians, and Ending with the Description of a smart, pretty machine of my own, which I am (almost) sure will do for the Longitude, and procure me the Twenty Thousand Pounds* (London, 1714), p. 2; cf. Taylor, *Tudor and Stuart England*, p. 306. Thacker's wit appeared in 1770 among the *Miscellaneous Works of the Late Dr. Arbuthnot. With an Account of the Author's Life* (London, 1770), II, pp. 295–310. See *New Cambridge Bibliography of English Literature*, II (Cambridge, 1971), p. 1051.

[44] Thacker, *Longitudes Examin'd*, pp. 3–4.

antagonistic remark of all directed against the Newtonians, especially in 1714, at a point when a Leibnizian as well as a Hanoverian succession were altogether possible. Perhaps Thacker was merely playing the politics of the longitude with considerable verve. He charged that it was only to Leibniz that the world could look to the improvement of astronomy if

> the great Mr. L—, with an Industry equal to his Candour, should join the Fluxions and Series, which he invented, to his known Skill in the Laws of Centripetal and Centrifugal Forces, and so vouchsafe to perform the Work for the Benefit of Mankind.[45]

Even the problem of the longitude could not escape the politics of the calculus. Thacker might have been gambling on the arrival of Leibniz in England in the entourage of George I. Whatever the reason, there is undoubtedly a recognition in Thacker's remarks that the Newtonians themselves might prove an obstacle to an interloper in the search for the longitude.

To circumvent the Board of Longitude, it was frequently the case that approaches were made directly to Newton in search of his approval. Although Newton was chronically irritated by these schemes, he did take time to show his petitioners where they had become confused. As far as he was concerned, improved clocks would be of little use without observations of the moon or of Jupiter's satellites.[46] A proposal by a Mr. Laurans to the Board of Admiralty for finding longitude at sea by means of an improved watch Newton dismissed out of hand, for "when the longitude at sea is once lost, it cannot be found again by any watch." In his view, the Act of Parliament related only to advancements in astronomy. Consequently, "this improvement must be made at land, not by watchmakers or teachers of Navigation or people that know not how to find the Longitude at land, but by the ablest Astronomers."[47] Newton's exclusive principles were to be maintained even in the practice of navigation. He told one petitioner that his proposal was "as impracticable as to make a perpetual motion." Thus, he demanded his correspondent "would not endeavour to concern me any further wth Projectors."[48]

[45] Thacker, *Longitudes Examin'd*, p. 8.

[46] Cambridge University Library, Add. MSS. 3972, fols. 36, 38., n.d., but in Newton's hand.

[47] Cambridge University Library, Add. MSS. 3972, f. 37, Newton to ?, n.d. On this draft, possibly to the Admiralty, Newton wrote but crossed out, "they that do not know how to find it at land will never be able to find it at sea."

[48] Cambridge University Library, Add. MSS. 3972, f. 38, Newton to ?, n.d.

Lobbying the prize

For desperate men, as were many of the Longitudinarians, there was hope in any lottery. With such a large purse, it is not surprising that politics should have played a profound role in the attempts to solve the longitude. When reading through the pamphlets debating the problem, one gets the impression that in the few years immediately following the Act, a good part of London, from the lowliest artisan to the most sophisticated of astronomers, was engaged in intense lobbying for support and patronage among the members of the Board of Longitude as well as Parliament and the Admiralty. Access to the Board might be smoothed by the support of the right patron or the Board could be circumvented entirely by an application for a patent. Robert Browne, an instrument maker who seems to have spent a good deal of his time in astrological forecasts, was a friend of Flamsteed, and although Flamsteed tried to discourage him, Browne pursued his reward mainly through improvements he claimed to have made in the observations of the sun and moon. Prior to the initial attempts by Whiston and Ditton to secure a reward, Browne had made representations to the South Sea Company on the grounds that his method would improve their coastal charts. His earlier petition was supported by Flamsteed, Halley, and James Atkinson, senior, a prominent teacher of mathematics at Rotherithe.[49] Nothing came of this and Browne withdrew his petition.

Ridiculing Whiston and Ditton proved much easier than seeking influential patrons. In 1714, one R.B., who seemingly knew Halley, claimed to have discovered a means far more effectual than that of Whiston and Ditton, who he noted, "may as well persuade our Country-Men, that White is Black, and Black's White: But never an Arrian in this Kingdom, shall seduce me to hold with him in so giant an Error."[50] Late in 1717, some time after the Act had passed, Robert Browne had occasion to solicit the assistance of General Stanhope, who had been instrumental in

[49] On Atkinson, see Taylor, *Tudor and Stuart England*, p. 254.

[50] E. G. R. Taylor suggests the author as the Robert Browne, *R.N.*, who published *Directions for sailing into the English Channel* in which the dangerous approach to the Scilly Isles is emphasized. *Tudor and Stuart England*, p. 301; *Hanoverian England*, p. 113. However, in 1714, Browne mentions "being not used to the Sea" and abandoning the opportunity to test his proposal. The member of the Royal Navy may be the *R.B.* who identified himself as the "secretary to the Honourable Sir Francis Wheeler, when Admiral and General in an Expedition to Martineco" and who published *Longitude To be found out with A new Invented Instrument, Both By Sea and Land... with a better Method for discovering Longitude, than that lately propos'd by Mr. Whiston and Mr. Ditton* (London, 1714). For the attack on Whiston and Ditton, see pp. 15, 26–7.

the passage of the Bill. Stanhope promised to aid Browne if Halley and Newton would be willing to present certificates testifying to the usefulness of Browne's work. Halley resisted on the grounds that he was about to print a proposal of his own. Moreover, Halley informed him that Newton was angry because Browne had printed something against his theory of the moon. At any rate, Stanhope refused to help, although he ordered Browne twenty pounds from the Treasury.[51] This was all the encouragement he received for his efforts. It was more than most achieved.

Judging from the number of books and pamphlets on the longitude, there were undoubtedly many such disappointments.[52] Occasionally, those seriously involved in the design of new instruments might find that their time and effort produced nothing but burdensome debt. Fatio de Duillier, the young mathematician who in his youth so impressed Newton, but whose reputation was effectively destroyed by his relations with a sect of millenarians – the so-called French prophets – carried on experiments in 1714 aimed at improving navigation, although these had little to do with the longitude. In 1724, he was attempting to secure the imprimatur of Newton for an advertisement of his project to manufacture of pierced rubies for use in watches.[53] By 1728, he complained of his "long and too unfortunate and ruinous Study of Mathematicks," and sought public support for an improved method of finding latitude.[54] In 1736, Fatio was still lamenting the loss of "several Thousands of Pounds for the Service of this Nation or of the Publick; and that, by my too unfortunate Invention of Jewell Watches, and by some other Discov-

[51] Robert Browne, *Methods, Propositions and Problems, For Finding the Longitude; with the Degree and Minute of the equator upon the Meridean. And the Longitude at Sea, by Coelestial Observations only. And also by Watches, Clocks, etc. and to correct them and know their Alterations* (London, 1714), pp. 18–20. This is to be found bound between pp. 90–7 of Browne, *A system of Theology Revealed from God by the Angels in the British Language herein is disclosed the mysteries of the Holy Scriptures, and the true Fundamentals of Religion, being a Lesson to the whole World. With any Curios and Occult Things in Church, State and Philosophy, All designed for the Public and Universal Good of Mankind, and a general Reformation of the present Times* (London, 1729). See also pp. 80–1, 89–90. Wallis and Wallis, *Biobibliography*, p. 62.

[52] Rousseau, "Science Books and their Readers," p. 226; and Wallis and Wallis, *Biobibliography, passim.*

[53] Cambridge University Library, Add. MSS. 4007(B), f. 662, Facio to Newton, April 1, 1724; Add. MSS. 3972, f. 37, Newton to ?, n.d..

[54] Nicholas Fatio de Duillier, *Navigation improved: Being chiefly the method for finding the latitude at sea as well as land, by taking any proper altitudes, Together with the time between observations: The Whole made easy by a new Trigonometric Rule*... (London, 1728). On the French Prophets, see Hillel Schwartz, *Knaves, Fools, Madmen, and that Subtile Effluvium. A Study of the Opposition to the French Prophets in England, 1706–1710* (Gainesville, 1978); and Schwartz, *The French Prophets. The History of a Millenarian Group in Eighteenth-Century England* (Berkeley and Los Angeles, 1980).

erys." He requested Sloane to open a "Book of Voluntary Subscriptions" at the Royal Society to enable the publication of his discovery of the sun's parallax.[55] As Denis Papin and Desaguliers knew, the business of experimental philosophy had become an expensive affair.

No more successful was a Welsh physician, Zachariah Williams, who came to London certain that he could obtain the longitude reward for a scheme based on magnetic variation, the trends in which he claimed to be able to predict. Unlike many aspirants to the prize, Williams was not an utter lunatic, although he was bitterly disappointed in his reception by the Board of Longitude. As a result, in 1729, he sought out the Duke of Chandos, the patron of Desaguliers. Chandos sent him on to Desaguliers at the Royal Society, where, as Desaguliers told it with some relish, Williams "took him for a conjuror," which was a reaction the philosopher encountered more than once. Nothing came of Williams' efforts, although he was well versed in electrical and magnetical experiments that he carried out at the Charterhouse with his friend Stephen Gray.[56] Williams' approach to Chandos was typical not only of his own predicament, but also of others whose reception by the Board was less than enthusiastic. What is more interesting is that a great investor like Chandos was not only pestered by claims of numerous inventors, but that he was frequently willing to listen – or rather, in his case, to ask Desaguliers to listen. It was not as though Chandos was unaware of the pitfalls and pretenses that surrounded him. In 1735, for example, after having been petitioned to promote a treatise that described an instrument to take the longitude at sea, he asked Desaguliers to examine it although "I know you are so tryed[?] with these pretended discoveries, that you have little inclination to give any attention to any thing of this sort. . . ."[57] No wonder.

These approaches were generally made to get around the obstacle of the Longitude Board and thereby secure a patent from the Crown. Such efforts had begun remarkably soon after the Act had been passed. In 1716, the mathematics teacher John French tried to circumvent the Board when he applied to the Prince of Wales for a patent under the Great Seal. French claimed to have perfected his "projection" by 1709 at considerable expense but in return to have had "nothing but hard and unjust Dealings,

[55] B.L., Sloane MSS. 4055, f. 27, N. Facio to Sloane, December 28, 1736.
[56] Taylor, *Hanoverian England*, pp. 148–9, 122; Parkinson, ed., *Private Journal of Byrom*, p. 383; and Albert J. Kuhn, "Dr. Johnson, Zachariah Williams, and the Eighteenth-Century Search for the Longitude," *Modern Philology*, LXXXII (August 1984), 40–52.
[57] Hunt. Lib., Stowe MSS. 57, XLVI, f. 83, Chandos to Desaguliers, May 16, 1735.

and Discouragement to the highest Degree." This was not quite true, as it may be that as early as 1706, Prince George, Queen Anne's consort, had ordered sea trials of French's notions. In any case, the institution of the reward in 1714 revived French and he approached the Royal Society with a notion that lighting a fire under a magnetic compass would somehow turn it from the direction of the pole at an angle equal to the latitude of the place of the compass. His request for a subscription to build the necessary instruments he could not afford because of the great charges and consequent "Great Loss of time for ye Publick Good" fell on deaf ears.[58] The report on French's petition in 1716 to George, Prince of Wales, merely pointed out that the Act for the discovery of the longitude was sufficient, and that such a patent would undermine the intent of the Act "to purchase the Discovery of the Longitude, for the Benefit of the Publick. . . ."[59] The Board was to be permitted to find its own way through the maze of petitioners.

The intensity of the competition induced amazing ingenuity – notably in finding ways to avoid the difficulties of convincing such a body. Naturally, the Royal Society became the focus of the swarm that sought its approbation. In 1730, a James Orr wrote to Hans Sloane for his assistance as he was a stranger to London and unfamiliar of the steps to be taken to secure the reward.[60] And in the early days of 1732, the Royal Society was provided with a series of references from mathematical practitioners, including William Pardon, an accountant to the Governor and Company for smelting down lead with pit coal and sea coal, and William Hodgkin, a writing master and accountant, who had evidently met for eight years at least in weekly gatherings on mathematical subjects. The subject of their petition concerned the invention of an instrument to find the longitude at sea by Stephen Plank, who had previously written pamphlets on the general problem.[61] Similarly, in 1736, Sloane was asked by William Burtsal of Surrey to communicate to the Society his discovery of yet

[58] *Correspondence of Isaac Newton* (Cambridge, 1973), VI, p. 94, John French to the Royal Society, December 9, 1714; Cambridge University Library, Add. MSS. 3972, f. 11. Cf. Taylor, *Hanoverian England*, pp. 118–19; Richard S. Westfall, *Never at Rest: A Biography of Isaac Newton* (Cambridge, 1980), pp. 836–7.

[59] Public Record Office, State Papers (Domestic), 35/5, no. 68(2). Report of Edward Northey, Attorney General. See also Christine MacLeod, *Inventing the Industrial Revolution. The English Patent System, 1660–1800* (Cambridge and New York, 1988), p. 38.

[60] B.L., Sloane MSS. 4051, f. 60, Orr to Sloane, June 24, 1730.

[61] Royal Society MSS., Classified Papers, VIII(2), no. 76, fols. 151–6. February 19–29, 1731/2; on Planck and Pardon, see Taylor, *Hanoverian England*, pp. 138–9, and Wallis and Wallis, *Biobibliography*, pp. 66, 219.

another machine to determine the longitude "by meer Inspection on the true Velocity of the Ship upon any Rumb. . . . "[62] Whiston's and Ditton's proposals for a reward had released a demon no one expected.

The passage of the Longitude Act in the summer of 1714 was testimony to the utilitarianism of those active in the propagation of Newtonian natural philosophy. Even before Whiston and Ditton had revealed their scheme, their patron Addison had suggested the potential public benefit from a project to discover the longitude.[63] Although these projectors reaped few rewards, their efforts nonetheless reflect the degree to which the mathematical practitioners of London, and especially the Newtonians, were engaged in the attempts to solve the riddle of the longitude. Whiston and Ditton represent the height of this practicality.

But the lust for the longitude did not attract universal praise. The notions of Whiston and Ditton were as controversial as they were innovative. Literary scholars have long since shown how they bore the brunt of bitter satire, and how Whiston especially became the target of much abuse. Many difficulties arise, however, in revealing the animus behind the nightmares of the Augustan critics. The effect on literature of the various projects to discover the longitude or to establish manufactures of wondrous variety would require a broader scope than this book intends. So much has been written of the personal motives, the political hostility, or the specific targets of the virtuosi to defy an simple summary. Pope, Gay, and Swift alone conjure up volumes of analysis. But beneath all of this lies the forceful reckoning of the social critic and satirist with those who were inexorably changing the early eighteenth-century world.[64]

Projectors and schism

The Tory visions of this tide of Longitudinarians confirmed the iconography of the early eighteenth century. New finance, duplicitous projectors and financiers, grasping Whigs, natural philosophy, and experiments in

62 B.L., Sloane MSS. 4054, f. 181, Burtsal to Sloane, February 16, 1735/6.

63 Marjorie Nicolson and G. S. Rousseau, "*This Long Disease My Life.*" *Alexander Pope and the Sciences* (Princeton, 1968), p. 167.

64 See, for example, Nicolson and Rousseau, "*This Long Disease My Life,*" pp. 131–235; Joseph M. Levine, *Dr. Woodward's Shield. History, Science, and Satire in Augustan England* (Berkeley, 1977), Part 1; Ernest Tuveson, "Swift and the World-Makers," *Journal of the History of Ideas*, XI (1950), 54–74; Pat Rogers, "Gulliver and the Engineers," *Modern Language Review*, LXX (1975), 260–70; Howard Erskine-Hill, *The Social Milieu of Alexander Pope* (New Haven, 1975), pp. 152–5; and Richard G. Olson, "Tory-High Church Opposition to Science and Scientism in the Eighteenth Century: The Works of John Arbuthnot, Jonathan Swift, and Samuel Johnson," in John G. Burke, ed., *The Uses of Science in the Age of Newton* (Berkeley, 1983), pp. 171–204.

religion were all linked in the collapse of Tory, High-Church power following the death of Queen Anne in the summer of 1714. For Swift, this was the nightmare of the new order for which the only appropriate response was satire or Bolingbroke's impotent *Craftsman*. The palpable hatred for the ubiquitous Whig projector became the foundation of Swift's attack on William Wood in the *Drapier's Letters*.[65] For Tory commentators, the vermin of Exchange Alley brought forth the financial crises that harmed so many unsuspecting investors.

These victims were ripe for Tory rhetoric, especially in the aftermath of such a great debacle as the South Sea Bubble in 1720. Belated Tory satirists like Alexander Pope or John Gay may be treated with some scepticism as their complaining followed their own losses. They too were heavily involved in South Sea Company loans.[66] But there were many who had a familiarity with the lament of the anonymous poet of *The Broken Stock-Jobbers: Or, Work for the Bailiffs*:

> With a fair flowing Fortune in my Hand,
> A clear Estate, a Borrough at Command.
> What mov'd me from my Country Seat to range
> And take a Lodging near the damn'd Exchange?
> . . . Why must my stupid Fancy e'er admire
> The Way of raising Water up by Fire?
> That cursed Engine pump'd my Pockets dry,
> And left no Fire to warm my Fingers by.[67]

There was much the same sense of events even in Whigs like White Kennett, the Bishop of Peterborough. It was Kennett's comment that "Enthusiasm in different shapes returns often to this poor nation: we have had religious enthusiasm, political enthusiasm, and this was mere secular enthusiasm."[68] Kennett was an astute observer of the political as well as of the psychological currents of his time.

The South Sea Bubble seemed proof that the Whig Junto had now produced monsters that could not be controlled. This was the era of the

65 J. M. Treadwell, "Swift, William Wood, and the Factual Basis of Satire," *Journal of British Studies*, XV (Spring 1976), 80; Isaac Kramnick, *Bolingbroke and His Circle. The Politics of Nostalgia in the Age of Walpole* (Cambridge, Mass., 1968), pp. 210–11.

66 P. G. M. Dickson, *The Financial Revolution in England, 1688–1756* (London and Melbourne, 1967), p. 195; McKendrick, Brewer, and Plumb, *The Birth of a Consumer Society. The Commercialization of Eighteenth-Century England* (Bloomington, 1982), p. 212.

67 [Anon.,] *The Broken Stock-Jobbers: Or, A Work for the Bailiffs. A New Farce. As it was lately Acted in Exchange-Alley* (London, 1720).

68 Quoted in G. V. Bennett, *White Kennett, 1660–1728. Bishop of Peterborough* (London, 1957), p. ix.

public emergence of Newtonianism that had a special relevance in the technical projects of some of Newton's most active proponents. This kind of development, which took place largely beyond the walls of the two universities, meant that natural philosophy was as much a public as an academic matter. This was precisely the point made by Thomas Watts in the 1722 edition of his *Essay on the Proper Method for Forming the Man of Business*:

> The proper Age for Education here is from about Thirteen or Fourteen upwards; and the Young Gentlemen are not only such as are immediately designed for Trades, Merchandize, the Sea, Clerkships in Offices, or to Attorneys, or any other Employments in Business at home or abroad; but those in general who are not designed for the Universities, or while they study there, intend to spend their Vacations in learning Accounts and Mathematicks, or in going through the Courses of Experimental Philosophy.[69]

The divisions of society, reflected in a university education, were no longer absolute. But, as the Schism Bill would lay bare, these divisions could still have great impact on the world of the mathematicians.

William Whiston's skirmishes with the High Church were part of a battle conducted simultaneously on numerous fronts. During 1713 and 1714, Whiston's campaign with Humphry Ditton for a reward for the discovery of the longitude was compromised by attacks on the Dissenters and their academies, which Sacheverell had defamed as seminaries of sedition. And these academies had established such an interest in natural philosophy that one Dissenting minister later complained that "among ministers and laity of the nonconformists, a taste for science prevails to a greater degree than a taste for literature and their clergy are better mathematicians and experimentalists than scholars."[70] It was out of such traditions that later came Joseph Priestley, among others. For example, we know that John Moore, who kept an academy at Bridgwater, and Joseph Hallett II, who had a similar institution at Exeter, had both subscribed to Harris's *Lexicon Technicum*. Of course, this tells us little about the teaching of science at these schools. However, the subscriptions suggest that these two academies maintained an active interest in science, as did the other academies at Bethnal Green, Findern, and Sheriffhales.

[69] Thomas Watts, *An Essay on the Proper Method for forming the Man of Business*, ed. Arthur H. Cole (Boston, 1946), p. 9. On Watts, see Wallis and Wallis, *Biobibliography*, p. 78.

[70] Olive M. Griffiths, *Religion and Learning: A Study in Presbyterian Thought From the Bartholomew Ejections (1662) to the Foundation of the Unitarian Movement* (Cambridge, 1935), pp. 36–7.

Samuel Clarke's translation of Rohault was required reading at Rathmell.[71] There was connection enough to fuel any Swiftian suspicions.

During 1714, the Dissenting Academies were shaken by the High-Church agenda. We have seen how, in the summer of 1714, the initial steps were taken to encourage the search for longitude, in part due to the persistent efforts of Whiston and Ditton. Also of some moment was a petition of May 25 from the wider marine community that pointed out how the solution of this navigational headache would not only increase safety but also improve trade. These events occurred during the deliberations of the House of Commons over matters of more immediate consequence in the reign of Anne. On May 12, 1714, the Tory William Wyndham was given leave to bring before the House a "Bill to prevent the growth of Schism; and for the further Security of the Church of England, as by law established." The genesis of this Bill is notably interesting. One apparent architect was Bolingbroke, who designed it as an effort to isolate his rival Harley and perhaps assuage the Jacobites. Playing both sides of the street, Bolingbroke had previously persuaded the Queen to issue a proclamation against James, which had delighted the Hanoverian Tories so much that they added a reward of £100,000 for his capture.[72] In this summer of prizes, it might have been easier to find the longitude.

Out of Bolingbroke's agenda, the Schism Bill materialized. The Bill was evidently drafted by Whiston's bête-noir Atterbury to have the same effect as the Act of Uniformity, that is, all teachers were to conform to the Church of England. Although Bolingbroke later denied it, the Bill was designed to effectively close the Dissenting Academies that certain High-Church circles considered a threat not only to the Church, but possibly, due to the electoral strength of some dissenting boroughs, a threat to the Tory party itself. To what extent this was merely the first stage of an open attack on the political rights of nonconformists was an open question.[73] In a comment that can be politely described as disingenuous and self-serving, Bolingbroke later claimed "that the persecution of dissenters entered into no man's head."[74] It certainly entered the head

[71] Griffiths, *Religion and Learning*, p. 37.

[72] D. Szechi, *Jacobitism and Tory Politics, 1710–1714* (Edinburgh, 1984), pp. 167–8.

[73] Geoffrey Holmes, *British Politics in the Age of Anne* (London, 1967), pp. 103–4; and G. V. Bennett, *The Tory Crisis in Church and State, 1688–1730* (Oxford, 1975), pp. 177–9.

[74] Henry St. John, Lord Viscount Bolingbroke, *A Letter to Sir William Windham* (London, 1753), pp. 22–3; and Geoffrey Holmes, *British Politics in the Age of Anne*, pp. 103–4.

of Whiston's patron, Richard Steele, that this assault upon schismatics was intended to undermine the supporters of the House of Hanover.[75]

The Schism Bill was presented to the Commons on May 24, the day before the petition of the marine community for the discovery of the longitude. Despite the hostility of Stanhope, Robert Walpole, and Peter King, the future Lord Chancellor and himself a pupil of Hallett at Exeter, the Bill passed the Commons on June 1.[76] Stanhope and Walpole had recently supported Whiston's longitude proposals before the House. Peter King was even more significant in Whiston's affairs as he had already served as one of Whiston's defense counsel, without fee, in a heresy trial before the Court of Delegates in 1713.[77] This Whig phalanx was not sufficient to stop Bolingbroke's Bill. For Dissenters, it was a catastrophe. As it stood, it would have destroyed the right of Dissenters to control the education of their own children.[78]

The summer of 1714 would experience a brutal alignment of politics and religion; and the Succession waited tentatively for the end of Anne and the Tories. The Dissenters seemed trapped between Jacobites and dismay. Between June 4 and June 9, 1714, seven petitions reached the House of Lords in a final effort to obstruct the Bill.[79] One of the first was from Dissenters who taught navigation, including James Atkinson, an associate of Robert Browne, and who was a well-known instrument maker, mathematical teacher, and writer of textbooks on navigation. Among other things, Atkinson taught algebra, navigation, surveying, merchants accounts, and astronomy as well as selling charts and all sorts of mathematical instruments from his shop in Rotherithe.[80]

[75] Richard Steele, *A Letter to a Member of Parliament Concerning the Bill for Preventing the Growth of Schism* (Edinburgh, 1714), pp. 12–13, 28–9.

[76] *Journals of the House of Commons*, XVII, pp. 640, 660; Edmund Calamy, D.D., *An Historical Account of Mine Own Life*, ed. J. T. Rutt (London, 1830), II, p. 282; Allen Brockett, *Nonconformity in Exeter, 1650–1875* (Manchester, 1962), p. 67.

[77] James E. Force, *William Whiston. Honest Newtonian* (Cambridge and New York, 1985), pp. 98–9.

[78] Josias Maultby, *God the Hope, and Saviour of his People, Apply'd to the Case of the Protestant-Dissenters: in a Sermon Preach'd at a Solemn Fast, June 2, 1714. On Occasion of the Bill for preventing the Growth of Schism, etc.* (London, 1714), pp. 10–11.

[79] The petitions are reprinted without signatures in Historical Manuscripts Commission, *Manuscripts of the House of Lords*, X, n.s. (1712–14), (London, 1965), pp. 341–3. The originals are in the House of Lords Record Office, Petitions, Box 195B, n.s. X, 3123.

[80] James Atkinson, *Epitome of Navigation: or, A Short and Easy Methodical Way to become a Compleat Navigator*, fifth edition (London, 1704); the petition was signed by both James Atkinson, senior and junior, who were both mathematical teachers. HLRO, Petitions 195B, n.s. X, 3123 (f); Taylor, *Tudor and Stuart England*, pp. 254, 296; Taylor, *Hanoverian England*, p. 109; Ralph Davis, *The Rise of the English Shipping Industry: In the 17th and 18th Centuries* (Newton Abbot, 1972), p. 126. See also Joyce Brown,

The mathematical practitioners had quickly recognized that the Schism Bill would have effectively eliminated all Dissenters from teaching of any kind, even those who were private tutors, and even those like Ditton at Christ's Hospital Mathematical School. To their credit, on June 9, several Anglican schoolmasters and schoolmistresses also added their complaints. Among this group was Ralph Snow, who had a school at the Hand and Pen in Little-Moorfields, and for whom Humphry Ditton had once advertised the teaching of "Mathematicks, with their Uses and Application to Natural Philosophy."[81] The petitioners pointed out that the Bill would be prejudicial to the nation as a whole, and especially to the teaching of navigation by those Dissenters who, as masters of ships, took apprentices.[82] This was a rather telling argument as several petitions had also been received calling for a reward for the discovery of the longitude. In the middle of concern over the Schism Bill, on June 11, the Commons Committee on the longitude was given leave to present a Bill for a reward, and on June 16, Stanhope brought forward such legislation.[83]

Quite apart from the other concerted parliamentary opposition to the Schism Bill, the reaction of the mathematical teachers had some effect. In view of the enthusiasm for the Longitude Bill throughout the London mathematical and philosophical community, including a significant number of nonconformists, it is interesting that one of the amendments made in the Lords to the Schism Bill stated:

> That this Act shall not extend, to be construed to extend, to any Person, who as Tutor or Schoolmaster, shall instruct Youth in Reading, Writing, Arithmetick, or any such Mathematical Learning only, so far as such Mathematical Learning relates to Navigation; or any Mechanical Art only. . . .[84]

Clearly, the opposition of the mathematical community proved important – especially as it could stand on the defense of trade and the Dissenting interest.[85]

The Schism Bill had the odor of desperation. Even among the High Churchmen there had been substantial nervousness that Atterbury, as

"Guild Organisation and the Instrument-Making Trade, 1550–1830: The Grocers' and Clockmakers' Companies," *Annals of Science*, XXXVI (1979), 16, 27.

[81] In Ditton's *An Institution of Fluxions* (London, 1706); HLRO, Petitions, Box 195B, n.s. X, 3123 (g).

[82] Dr. Williams's Library, anon., *A Letter to a Member of Parliament*, in *A Collection of all the Papers that have been Giv'n to the Lords and Commons, in Relation to the Bill to Prevent Schism* (London, 1714), p. 9.

[83] *Journals of the House of Commons*, XVII, pp. 677–8, 686.

[84] *Journals of the House of Commons*, XVII, p. 698 (June 23, 1714).

[85] De Krey, *A Fractured Society*, pp. 99–102.

Prolocutor, was carried away in his hunt for heresy – and that his motives were less than pure. The Tory Lord Treasurer Oxford, with his Dissenting background, was horrified.[86] But Low-Church, Latitude, and Dissent had formed a powerful force to deflect the bile of zealous and ambitious clergymen like Atterbury. For the High-Church, time was running out and Tory intrigues between Bolingbroke and Harley threatened to erupt into open warfare. As Anne's life slowly slipped away in the heat of the summer, the specter of a Hanoverian, and Whig, Succession was more real than ever. For Whiston, this finally promised rescue from the maws of the beast. But even for the Newtonians, this might be a mixed blessing, for Hanover meant not only George, but also Leibniz.

Jonathan Swift and the Tory compass

The scientific spirit that Newtonian lecturers so happily exploited was not immune from the politics Swift also neatly dissected. For Swift, "party" was the real ground of the projects he had seen unfold; in this, the Longitude Act was no exception. It had been the Whigs who had foist it on a House too preoccupied with other, more obviously political, matters. But Swift saw the pattern, and sensed a plot. The failure of Desaguliers' teacher John Keill to secure the Sedelian Professorship at Oxford in 1704 had been attributed to Whig, Low-Church interference.[87] The tale was repeated a few years later upon the death of David Gregory, the Savilian Professor of Astronomy at Oxford. Although Keill made an attempt at it, Gregory was succeeded by Edmund Halley despite what Flamsteed called Halley's "vile character." It was Swift's conclusion that "... Party reaches even to Lines and Circles, and he [Keill] will hardly carry it being reputed a Tory, wch yet he wholly denyes."[88] Keill had to wait until 1712 and, as though Swift was right, the zenith of Tory influence.

Swift cannot be seen as an unregenerate opponent of science. What troubled him was not purely the politics, but the corrosion of society at large that inevitably followed the agonizing collapse of projects that crushed the unsuspecting. When natural philosophers like Whiston were connected with schemes that seemed never to bear fruit, Swift gave no

86 Bennett, *The Tory Crisis in Church and State*, p. 177.

87 Anita Guerrini, "Newtonian Matter Theory, Chemistry, and Medicine, 1690–1713" (Ph.D. thesis, Indiana University, 1983), p. 158.

88 Cudworth, *Life and Correspondence of Sharp*, p. 96. Flamsteed to Sharp, March 24, 1709; Williams, ed., *Correspondence of Jonathan Swift*, I, p. 121. Swift to Robert Hunter, January 12, 1708/9.

quarter.[89] This was especially the case in *Gulliver's Travels*, published in the aftermath of the South Sea Bubble. But even there, Swift's alarm did not entirely sound true to Tory ears. Dr. John Arbuthnot, who had met Swift in 1711, had also seen the link between notions of improvement in the Baconianism of the eighteenth century and the influence of Dissenters. But he was nervous about the caricature in *Gulliver's Travels*. This was something about which Arbuthnot had more than a passing knowledge, having become active in the affairs of the African Company through his connection with the Duke of Chandos.[90] His interest in such financial matters, even in the aftermath of the South Sea Bubble, in addition to his familiarity with the Royal Society, induced a more dispassionate view as he wrote to Swift in 1726, "I tell yow freely the part of the projectors is the least Brilliant."[91] Nevertheless, it is clear that Swift reflected the views of many Tory sympathizers such as Thomas Baston, who, in 1716, had complained in his *Thoughts on Trade and a Publick Spirit* of the vast number of schemes, "numerous as Egyptian Locusts . . . as if the Devil had a mind to deal with us by wholesale. . . ."[92]

No more remarkable example can be found of the satirical weapon as social criticism than in the hands of Jonathan Swift. His railing against the pestilence of projectors, freethinkers, and philosophers had a broad horizon that took in the breadth of Jeremy Thacker's Latitudinarians and Longitudinarians. What makes Swift so powerful is the paranoic edge of a man personally embittered by the forces upon which Whig philosophers like Whiston had come to rely. By the time of the Longitude Act, such special interests were already a target for ridicule. In 1712, Swift had complained to Archbishop William King about a projector who had asked him to assist in presenting a petition to Bolingbroke for a means of locating longitude. As far as Swift was concerned, it was "a Thing as improbable as the Philosopher's Stone, or perpetual Motion."[93] Newton might have agreed. There is no reason to suppose that Swift was referring directly to Whiston at this point; there were numerous such projectors. Never-

89 G. S. Rousseau, "Science," in Pat Rogers, ed., *The Eighteenth Century* (London, 1978), pp. 160–2. Cf. Olson, "Tory High-Church Opposition to Science," pp. 185–91.

90 Olson, "Tory High-Church Opposition to Science," pp. 177–9; L. Stewart, "The Edge of Utility: Slaves and Smallpox in the Early Eighteenth Century," *Medical History*, XXIX (1985), 63.

91 Williams, ed., *The Correspondence of Jonathan Swift*, III, p. 179. Arbuthnot to Swift, November 5, 1726; Marjorie Nicolson and Nora M. Mohler, "The Scientific Background to Swift's *Voyage to Laputa*," in Marjorie Nicolson, *Science and Imagination* (Ithaca, N.Y., 1956), pp. 152–3.

92 [Thomas Baston], *Thoughts on Trade and a Publick Spirit* (London, 1716), p. 10.

93 Harold Williams, ed., *The Correspondence of Jonathan Swift* (Oxford, 1963), I, p. 295. Swift to King, March 29, 1712.

theless, Swift had long associated the Royal Society with ephemeral schemes and found them as threatening as notions of latitude in religion with which the Society had long been tarred. He had no use for Society apologetics, noting once that if Joseph Glanvill's "Scepsis Scientifica comes to me I'll burn it for a fustian piece of abominable curious Virtuoso stuff."[94] Swift was hardly alone in his views of the Society; in 1700, Hans Sloane had been the target of a satirical pamphlet written anonymously by the controversialist William King, who had already attacked Richard Bentley in the so-called "Battle of the Books."[95] By this time, Whiston had become infamous for his efforts in the conflict that had followed the publication of Thomas Burnet's *Sacred Theory of the Earth*. Whiston and Halley had both suggested that changes in the history of the earth could be accounted for by the near approach of a comet.[96] Comets overturning Scripture and the lunacy of the longitude was all too much for the mind to bear.

Swift represented a rear guard action against those who would change his world. And it soon developed into full-scale war. The remarks upon Whiston's proposals for the longitude do not represent a narrow or specific attack, but rather one that was based on the defense of revelation against the purveyors of natural religion and the protection of society against scurrilous projectors. Nor would the politics of the Act have escaped Swift. On the day that it passed into law he wrote to Arbuthnot, "the Commons offer 100000 [for the capture of James] If I was the Pretender I would come over my self, and take the money to pay my Troops. They had better Put out a Proclamation that whoever discovers the Pretender or the Longitude shall have 100000. . . ."[97]

It was Swift's particular genius to be able to sense the madness of his age in the lust for projects coexisting with the crumbling foundations of revealed religion. Here was a real awareness that there were tides building in the political world that had not only shaken James from his throne, but that would now alter the foundations of the social world. His alarm was deeply rooted, perhaps in personal fears still ephemeral, but given substance by the projectors and the Low Churchmen whose tolerance extended to Dissent and Occasional Conformity. So Whiston was both a Longitudinarian and a Latitudinarian, and a symbol of alarm. Much

[94] *Correspondence of Jonathan Swift*, I, p. 30. Swift to Rev. John Winder, January 13, 1698/99; see also Tuveson, "Swift and the World-Makers," p. 73.

[95] Levine, *Dr. Woodward's Shield*, pp. 86–91, 247–50; Dorothy Stimson, *Scientists and Amateurs. A History of the Royal Society* (New York, 1968), p. 143.

[96] Tuveson, "Swift and the World-Makers," p. 67.

[97] *Correspondence of Jonathan Swift*, III, p. 47. Swift to Arbuthnot, July 3, 1714.

has been written of the specificity of Swift's satire – a blunter weapon would have been much less effective – but it cannot be denied that Swift saw the enemy in every alley. In Exchange Alley, especially, the swarms of locusts were bred that would plague a Christian nation. But he was no Lear, howling into a hurricane of lunacy and vice. His remarkable *Gulliver's Travels* with its nuances about the mechanics and projects of the Academy of Lagado was a broadside against both the Royal Society and the wider commercial world.[98]

At the end of the reign of Anne, public science was of sufficient weight to draw attention to the utility Thomas Sprat had once employed in defense of the Royal Society. Practicality and comprehension had met in the coffeehouses. The Newtonians came to reflect the broad changes within British society in the early eighteenth century – and also to give substance to the fears of Jonathan Swift and Roger North. Latitude had clearly not meant lassitude in matters of religion or politics. The sympathies of Newtonian theologians as William Whiston and Samuel Clarke with the Dissenting community reflected a relationship already recognized among the instrument makers, mathematical teachers, and lecturers of experimental philosophy.

[98] Rogers, "Gulliver and the Engineers," pp. 260–1; Philip Harth, "The Problem of Political Allegory in *Gulliver's Travels*," *Modern Philology*, LXXIII (May 1976), Supplement, pp. s.40–s.47. On similar views of Pope, see Brean S. Hammond, *Pope and Bolingbroke: A Study of Friendship and Influence* (Columbia, 1984), p. 83.

7

Desaguliers and the usefulness of philosophers

In the natural philosophy of Newtonians like Desaguliers, there was a residual Baconianism hidden and submerged by the contemporary reputation of Newton himself. Newton's philosophical grandeur obviously represents a pantheon at which more than his contemporaries have since worshiped. At first glance, all else seems in eclipse in the early eighteenth century. But we should not assume that other philosophers were dismissed out of hand by the early lecturers of natural philosophy. In 1731, Peter Shaw gathered together a large number of booksellers to publish by subscription a new edition of Bacon's works. Here Shaw declared what many had obviously come to accept: that " . . . Theory and Practice, as we somewhat abusively distinguish them, are not two, but one inseparable Thing."[1] The Baconian polemic remained impressive in the eighteenth century and it became more meaningful as the century wore on.

Making connections: The early career of Desaguliers

The career of John Theophilus Desaguliers straddled the realms of landowners and industrialists. A son of the Huguenot diaspora, educated by

[1] B.L., Add. MSS. 4258. Peter Shaw, "Proposals for Printing, by Subscription, the Philosophical Works of Francis Bacon," p. 4, March 25, 1731. See also Peter Shaw, M.D., *Three Essays in Artificial Philosophy, Or Universal Chemistry . . .* (London, 1731), p. 9. Shaw was firmly rooted in the eighteenth-century link between philosopher and artisan where " . . . if Things were in their proper Chanel; all Arts, Inventions, and Works, should flow from natural Philosophers." Shaw, ed., *The Philosophical Works of Francis Bacon*, quoted in Golinski, "Peter Shaw: Chemistry and Communication in Augustan England," *Ambix*, XXX (March 1983), p. 24. On the rhetorical Baconianism of Newton's editors, see Antonio Perez-Ramos, *Francis Bacon's Idea of Science and the Maker's Knowledge Tradition* (Oxford, 1988), pp. 16–18.

his clergyman father, who opened a school at Islington, near London, Desaguliers went up to Oxford, where he came under the influence of John Keill. Too much the product of the Protestant cause to be affected by the High-Tory cant of Christ Church, Desaguliers had no truck with the Oxford Jacobitism of the latter part of Anne's reign. His sentiments were entirely with the supporters of the Hanoverian Succession. Desaguliers took deacon's orders from Henry Compton, the Bishop of London, in 1710, although he remained at Oxford until his marriage in 1712.[2] His added responsibilities caused him to loosen the bounds that tied him to an academic life and he risked all in the rage for natural philosophy in London. His move into the metropolis was a dramatic change in the direction of his life and it was here that Desaguliers faced the dialectic that reshaped British society.

As a lecturer, Desaguliers was enormously successful, a remarkable feat when one considers how many attempted to secure at least part of their livelihood from the practice. Of course, like John Harris, with whom he collaborated from time to time, Desaguliers was also a clergyman and ever a small living promised to provide for body and soul. But the last thing Desaguliers wanted was to be expected to provide much in the way of religious service. He understood, as Harris had also recognized, that patronage was the predecessor of religious preferment. Indeed, it is conceivable that Desaguliers' presentation in 1717 to the living of Bridgeham in Norfolk in the gift of William Lord Cowper, then Lord Chancellor, was the result of his association with Harris, whom Cowper had long patronized. By far the most important of his clerical appointments, however, came from the largess of James Brydges, F.R.S., later the Duke of Chandos, who, as Earl of Carnarvon, appointed Desaguliers his personal chaplain in 1716. This made a great deal of sense for Brydges, who had an extensive interest in the affairs of the Royal Society and an increasing passion for things mechanical, especially if they might turn a profit. In 1718, Desaguliers obtained his LL.D. from Cambridge, seemingly indicating a hope for further ecclesiastical advancement. He was rewarded in 1719 when Brydges appointed Desaguliers rector of Whit-

[2] There is a biographical sketch corrected by Desaguliers in 1736 in Bodleian Library, MS. Rawlinson J, fol. 3. See also A. Rupert Hall, "Desaguliers," *Dictionary of Scientific Biography*, IV (New York, 1971), p. 45. The standard sources on Desaguliers' life are badly outdated. See Paul Richard Major, "The Physical Researches of J.T. Desaguliers" (M.Sc. thesis, University of London, 1962); W. R. Hurst, *An Outline of the Career of John Theophilus Desaguliers, M.A., LL.D., F.R.S.* (privately printed, 1928), which seems largely based on the Bodleian MS.; and Margaret E. Rowbottom, "John Theophilus Desaguliers (1683–1744)," *Proceedings of the Huguenot Society of London*, XXI (1968), 196–218.

church or Stanmore Parva in Middlesex. By this point, it must have become obvious that it was not the religious employment that interested Brydges, who was convinced that Desaguliers was "certainly the best Mechanick in Europe." Like Robert Hooke before him, Desaguliers' contemporary reputation was to be made not so much as curator of experiments, but as upon his mechanical skill.[3]

We cannot now be certain why Desaguliers originally commended himself to James Brydges, although by 1716, the philosopher had made himself useful in the presentation of experiments before the Royal Society. In the summer of 1714, when he was elected a Fellow as a protege of Keill, Desaguliers had reproduced experiments to verify Newton's theory of colors and since then had obtained some income from the Society as had his predecessor Francis Hauksbee. In November 1716, shortly after Brydges had appointed him chaplain, Desaguliers was once again making important experiments at the request of Newton.[4] Clearly, by 1716, Desaguliers was a man of increasing reputation. But he was employed by Brydges not in the making of crucial experiments and only marginally in his function as a clergyman.

Understanding where Desaguliers fit in the rush to the improving spirit is best obtained by seeing where his patron stood at the end of the seventeenth century. By 1694, Brydges was a Fellow of the Royal Society and, in 1698, elected M.P. for Hereford, a promising and ambitious descendant of both landed gentry and the merchant class.[5] The brief diary we have of his early years in London from 1697 to 1702 reveals the range and the convergence of his interests. Here he followed, quite literally, in the footsteps of Hooke. He often dined with the Royal Society at Pontack's in Abchurch Lane between visits to Garraway's Coffee House in Exchange Alley to follow the progress of stocks before proceeding to the meetings of the Society in Gresham College with Hans Sloane or Dr. John Woodward. Probably in Garraway's, he became an investor in Dr. Hugh Chamberlen's land bank scheme, which was only one of the variety of financial fancies that he experienced during his lifetime.[6] It was into

[3] Hunt. Lib., Stowe MSS. 57, XV, f. 252, June 16, 1718. Chandos to Mr. Wm. Mead (his banker); Bodl. MS., Rawl. J., f. 3. See John Aubrey's comment on Hooke, quoted in Steven Shapin, "Who was Robert Hooke?" in Michael Hunter and Simon Schaffer, eds., *Robert Hooke. New Studies* (Woodbridge, 1989), p. 253.

[4] Henry Guerlac, *Essays and Papers in the History of Modern Science* (Baltimore, 1977), p. 124.

[5] The most complete examination of his life is C. H. Collins Baker and Muriel I. Baker, *The Life and Circumstances of James Brydges First Duke of Chandos* (Oxford, 1949).

[6] Hunt.Lib., Stowe MS. 26. James Brydges, First Duke of Chandos, "A Journal of My Daily Actions, 1697–1702." 2 vols., *passim*; Shapin, "Who was Robert Hooke?" p. 261.

this world that John Theophilus Desaguliers was soon drawn. There would emerge a history of much greed and just as much skullduggery as financial imaginations could devise.

From time to time, Desaguliers had need to set his course through Brydges' tangle of mechanics and finance. By 1717, Desaguliers was lecturing before the Royal Society on the breadth of the experimental program, from electricity and the barometer to improvements in the Savery engine. At this point, Brydges displayed an active interest in natural philosphy. He borrowed the Society's 120-foot object glass in order to construct a tube to house it and a mechanism whereby it might be raised. At the same time he clearly saw possibilities for his own estates in employing the talents of Desaguliers. By the end of 1718, Desaguliers, who had established a close friendship with the engineer Henry Beighton, erected his new improved engine at Brydges' estate at Canons.[7] In these circumstances, Desaguliers' receipt of the living of St. Lawrence's, Whitchurch in 1719 from the new Duke of Chandos, did not relieve the philosopher of what came to be the most essential aspects of his association with the Duke. On the contrary, the security of a clerical sinecure allowed him to attend to the matters that most interested him – natural philosophy and the possibilities for profit. Functioning as a scientific or technical consultant to the Duke, Desaguliers became deeply embroiled in the numerous investments of his benefactor.

During the early 1720s, Chandos was much concerned to find ways to rescue the failing fortunes of the African Company. The slave trade had not met investors' expectations. One of his notions was the importation of African potash to lay the foundation for a soap manufactory. Desaguliers was frequently asked for advice on the likelihood of success of the project and was assigned the task of liaison with Colonel Samuel Horsey, the Duke's agent in the York Buildings Company.[8] At this point, the York Buildings Company was also giving some thought to exploiting pine forests on estates recently purchased in Scotland. Production of potash in Scotland became the focus of much of their attentions to ensure the success of the soap project and Desaguliers was given the task of assisting or, at least, monitoring the progress with Col. Horsey. Chandos put great store in these possibilities; at least he wanted to try the Scottish experiment before he went to the expense of sending a ship to Gambia for the African potash.[9]

[7] Royal Society MSS., Journal Book (C), XI (1714–20), pp. 162ff. (February 28, 1716/17); p. 178 (May 2, 1717); pp. 269–70 (December 4, 1718).

[8] Hunt. Lib., Stowe MSS. 57, XXIII, f. 33, Chandos to Desaguliers, October 24, 1723.

[9] Baker and Baker, *Life and Circumstances of James Brydges*, pp. 230, 290; see also Hunt. Lib., Stowe MSS. 57, XXIV, f. 256. Chandos to Desaguliers, September 27, 1724.

The patronage of a Duke

Desaguliers' role in the schemes of the York Buildings Company was characteristic of the kind of relationship he had with the Duke. The philosopher was regarded as a person upon whom Chandos could rely for sound technical advice, especially on the many possibilities that confronted a man of great wealth. One of the Duke's major concerns was with the mining and refining of ores not only on his estates, but on the lands owned or leased by some of the companies in which he invested.[10] During the summer of 1732, while Desaguliers was in Holland, he was instructed to inquire into a method used in the refining of gold for fixing quicksilver that had been discovered in the Netherlands by Baron Silburgh, who lived near the Hague. Although Desaguliers was warned that the Baron "drinks prodigious hard," he was asked to try to convince the Baron "either to sell the Secret, or admit a partner into the participation of it."[11] This was a typical approach of Chandos. Whatever transpired, it didn't meet expectations, for the Duke continued his search for the elixir. In 1734, Desaguliers was asked to consider what would happen to quicksilver surrounded by a ball of bone ash in the extreme heat of a furnace. Evidently, Chandos wished to undertake the experiment with hope of finding a way of best utilizing mercury in the refining of gold ores. Desaguliers even referred to him an account of the methods used by the Spanish in Chile that had been printed in the *Philosophical Transactions*.[12]

The relationship was, fundamentally, a marriage of convenience. Desaguliers was a frequent visitor at the Duke's estate at Cannons in Middlesex. In the early years of their arrangement, when Desaguliers' advice was frequently sought over the improvement of the estate, the Doctor was at Cannons at least once a week.[13] For various reasons, over the more than a quarter century they knew one another, their association became more and more stormy. This is not surprizing. A major cause of friction was the constant financial difficulties in which Desaguliers found himself.

Despite these troubles, Desaguliers was enormously successful as a public lecturer. This provided him with a considerable income that was supplemented by the demonstrations for which he was paid by the Royal

[10] It was likely because of the efforts of the York Buildings Company that Chandos consulted his solicitor about the rights of a lessee in the event of the discovery of copper or coal in sufficient quantity to mine. Hunt. Lib., Stowe Brydges MSS., Legal, Box 1. Opinion of J. Jones, March 9, 1725.

[11] Hunt. Lib., Stowe MSS. 57, XL, f. 9, Chandos to Desaguliers, August 25, 1732.

[12] Hunt. Lib., Stowe MSS. 57, XLIV, f. 229, Chandos to Desaguliers, July 4, 1734; f. 264, Chandos to Desaguliers, July 19, 1734.

[13] Hunt. Lib., Stowe MS. 59, "A Book of Strangers" dining at Cannons, 1721–2.

Society, although not without some controversy. Nevertheless, this does not appear to have allowed Desaguliers to maintain the style he wished, especially, perhaps, because of his desire for a presence at Court. Whether he obtained additional support from his Royal patrons is not certain, but it would seem likely. In any case, Desaguliers evidently had an income that often failed to meet his needs. The constant reach for additional revenue was responsible for the many projects in which Desaguliers became involved, and it also explains why he was so unsuccessful as a clergyman. The problem was not mammon, but his inability to actually meet all of the demands on his time.

The living at Whitchurch created constant headaches, some of them certainly of Desaguliers' own making. Disputes with the tenants festered for years, with Desaguliers attempting to lease the tithes and ultimately threatening to take legal action against the delinquent. This was not exactly what Chandos had hoped for. By 1733, the affair had degenerated into a comic opera. In March, in obvious anger, Chandos blasted Desaguliers for his conduct in the dispute:

> ye Sum is far from being so considerable as one would have imagined, from ye clamour you have made; & considering ye neglect of Duty on your part wch: they complain of, I wonder that they pay you so well; for I am credibly informd, that a dead Body has lain in ye Church this last week for three days together without Christian Burial, & tho notice was sent to you of it last Wednesday neither you nor your Curate thought fitt to attend till today.[14]

For a few years, probably because they were both so heavily engaged in one technical project after another, relations between the two improved. But these ventures produced other tensions. Nevertheless, Desaguliers was clearly preoccupied with technological ventures and with his public lectures. The result was predictable.

The erosion of the relationship between Desaguliers and Chandos followed increasing frustration at Whitchurch. By the spring of 1739, it was evident that Desaguliers had done very little to rectify the situation as Chandos frequently heard complaints of the Doctor's neglect. In March 1739, Chandos informed Desaguliers that he would take legal action in the Church courts if the Doctor did not immediately appoint a curate to take care of the parish. He was incensed by complaints of the Church

[14] Hunt. Lib., Stowe MSS. 57, XLI, f. 185, Chandos to Desaguliers, March 11, 1732/3; Baker and Baker, *Life and Circumstances of James Brydges*, p. 152; Hunt. Lib., Stowe MSS. 57, XXXIX, f. 207, Chandos to Desaguliers, May 10, 1732; XLI, f. 203, Chandos to Desaguliers, March 20, 1732/3; and XLVIII, fols. 196–7, Chandos to Desaguliers, February 1, 1736/7.

wardens that "ever since the 6th. day of Novr. there has been no settled Minister to officiate in the Parish insomuch that the Inhabitants & Officers of it have been forced to go a begging to other Ministers to bury their dead," and lectured, "This is a very shameful neglect & what I have more than once complained to you of."[15] Chandos had reached his limit, or so it seemed. Desaguliers quickly agreed to remedy matters by appointing a reliable curate.[16]

This resolution did not, however, solve the essential problem of Whitchurch – which was Desaguliers' neglect. Apart from an enquiry made by the Duke's secretary in 1740 about the disposal of some iron railings from the estate at Cannons, the only direct communication that survives is the polite refusal in 1741 of Chandos to have anything to do with some obscure scheme that Desaguliers had apparently devised.[17] It was a sad, and rather tired, end to an association that had been at the focus of some of the most remarkable ventures of preindustrial Britain.

Desaguliers' enthusiasm for natural philosophy obviously obstructed his clerical duties. In some circumstances, this mattered very little; there seems to be little indication of religious interests even after his presentation to livings at Bridgeham, Norfolk, and to Little Warely in Essex. Likewise, his appointment as chaplain to Frederick, Prince of Wales as well as to Bowle's Regiment of Dragoons, speaks more of important patrons than of theological virtuosity. Of Desaguliers' religious passion, there is little evidence except for the preface he wrote to John Chamberlayne's translation of *The religious philosopher*, a work of natural religion by the Dutch philosopher Niewentyt.[18] Natural philosophy was the central fact of Desaguliers' life. This had been the case since 1713, when he had come up to London and given courses for Hauksbee's widow. Through his public lectures and his role as curator of experiments to the Royal Society from 1714 he was able to establish the reputation that brought him to the attention of wealthy and powerful patrons like William Cowper and

[15] Hunt. Lib., Stowe MSS. 57, XLI, fols. 131–2, Chandos to Desaguliers, March 20, 1738/9.

[16] Hunt. Lib., Stowe MSS. 57, XLI, fols. 137–8, Chandos to Desaguliers, March 22, 1738/9; f. 145, Chandos to Revd. Mr. Perkins, March 26, 1739; fols. 152–3, Chandos to Revd. Dr. Leigh, April 6, 1739; fols. 230–1, Chandos to Leigh, June 3, 1739; LI, fols. 249–50, Chandos to Desaguliers, June 14, 1739; also XLVII, f. 227, Chandos to Desaguliers, June 3, 1736

[17] Baker and Baker, *Life and Circumstances of James Brydges*, p. 161; Hunt. Lib., Stowe MSS. 57, LI, f. 19, John Pudsey to Desaguliers, October 25, 1740; LV, f. 212, Chandos to Desaguliers, August 28, 1741.

[18] Rowbottom, "John Theophilus Desaguliers (1683–1744)," 196–218; see also *Gentleman's Magazine*, VIII (1738), 325.

James Brydges.[19] The clerical livings he thereby obtained merely provided the opportunity to engage the attractions of natural philosophy.

In Desaguliers' mind, there was little barrier between the subjects that he pursued in his public lectures and those that he examined before the Royal Society. Desaguliers involved himself in the pursuit of answers to some of the fundamental philosophical issues raised by Newton. His explorations of the theory of the composition of light, or of electricity, would substantiate that. The overwhelming presence of Newton and his lasting influence tends to mask the fact that disciples like Desaguliers took delight in the Queries to the *Opticks* and, after 1713, expanded the experimental doctrine of the *Principia*. The records of the Royal Society reveal the exceptionally large number of experiments that Desaguliers performed. Some were experiments of importance in the defense of the Newtonian philosophy, such as those regarding the refraction of light.[20] Generally, his public lectures and the demonstrations suggest only an occasional difference in sophistication from those given before the Royal Society.

Between experiment and invention

If there was a gap between theory and practice, Desaguliers bridged it. Desaguliers' public philosophy, demonstrated by his numerous popular lectures, moved into areas of more direct mechanical interest than required by his Royal Society obligations. It is interesting, for example, that in 1716, when Willem s'Gravesande attended his course of experimental philosophy, Desaguliers was using a copy of John Harris's *Lexicon Technicum* in which there was described a Savery engine. Desaguliers later claimed that their discussions led to improvements in the engine, although by the time he made the claim, it was the Newcomen engine that was much sought after.[21] Moreover his experiments on friction and weight, witnessed by the King and the gentlemen of the Court in 1716, might reveal the imperfections and impossibilities of machines. Obviously, experimental philosophy, in the mind of Desaguliers, might be closely tied

[19] Rowbottom, "The Teaching of Experimental Philosophy in England, 1700–1730," *XIe congres international d'histoire des sciences (1965)*, Actes IV (1968), 51.

[20] J. L. Heilbron, *Physics at the Royal Society during Newton's Presidency* (Los Angeles, 1983), pp. 92–3. For Desaguliers' experiments, see Royal Society, MS. Register Book; B.L., Sloane MSS. 4049, f. 215, Desaguliers to Sloane, August 28, 1728; and Simon Schaffer, Simon, "Glass Works: Newton's Prisms and the Uses of Experiment," in David Gooding, Trevor Pinch, and Simon Schaffer, eds., *The Uses of Experiment. Studies in the Natural Sciences* (Cambridge, 1989), pp. 92–8.

[21] Desaguliers, *A Course of Experimental Philosophy*, II (London, 1744), pp. 484–5.

to technological improvement. By the early eighteenth century, it is difficult to imagine a Royal Society that was entirely averse to explorations of technological matters.

Nor was experimental philosophy an obstacle to entertainment. While Desaguliers and Harris jointly performed experiments on Villette's burning glass for the benefit of the Royal Society, it was later shown to the general public for a small fee before Villette returned to Liege. Francois Villette was first introduced to the Royal Society in the spring of 1718. For a year and a half, his concave mirror was used by a number of the Fellows in experiments. Desaguliers burned chalk and showed the remains of an emerald that had been melted before an audience at Whitehall.[22] Likewise, at the request of the King in 1716, Desaguliers performed experiments to explain feats of strength in the markets, which some noblemen were also inclined to attempt.[23] All of these efforts served to unlock the door of the laboratory.

Desaguliers was a man forever in search of an experiment. Once on a barge on the Thames during a day of celebration in the company of Colonel Samuel Horsey, another crucial figure in the Chandos connection, Desaguliers was lighting water rockets. These were designed to burn under water and then to surface with spectacular effect. Unfortunately, a barge of overly curious gentlemen and ladies came close by and one of Desaguliers' rockets rose underneath their vessel, blowing a hole in the stern. The result was a frantic rescue as the oarsmen desperately rowed for shore near Mortlake and the barge filled with water. In near disaster, Desaguliers saw opportunity surface. He had the notion to try an experiment on the resistance of water to compression by exploding charges

[22] These public displays induced the image of the "catoptrical Victuallers," who captured the sun to roast an ox. Hence, even cooks and maids, it was suggested, would soon be required to study optics and astronomy. So as not to undermine the price of coal, a tax was proposed upon those confining sunbeams for cooking. [John Arbuthnot], *To the Right Honourable the Mayor and Alderman of the City of London: The Humble Petition Of the Colliers, Cooks, Cook-Maids, Blacksmiths, Jackmakers, Brasiers, and Others* [1716], in [Jonathan Swift], *Miscellanies*, Vol. III (London, 1732), pp. 72–8.

[23] Royal Society MSS., Journal Book (C), XI (1714–20), p. 241 (May 5, 1718); p. 266 (November 20, 1718); pp. 285ff. (January 29, 1718/19); p. 344 (May 7, 1719); p. 371 (June 25, 1719). In 1719, Desaguliers reported to the Society on Villette's invention of an artificial magnet. Journal Book (C), XI (1714–20), pp. 393–4 (November 5, 1719). See *A description of the great burning-glass made by Mr Villette and his two sons* (London, 1718); Desaguliers, *Course of Experimental Philosophy*, second edition (London, 1745), I, pp. 281–2; see also pp. 70–1, 138–43; *Daily Courant* (Thursday, July 16, 1719). These experiments on combustion were generally regarded to establish the providence of God in the provision of heat from such a great distance. John Laurence, *A New System of Agriculture. Being a Compleat Body of Husbandry and Gardening* (London, 1726), pp. 18–19.

of gun powder beneath the surface of a pond, causing a great shock to those recruited to stand nearby.[24] In the three decades since his election as a Fellow in 1714, Desaguliers constantly provided the Society with such experiments. This was the case except during the early 1730s, when he was in Holland or when he was racked by the successive fits of gout that often plagued him during winter.[25]

The range of Desaguliers' experiments ran from electricity to innumerable mechanical contrivances. Many of these were reported in the *Philosophical Transactions*, in which, frequently, the practical consequences of the various mechanical and hydrostatical problems were discussed. He reported on difficulties encountered in water pipes from a reservoir on Chandos's estate, upon experiments that revealed the constriction resulting from air, and a new machine to clear the air from pipes, called a Jack in a Box, that he claimed had been invented by a group including himself and, among others, Thomas Newcomen.[26] Throughout his career in the Royal Society he published various reports on his mechanical inventions such as a machine to clear damp and foul air out of mines, a contrivance that he invented with the Reverend Stephen Hales for measuring depth at sea, or a calculation of the worth of centrifugal bellows, as he called them, that might be of use in mines or in ships to clear the air below decks.[27]

[24] Desaguliers, *A Course of Experimental Philosophy*, second edition (London, 1745), I, pp. 420–4. This might be seen within the broader context of the resistance of fluids upon which Desaguliers offered experiments. See *Philosophical Transactions*, no. 367 (Jan.–April, 1721), 142–4.

[25] Royal Society MS., Letter Book D, no. 71, Desaguliers to Cromwell Mortimer, November 8, 1735; no. 73, Desaguliers to Mortimer, October 29, 1733; Hunt. Lib., Stowe MSS. 57, XXXIX, fols. 265–6, Chandos to Senex, June 28, 1732; f. 272, Chandos to Sir John Meres, July 6, 1732; XL, f. 9, Chandos to Desaguliers, August 25, 1732; W. D. Hackmann, "The Growth of Science in the Netherlands in the Seventeenth and Eighteenth Centuries," in Maurice Crosland, ed., *The Emergence of Science in Western Europe* (New York, 1976), p. 100; and Johan A. Van Reijn, "John Theophilus Desaguliers, 1683–1983," *Thoth*, V (1983), 165–204.

[26] Desaguliers, "An Account of several Experiments concerning the Running of Water in Pipes, as it is retarded by Friction and intermixed Air," *Philosophical Transactions*, XXXIV, no. 393 (March-April 1726), 77–82. For a discussion of his early experiments with Lowthorp, see his amendments to his translation of *The Motion of Water, and other Fluids. Being a Treatise of Hydrostaticks, Written originally in French, by the late Monsieur MARRIOTTE, Member of the Royal Academy of Sciences at Paris* (London, 1718), p. 289.

[27] Desaguliers, "An Attempt made before the Royal Society, to shew how Damps, or foul Air, may be drawn out of any Sort of Mines, . . . ," *Philosophical Transactions*, XXXV, no. 399 (July–Sept. 1727), 353–6; "An Account of a Machine for measuring any Depth in the Sea, with great Expedition and certainty . . . ," *Philosophical Transactions*, XXXV, no. 405 (November 1728), 559–62; "A Calculation of the Velocity of the Air moved by the new-invented Centrifugal Bellows . . . ," *Philosophical Transactions*, XXXIX, no. 437

The temptation to reduce invention to the work of mere mechanics, as one might describe Thomas Newcomen or his associates, must be resisted. More significantly, this presumed distinction was one that Desaguliers himself sought to overcome. Neither Desaguliers nor Stephen Hales were mere empirics; they were both sophisticated experimentalists who had been deeply influenced by the work of Newton, most especially by his *Opticks*.[28] Neither felt invention to be something to disdain; in fact, they gloried in the possibilities of the solution of technical problems such as those that inflicted the mining industry.

This attitude was widespread and it is reflected in the view that natural philosophers had much to offer to mechanical invention. To focus on this attitude rather than upon the infrequent successes of early eighteenth-century inventors would seem to be an improvement on the smug disparagement of their many failures. For example, Desaguliers' friend Henry Beighton, who was involved in mining though better known as the editor of the *Ladies Diary*, saw great possibilities in the link between natural philosophy and mechanics:

> We generally see, those who pretend to be Engineers, have only guess'd, and the Chance is, they sometimes succeed; else they have made them like others that have done pretty well. But he who has skill enough in Geometry, to reduce the Physico-Mechanical Part to Numbers, when the Quantity of Weight or Motion is given, and the Force designed to move it, can bring forth all the Proportions, in a Numerical Calculation, so as it may be almost impossible to Err.[29]

Such ideals were increasingly common. Even a mere tradesman, like the exceptionally well-read gardener Stephen Switzer, lamented that some had distained those mechanics who might seek enlightenment in philosophy. It was the separation of theory and practice, he maintained, that made it possible for gentlemen to be overcome by the interminable frauds and fancies even in the excessive claims of some gardeners. And, he acknowledged, "there is too often some awkward, ignorant, I might add dishonest Person and Pretender to a great deal of Knowledge inn this Matter at hand, and ready to offer their Advice, tho' never so weak and

(April-June 1735). On Desaguliers and Hales, see D. G. C. Allen and R. E. Schofield, *Stephen Hales: Scientist and philanthropist* (London, 1980), *passim*; and Arnold Zuckerman, "Disease and Ventilation in the Royal Navy: The Woodenship Years," *Eighteenth Century Life*, XI (November 1987), 81–4.

28 Allen and Schofield, *Stephen Hales*, pp. 39–41; and I. Bernard Cohen, *Franklin and Newton: An Inquiry into Speculative Newtonian Experimental Science and Franklin's Work in Electricity as an Example Thereof* (Philadelphia, 1956), pp. 246ff.

29 Quoted in A. E. Musson and Eric Robinson, *Science and Technology in the Industrial Revolution* (Toronto, 1969), p. 49.

false."[30] Now almost entirely ignored, Switzer was a sophisticated observer of the tension between the trades and natural philosophy so that in his treatise on hydraulics in 1729, he argued that "it must be own'd . . . , that how greatly soever Gentlemen and others, that have often Occasion of these Things, are vers'd in the Theory; yet the Practice is somewhat difficult, nor is it so well known to them, as to judge when they are well or ill us'd, or whether their Undertakings are likely to succeed or not."[31]

These pronouncements, of Beighton and of Switzer, demonstrate the influence of Desaguliers, whose intent was not to censure the mechanics and so-called engineers of his day. Rather, they were determined to see mechanics and their methods assisted by the knowledge that natural philosophy could provide. Undoubtedly, Desaguliers would have been the beneficiary if his lectures could be demonstrated to be of value to mechanics and tradesmen. But his purpose was also to demonstrate how damaging the lack of such natural philosophy might be – not only to mechanics, but to those who were drawn into their multitude of proposals. Desaguliers sought out schemers like a guardian of philosophic virtue as much as a protector of the Royal Society.

Desaguliers certainly knew enough examples of engineering frauds and errors to be able to assert the virtue of natural philosophers in mechanical projects. His contact with a large number of contemporary engineers like Henry Beighton, John Grundy, senior and junior, Martin Triewald, and Bernard Forest de Belidor reflected this belief.[32] His connections with inventors might even mean improvement. At Desaguliers' home in Channel Row, the inventor Joshua Haskins set up a model of his mercury engine that raised water by creation of a vacuum. Desaguliers puzzled over its limitations and later protected the claims of Haskins' heirs for a patent for the improved engine, which was shown to the Royal Society.[33]

[30] Stephen Switzer, *Ichnographia Rustica: Or, The Nobleman, Gentleman, and Gardener's Recreation. Containing Directions for the general Distribution of a country Seat, into Rural and Extensive Gardens, Parks, Paddocks, &c. And a General System of Agriculture* (London, 1718), Vol. I, pp. xii–xiv; and Switzer, *An Introduction To a General System of Hydrostaticks and Hydraulicks, Philosophical and Practical* (London, 1729), Vol. II, "Advertisement."

[31] Switzer, *Introduction to a General System of Hydrostaticks and Hydraulicks*, Vol. I, p. xix.

[32] Royal Society MSS., Letter Book A, no. 90. Belidor to Cromwell Mortimer, November 4, 1731; see also Musson and Robinson, *Science and Technology*, p. 45. On Grundy, see Margaret C. Jacob, "Scientific Culture in the Early English Enlightenment: Mechanisms, Industry, and Gentlemanly Facts," in Alan Charles Kors and Paul J. Korshin, eds., *Anticipations of the Enlightenment in England, France, and Germany* (Philadelphia, 1987), pp. 157–8; and Jacob, *The Cultural Meaning of the Scientific Revolution* (Philadelphia, 1988), pp. 151–4.

[33] Desaguliers, *Course of Experimental Philosophy*, II, pp. 491–7. See Royal Society MSS.,

Likewise, before the Royal Society he explored the benefits of many machines of his own invention. Such was the centrifugal bellows with a diameter of eight feet that he built at Channel Row. Here, he hoped, was a contrivance by which air might be cleared in a ship's hold, in sick rooms, or damps dispersed from mines.[34] Despite his reservations about the vulgar projector's alchemy, Desaguliers did not shirk from the effort to apply philosophical principles to the design and construction of machines, models, and experimental apparatus. Channel Row, where he frequently held his lecture courses, must have been a jungle of instruments and machines. It is a great pity that we do not know more about the conditions in which Desaguliers exercised his particular enthusiasm; we can at least imagine why he did not want to desert London for the morbid confines of a country church.

Compared to the business of experimental philosophy, Desaguliers saw little benefit, beyond a small and uncertain income, from his clerical duties. He therefore chose to focus the great part of his attentions on London, on the Royal Society and the public lectures, and especially on the community of lecturers and instrument makers with whom he came daily into contact. The instrument makers were essential to the experiments that he carried out from the spring of 1713 in the gap left by the death of Francis Hauksbee, the elder.

One of those on whom he soon came into rely was William Vream, whom Desaguliers later described as his "Operator for Philosophical Machines."[35] Vream had likely been employed by the elder Hauksbee, and later opened his own shop for selling instruments in Earl Street, near Seven Dials. Vream was first mentioned by Desaguliers in 1715, when Vream was apparently living in the vicinity of the George Inn, in Hedge-Lane, near Leicester Fields. Vream, an artisan known for his brass work, had his model of a double air pump exhibited at the Royal Society by Desaguliers in November 1715. A year later, he was introduced to the Royal Society as a "Maker of Instruments for Experimental Philosophy" when he showed his device to make the pistons of a double pump work alternately. This design crossed over from experimental apparatus to an

Classified Papers, VI (1660–1740), no. 64 (April 19, 1722); and *Philosophical Transactions*, XXXII (Jan.–March 1722), 5–15.

[34] Royal Society MSS., Letter Book D.2, no. 71, Desaguliers to Mortimer, November 8, 1735. This was the bellows that he reported in the *Philosophical Transactions*; see above, note 27.

[35] Desaguliers, *Course of Experimental Philosophy*, II, pp. 377. On Vream, see E. G. R. Taylor, *The Mathematical Practitioners of Tudor and Stuart England* (Cambridge, 1954), p. 303; and E. G. R. Taylor, *The Mathematical Practitioners of Hanoverian England* (Cambridge, 1966), p. 145.

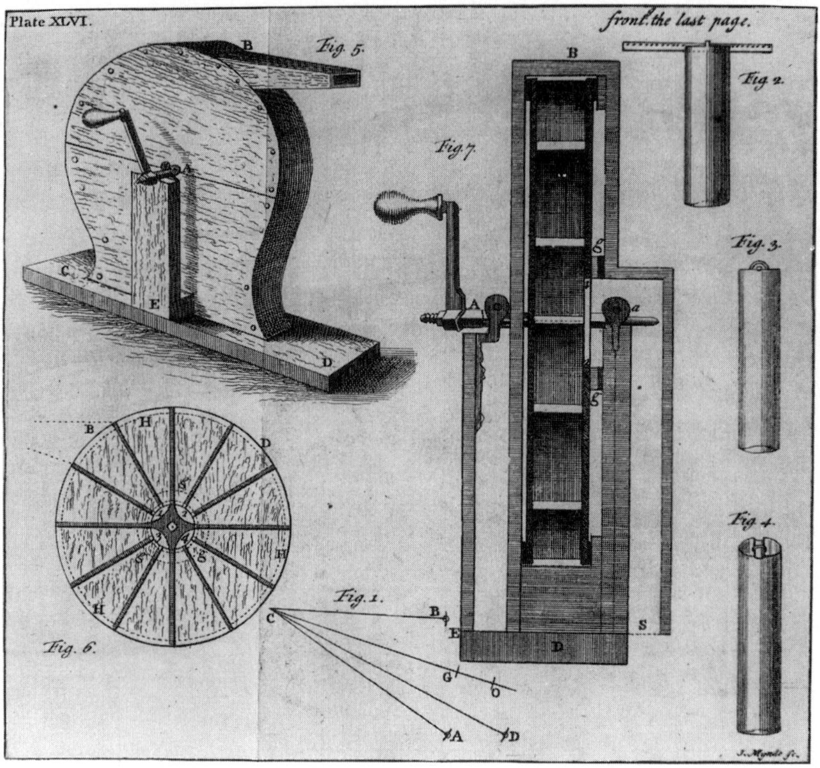

Figure 8. Desaguliers' centrifugal bellows. By courtesy of Thomas Fisher Rare Book Library, University of Toronto.

industrial use; this was a machine thought "very proper to be Apply'd to Water Works."[36] While promoting new-model chimneys, Desaguliers also recommended Vream in the translation of a book by Nicolas Gauger. The work was *Fires Improv'd* and the promises of Desaguliers and Vream to dispel the smoke of faulty chimneys opened many aristocratic drawing rooms to the two men.[37]

[36] Royal Society MSS., Journal Book (C), XI (1714–20), pp. 84–5 (November 3, 1715); pp. 142ff. (November 15, 1716).

[37] Desaguliers, trans., *Fires Improv'd: Being a New Method of Building Chimneys, So as to prevent their Smoaking: in which A Small Fire, shall warm a Room better than a much Larger made the Common Way. With the manner of altering such Chimneys as are already Built, so that they shall perform the same Effects*, written in French by Monsieur Gauger (London, 1715). See Royal Society MSS., Journal Book (C), XI (1714–20), p. 86 (November 10, 1715).

Shivering before an open fire was too frequent an experience in many country homes. In a very short time, Desaguliers and Vream were in much demand to solve the perpetual problems of smoke and lack of heat that many buildings experienced. In the winter of 1717–18, Desaguliers was asked for advice on repairs that were to be made to one of the towers of Parliament in order to improve the heating of the House of Commons. It soon reached the point where anyone with a claim to improve chimneys was taken seriously by the King's advisors, Baron Bothmer and Secretary Robethon. When one French mechanic sought a patent on knowledge of Gauger's method, Desaguliers objected. This simply resulted in an offer to Desaguliers of half the patent on the basis of his own improvements. Desaguliers thought the whole business a fraud.[38] By 1719, it was the Royal Society that sought Desaguliers' newly invented designs to cure the damp inflicting their repository at Crane Court.[39] His method essentially was to place plates in the chimneys to create a proper draft. Such activities argue persuasively that Desaguliers' interests lay as much in the resolution of practical difficulties as in the sophisticated experimentation of natural philosophy. This is further confirmed by the assistance he provided to the Duke of Chandos in the winter of 1720. The library at Cannons had become so filled with smoke from the fire that it was impossible to sit in the room. Desaguliers was asked to hurry along William Vream, who was to make the plates to fit into the chimney.[40]

The partnership between Vream and Desaguliers was a pragmatic and a philosophical venture. Vream's involvement with Desaguliers' lectures was extended into Thomas Watts' Academy when Desaguliers began to lecture there.[41] Their cooperation also drew others into the promise of their inventions. In 1720, Vream and Desaguliers joined with Daniel Niblet, a coppersmith, to produce a device that would use the heat of steam in manufacturing. Desaguliers was convinced that the principle of the expansive power of steam would permit it to be used in applying heat in the boiling or evaporation of volatile substances. His theoretical explanation that steam maintained its heat and elasticity was introduced to the Royal Society in the summer of 1720 and was clearly based on the notions of elastic fluids found in the 1717 edition of the *Opticks*.[42] De-

[38] Desaguliers, *Course of Experimental Philosophy*, II, pp. 559–60.

[39] Public Record Office, Treasury Papers, T1/255/259. January 9, 1718; and Royal Society, MSS. Council Minutes, Vol. II, f. 315, April 14, 1719.

[40] Hunt. Lib., Stowe MSS. 57, XVII, fols. 11–12, Chandos to Desaguliers, February 15, 1719–20.

[41] Vream later became an assistant to James Stirling, the mathematician, at Watts' Academy. Musson and Robinson, *Science and Technology*, p. 41n6.

[42] Desaguliers, *Course of Experimental Philosophy*, II, pp. 310ff.

saguliers argued that by confining the heat of steam in a copper vessel it could be brought to act upon compounds where the application of fire might be excessively dangerous such as with turpentine, varnishes, in the distillation of liquors, or even in the drying of gunpowder. This was to be done by means of a copper boiler connected by pipes to other containers with double bottoms, which would enable the indirect application of steam, and thus transfer the heat, the temperature being regulated by the amount of steam allowed to each of the secondary vessels.[43]

This inventive partnership did not secure immediate advantages. With some disgust, they were drawn headlong into a crass and competitive world for which philosophers were seldom prepared. They did receive a warrant for a patent in June 1720 for their dryer, the uses of which had been extended to the drying of malt, hops, and other humid substances. But they also encountered the cynicism of a Buckinghamshire gentleman who assisted in making some trials, but whom Desaguliers soon accused of stealing his ideas. The culprit, a Captain John Busby of Marsh Gibbon, then had the effrontery to offer Desaguliers shares in a joint stock that he intended to set up. While the philosopher lamented that Busby had made as much as twenty thousand pounds by the project – and this may not have been much of an exaggeration given the hysteria to invest in any scheme offered to the public – the culprit lost his money in the South Sea Bubble.[44]

Desaguliers' dismay over the affair was obviously self-serving. Yet, like much else that he later recounted, it appears remarkably accurate, at least insofar as we can now establish. His problem with Busby affords a useful insight into the way in which the competition for patents made it difficult to overcome the suspicions that inevitably followed. By 1722, Busby had also decided that machines to clear the air in mines would be extremely lucrative. This was something that Desaguliers had spent years trying to perfect. But Busby made an important and pointed claim to clear the "Damps in Mines, Pitts, or Wells by an Engine which is an Improvement from Mr Boyls Expermts of the pressure of the air, And which has never had the least Imitation of it, but in an Minute Degree to show the pressure of the Air, *by those persons who read Lectures in Experimental Philos-*

[43] Royal Society MSS., Classified Papers XVIII (2), 40. "Steam Confined, A Description of an Invention (contriv'd by J. T. Desaguliers, LL.D. F.R.S., Daniel Niblet Copper-Smith, and William Vream Instrument-maker) for performing several operations by the steam of Boyling liquors, sooner, better, and cheaper then is usually done by the common way of Applying Fire" (July 1720; read April 6, 1721).

[44] Desaguliers, *Course of Experimental Philosophy*, II, pp. 560. The connection with Busby was likely through Gun-Jones; Busby was made F.R.S. in 1719. See Royal Society MSS., Journal Book (C), XI (1714–20), p. 311 (March 19, 1718/19) and p. 326 (April 16, 1719).

ophy."[45] Desaguliers was driven to utter distraction by such interlopers. He complained bitterly about those projectors who "go to the charge of 70 or 80 l. for a Patent for their new Invention; then divide it into Shares, and draw persons more ignorant than themselves to contribute towards this (suppos'd advantageous) Undertaking."[46] This, he argued, was precisely why a public experimental philosophy was necessary.

Desaguliers' early experiences as an entrepreneur, like those of Nehemiah Grew a generation earlier, were not happy ones. The collapse of numerous enterprises in the 1720 Bubble could not have made conditions very easy for any new venture. Yet his association with Niblet, the coppersmith, continued throughout the next two decades. In 1726, for example, Niblet was employed by the Duke of Chandos to make 200 yards of copper pipe to be used in the irrigation of the garden at Cannons.[47] A decade later, Niblet was involved in a peculiar incident of which only the barest trace now remains. Chandos had been approached by a Robert Williams who claimed a new invention that enabled the production of steam at about one-third less fuel than was usually required by a Newcomen engine and that enabled the engine to operate with a greater force. Chandos was told that Niblet had approved of the invention. Intrigued, if sceptical, the Duke decided that Desaguliers should be consulted. The initial performance in 1736 was evidently so impressive as to suggest further trials be undertaken at Desaguliers' house in Channel Row, "where there were sufficient conveniencies for making all the experiments requisite." But Williams complained that he was never told the results of these trials, and Chandos undertook to write to Sir George Beaumont, M.P., to ensure that Williams not "be entirely defeated of a Share of the benefit (if any there should be) of an Invention that was originally wholly his own."[48] Williams, however, was left in the cold. Chandos understood there were few pleasant tales in the brutal scramble for patents. Of course, he did not trust Desaguliers, and by 1736, he was also distinctly unhappy with Desaguliers' performance at Whitchurch.

What George Beaumont made of this affair is not entirely clear. Desaguliers claimed that in 1736, Beaumont, along with other members of the House of Commons, was interested in his inventions. This had come

[45] Public Record Office, State Papers (Dom.) 35/30/45. Busby Petition, March 2, 1721 [/2]. My italics.
[46] Desaguliers, *A Course of Experimental Philosophy*, second edition (London, 1745), I, p. 138.
[47] Hunt. Lib., Stowe MSS. 57, XXVIII, f. 33, Chandos to Desaguliers, March 30, 1726.
[48] Hunt. Lib., Stowe MSS. 57, XLVII, fols. 171–2. Chandos to Sir George Beaumont, March 26, 1736.

about because of a machine that he had designed in 1727 for the Earl of Westmorland, formerly Governor of the York Buildings Company, to clear mines of foul air. Desaguliers' experiments on the device had been reported in the *Philosophical Transactions* and Beaumont asked the philosopher if he could construct a machine to "draw the hot and foul Air out of the House."[49] Once again, the Doctor had made the best use of his connections.

Air, water, and steam

The interest of Desaguliers in ventilation provides clear evidence of his committment to the resolution of technical problems. It also places Desaguliers in a much wider community of practical philosophers than a focus on the Royal Society might suggest. When he published his proposals in 1727 for a machine that could clear mines of foul air, he was part of a process that involved mining engineers as well as physicians throughout the century.[50] For Desaguliers, the impetus was undoubtedly the burgeoning mining developments of the early eighteenth century. This interest followed the investments of the Duke of Chandos as well as the new steam engines then being deployed in British and continental mines.

A parallel case can be seen in the concerns of the Swedish engineer, Martin Triewald, with whom Desaguliers became familiar. Triewald actually made his start in the Newcomen technology in the Ridley collieries around Newcastle, which were among the first to employ the new engines. Recent evidence indicates that in 1724, the Scottish traveler Sir John Clerk discussed with Triewald techniques being used to "draw ill aire out of Alderman Ridley's works." Triewald in the north, like Desaguliers, also lectured on natural and experimental philosophy during the winter of 1724–5 and later in Sweden. Whether any of his lectures made reference to the new technology of mining, including ventilation, is uncertain. However, after he had returned to Sweden he published several papers on coal mining, and in 1742, did discuss the application of ventilators in the navy.[51] The transfer of technology in the 1740s to

[49] Desaguliers, *Course of Experimental Philosophy*, II, pp. 561, 563. On Beaumont's interest in mining, see A. W. A. White, "Early Newcomen Engines on the Warwickshire Coalfield, 1714–1736," *Transactions of the Newcomen Society*, XLI (1968–9), 205; Colin Owen, *The Leicestershire and South Derbyshire Coalfields* (Ashbourne, 1984), p. 113.

[50] J. T. Desaguliers, "An Attempt made before the Royal Society, to shew how Damps, or foul Air, may be drawn out of any Sort of Mines, etc. by an Engine contriv'd by . . . ," *Philosophical Transactions*, XXXV, no. 399 (July–Sept. 1727), 353–6.

[51] Svante Linqvist, "The Work of Martin Triewald in England," *Transactions of the Newcomen Society*, L (1978–9), 167–70.

wider uses than those that had originally been envisaged involved many of the experimentalists.

Desaguliers' practical interests in the 1720s, therefore, did not spring from a solitary concern. His explorations and eventual designs of ventilating apparatus had a number of competitors that constantly haunted him. Judging from the entries in the Register Book, the philosophical inclinations of the Royal Society included the possibilities such devices might have promised.[52] Desaguliers' experiments were further stimulated, not only by the possibilities of a patent, but also by the necessities imposed on him by his aristocratic patron. Connection, in philosophy as in politics, might make all the difference.

In the 1730s, Chandos was hot on the trail of veins of ore on estates for which he had established leases. Just as the draining of mines might render old shafts workable, the ability to clear the air of danger to miners made the idea of deeper mines conceivable. Therefore, there were evident economic benefits to be derived by the application of the new technology. In the spring of 1733, Desaguliers had convinced Chandos such was the value of the apparatus "for sucking the bad air, and driving in fresh to any depth" that the duke wrote to Lewis Baden, his mining agent, to expect a model to be sent to him for trial.[53] It is likely, however, that the tests were never made because Baden failed in his digging and was only one of many whom Chandos decided was a swindler.[54]

In the improvement of devices to heat rooms or to clear away foul air, the Newtonian experimentalists were aware not only of the possibilities of economic improvement, but also of fundamental issues of good health. One such was the Reverend Stephen Hales, the vicar of Teddington, who is perhaps best known for his experimental *Vegetable Staticks*, which was deeply influenced by Newton's queries to the *Opticks*.[55] By 1727, Hales was convinced by experiment that fresh air was necessary for the preservation of healthy life:

> And thus many of those who have weak lungs, but can breath well enough in the fresh country air, are greatly incommoded in their breathing, when they come into large cities where the air is full of fuliginous vapours, arising from innumerable coal fires, and stenches from filthy lay-stalls and sewers.[56]

[52] Royal Society MSS., Register Book (C), XIX, 89–92, 193–4, 288–9.

[53] Hunt.Lib., Stowe MSS. 57, XLI, fols. 213–4, Chandos to Baden, March 24, 1713.

[54] Baker and Baker, *Life and Circumstances of James Brydges*, pp. 341–3.

[55] On Hales' experimentalism, the best source remains I. Bernard Cohen, *Franklin and Newton*, pp. 266ff.

[56] Stephen Hales, *Vegetable Staticks*, ed. M. A. Hoskins (London, 1969), pp. 146–7.

Hales went on to explain the grave dangers of air in mines, accounting for the almost certain death that followed explosions and the immediate destruction of the elasticity of air by the "fuliginous vapours."[57] Thus, Hales' experimentalism, like that of Desaguliers, often had a specific practical application.

To Desaguliers, among others, the disasters in mines suggested experimental possibilities. It had become increasingly obvious that the attempts by landowners to establish mines on their estates was creating greater demand for a solution to the well-known dangers of mine damps. This was a concern that had frequently surfaced among the Fellows of the Royal Society. In the early eighteenth century as shafts were sunk deeper, as the search for ores became more urgent, the threat of explosive gases to both life and investment meant more pressure on the experimental agenda of the Royal Society.

Sir James Lowther was another Fellow of the Royal Society and great landowner who, like the Duke of Chandos, felt the frustrations induced in mining by the powers of nature. In Cumbria, where the hills dissolved away to the sea near his seat at Whitehaven, Lowther had inherited collieries that turned the port town into one of the most prosperous in Britain. While absent in London during most of the fifty-five years he sat as an M.P., the affairs of the collieries were largely in the hands of John Spedding and his brother Carlisle, who became especially well-known for his knowledge of mining technology.[58] But explosions and fires remained a chronic fear that required great skill and courage in their handling. By the 1730s, when Carlisle Spedding oversaw the sinking of deeper mines near the Irish Sea, he and Lowther set out to determine what was to be done about the damps. Saltom Pit, which reached under the sea, was very probably one of the richest and deepest in Britain, but it was also especially plagued by the threat of burning gas and sudden explosions. In 1733,experiments were tried on foul air collected from the pit in a bladder, and early in 1734, Lowther read an account to the Royal Society of the mine damps found at Whitehaven. It was likely this that occasioned Desaguliers' further experiments on the effects of simply burning candles underground.[59] By 1740,Carlisle Spedding was still struggling

[57] Hales, *Vegetable Staticks*, p. 148; Cohen, *Franklin and Newton*, p. 268.

[58] Hugh Owen, *The Lowther Family. Eight Hundred Years of "A Family of Ancient Gentry and Worship"* (Phillimore, 1989), pp. 243, 248–54.

[59] Royal Society MSS., Register Book (C), XVIII, 129–32. Sir James Lowther, "An Account of the damp-Air at Saltom Pit" (read February 8, 1732[/3?]; Classified Papers (1660–1740), IV (2), no. 12. "An Accot. of the Damp Air in a Coal Pitt of Sir James Lowther Bart. sunk within 20 Yds of the Sea," dated December 20, 1733 (read February 8, 1734); see *Philosophical Transactions*, no. 429; and Register Book (C), XIX, 193–4. Desaguliers,

to make mines workable. In sinking one shaft, he found the air so vile that he collected it and was able to light his nearby shed by burning the gas piped from the pit. Spedding's survey for Lowther formed the basis of an extensive report to the Royal Society and of the explorations of subterranean gases by Lowther's physician, William Brownrigg.[60] There was great risk even for those used to the damps; Spedding died in a gas explosion at Whitehaven in 1755.[61]

Obviously, the problem of ventilation discussed by Desaguliers, Hales, and Triewald had potentially widespread consequences. The air in mines, ships' holds, or the Houses of Parliament was all sufficiently foul to warrant continued attention throughout the first half of the century.[62] By the 1740s, there was a raging competition amongst inventors like Hales, Desaguliers, and Samuel Sutton, a London brewer who had the support of Dr. Richard Mead.[63] Desaguliers complained that while he had been asked in 1740 by the Lords of the Admiralty to outfit a ship with a smaller version of his centrifugal wheel that had been installed in the House of Commons, he had been obstructed by Sir Jacob Ackworth, Surveyor of the Navy. There was likely some truth to this, for Hales encountered the same obstacle even after he had used his influence as a trustee of the Georgia colony to have merchant ships and slavers outfitted with his device.[64] But Hales persisted. He continued to send descriptions of the experiments he made on his machine to the Royal Society for their consideration. In 1742, for example, he suggested the usefulness of his apparatus not only in ships' holds, but also possibly in the drying of

"Experiments to shew that some damps in Mines may be occasioned only by the burning of Candles under ground without the addition of any noxious Vapour, even when the bottom of the Pitt has Communication with the outward Air: unless the outward Air be forcibly driven at the said Communication or Pype," ca. 1734–5.

60 Royal Society MSS., Letters and Papers. Decade I, vol. I (Oct. 1741-April 1742), no. 78, Carlisle Spedding to Sir James Lowther, November 14, 1740 (read January 8, 1740/1); J. V. Beckett, "Dr William Brownrigg, F.R.S.: Physician Chemist and Country Gentleman," *Notes and Records of the Royal Society of London*, XXXI (Jan. 1977), 255–71.

61 *The Newcastle General Magazine* (August 1755), p. 432. This obituary was probably written by Isaac Thompson, who edited the paper.

62 See, for example, Stephen Hales, *Philosophical Experiments: Containing Useful and Necessary Instructions for Such as undertake Long Voyages at Sea . . .* (London, 1739). See also Arnold Zuckerman, "Dr. Richard Mead (1673–1754). A Biographical Study" (Ph.D. thesis, University of Illinois, 1965), pp. 154–9.

63 Allen and Schofield, *Stephen Hales*, pp. 82–8; Daniel A. Baugh, *British Naval Administration in the Age of Walpole* (Princeton, 1965), p. 244; and Arnold Zuckerman, "Disease and Ventilation in the Royal Navy: The Woodenship Years," *Eighteenth Century Life*, XI (Nov. 1987), 84–6.

64 Allen and Schofield, *Stephen Hales*, p. 84; Desaguliers, *Course of Experimental Philosophy*, II, p. 560; and Zuckerman, "Disease and Ventilation," pp. 81–2.

gunpowder and in preventing not only increasing dampness over time as well as spontaneous explosions.[65]

By 1744, the prevention of explosions in mines had also suggested other possibilities. Martin Triewald wrote from Stockholm to the Royal Society with a proposal that might be of benefit to the bakers of London. Triewald explained that he had outfitted a Swedish man of war with a ventilator quite different from that of Desaguliers. Having the idea to bring to England some of his inventions, Triewald asked the Society to enquire of the Bakers' Company whether it might be interested in providing a reward for the machine for which he held a Swedish patent. He claimed the machine actually would permit the bakers to save over half their fuel and thus, with the large number of bakers in London, a charge of two guineas each would be relatively small, but would obviously provide a handsome reward to the inventor.[66] What came of this is uncertain, although the similarity of such an effort with the proposals of Denis Papin forty years earlier is quite remarkable.

Throughout the eighteenth century, the propagation of Newtonianism was not divorced from the world of mechanics and instrument makers. In the tendency to accept that Newton's natural philosophy eclipsed all competitors in the early eighteenth century, there has also been the capacity to ignore the activities of natural philosophers that brought them directly face to face with the technological and commercial concerns of their age. Desaguliers, for example, even supervised the manufacture of copper pipe by Niblet in 1726 and had a role in the construction of a system of irrigation on Chandos's estate at Cannons. Desaguliers had long been interested in the problems of hydraulics. Hydrostaticks, as he called it, became a consistently prominent part of his public lectures. Undoubtedly, this had a great deal to do with the demands that Chandos placed on him. But Desaguliers also saw this as an opportunity to consider some of the technical issues involved in water supply in general. In 1718, he had published, with numerous annotations, a translation of *The Motion of Water and Other Fluids* by Edme Mariotte of the Royal Academy in Paris. In his additions, Desaguliers described experiments he had undertaken on James Brydges' estate, before his elevation as Duke of Chandos,

[65] Royal Society MSS., Letters and Papers, Decade I, vol. I, no. 38, Hales to Cromwell Mortimer, January 23, 1741/2.

[66] Royal Society MSS., Letters and Papers, Decade I, vol. II, no. 93, Triewald to Mortimer, October 12, 1744. Triewald had hit upon a resolution to a severe problem for some trades. For example, in 1721, the Sun Fire Office determined not to offer policies to brewers, distillers, sugar and bread bakers, soap boilers and other trades where explosions and fires were common. Guildhall Library, MS. 11931, f. 54, Sun Insurance and General Committee Minutes, May 31, 1721.

with John Lowthorp, an engineer and Fellow of the Royal Society. These experiments disproved the assertion of Mariotte that the flow of water in pipes was constricted by friction with the walls of the vessel. Desaguliers and Lowthorp were convinced that the real problem was the air in the pipes.[67]

For Desaguliers, the wonder of water supply was a curse and an opportunity. As early as 1721, when he was in Edinburgh on matters relating to his beloved Freemasonry, he appears to have given instructions that enabled the city to maintain water flow by discharging air in the pipes.[68] He also continued to advise on the manufacture of pipes that were intended for the estate at Cannons.[69] By 1725, these efforts expanded, following the growth of the Duke's interests in London, to consideration of the pipes necessary to bring water from a new reservoir to his properties in Cavendish Square. These ventures, however, only scratch the surface of Chandos's financial follies and especially of the demands that Desaguliers increasingly had to face.[70] It was in this context in the spring of 1726 that Desaguliers reported before the Royal Society on his experiments concerning running water in pipes and published a description of his Jack in a Box.[71]

[67] J. T. Desaguliers, ed., *The Motion of Water, and Other Fluids. Being a Treatise of Hydrostaticks. Written originally in French, by the late Monsieur Marriotte, Member of the Royal Academy of Paris* (London, 1718), p. 289; Desaguliers, *Course of Experimental Philosophy*, II, p. 124; Stephen Switzer, *An Introduction to a General System of Hydrostaticks and Hydraulicks* (London, 1729), I, pp. xxv, 107. The first report on these experiments was read to the Royal Society on May 15, 1718. Royal Society MSS., Journal Book (C), XI (1714–20), p. 243. Lowthorp's calculations date from 1717–20 can be found in Hunt. Lib., Stowe MS. 28, "A Collection of Remarkable Papers," pp. 9–10. Another copy is Bodleian MS., Gough 7.

Of John Lowthorp, little is known. He was made a Fellow in 1702, and shortly thereafter, he was in the employ of the New River Company as an engineer attempting to solve some difficulties of water supply from Islington, where Desaguliers may have met him. B.L., Loan 29/2. New River Company. Miscellaneous Papers. These reports, evidently from Lowthorp, date from 1704–6. On the company and its engineers, see H. W. Dickinson, *The Water Supply of Greater London* (London, 1954), pp. 34ff. Lowthorp continued to be engaged in Chandos's affairs at least until 1727 at which time he appeared to be employed by the York Buildings Company. See Hunt. Lib., Stowe MSS. 57, XXIX, f. 133, Chandos to Samuel Horsey, December 24, 1726; XXX, f. 5. Chandos to Horsey, May 10, 1727.

[68] Desaguliers, *Course of Experimental Philosophy*, II, p. 126; Calhoun Winton, *Sir Richard Steele, M.P. The Later Career* (Baltimore, 1970), pp. 203–4. On Desaguliers and Freemasonry, see Margaret C. Jacob, *The Radical Enlightenment: Pantheists, Freemasons, and Republicans* (London, 1981), pp. 123–8.

[69] Hunt. Lib., Stowe MSS. 57, XXII, f. 81, Chandos to Desaguliers, October 6, 1722.

[70] Hunt. Lib., Stowe MSS. 57, XXII, fols. 158–9. Chandos to Desaguliers, July 6, 1725. On the Cavendish Square development, see Baker and Baker, *Life and Circumstances of James Brydges*, pp. 289ff.

[71] Royal Society MSS., Register Book (C), XII, pp. 744–50.

The lack of any formal barrier between natural philosophy and the technical interests of the early eighteenth century is revealed in Desaguliers' experimental attitude. In his public lectures, he had been accustomed to giving demonstrations of models of the Savery engine. Among those whom he instructed were s'Gravesande, the engineer Martin Triewald, who became expert in the construction of steam engines, and Joseph Fischer von Erlach. In 1721, when discussions were taking place for the building of a Newcomen engine for the Land-Grave of Hesse, Desaguliers recommended von Erlach to s'Gravesande to undertake the task.[72] Desaguliers' influence extended to the international commerce of the new science and technology.

A good example of the extent of Desaguliers' technical associations occurred in Liege. The roots of this particular incident are far from clear, although we do know that he had a connection with Villette at Liege, who had recently left London after the experiments with his burning glass. In September 1720, there arrived in Liege John O'Kelly, an engineer, probably an acquaintance of Henry Beighton. O'Kelly had struck a partnership with Canon Wanzoulle, Baron d'Eynstatten d'Aubay, and Lambert van den Steen to construct a steam engine to drain their coal mines about Liege. Within a year, as was frequently the case in such enterprises, the partners had a falling out and O'Kelly and his family were in grave danger of being imprisoned for debt. Although the engine may have been ready for work early in 1721, it stood idle while O'Kelly's resources were drained rather than the mines.[73] It was into these dire circumstances that Desaguliers may have been drawn more than has been supposed. It was Chandos who came to the rescue. Having heard of the plight of O'Kelly, Chandos requested details of the patent granted by the Bishop of Liege, as well as of the articles of association of the partners.[74] As soon as a petition of relief was prepared, the Duke began to manipulate the strings of connection. Chandos wrote to Secretary of State Carteret to try to obtain the support of the King in seeking for the engineer the protection of the Elector.[75] O'Kelly was saved from the edge of disaster, although by the time the engine was set to work in 1723, he had sold his share in the partnership.

[72] G. J. Hollister-Short, "The Introduction of the Newcomen Engine into Europe," *Transactions of the Newcomen Society*, XLVIII (1976–7), 11–12, 15.

[73] Hollister-Short, "Introduction of the Newcomen Engine," p. 12.

[74] Hunt. Lib., Stowe MSS. 57, XVIII, f. 363, Chandos to Mr. Bridges, January 19, 1721/2. On O'Kelly, see L. T. C. Rolt and J. S. Allen, *The Steam Engine of Thomas Newcomen* (New York, 1977), pp. 45, 71.

[75] Hunt. Lib., Stowe MSS. 57, XX, Film 417, f. 206, Chandos to Lord Carteret, April 28, 1722.

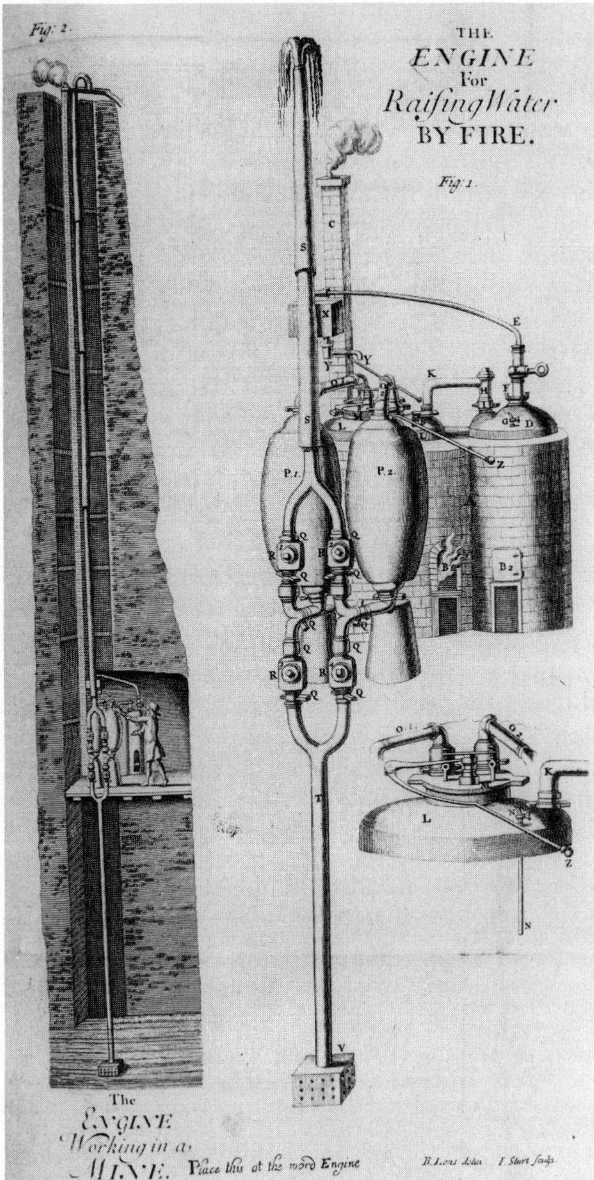

Figure 9. Savery Engine from Harris's, *Lexicon Technicum* (1704). By courtesy of Thomas Fisher Rare Book Library, University of Toronto.

Of O'Kelly only a few meager traces remain. In 1725, he was variously reported to be in London, or in Amsterdam "sur le point de s'embarquer pour L'Espagne ou il compte etre charge d'assecher les mines de Biscaye."[76] What is more certain is that in March 1725, O'Kelly was in The Hague, where he visited the Swedish Resident, Joachim Frerik Preiss, with a proposal to erect an engine in Sweden. Although he could not go himself, O'Kelly offered to send his chief workman, George Saunders.[77] Although this, too, fell through, Saunders, unlike O'Kelly, continued to surface, being employed as "directeur de la machine a feu" at Vedrin, near Namur from 1730.[78] What is especially curious is that when O'Kelly approached the Swedish government in 1725, he provided them with a drawing of his proposed engine that was a copy from a print of the York Buildings Company engine still idle near the Strand. This print was not sold in London until a year later in an effort to restore the tatters of the Company's reputation. By then, the York Buildings engine had experienced a number of difficulties about which Desaguliers was consulted. In the crosscurrents of the international exchange of engineers, it is conceivable that O'Kelly may have been connected with the controversial "Committee of Proprietors of the Invention for Raising Water by Fire" created to market the Newcomen improvement.[79]

It is hardly surprising that the developments of the steam engine in the early eighteenth century should have become a matter of dispute. There were enormous profits to be made from the draining of mines that would otherwise have been unworkable. Consequently, the Royal Society and the Crown were frequently inundated with plans for machines that would

[76] G. Hansotte, "L'Introduction de la machine a vapeur du Pays de Liege (1720)," *La vie wallonne*, XXIV, 55; Hansotte, "Machines a vapeur dans la region Liegeoise au XVIIIe siecle," *Societe royale le vieux Liege* (1955), 492–3; Pierre Delree, "A propos de l'introduction des machines a vapeur dans la region liegeoise," *Chronique archaeologique de pays de Liege*, LIII (1962), 109–16; Herve Hasquin, *Une mutation. le "Pays de Charleoi" aux XVIIe et XVIIIe siecles. Aux origines de la Revolution industrielle en Belgique* (Brussels, 1971), pp. 135–8; and Svante Lindqvist, *Technology on Trial. The Introduction of Steam Power Technology into Sweden, 1715–1736* (Stockholm, 1984), pp. 154–5.

[77] Carl Bjorkbom, "A Proposal to Erect an Atmospheric Engine in Sweden in 1725," *Transactions of the Newcomen Society*, XVIII (1937–8), 75–8.

[78] Hasquin, *Une mutation*, pp. 138–9; and Hollister-Short, "Introduction of the Newcomen Engine," pp. 12–14.

[79] It is also not beyond possibility that Desaguliers may have been the source of the Nicholls print. *Daily Courant* and *Daily Post*, March 28, 1726; Bjorkbom, "A Proposal to Erect an Atmospheric Engine," pp. 75, 77, 84–5; and Rolt and Allen, *The Steam Engine of Thomas Newcomen*, pp. 80–4. On the proprietors, see Alan Smith, "Steam and the City: the Committee of Proprietors of the Invention for Raising Water by Fire, 1715–1735," *Transactions of the Newcomen Society*, XLIX (1977–8), 5–20; see also H. W. Dickinson, *A Short History of the Steam Engine* (Cambridge, 1939), pp. 45–50.

raise water.[80] Great landowners were likewise not immune to the prestige of fabulous fountains the engines could promise. The result was an amalgamation of old and new technology, of water wheels and steam engines, used for the raising of water and for fountains on lordly estates. This market in the delights of jet d'eaux further compounded the demand – and disappointments – for those familiar with the technology.

Practitioners and defending public philolophy

The Reverend Thomas Holland, the minister of Amesbury, near Stonehenge in Wiltshire, was another such embittered expert. Holland has received very little credit for his efforts although his work was once well known. He had created fountains using water wheels on the estate at Wilton of Thomas Herbert, the Earl of Pembroke and former President of the Royal Society, and at Lord Tinley's at Wanstead. Holland's technique allowed jet d'eaux to work continuously without the need of a reservoir. He was also Engineer General to the Mines Royal and Mineral and Battery Works in 1709, at the same time that John Harris and the engineer George Sorocold were involved.[81] Holland, thus, had some considerable reputation amongst his contemporaries, although Desaguliers claimed that Holland was ruthlessly exploited by those who learned of his machines. Inventions and patents proved easy prey for the unscupulous.

One for whom Desaguliers had no use whatever was William Benson, who succeeded Wren as Surveyor-General. In order to secure his election as M.P. for Shaftesbury, Benson had arranged that the town should be supplied with water by means of an engine constructed by Holland although Benson got the credit. Benson's later fury at his failure at the polls in 1727 caused him to shut down the engine. But his early rise had allowed him to convince George I to permit him to construct an engine at the summer residence in Herrenhausen. Benson seemingly had little knowledge in such matters, and Desaguliers claimed that he had simply taken away Holland's chief workman, John Cleve, and employed him at Herrenhausen. The whole affair was bungled from the begining. With

80 For example, see B.L., Sloane MSS., 4058, f. 178, Godfrey Copley to Hans Sloane, n.d.; 4034, f. 92, Charles Leigh to Sloane, December 8, 1733.

81 B.L., Loan 16. Records of the Mines Royal and Mineral and Battery Works, fols. 233–243; see also Switzer, *An Introduction To a General System of Hydrostaticks and Hydraulicks*, I, pp. xix–xx; Desaguliers, *Course of Experimental Philosophy*, II, pp. 431–6, 520–522; Terry S. Reynolds, *Stronger Than a Hundred Men: A History of the Vertical Water Wheel* (Baltimore, 1983), p. 193

George and the entire court assembled to watch the first test, the spout of water only rose ten feet because Cleve had managed to produce a pipe with a right angle and thereby created a severe constriction of the flow. The chance advice of a French watchmaker saved the day.[82] It was perhaps just as well that Holland received no credit at all for the design. Cleve, evidently, survived as an ironmonger.[83]

The episode is an interesting one because Desaguliers later used it as an illustration of how easily people could be defrauded by extravagant claims. If George I could be misled, despite all those like Desaguliers who might have provided more sound warnings, how could an unschooled investor or landowner possibly survive the blandishments of the packs of projectors who hunted in Exchange Alley? But it was also evidence of how important the advice of a natural philosopher could be. Improvements in the steam engine were particularly susceptible to financial legerdemain, even when claims were made by inventors of engines that would have required the repeal of the law of gravity.[84] An excellent example of such hopes occurred in Chandos's own household in 1724 involving a Signor Ghirardo, one of the violinists. Chandos passed on his proposals to Desaguliers with instructions to assist the musician:

> you'l see by reading it [the letter] what hopes he has of making himself a vast Fortune & as nobody knows what may happen since stranger things have come to pass I desire you'l be pleased to order such a Model of the Fire Engine to be made as he writes for & the Description necessary for the use of it drawn up.[85]

Nothing much came of this, but Chandos and Desaguliers were often preoccupied with issues deriving from the adaptation of steam engines to water supply. This involved not only the needs of his estates, but also some of the companies in which Chandos had a long-term and extensive financial interest. Desaguliers was, therefore, often consulted on the difficulties that seemed to plague companies like the Thames Water Supply

82 Desaguliers, *Course of Experimental Philosophy*, II, pp. 521–6; Ragnhild Hatton, *George I: Elector and King* (Cambridge, Mass., 1978), pp. 262–3, 364n1. For Benson's reputation, see H. M. Colvin et al., eds. *The History of the King's Works*, V (1660–1780) (London, 1976), pp. 57–65; Romney Sedgwick, *The House of Commons, 1715–1754* (New York, 1970), I, p. 455; and A. J. G. Cummings and Larry Stewart, "The Case of the Eighteenth-Century Projector: Entrepreneurs, Engineers, and Legitimacy at the Hanoverian Court in Britain," in Bruce T. Moran, ed., *Patronage and Institution* (Woodbridge and Rochester, 1991), pp. 236–7.

83 Hunt. Lib., Stowe MSS. 85, James Brydges, accounts with various tradesmen for work at Canons (1721–3), March 29, 1723.

84 Desaguliers, *A Course of Experimental Philosophy*, II, p. viii.

85 Hunt. Lib., Stowe MSS. 57, XXIII, f. 280, Chandos to Desaguliers, February 11, 1723/24; Baker and Baker, *Life and Circumstances of James Brydges*, pp. 130–1.

in York Buildings which briefly used both Savery and Newcomen engines.[86]

These engines invited disaster. During the 1720s, there was a rash of proposals put before the Royal Society that gave Hans Sloane more than a little pause for thought. An unfortunate Mr. Thinn, surgeon at Mansfield, placed his bets on the losing propositions of both the longitude and perpetual motion.[87] In the summer of 1729, claims were made, as usual, for a new engine for raising water along with the discovery of a perpetual motion.[88] A few months later, Desaguliers was faced with the same aggravation when he suggested experiments to disprove a French claim, in a translation of Vitruvius, to have invented a machine that worked without friction. It is notable that Desaguliers' argument for such a demonstration was that "it is of Service to the Public to shew where the Error is." He wrote to Sloane, "I am sure he is mistaken & I can both demonstrate mathematically, and experimentally by Models wherein he is deceiv'd, in such manner as to prevent all Faults of that kind in Practice by shewing what it is that leads People in to those kinds of mechanical Mistakes."[89] Desaguliers' enthusiasm (possibly to capture the right to perform the Copley experiment, which carried with it a tidy reward) certainly overcame any hesitation he might have had about correcting the misguided. Experimental answers to proposals of perpetual motion were nothing new in the Society.[90]

Desaguliers was no more successful than others had been to overcome the faith in the existence of a holy, and frictionless, grail. Late in his life, he reflected on the many such frauds he had witnessed:

[86] Hunt. Lib., Stowe MSS. 57, XXV, f. 69, Chandos to Desaguliers, December 13, 1724; XXV, f. 323, Chandos to Desaguliers, April 7, 1725.

[87] Royal Society MSS., Letter Book (C), XIX, 182–3, Thinn to Sloane, August 9, 1727, and October 17, 1727; 283. Thinn to James Jurin, January 6, 1728/9; 303–4. Thinn to James Jurin, January 18, 1729; 309–10. Thinn to Jurin, May 24, 1729; Register Book (C), XIV, 247. Thinn to Dr. Rutty, April 6, 1729.

[88] B.L., Sloane MSS. 4050, f. 175, Geo. Cunningham to Hans Sloane, August 19, 1729. See also f. 114, Cunningham to Sloane, May 24, 1729. Desaguliers made a career of dealing with these deceptions, having corrected the enthusiasm of John Rowley, the celebrated instrument maker, in 1720–1. Desaguliers, *A Course of Experimental Philosophy*, second edition, corrected (London, 1745), II, pp. 183–6. This posthumous edition was published by the booksellers Innys, Longman, and Senex without need of a subscription.

[89] B.L., Sloane MSS. 4050, f. 244, Desaguliers to Sloane, December 12, 1729; Royal Society MSS., Register Book (C), XIV, 314–20. Desaguliers, "An Examination of M. Perault's new-invented Axis . . . " (December 18, 1729). See the *Philosophical Transactions*, XXXVI (Jan.–Feb. 1730), 222–30.

[90] J. L. Heilbron, *Physics at the Royal Society during Newton's Presidency* (Los Angeles, 1983), pp. 70–1.

> I know five or six Persons who have been taken in . . . even after I had told them that the Persons applying to them were ignorant Pretenders. What they lost by them, and reading this, will make them remember it. These bold Undertakers, who are generally Perpetual-Motion Men, as so ignorant as not to understand the Language whereby they should be shewn their Error; and I have known but one of five and thirty who has been convinced that he was in the wrong in his Scheme for a perpetual motion.[91]

True philosophy might promote the public good.

The adjustments to the world of mechanics and traders that the Royal Society had been unable to make in the seventeenth century were now thrust upon the philosophers. Swift and Pope were right; the universe was changing beyond recognition and the slight of Stuart claims to a throne was not the half of it. The growth of a powerful bureaucracy that laid the foundation of connection between Court and Commons manipulated so ruthlessly by Walpole was only the surface of a swell of opportunity linked to commerce as well as to government. It was not necessary for the Royal Society to acquiesce to these forces. Philosophers were doing it themselves in their rush to exploit the public excitement over technical and scientific expertise that still offered more than it could deliver. The same search for advantage, the forces of ruthless innovation, that fed the South Sea Bubble might save a philosopher from ruin, or lead him to it. Lectures could be given, broadsheets or major works could be sold to a public that ranged from meager tradesmen to the most fashionable of lords. So it was that the ideal of Robert Hooke or Robert Boyle for a Society that could support the technical proposals of its Fellows never really died; it was merely transformed.

Henry Beighton: The career of the philosopher-engineer

There may be some temptation to reduce technical advantage to rudimentary mechanics, as in the warnings Desaguliers gave to workmen who would dare to build pressure by overcoming the workings of the safety clack.[92] But its inventor, Henry Beighton, was another of those Fellows of the Royal Society who found gainful employment in the ventures of builders of engines. With his friend Desaguliers, Beighton appears to have made several changes to the safety valves on the Newcomen engines and

[91] Desaguliers, *Course of Experimental Philosophy*, II, p. 416.
[92] Desaguliers, *Course of Experimental Philosophy*, II, p. 489; and, Rolt and Allen, *The Steam Engine of Thomas Newcomen*, pp. 89, 92–6.

was likewise regarded as something of an authority on the principles of the engines. John O'Kelly appealed to the authority of Beighton in 1721 when his own knowledge was questioned. It seems likely that O'Kelly and Isaac Potter had both gained experience with Beighton through his work on a Newcastle engine in 1717.[93]

Beighton moved with great ease from mines along the North Sea coast to the coffeehouses of the London experimentalists. But his foundation in the world of artisans and engineers can now be partially reconstructed. As early as 1714, in a brief reference in the letters of the northern mining baron William Cotesworth, there is an acknowledgment of Beighton's willingness to buy and sell lead extracted from Cotesworth's mines.[94] This suggests the intriguing possibility that Beighton may have been connected with the London Lead Company and Richard Peck although this is pure conjecture at this point. By the 1720s, Beighton's reputation was founded largely upon his association with Sir Richard Newdigate and his steam engines in Warwickshire. We now know that Beighton held a share in the Heaton Colliery near Newcastle by 1717 and that subsequently he was engaged as engineer with Stonier Parrott, who constructed a number of engines in the coal fields.[95] Beighton's expertise brought his introduction to the Royal Society by Desaguliers in December 1717. The immediate interest was in a draft of Beighton's improved steam engine. In the spring of 1718, Desaguliers revealed that his experiments upon Beighton's engine established a remarkable reduction in the amount of fuel consumed to raise water. By that point, Beighton was evidently at Newcastle and reported to Desaguliers, somewhat optimistically, on the construction of an engine, the stroke of which exceeded all previously built.[96] Beighton was elected a Fellow of the Royal Society in 1720

[93] Desaguliers, *Course of Experimental Philosophy*, II, p. 532; Rolt and Allen, *The Steam Engine of Thomas Newcomen*, p. 45. This may have been the Oxclose engine, Washington Fell, county Durham. On Isaac Potter and the involvement of his family with the Proprietors of the Fire Engine Patent, see J. S. Allen, "Bromsgrove and the Newcomen Engine," *Transactions of the Newcomen Society*, XLIII (1970–1), 183–98; and Mikulas Teich, "The Early History of the Newcomen Engine at Nova Bana (Konigsberg): Isaac Potter's Negotiations with the Hofkammer and the Signing of the Agreement of 19 August 1721," *East-Central Europe*, XI (1982), 29n18

[94] Northumberland County Record Office (NCRO), Gosforth Park, Newcastle-upon-Tyne, MS ZDE/10/2. William Cotesworth letter books.

[95] Rolt and Allen, *The Steam Engine of Thomas Newcomen*, p. 54; and J. S. Allen, "Some early Newcomen Engines and the Legal Disputes surrounding them," *Transactions of the Newcomen Society*, XLI (1968–9), 198–201.

[96] Royal Society MSS., Journal Book (C), XI (1714–20), pp. 208–9 (December 5, 1717); p. 211 (December 12, 1717); p. 220 (February 20, 1717/18); p. 227 (March 6, 1717/18); p. 232 (March 7, 1718).

and subsequently leased coal mines from Sir Richard Newdigate in Warwickshire.

Beighton, who lived mostly at Griff, in Warwickshire, became linked to the Cotesworth enterprises through his involvement with the engineers Stonier Parrott and George Sparrow. Parrott and Sparrow had already leased some of the coalworks of Sir Richard Newdigate at Griff sometime before 1713. By the spring of 1715, Parrott and Sparrow obtained a license to construct Newcomen engines from the Proprietors of the patent in London. Possibly pursuing an intial contact made by Savery, they had also approached George Liddell of Ravensworth, one of Cotesworth's major allies in the coal trade, offering to build such engines as necessary to resurrect their mines.[97] The first engine built by Sparrow and the Proprietors throughout 1716–17 was for the lead mine at Yatestoop, Winster, in Derbyshire, which raises further possibilities for their contact with Beighton.[98] There is no doubt that Cotesworth and Liddell were intrigued, although they soon entered into negotiations with Sir John Meres and the Proprietors directly.[99]

By the winter of 1717–18, Beighton had become a major player of the progress toward construction of engines although his role has not, I think, been clearly understood.[100] It is difficult to accept the notion that his function was "rather as a theoretical consultant than a practical engineer."[101] Such a distinction was inappropriate in these circumstances. In fact, the evidence shows that Beighton was directly concerned in the technical difficulties in setting up the engines as well as acting on behalf of the Proprietors in London, and not simply as an agent of Sparrow or Parrott. In early 1718, George Liddell had heard from Parrott of the activities of Beighton who was then at Griff:

> he tells me that Mr. Beyton was come from the founders & tells him [Parrott] yt one of the Cylinders is Cast & boring for ye Parke [Colliery] & yt all the pitt Barrells are Cast & that the Moulds are made for Casting 3 more by & that the first of them shall be for the parke. That the Pumps are Shipt fr: London this

[97] J. M. Ellis, ed., *The Letters of Henry Liddell to William Cotesworth*. Surtees Society Publications, CXVII (1985), 174–5, 240, 242, 250 (April 21, 1715; June 19, 1716; June 26, 1716; October 2, 1716); Marie B. Rowlands, "Stonier Parrott and the Newcomen Engine," *Transactions of the Newcomen Society*, XLI (1968–9), 49–51; and Edward Hughes, "The First Steam Engines in the Durham Coalfield," *Archaeologia Aeliana*, XXVII (1949), 29–45.

[98] Rolt and Allen, *The Steam Engine of Thomas Newcomen*, p. 61.

[99] Gateshead Public Library, Cotesworth Correspondence, CP1/1, George Liddell to William Cotesworth, December 6, 1717; CP1/2. Liddell to Cotesworth, December 8, 1717.

[100] Gateshead Public Library, CP1/3. Liddell to Cotesworth, January 5, 1717[/18].

[101] Rowlands, "Stonier Parrott and the Newcomen Engine," p. 57.

Christmas. That the Smith comes downe to make the Boylers the latter End of Jany., & that Mr. Beighton comes at the Same time to get the Wood Squared & Shipt from Yorkshire & to get the Pumps placed here.[102]

This is hardly the itinerary of someone who thought the theoretical aspects of the Newcomen engine would impress mine owners like Cotesworth or Liddell.

The practicalities of actually getting the engines operating must at times have seemed insurmountable. And it was precisely this feeling that led to difficulties with the engineers. By the middle of February 1718, no word had come of Beighton and the work stood idle. On top of those frustrations, Liddell and Cotesworth were finding Sparrow to be "Refractory not to call it worse" and ultimately their opinion of Parrott was no better.[103] One of the reasons why there was such an early falling out between the agents of the London Proprietors and the mine owners was the overwhelming demand for the services of the engineers and their mechanics. The dispute gathered momentum over the agreement to erect an engine at Farnacres, west of the Gateshead Park site near the River Derwent. There were major problems securing the services of a smith to construct a boiler, and Beighton, who was evidently in charge of construction on the site, complained that the smith could not stay because "it was wth great Courting they got him to come over to make yt for the Parke & Washington & Mr Parrott & Mr Sparrow have so many bespoke Elsewhere that they would not consent to his Staying any longer."[104] This was the problem in a nutshell.

The demand on the engineers made it difficult for them to meet all requests, but it also made them an exceedingly valuable commodity. At a meeting with Liddell, Beighton was read part of a letter from the smith who complained that not only had he to repair all the salt pans in Cheshire, but "he had one Engine to set up for Mr. Meeres in Yorkshire: Another at Griff & a third in Derbyshire before he could Either furnish the other Boyler for the Parke or that for farnacres."[105] Beighton was held partly to blame for the lack of progress at Farnacres, although throughout the spring of 1718, he continued to advise on the details of construction of

[102] Gateshead, CP1/8, Liddell to Cotesworth, January 5, 1717[/18]; see also Rowlands, "Stonier Parrott and the Newcomen Engine," p. 53.

[103] Gateshead, CP1/15, Liddell to Cotesworth, February 20, 1717[/18]; also CP1/12. Liddell to Cotesworth, February 8, 1717/8. See Rowlands, "Stonier Parrott and the Newcomen Engine," p. 53. On the great difficulties in making these engines work see Richard L. Hills, *Power from Steam. A History of the Stationary Steam Engine* (Cambridge, 1989), pp. 16–29.

[104] Gateshead, CP1/21, Liddell to Cotesworth, March 5, 1717/18.

[105] Gateshead, CP1/23, Liddell to Cotesworth, March 15, 1717/18.

the engine house and the pumps. When he tried to defend the role of Parrott in the prolonged negotiations, however, he was met with a torrent of abuse from George Liddell.[106] Parrott was desperately sought after as he had devised a procedure for replacing expensive copper boilers with ones of iron. Although the Parke engine was in place by 1719, Cotesworth was complaining to London over its failure to be put to work to overcome the flooding of the shafts: "Mr Sparrow owns the Affaire no more than if he had no Concerne in it. – Mr Parrot was here but since he has been gone nothing has been done tho Mr Beighton lives within 800 Yards of the Engine – the Slow progress that is made in wynning the Parke Colliery brings a great discredit upon the Engine."[107]

Nerves and pocketbooks were frayed by the continuing expenses and the marginal results. In addition, Beighton was also being pursued by Cotesworth's rival Alderman Ridley, who sought an engine at Newcastle for his own mines. This was not a risk Beighton was prepared to take. The London Proprietors were less circumspect and fashioned an independent agreement with the Ridleys for the construction of engines for "the draining of water only according to the method now used at the Griff near Coventry in the County of Warwickshire."[108]

Henry Beighton found himself in the thick of the conflicts surrounding the efforts of the London Proprietors to control the building of the Newcomen engines. As such, he was frequently on the move from one mine to the next, to the foundries and the smith's forges, and to London, where he sometimes met Parrott and Sparrow or the Proprietors themselves. His visits to London must have been frequent and would also have afforded him the opportunity to consult with the scientific lecturers like Desaguliers, whose courses he attended in 1718.[109] Beighton's interest in natural philosophy was confirmed continually throughout a life dedicated to solving practical problems of the mining engineers.

[106] Gateshead, CP1/31, Liddell to Cotesworth, April 4, 1718; CP1/45, Liddell to Cotesworth, May 4, 1718.

[107] Desaguliers, *Course of Experimental Philosophy*, II, pp. 480–1; NCRO, ZDE 10/3. Cotesworth's secretary to Meres. February 19, [1718[/19]. On these difficulties, see Hughes, "The first steam engine in the Durham Coalfield," pp. 37ff. On the problems of the boilers, see Hills, *Power from Steam*, pp. 120–4.

[108] NCRO, Ridley (Blagdon) MSS., ZRI/23/1, Articles of Agreement, November 4, 1718; and Rowlands, "Stonier Parrott and the Newcomen Engine."

[109] B.L., Add. MSS. 4432, f. 198, Henry Beighton to [Hauksbee?] the Royal Society, November 15, 1732. Desaguliers gave at least two courses in 1718. See *Daily Courant* (January 13, 1718 and November 15, 1718). Beighton had been in London in December, 1717, but by the end of March 1718, he was in Newcastle. Royal Society MSS., Journal Book (C), XI (1714–20), p. 208 (December 5, 1717) and p. 232 (March 27, 1718).

Seeking the Royal Society

When he took over the *Ladies Diary* as editor in 1714, Beighton was able to respond to public fixation with natural philosophy. Even from Griff in Warwickshire, like numerous other provincial correspondents, he maintained his contact with the Royal Society. In 1730, he presented some papers to the Royal Society, which included a diary of the weather that he had kept in 1707. These meteorological records were common amongst eighteenth-century naturalists; even the *Athenian Oracle* proposed such a project "to all Ingenious Gentlemen."[110] Such diaries, of the kind that we might have associated with William Derham, Beighton had maintained over a long period of time although his fascination began with the dramatic. Such, for example, was his report of a wild hurricane in Warwickshire in 1712. Perhaps under the influence of Derham, these diaries were generally much more systematic and continued with detailed barometrical observations for over twenty years, mainly at Griff. They also included notations for the three years, from 1725 to 1727, that he spent in Derbyshire, probably with the Yatestoop lead mine of George Sparrow, and in the fens of Lincolnshire.[111] Neither a provincial philosopher nor a peripatetic engineer need be completely isolated from London science. Beighton continually maintained his observations, providing them to the Royal Society at least up to 1739.[112]

This did not mean that Beighton's mechanical concerns were of lessening importance. In 1730, he sent to the Society a description of the London Bridge water works, where the pumps were worked by great wheels. This sketch Beighton described as "only a Specimen at what I have for many years been Collecting, in order for a Treatise of Engines

[110] [R. Sault], *A Supplement to the Athenian Oracle: Being a Collection of the Remaining Question and Answers in the Old Athenian Mercuries* (London, 1710), pp. 153ff.

[111] Royal Society MSS., Classified Papers V (1660–1740), no. 12, "A Diary of the Weather ... Observ'd at the City of Coventry ... by H. Beighton" (n.d.); no. 14. "A Table Shewing the Height of the Mercury in the Barometer, the Coast & Strength of the Winds and the Weather on the first day of 8 Months in the Years 1707 & 1708. Observed at Coventry in Warwickshire by Mr. H. Beighton, F.R.S.; & at Upminster in Essex by W: Derham F.R.S." November 11, 1731. See *Philosophical Transactions*, XXXVII (April-June 1732), 269–73; B.L., Add. MSS. 4432, fols. 200–1, Beighton to Royal Society, March 1, 1732/3. "An Account of a most Terrible Hurricane at Nun-Eaton at Warwickshire. the 29 Augt. 1712 and the Next morning"; Add. MSS. 4433, f. 273, Beighton to Sloane?, June 10, 1736.

[112] B.L., Add. MSS. 4433, f. 276, Beighton to Francis Hauksbee, November 25, 1736; Add. MSS. 4436, ff. 19–24b, Beighton to Hauksbee?, Diary of the Weather & Height of ye Barometer, November 1736–December 1739.

Figure 10. Henry Beighton's view of the London Bridge Water Works. From Desaguliers, *A Course of Experimental Philosophy*. By courtesy of Thomas Fisher Rare Book Library, University of Toronto.

of Use & Service in a Method as I conceive yet unpresidented, the Whole being reduced to a Mathematicall Calculus."[113] This was an objective that the engineer only partially fulfilled.

Like many engineers of the period, his interests extended to philosophical instruments. In the company of John Grundy, senior, at Ashby-de-la-Zouche in January 1734, Beighton examined and commended to the Spalding Gentlemen's Society a newly perfected barometer. Made by Charles Orme, these barometers, which were treated to expel any air trapped in the mercury, so impressed Beighton and Grundy because they

113 Royal Society, MSS. Letter Book, B.3, no. 17, Henry Beighton to the Royal Society, March 4, 1730; Register Book (C), XV, 314–21. A Description of the Water works on London Bridge, explaining the Draught here annex'd. This was later published in the *Philosophical Transactions*, XXXVII, (April-June 1732), 5. See also Guildhall MS. 537, "A short account of the London Bridge Water Works & memorandum relative to Their Business." In 1736, Beighton offered to the Royal Society "A Mechanical, Mathematical & Philosophical acct of the Engine for Raising Water by Fire done in the Year 1718. And on several other Engines In the Manner I described London Bridge Water Works," B.L., Add. MSS. 4433, f. 273, Beighton to Sloane?, June 10, 1736.

appeared to give the most precise measurements of the weather yet possible. Beighton, whom Grundy described as "an ingenious Mathematician," likewise proposed to the Royal Society in 1739 the construction of a large wall quadrant that could be made cheaply of wood.[114] It is significant that Beighton sought to integrate interest in natural philosophy or natural history with his technological ventures. In 1733, in a curious *Essay towards Assigning a Reason why Persons falling into a deep Coalpitt are Seldom killed*, he declared that "Having from my Younger Years been conversant amongst many Coal mines, and seen & observ'd most In England & Scotland, my Curiosity led me to make a Great many observations in the Works of Nature."[115] Beighton was one of the many who had bridged the cultures of philosophers and mechanics of the early eighteenth century.

Henry Beighton was also an entrepreneur who had lived and worked among the improvers and adventurers who recognized the value in men who were capable of signaling a necessary caution. That they might not always have done so is another issue, linked more directly to their lust for personal advantage than to the limitations of their knowledge of machines. Like many, Beighton was also a victim. In 1736, his circumstances forced him to write to Desaguliers to beg that the Society might excuse him the arrears of his dues as had been promised if he continued to make suitable communications. His adventures in mining had followed the desolate track of numerous investors. Risk turned to ruin, as it was to do for many, in the misfortunes of the Mine Adventurers Company.[116]

When Beighton and his son died suddenly in 1744, he left his widow destitute. For Elizabeth Beighton, there remained only a disputed claim to the editorship of the *Ladies Diary* with which she had assisted since her husband had become involved in it in 1714. Her role as editor continued, although she had to rely on friends who taught mathematics nearby because she had insufficient expertise to calculate eclipses of Jupiter's satellites or the moon's rising and setting.[117] One upon whom she

[114] Spalding Gentlemen's Society MSS., Henry Beighton and John Grundy to the Spalding Gentlemen's Society, March 20, 1734; John Grundy, *A Description of the Improvements that have been made in the Barometers by Mr. Charles Orme of Ashby-de-la-zouch in the County of Leicester* (1734), n.p.; Royal Society MSS., Journal Book (C), vol. XVI (1736–9), p. 442 (May 24, 1739).

[115] B.L., Add. MSS. 4432, fols. 199–200, Beighton to the Royal Society, March 1, 1732/3.

[116] Royal Society MSS., Council Minutes, III (1727–47), pp. 182–3. Beighton to Desaguliers, January 12, 1735[/6]; also, B.L., Add. MSS. 4433, f. 276, Beighton to Hauksbee, November 25, 1736.

[117] B.L., Add. MSS. 43741, f. 6, Elizabeth Beighton to Mr. Cooper, May 10, 1744. This

came to depend was her husband's former assistant Robert Heath, who was partly responsible for encouraging mathematicians like William Emerson to contribute as well as helping fend off what Elizabeth Beighton called the "base dealings of Holliday and the Nottingham Quacks" who wanted to take over the *Diary*.[118] Emerson was a remarkable eccentric who took interest in publishing practical works for mechanics. The activities of Emerson, Heath, and Elizabeth Beighton confirm what the work of Peter Wallis has already suggested – that we know precious little of the world of the provincial mechanics and their interests in natural philosophy and mathematics.[119] Increasingly, it seems to have been a world that no longer looked for signs of success in the great metropolis of London. Emerson once barked that "when a man is eminent for his discoveries, perhaps he is dignified by the title of F.R.S. but he has to pay a quarterly cess for the honour of it. This is the way that ingenuity is rewarded in England."[120]

Benjamin Martin, in seeking his own F.R.S., thought it might be of advantage to him in the provinces as an itinerant lecturer. Such benefit does not appear to have occurred to others, but Martin's reputation as a "country showman" was to haunt him in a later dispute with John Freke, F.R.S. Freke's dismay of books "written to pay a landlady; or an ale-house score" clashed dramatically with Martin's vision of a commonwealth where "the Chemists, the Anatomists, the Physicians, and Divines, everywhere read lectures for money."[121] The principle of exclusion of which Martin had been victim could not be disguised as a principle of morality directed against public experimentalists. This, perhaps, was what lurked behind the reflections of the engineer John Smeaton upon the cost of election to the Royal Society. Undoubtedly, there might be great suspicion of what Smeaton called the "common Herd of conjuring Philosophers about Town."[122] But in 1747, Smeaton complained to the ex-

was Thomas Cooper at Willingborough in Nottinghamshire. On Elizabeth Beighton, see Ruth and Peter Wallis, "Female Philomaths," *Historia Mathematica*, VII (1980), 57–64.

118 B.L., Add. MSS. 43741, f. 8, Elizabeth Beighton to Thomas Cooper, May 29, 1746; see also Taylor, *Hanoverian England*, pp. 157, 180–1; and Robert R. Bataille, "Robert Heath, Thomas Cowper, and Eighteenth-Century Mathematical Journalism," *Papers of the Bibliographical Society of America*, LXXXI (Sept. 1987), 339–43. On the history of the *Ladies Diary*, see Bernard Capp, *Astrology and the Popular Press. English Almanacs 1500–1800* (London and Boston, 1979), pp. 242–3, 245–7.

119 P.J. Wallis, "British Philomaths – Mid-eighteenth Century and Earlier," *Centaurus*, XVII (1973), 301–14.

120 Taylor, *Hanoverian England*, pp. 34–5, 157. On Emerson, see P. M. Horsley, *Eighteenth-Century Newcastle* (Newcastle, 1971), pp. 34–42.

121 John R. Millburn, *Benjamin Martin: Author, Instrument-Maker and "Country Showman"* (Leyden, 1976), pp. 35–8, 54–7.

122 B.L., Add. MSS. 30094, f. 41, Smeaton to Benjamin Wilson, August 24, 1747.

perimentalist Benjamin Wilson, years before his own election, that the price of admittance of 22 guineas was so great "rich Philosophers can only afford to pay so that I suppose ye popish proverb will fitt them *no penny no pater noster.*"[123]

Natural philosophers did not await the approval of the Royal Society. Like the most eager projectors, they had launched themselves into the exploitation of the market that lay before them. Some took to showing the wonders of nature to subscribers in the coffeehouses; others hunted patronage and projects with equal desperation, like "technical Harris," or James Stirling, the mathematician of Watts' Academy, who abandoned philosophy for commerce in the management of the Scotch Mines Company at Leadhills in Lanarkshire.[124] It was no surprise that the link between the teachers of natural philosophy and the commercial enterprises of the early eighteenth century emerged from the schools that taught commercial practice alongside navigation, mathematics, and experimental philosophy.[125]

Philosophy and improvement

The first three decades of the eighteenth century saw the transformation from the patronage of a single lord to that wider public philosophers had first acknowledged in the coffeehouses and then in the academies. So it was that many book subscriptions were given birth. In the coffeehouses and in newspapers, subscriptions could be announced for the latest work of astronomy or for recent conglomerates to exploit mines in Jamaica. Natural philosophers became central to the transition "from the client economy to the mass market," their subscriptions providing "a degree of security in a period of change, enabling the small tradesman or producer to move outside the protected realm of patrician patronage."[126] With this,

123 B.L., Add. MSS. 30094, f. 42, Smeaton to Benjamin Wilson, September 7, 1747. Wilson was elected F.R.S. in 1751; Smeaton was chosen in March 1753 on his reputation as a "maker of Philosophical Instruments, a Gentleman well known to the Society for his great skill in the Theory and Practise of Mechanicks. . . . " Royal Society MSS., Certificates, II (1751–6), nos. 442 and 7.

124 See Stewart, "The Selling of Newton: Science and Technology in Early Eighteenth-Century England," *Journal of British Studies*, XXV (April 1986), 190–1; T. C. Smout, "Lead-mining in Scotland, 1650–1850," in Peter L. Payne, *Studies in Scottish Business History* (Edinburgh, 1967), pp. 120, 124ff.

125 Sidney Pollard, *The Genesis of Modern Management. A Study of the Industrial Revolution in Great Britain* (London, 1965), pp. 112–5.

126 Neil McKendrick, John Brewer and J. H. Plumb, *The Birth of a Consumer Society. The Commercialization of Eighteenth-Century England* (Bloomington, 1982), pp. 224–5.

the Newtonians had to come to terms. Useful knowledge, as Bacon, Hooke, and Boyle had all recognized in their own way, was a key to the transformation of which the natural philosophers were inevitably a part.

The adjustment of philosophers was made easier by the fact that landowners were forced to make their own accommodations with the power of commercial and financial revolution. Improvement was not only increasingly desirable on estates, it some cases it could be essential. The technical expertise of philosophers made them much sought after on the estates of the great landlords. This was manifest when the market for books and schemes of agricultural improvement, heirs of a long tradition, grew rapidly in the early eighteenth century.[127] Those with ideas or expertise to sell soon found a ready market. The union of astronomy and agriculture in Fatio de Duillier's schemes for sloping fruit walls to take full advantage of the obliquity of the sun's rays was not altogether unique. Edmund Halley made calculations of the quantity of the sun's rays at different latitudes, and William Whiston undertook similar calculations in 1726 on behalf of his old friend John Laurence, the rector of Bishop Wearmouth.[128] In Laurence's *New System of Agriculture* – in which he also advertised the value of the Savery engine in the draining of mines – great play was made of the force of utility inherent in natural knowledge. Likewise, in 1729, Stephen Switzer affirmed the importance of these early engines in the development of estates. Laurence had a broad view of such utility-making claims, not only for the improvement of estates, but also for the value of mathematics, which led to the "glorious Discoveries of a Newton."[129]

John Laurence's attitude toward the practical value of natural philosophy represented an increasingly common conviction. Yet he is now long forgotten amongst the innumerable agricultural writers of the eighteenth century. Utility was all – and he had no wish to glorify philosophers who failed to appreciate the potential. Of fruit walls he once remarked, "To so little purpose is it for Men of Theory to Philosophize about these

[127] Joan Thirsk, "Plough and Pen: Agricultural Writers in the Seventeenth Century," in T. S. Aston, P. R. Cross, Christopher Dyer, and Joan Thrisk, eds., *Social Relations and Ideas. Essays in Honour of R.H. Hilton* (Cambridge, 1983), pp. 295–318.

[128] John Laurence, *A New System of Agriculture* (London, 1726), pp. 16–17. Whiston to Laurence, February 10, 1725/6. On Laurence, see G. O. Bellewes, *Notes on the Life and Works of John Laurence, Rector of Bishop Wearmouth, 1721–1732* (Sunderland, 1905); and Blanche Henrey, *British Botanical and Horticultural Literature Before 1800*, II, pp. 415–24.

[129] Laurence, *New System of Agriculture*, "Preface." See Stephen Switzer, *An Introduction to a General System of Hydrostaticks and Hydraulicks, Philosophical and Practical*, Vol. II (London, 1729), pp. 402ff.

Matters without having had some Experience and Knowledge in the practice."[130] Later, the popular lecturer Benjamin Martin pointed out how philosophers had been important to the development of the steam engine and its improvement.[131] The unity of theory and practice was rapidly becoming a commonplace view. It is certainly true that Desaguliers lamented the fact that many who proclaimed themselves men of practice had little knowledge of natural philosophy, but his complaint has been much misunderstood. His dismay in 1744 that "Now we are over run with Engineers, (not Ingeneurs of a Proper Education for the Science) and Projectors, especially in Water-Works" should not be taken for a criticism of mechanics.[132] Desaguliers had more than enough experience with the schemes of his age to know whereof he spoke. It was the convergence, and not the isolation, of practice and philosophy that he sought.

The activities of the Newtonians thus reflected crucial changes in the social and economic structure of British society since the Revolution. More important, their intellectual credentials provided legitimacy to a process of mechanization that would overwhelm the century. The impact of natural philosophy cannot be appreciated by a focus that is limited to the experimentalism of the Royal Society – even when it did reach into the community of instrument makers and mathematical practitioners. Natural philosophy was part of a widely based epistemological shift, to be sure, but one not confined to philosophical experiment. It had much to owe to the foundations of a providential view of nature of which Desaguliers and Hales were proponents. And it had a great deal to do with financial and commercial innovation.

The rhetoric of experiment – in its broadest sense – had merged, just as Peter Shaw said it should, with the rhetoric of power. The ideology of improvement, of the capacity and the necessity to alter nature, was founded on a powerful epistemology – of the present and the actual – that resided in the authority of experiment. Hence, alarm easily arose about the consequences of natural philosophy from critics of the Royal

130 Laurence, *The Clergy-Man's Recreation*, quoted in Peggy Kidwell, "Nicholas Fatio de Duillier and *Fruit-walls Improved*: Natural Philosophy, Solar Radiation, and Gardening in Late Seventeenth Century England," *Agricultural History*, LVII (1983), 414.

131 Benjamin Martin, *Philosophia Britannica: Or, A New and Comprehensive System of the Newtonian Philosophy, Astronomy and Geography, in a Course of Twelve Lectures*, second edition (London, 1759), II, pp. 71–5. On Martin's association with Desaguliers, see Augustus de Morgan, *A Budget of Paradoxes* (London, 1872), p. 91; and the criticism of this view in John R. Millburn, *Benjamin Martin: Author, Instrument-Maker and "Country Showman"* (Leyden, 1976), pp. 45–6.

132 J. T. Desaguliers, *A Course of Experimental Philosophy*, II (London, 1744), p. 415; cf., Musson and Robinson, *Science and Technology*, p. 38.

Society. In this respect, at least, Jonathan Swift was not alone. When philosophers like Desaguliers, or Denis Papin, or James Stirling, or Henry Beighton, and many others reached into the world of the entrepreneurs, they were merely extending their investigations into exploitation of a world, they believed, they had been given to make the best of that they could. The ideology of improvement was founded upon the same epistemology that formed the ground for experimentalism.

PART III

For use

Who sees with equal eye, as God of All,
  A Hero perish, or a Sparrow fall,
  Atoms, or Systems, into ruin hurl'd,
And now a Bubble burst, and now a World!

– Alexander Pope, *Essay on Man. In Four Epistles to H. St. John, Lord Bolingbroke* (London, 1733–4), Epistle I, 11:87–90

They who find themselves inclined to censure new undertakings, only because they are new, should consider that the folly of Projection is seldom the folly of a fool; it is commonly the ebullition of a capricious mind, crowded with a variety of knowledge, and heated with an intenseness of thought; it proceeds often from the consciousness of uncommon powers, from the confidence of those who, having already done much, are easily persuaded that they can do more.

– Samuel Johnson, *The Adventurer*, no. 99 (Tuesday, October 16, 1753)

8

The culture of enterprise before 1750

For those in the eighteenth century who sought to improve their estates or their fortunes, it was by providence and the bounty of nature that they set their course. Natural philosophers necessarily confronted metaphysical issues like those analyses of providence with which the century abounds. From Sprat to Desaguliers, there was a sense of mission, "To contemplate the Works of GOD, to discover Causes from their Effects, and make Art and Nature subservient to the Necessities of Life, by a skill in joining proper Causes to produce the most useful Effects."[1] After Newton, experimental method had melded into material purpose. Philosophers might not only justify the exploitation of nature, but even enrole in the grand schemes to build fortunes in a more prosperous Great Britain. From this could be drawn all kinds of ideological conclusions – from the general defense of trade to necessity of the reform of the coinage and the construction of a National Debt. As a result, any analysis of the relation between the entrepreneurs and the philosophers of the early eighteenth century inevitably encounters the political rhetoric of party and corruption. Accusations of shady practices abound in politics and they form the background of the attacks on innovation in the early eighteenth century. To Jonathan Swift, this world seemed deeply fragmented, shattered into a spectrum of Revolution and financial chaos, compounded by innovations in theology and philosophy. But as with the Revolution, philosophy, too, needed defenders. It could achieve credibility if it could represent itself as a champion of the public interest. For numerous reasons, Newtonians could not avoid this deeply polemical conflict.

[1] Desaguliers, *Course of Experimental Philosophy*, second edition, corrected (London, 1745), I, p. iii. Dedication to Frederick, Prince of Wales. See also John McVeagh, *Tradeful Merchants. The Portrayal of the Capitalist in Literature* (London, 1981), pp. 45–9.

This tension has much to do with social sanction – founded on the experimentalists' power to understand natural law and, as Desaguliers obviously proposed, on the claim to objectivity as well as to a power to transform nature. In the early eighteenth century, this claim to objectivity had a special urgency; not only was it antimetaphysical, but there was also emerging in Britain the kind of social and economic structure that made the manipulation of nature possible and desirable. Who better to do so than those who claimed to understand nature best? Who better to encourage them than improving landlords and financiers who were spurred to act by a financial revolution? Newtonian science, I submit, fit nicely into the power relations of a society that was just constructing the incipient framework of capitalism that was to make the Industrial Revolution possible. Subscription to a creed, whether in religion or in science, was the ultimate sanction. I would not care to argue, however, that Newtonianism was responsible for either industrialization or for a capitalist ideology in the eighteenth century. Nor do I wish to oversimplify that argument. It is my view that the relationship between natural philosophy and the emergence of an industrial England was a subtle one tied to the forces of land as much as to the marketplace, to the interests of the aristocracy as to those of radical materialists and heterodox theologians.[2]

To be clear, this is not a simply a matter of context without which an ideology cannot exist. It is a matter of the meaning and place of natural philosophy as well as the level of comprehension. Where ideology begins, I do not know. The recognition that science is partly a social phenomenon has led to the remarkable and controversial studies of Margaret C. Jacob, whose examinations of the Boyle Lectures at the interface of science and religion have provided a much needed inroad into the perceptions of science by Newton's contemporaries.[3] But if we are to approach the

[2] This matter is discussed in Margaret C. Jacob, "Early Newtonianism," *History of Science*, XII (1974), 145; Jacob, *The Newtonians and the English Revolution, 1689–1720* (Ithaca, 1976), pp. 67–8, 180–4, 272; Jacob, *The Radical Enlightenment: Pantheists, Freemasons, and Republicans* (London, 1981), pp. 89–96. A brilliant exposition of the role of science and social context is given in Roger Cooter, *The Cultural Meaning of Popular Science. Phrenology and the Organization of Consent in Nineteenth-Century Britain* (Cambridge, 1985), pp. 9–19.

[3] Jacob, *The Newtonians and the English Revolution*, Chaps. 4–5; Jacob, "John Toland and the Newtonian Ideology," *Journal of the Warburg and Courtauld Institutes*, XXXII (1969), 321–4; Jacob, "The Church and the formulation of the Newtonian world-view," *Journal of European Studies*, II (1972), 1–21; with Henry Guerlac, "Bentley, Newton and Providence," *Journal of the History of Ideas*, XXX (1969), 307–18. See also her review article, "Science and Social Passion: The Case of Seventeenth-Century England," *Journal of the History of Ideas*, XLIII (April–June 1982), 331–3. Cf. John Gascoigne, "From

meaning of Newtonianism for the early eighteenth century, we have to recognize that Newton's followers created from his natural philosophy a social movement of many facets within the particular set of intellectual and political circumstances of late Stuart and early Hanoverian England. We must deal, therefore, not simply with the expression of an eristic manifesto of experimental philosophy – whether of Boyle in the Restoration or of Newton's "General Scholium" in 1713 – but with an *experiential* culture that emerged with the market economy of the eighteenth century. In relation to the eighteenth century, J. V. Golinski remarked perceptively about the "social-cultural construction" of useful science.[4] Utility was not only a response to crude perceptions of need, but also to seductive assertions of the ability to control nature. Herein was to lie a broad acceptance of the truth of scientific claims generally – and especially of those claims of utility repeatedly made from the Restoration to the mideighteenth century.

The issue of the association between science and technology prior to the great Industrial Revolution of the early nineteenth century has produced less light than sound and fury. And the debate has been obscured by some nicely considered prose. Twenty years ago, A. E. Musson and Eric Robinson took issue with what they regard as exaggerations and "tilting at windmills" by Rupert Hall in his effort to discount any causal link between the scientific revolution of the seventeenth and the industrialism of the eighteenth century.[5] Hall has forcefully replied that the success of science is a matter of "esteem from fellow-scientists," and has nothing whatever to do with "popular acclaim" or "practical usefulness."[6] This is historically naive; one cannot agree that "to find a causal relationship between the popularity of science and innovation in technology in eighteenth-century Britain requires a degree of faith."[7] However, the question has been often reduced to a matter of faith. We have only begun

Bentley to the Victorians: The Rise and Fall of British Newtonian Natural Theology," *Science in Context*, II (Autumn 1988), 224–5.

4 J. V. Golinski, "Utility and Audience in Eighteenth-Century Chemistry: Case Studies of William Cullen and Joseph Priestley," *British Journal for the History of Science*, XXI (March 1988), 1–31, esp. 3–4.

5 A. E. Musson and Eric Robinson, *Science and Technology in the Industrial Revolution* (Toronto, 1969), pp. 27–9; cf. A Rupert Hall, "On Knowing, and Knowing how to . . . ," *History of Technology*, III (1978), 99–100.

6 A. Rupert Hall, "What did the Industrial Revolution in Britain Owe to Science?" in Neil McKendrick, ed., *Historical Perspectives: Studies in English Thought and Society in Honour of J. H. Plumb* (London, 1974), p. 131.

7 Hall, "What Did the Industrial Revolution in Britain Owe to Science?" p. 143. A more recent analysis of this problem is Michael Fores, "Constructed Science and the Seventeenth Century 'Revolution,'" *History of Science*, XXII (1984), 217–44.

to explore the wide social interest in science and the enormous demand for its dissemination.[8] Therefore, we need to know more about the historical reality behind what Margaret Jacob has asserted as the role of Newtonianism in "the underpinning of a social and political ideology" associated with the emergence of "capitalistic enterprise."[9]

As far as the early eighteenth century was concerned, what converged was a whole series of economic, religious, political, and technological influences beyond those that formed in the narrow crucible of the early Royal Society.[10] Among the practitioners of natural philosophy, a convincing epistemology founded in experiment – one, therefore, of public display – would ultimately make the claims of utility convincing. The association between practical men and the Newtonians, evident in the lists of subscribers to Newtonian publications, in the designs for engines and ships to be found in Harris's *Lexicon Technicum*, for example, or in the plans to solve the difficulties of navigation suggests that the new philosophy had great appeal to men of trade and finance. This presents the problem of a public, and frequently anonymous, technological culture, the origins of which remain often hidden in the fragments of obscure lives. If Sprat's rhetoric on the importance of artisans, or later Jonathan Swift's satire on commercial or technological projects, is to have any meaning, we must lift the veil on the lives of those who would unite natural philosophy with the world of machines and their makers.

Encountering the stock-jobbers

The involvement of natural philosophers in the plans of projectors was not likely to endear them to the social critics of the eighteenth century. Swift's well-aimed satire on the Royal Society was not so much an attack on the Society as a broader slash of the sword against all projectors who more often than not duped the unwary. Those natural philosophers like John Harris or Desaguliers, among others, who were drawn to the plans of some of the early eighteenth-century innovators, were all tarred with

[8] Edwin T. Layton, Jr., "Technology as Knowledge," *Technology and Culture*, XV (1974), 40–1.

[9] Margaret C. Jacob, *The Newtonians and the English Revolution, 1689–1720* (Ithaca, 1976), p. 272; also Jacob, "Early Newtonianism," *History of Science* (1974), 142–6; Jacob, *The Radical Enlightenment: Pantheists, Freemasons and Republicans* (London, 1981), pp. 91, 94–5, 136. Cf. Colin Russell, *Science and Social Change 1700–1900* (London, 1983), pp. 38ff., 58.

[10] J. L. Heilbron, *Physics at the Royal Society during Newton's Presidency* (Los Angeles, 1983), *passim*.

the same brush. And in a society deeply agitated by questions of legitimacy and corruption, they were forced to face sometimes unpleasant attacks. The fact that many of them were clergymen would also not escape the attention of those incensed by a surge of irreligion.

The projectors of the early eighteenth century represent the main stream of entrepreneurial life. They were not merely a subculture whose questionable respectability prohibited them access to polite and proper society. On the contrary, they were held in low regard precisely because they had duped so many prominent individuals. Of course, effort to revile projectors for the carnage they left behind their plunders was well enough founded. Those looking at the South Sea Bubble had no trouble concluding who was to blame. The problem was that such views were often after the fact, when doomsayers delighted in the confirmation of disaster. In fact, the projectors had a willing clientele; that was exactly what made them so effective, and often dangerous.

Financial chaos inevitably induced in the public mind broader social questions about the emergence of a new kind of entrepreneur. They were men whose roots in land seemed very slight – but that was only at first glance. One must also be aware of cases where the boundary between land and industry may be more rhetorical than historically real. Prominent landowners were happily drawn to the hunt, for silver or for coal on their estates, willing to explore the possibilities of new technology, and certainly deeply involved in the machinations of investment in some of the more adventurous companies. As one anonymous pamphleteer argued in 1695, "the interest of Trade and Land are the same, for the mutation frequently happens; the Moneyed Man today is a Landed Man to morrow; and the Landed Man to day becomes a Moneyed Man to morrow."[11]

The gulf between pristine nature and disruptive industry may have more romantic appeal than actuality. The management, for example, of the coal mines near Dudley Castle certainly must have had much in common with the control of a large estate.[12] Likewise, there were improving landlords like the Duke of Chandos or Thomas Coke, later first

[11] *Some thoughts concerning the better security of our trade and navigation*, quoted in Joyce Oldham Appleby, *Economic Thought and Ideology in Seventeenth-Century England* (Princeton, 1978), p. 177. See Julian Hoppit, "Attitudes to Credit in Britain, 1680–1790," *The Historical Journal*, XXXIII (June 1990), 305–22, esp. 308–11.

[12] Sidney Pollard, *The Genesis of Modern Management. A Study of the Industrial Revolution in Great Britain* (London, 1695), p. 26; cf. Francois Crouzet, *The First Industrialists. The Problem of Origins* (Cambridge, 1985), pp. 22–5, 68–9; and K. G. Davies, "Joint-Stock Investment in the Later Seventeenth Century," *Economic History Review*, second series, IV (1952), 286

Earl of Leicester, who learned many a bitter lesson in the myriad of stocks that disappeared in the vapors of the South Sea Bubble.[13] If there were barriers, they broke them often, frequently to their regret.

Likewise, the ambitions of projectors might only be realized with the success of technical innovations.[14] The careers of these remarkable innovators, who skirted the fringes of the great chartered companies, would certainly be worth further examination. If they sometimes disappear into the shadows of King's Bench Prison, or leave vague traces of their flight to the Continent, this should not give reason to ignore the exploits that captivated the London press. In the realm of politics or natural philosophy, they were mostly marginal men. But those who relied on patronage, like Desaguliers, are as revealing as those, like Newton or Chandos, who dispensed it.

The concept of corruption in the early eighteenth century was born out of a widespread feeling of injustice. This had been created, in part, in the post-Revolutionary experiments with the coinage, but also out of the direct link between the chartered companies like the African Company or the South Sea Company, which seemed the chief beneficiaries of British foreign policy and the National Debt.[15] This real sense of grievance by those forced to pay taxes to fund such policies was even more deeply aggravated when many became entangled in the intrigues of stock-jobbers in the early years of the Hanoverians. Trade seemed increasingly disreputable. Daniel Defoe complained loudly in *Applebee's Journal* about the "Knaves of Projectors" who sought out "high Patrons; not only to Disguise their Frauds . . . but to bear the weight of popular Clamour, when that Fraud is Detected."[16] By the time Bolingbroke organized the *Craftsman* and its attack on the Robinocracy, there was a long-established fear of the influence of ill-gotten wealth. In 1727, for example, the Tory William Statford complained about stock-jobbers buying the votes of freemen to secure a place in Parliament.[17] From the Exchange and the pawnshop to the House of Commons, they wreaked their havoc.

[13] R. A. C. Parker, *Coke of Norfolk. A Financial and Agricultural Study* (Oxford, 1975), Chapter 2.

[14] Such was the case, for example, with William Wood. See J. M. Treadwell, "William Wood and the Company of Ironmasters of Great Britain," *Business History*, XVI (1974), 93–112; cf. Crouzet, *First Industrialists*, pp. 24–5.

[15] P. G. M. Dickson, *The Financial Revolution in England, 1688–1756* (London and Melbourne, 1967), pp. 17–18; and Michael John Jubb, "Fiscal Policy in England in the 1720s and 1730s" (Ph. D. thesis, University of Cambridge, 1978), pp. 221, 408.

[16] William Lee, ed., *Daniel Defoe: His Life, and Recently Discovered Writings* (London, 1869), III, pp. 101, 175–6.

[17] H. T. Dickinson, *Bolingbroke* (London, 1970), p. 188; and Isaac Kramnick, *Bolingbroke*

It was virtually impossible even for those disposed to trade to isolate the mischief of stock-jobbers from the legitimacy of the merchant. Even pretenders to country virtue like Pope or Gay had trouble with the segregation.[18] This was especially true in the aftermath of the collapse of the numerous companies that failed in the early eighteenth century. As one anonymous critic reflected in the chaos that surrounded the scandal of the Charitable Corporation in the 1730s, "Stock Jobbing is at once so great and so known an Evil, that like Atheism, though many Hazard their Safety in its Practice, yet few care to risque their Reputation in defending it in theory."[19] In the new property of stocks resided not only a fear of corruptibility of the political nation, but also a fair share of fantasy that was the engine that moved so many projects. And in such dreams, there was a susceptibility to claims of invention, and to manipulation by inventors whose claim to a patent received only a rudimentary legal scrutiny.

Inventors were often at the center of the storm, the focal point of the energy that generated the financial patronage that enabled projects to take shape. At the end of the seventeenth century, patents and joint-stock companies were inseparable. The large number of patent applications for engines to drain mines, for example, was a reflection not of mechanical success, but of industrial intention.[20] And in that purpose lurked the best promotion of natural philosophy that the lecturers, instrument makers,

and His Circle. The Politics of Nostalgia in the Age of Walpole (Cambridge, Mass., 1968), pp. 42–8. It is entirely possible that Statford was thinking of Chandos's confederates.

18 Appleby, *Economic Thought and Ideology*, p. 258; J. G. A. Pocock, *The Machiavellian Moment. Florentine Political Thought and the Atlantic Republican Tradition* (Princeton, 1975), pp. 465–6. See also Hector Monro, *The Ambivalence of Bernard Mandeville* (Oxford, 1975), Chapter 7.

19 Quoted in Armand DuBois, *The English Business Company after the Bubble Act, 1720–1800* (New York, 1938), p. 330; cf. Pocock, "Early Modern Capitalism – The Augustan Perception," in Eugene Kamenka and R.S. Neale, eds., *Feudalism, Capitalism and Beyond* (Canberra, 1975), p. 80.

20 Pocock, "Early Modern Capitalism," pp. 72–9; and K. G. Davies, "Joint-Stock Investment in the Later Seventeenth Century," *Economic History Review*, second series, IV (1952), 285. Note the glorification of joint-stocks in John Houghton, *A Collection for Improvement of Husbandry and Trade*, V, no. 103 (Friday, July 20, 1694). He refers directly to Halley's diving bell, as well as the New River and City Conduits. The joint-stock companies became increasingly important not only for their financial impact, but also as a means of entry into trade. Richard Grassby, "Social Mobility and Business Enterprise in Seventeenth- century England," in Donald Pennington and Keith Thomas, eds., *Puritans and Revolutionaries: Essays in Seventeenth-Century History, Presented to Christopher Hill* (Oxford, 1978), p. 34. On the rapid growth of such concerns after the Revolution, see W. R. Scott, *The Constitution and Finance of English, Scottish and Irish Joint-Stock Companies to 1720*, Vol. I (reprint edn., Gloucester, Mass., 1968), pp. 327–33.

and inventors could possibly have wished. In 1697, Daniel Defoe remarked in his *Essay on Projects* that "There is, 'tis true, a great difference between New Inventions and Projects, between Improvement of manufactures or Lands, which tend to the immediate Benefit of the Publick, and Imploying of the Poor; and the Projects fram'd by subtle Heads, with a sort of Deceptio Visus, and Legerdemain, to bring People to run needless and unusual hazards."[21] To some, this was a deeply rooted confrontation of vice against virtue. This opinion, with more or less degrees of dismay, became the currency upon which many meditated after they had been swindled.

Such alarm also formed the basis almost half a century later for Desaguliers' defense of the employment of natural philosophers: "This is a very necessary Caution; for there are several Persons who have Money, that are ready to supply boasting Engineers with it, in hopes of great Returns; and especially if the Project has the Sanction of an Act of Parliament to support it – and then the Bubble becomes compleat, and ends in Ruin."[22] In this caveat, a case was made for the continued exploration of nature – so that deceit might be defeated, and the virtue of trade defended. In 1727, in his *History of the Principal Discoveries and Improvements, In the several Arts and Sciences*, Defoe compared the failures that had gone before with the current spirit of enquiry and improvement:

> Where were the Men that arriv'd to Characters, to Fame, and to Distinction, by Trade, by the Mathematicks, by the Knowledge of natural or experimental Philosophy? Where was the Sir Walter Raleighs, the Verulams, the Boyls, or Newtons of those Ages? Nature being not enquir'd into, discover'd none of her Secrets to them, they neither knew, or sought to know, what now is the Fountain of all human Knowledge, and the great Mistery for the Wisest Men to search into, I mean Nature.[23]

There was no real ambivalence in Defoe, or in Addison, Steele or the Newtonian James Thomson, for whom the merchant could match the

[21] Defoe, *An Essay on Projects* (London, 1697), p. 15. See Simon Schaffer, "Defoe's Natural Philosophy and the Worlds of Credit," in John Christie and Sally Shuttleworth, eds., *Nature Transfigured. Science and Literature, 1700–1900* (Manchester and New York, 1989), pp. 13–44.

[22] Desaguliers, *A Course of Experimental Philosophy*, II (London, 1744), p. viii.

[23] Daniel Defoe, *The History Of the Principal Discoveries and Improvements, In the Several Arts and Sciences: Particularly the great Branches of Commerce, Navigation, and Plantation, In all Parts of the known World* (London, 1727), pp. 238–9. See also McVeagh, *Tradefull Merchants*, pp. 47, 59–61; and Neil McKendrick, "'Gentleman and Players' revisited: the gentlemanly ideal, the business ideal and the professional ideal in English literary culture," in McKendrick and R. B. Outhwaite, eds., *Business Life and Public Policy* (Cambridge, 1986), pp. 108–11, 118–19.

aristocrat. Placing the philosophers in this universe of stock-jobbers and projectors provides one clue to appreciating the rage for natural philosophy. Public science had the capacity to supply the necessary prudence.

Stocks, Steele, and the Duke of Chandos

By 1744, when Desaguliers made his complaint about projectors, he had not only the warnings of Defoe and the lamentations of the Tory satirists, he had a wealth of experience. One of the most remarkable, and most overlooked, aspects of Desaguliers' life was his extensive personal knowledge of the universe of the eighteenth-century inventor. In 1697, Defoe had remarked that "Stock-Jobbing nurs'd Projecting, and Projecting in return has very diligently pimp'd for its Foster-parent, till both are arriv'd to be Publick Grievances; and indeed are now almost grown scandalous."[24] At least, he was prophetic. When Desaguliers later reflected upon how true this warning had turned out to be, his reaction to the support for ludicrous mechanical projects dripped with scorn: "Our Legislators may make Laws to govern us, repeal some and enact others, and we must obey them; but they cannot alter the Laws of Nature; not add or take away one iota from the Gravity of Bodies."[25] There was some considerable irony in the fact that Desaguliers had been implicated in some of the schemes that went seriously awry, and thus had earned the approbation of project. And he had seen first hand many of his associates drawn by the magnetic promise of sudden wealth. All too often hope became a dismal spiral into poverty.

The rage in stocks spread like the ubiquitous smallpox of Desaguliers' time. In one sense, smallpox might be preferable to the madness of trading in Exchange Alley; at least the organic disorder left its survivors with an effective immunity, which was not the case with the roulette of eighteenth-century investment. But both disorders could be very difficult to avoid, and no one had yet discovered an inoculation against greed.

Desaguliers was drawn very early to the machinations of investors. In 1717, he had been involved in advising Chandos about an investment that cannot now be identified with certainty.[26] It is possible that the scheme

[24] Defoe, *An Essay on Projects*, p. 30.

[25] Desaguliers, *Course of Experimental Philosophy*, II, p. viii. His sentiment was not uncommon. In 1667, Thomas Sprat argued that to overcome idleness and improve industry "the true method of increasing industry, is by that cours which the Royal Society has begun in Philosophy, by Works, and endeavours, and not by the praescriptions of words, or Paper Commands." Sprat, *History*, p. 422.

[26] Hunt. Lib., Stowe 57, XIV, f. 377, Chandos to Desaguliers, March 18, 1716/17.

to which their correspondence refers was the fish pool of William Whiston's patron, and Whig M.P., Richard Steele. At the time of the consultation between Desaguliers and Chandos, Steele was attempting to have constructed a model of a vessel designed to transport live fish to the London market. In this, he was assisted by the mathematician William Gillmore and had taken in his Whig friend, William Benson, who was lobbying for one of the lucrative auditorships of the Imprest.[27] Steele was always strapped for funds and it was Benson who appears to have put up the initial investment, although within the year, he no longer seems to have had much to do with the project, perhaps then concentrating his attentions on the patronage of the Court.[28] Steele seduced a patent from the Crown and work on the construction of a full-scale vessel followed from the more substantial backing of the Duke of Chandos. Chandos loaned Steele enough funds to keep the project going throughout 1718, although toward the end of 1719, Chandos was beginning to have his doubts. In August, he told Steele that he was willing to remain a subscriber, but that he would take no further direct role in the affair.[29]

Chandos's reference to a subscription or to shares in the stock of the company to operate the Fish Pool was an interesting one. In the following spring, Steele explored the possibilities of a joint-stock consulting Chandos, along with Robert Knight, the treasurer of the ill-starred South Sea Company, and with John Law in Paris, the originator of the infamous Mississippi Scheme.[30] By the end of 1720, not only had shares in the Fish

[27] This was the same William Benson whom Desaguliers came to loathe as a great fraud in his building of steam engines. There are several extensive treatments of Steele's venture. See George A. Aitken, *The Life of Richard Steele*, Vol. III (London, 1889), pp. 157–79; and Calhoun Winton, *Sir Richard Steele, M.P. The Later Career* (Baltimore, 1970), Chapter VI. On William Benson, see Winton, *Sir Richard Steele*, p. 103; and H. M. Colvin et al., eds., *The History of the King's Works*, Vol. V (1660–1720) (London, 1976), esp. Chapter V, which describes Benson's skulduggery in succeeding Wren as Surveyor-General. Benson was dismissed in 1719 after falsely claiming that the House of Lords was in danger of collapse and in need of rebuilding. Desaguliers' remarks about Benson and the building of fountains of Herrenhausen were evidently accurate. See Desaguliers, *Course of Experimental Philosophy*, II, p. 526.

[28] Aitken, *Life of Richard Steele*, II, p. 135; and Winton, *Sir Richard Steele*, p. 136n27.

[29] Hunt. Lib., Stowe MSS. 57, XVI, f. 243, Chandos to Steele, August 11, 1719. On Chandos's involvement, see XV, f. 245, June 6, 1718; XV, f. 305, August 19, 1718; and XVI, f. 9, October 13, 1718. This exchange is reprinted in Rae Blanchard, ed., *The Correspondence of Richard Steele* (Oxford, 1941). Between 1710 and 1719 Chandos claimed to have loaned Steele in excess of £1000. Blanchard, *Correspondence*, p. 128n2. On Steele's patent, see PRO, SP. (Dom.) 35/11/23. It did not take long before rivals appeared, such as the invention of Joseph Avery "Mathematician and Shipwright." PRO, SP (Dom.) 35/12/92. Oath, October, 1719. Avery's claims to have improved on Steele's invention were reported to the Royal Society by Halley. Royal Society MSS., Journal Book (C), XI (1714–20), 466 (March 17, 1719/20).

[30] Winton, *Sir Richard Steele*, p. 155. By this point, Steele was no stranger to joint stocks,

Pool surfaced on the Exchange, but he had also sold his patent to William Dale, the wealthy upholsterer.[31] This resulted in a storm amongst the shareholders who held a violent meeting in January 1721 in Steele's Censorium in York Buildings. Given Desaguliers' contact with Steele and with Chandos in 1719, it is inconceivable that he could have been unaware of the numerous opportunities in joint stocks. In any case, like so many other grand designs, the Fish Pool floundered on the competition and disappeared into the maw of Brunsden's Dock in November 1722.[32]

The slightest scent of difficulty could scuttle the most worthy of projects. Even Thomas Watts had seen the possibilities and was trying to revive the Royal Fishery, which had received its charter under Charles II.[33] Early in 1720 an attempt was made to interest Chandos in the project, although by that time, it seems the Duke disliked the smell of fish. He declined to be concerned in the venture, as he told the former South Sea Director Arthur Moore, unless "there was a probability of its ending more advantageous than a Stock-jobbing Voyage generally proves."[34] Chandos was undergoing one of his infrequent bouts of caution.

having sold partnerships in his rights to the patent to Drury Lane theater in order to clear his away debts. See John Loftis, *Steele at Drury Lane* (Berkeley and Los Angeles, 1952), pp. 91ff.

31 On Dale, see Rae Blanchard, *Correspondence*, pp. 157–8. Steele to Joseph Gillmore, December 10, 1720; and *ibid.*, pp. 502–3. The identification of William Dale offers fascinating possibilities. In 1720 or 1721, a William Dale was engaged in a transaction to purchase leases held by others from the Mines Royal and Mineral and Battery Works. As it happens, he was involved in the enterprise with Case Billingsley of the Longitude hunt, as well as in the numerous lawsuits that followed. PRO, Chancery Exhibits and Answers, C11/1455/8. John Essington v. Case Billingsley. The plan had been to improve the production of copper and brass. How Dale became concerned in Steele's Fish-Pool is not clear, but at Temple Mills, near Hackney, where the Fish-Pool was first tested, there was a large brass works. It is also interesting that in December 1720, Dale was reported to have purchased the forfeit estate of Viscount Bolingbroke. If true, then it suggests connections not only with the York Buildings Company – a Billingsley scheme – but also with Chandos, who had acted as Bolingbroke's financial agent after his flight in 1715. See Hunt. Lib., Stowe MSS. 57, XVII, fols. 206–7, Chandos to Bolingbroke, October 5, 1720. Steele was a member of the Forfeited Estates Commission from 1716 to 1723, the body charged by the Crown with the disposal of the Jacobite estates. On Dale, see Winton, *Sir Richard Steele*, pp. 192–3; and J. R. Ward, *The Finance of Canal Building in Eighteenth-Century England* (Oxford, 1974), pp. 11–12.

32 Aitken, *Life of Richard Steele*, II, p. 255n4.

33 Scott, *Constitution and Finance*, II, pp. 372–6. On Watts, see Spencer Library, University of Kansas, Lawrence, Kansas, MS. 83/3/19. Watts to Arthur Moore, March 5, 1719. Moore had succeeded Brydges as paymaster of the armed forces, was a former director of the South Sea Company and M.P. for Great Grimsby. *D.N.B.*, XXXVIII, 340; Sedgwick, *History of Parliament*, II, pp. 270–1

34 Hunt. Lib., Stowe MSS. 57, XVII, fols. 3–4, Chandos to Arthur Moore, February 9, 1719/20. In the aftermath of the Bubble, the fishery encountered severe legal difficulties. See DuBois, *The English Business Company*, p. 19.

The Fish Pool was only one of that great rage of investment that inflicted itself on an all-too-eager Exchange Alley soon after the Hanoverian Succession. For a while, the mists of the Thames may have masked the excesses of the stock-jobbers near the Royal Exchange. But there was far too much clamor for stock and a frenzy for the subscriptions to the great organizations like the South Sea Company to prevent disaster.

Parliament and the South Sea Bubble

At first, the sense of danger was misplaced. Parliament was initially concerned about those smaller enterprises that sailed far too close to the political wind to make an independent City Member of the House of Commons like John Hungerford very comfortable.[35] Rumors of bribes in Exchange Alley were increasingly frequent. But by the spring of 1720, questions were being asked in the House of Commons about the proliferation of insurance schemes, most especially about those that had found it possible to operate as joint stocks on the basis of the purchase of obsolete charters. A committee of the House was set up to examine these and particularly the interlocking directorships of some of the more controversial schemes such as the Royal Exchange Assurance or the London Assurance.[36] As it turned out, suspicions were well founded about an extensive manipulation of charters and the financial markets. Not only did the committee find enormous sums of money credited to the various subscriptions, but they also took note that many of them were organized from the numerous coffeehouses in the vicinity of the Exchange. The British Fishery, for example, was directed from Garraway's, a subscription for £2,000,000 was set up in Cooper's Coffee House, and the Company for Raising Thames Water in York Buildings bubbled merrily away in the Marine Coffee House. All were seen to prejudice public trade and commerce.[37] They had nothing to say, however, about the constant subscriptions of the South Sea Company. The timing was hardly propitious

[35] Lists of the numerous bubbles are provided in Adam Anderson, *An Historical and Chronological Deduction of the Origin of Commerce*, Vol. III (London, 1801), pp. 108–12; [Anon.,] *The Battle of the Bubbles* (London, 1720), pp. 18–24. See also Scott, *Constitution and Finance*, Vol. III, pp. 445–58.

[36] See *The Special Report, from the Committee Appointed to Inquire into, and Examine the several Subscriptions for Fisheries, Insurances, Annuities for Lives, and all other Projects carryed on by Subscription, in and about the Cities of London and Westminster; and to Inquire into all Undertakings for Purchasing Joint-Stocks, or Obsolete Charters* (London, 1720). This report of the Hungerford Committee is reprinted in Sheila Lambert, ed., *House of Commons. Sessional Papers of the Eighteenth Century* (Wilmington, 1975). See also John Carswell, *South Sea Bubble* (London, 1960), pp. 116–7, 139–40.

[37] PRO, SP (Dom) 35/21. Resolution of the Hungerford Committee.

for any new venture like those of Richard Steele or even for the visions of water supply that danced in Desaguliers' head.

In 1718 and 1719, a flood of projects swept the public away in a hysteria of hyperinvestment. Some were based on proposed solutions to problems that had long vexed philosophers. We are already aware of the effort of the Government to secure a solution to the longitude problem when it provided a reward in 1714. At that time, such an initiative was only the briefest hint at a way to extend the rewards that might exist in the receipt of a patent. But patents were sometimes highly contentious affairs. Whiston and Ditton had convinced the Government that the problem was such a fundamental one for a maritime nation that urgent measures had to be taken. Only a bounty might produce a quick solution. Innovation, however, took an unexpected turn with the initiation of organized marine insurance.

The locus of this world of mariners and merchants was in the coffeehouses surrounding the Royal Exchange. The center of the early private underwriting of ships and cargoes was at Lloyd's in Lombard Street. However, the creation of joint-stock companies to undertake marine insurance dates from 1717, in part through the idea of a self-styled solicitor, Case Billingsley of the firm of Bradley and Billingsley. This is the same gentleman who had made an attempt to solve the longitude, and claim the reward, in 1714 with the publication of a tract, *The Longitude at Sea*.[38] Shortly thereafter, the burden on the longitude was deflected by notions of marine insurance.

Billingsley initiated a plan to purchase the charter of the Thames Water Supply Company at York Buildings, a process that would be far cheaper than bidding for a new charter. Patented in 1675, the old company was sold in 1719 on the eve of the South Sea speculations to a group of shareholders under the direction of Case Billingsley, who was one of those who saw the financial possibilities in the application of scientific discovery. Billingsley's schemes eventually connected this York Buildings Company, as it became known, to the Royal Exchange Assurance concerned with maritime insurance. These companies were to become the target of a Parliamentary investigation.

A similar example of the emergence of the insurance market in London was the foundation of the Sun Fire Office. This was a creature of Charles Povey, who published *The General Remarks on Trade* from the Traders

[38] E. G. R. Taylor, *The Mathematical Practitioners of Tudor and Stuart England* (Cambridge, 1954), p. 304; and E. G. R. Taylor, *The Mathematical Practitioners of Hanoverian England* (Cambridge, 1966), p. 111.

Exchange-House in Hatton-Garden. By 1705 Povey was touting the notion of life insurance, which was soon transformed by 1708 into insurance against losses by fire. The result was the Company of London Insurers, or the Sun Fire Office, after the symbol chosen by Povey. It is significant that of the twenty-four directors chosen in 1709 to run the company, several were among the Fellows of the Royal Society. These included John Harris, Richard Richardson, M.D., Dr. Benjamin Pratt, the Dean of Rochester, and possibly John Billers.[39] Harris soon appeared in the list of directors of the Union, or Double Hand in Hand, Fire Office founded in 1714 for the insurance of goods, along with his patron Sir Charles Cox with whom he had initiated public lectures at the Marine Coffee House.[40] In any case, by 1710, the Sun was operating from Paul's or George Causey's Coffee House in St. Paul's Churchyard, although they soon took a new office near the Royal Exchange in rooms next door to the Amsterdam Tavern. This may also have had some connection with Garraway's Coffee House, which was to play a role in the establishment of the new insurance schemes in 1719. We find yet another strong link with the Newtonian community in the involvement of Thomas Watts in the Sun Fire in 1720, notwithstanding his interest in the Academy in Little Tower Street.[41]

Newtonians like Desaguliers and Harris exploited the connections established during the lectures on Newton's philosophy. They and many others, including Thomas Watts, had proved important in widening the cultural impact of Newtonianism through their lectures and publications. But they were also important in bridging what now must appear as a very narrow gap between the world of natural philosophy and the entrepreneurs of the eighteenth century. This presents us with the very problem that so tantalized the Tory satirists: did natural philosophy inevitably lead to the kinds of disruptive experiments that were evident in Exchange Alley? In one sense, the answer must surely have been yes, for

[39] Edward Baumer, *The Early Days of the Sun Fire Office* (London, 1910), pp. 8–9; P. G. M. Dickson, *The Sun Insurance Office, 1710–1960* (London, 1960), pp. 17–31. Among the Committee, several were to be subscribers to the *Lexicon Technicum* (1710). The inclusion of Dr. Pratt is interesting as in 1708 Harris had been named a Prebend of Rochester.

[40] *Daily Courant* (October 17, 1716). In the following year, Harris was no longer on the list, but Cox remained. There had also been added John Payne who may well be the Duke of Chandos's agent at Bridgewater, on whom see below. *Daily Courant*, October 11, 1717.

[41] Dickson, *Sun Insurance Office*, pp. 31, 272; Bryant Lillywhite, *London Coffee Houses* (London, 1963), p. 439; Baumer, *Early Days of the Sun Fire*, p. 50. Cf. Aytoun Ellis, *The Penny Universities: A History of the Coffee-Houses* (London, 1956), pp. 112, 126–7.

it was clear that many like Desaguliers saw the merit in philosophers lending their expertise to the improvement of some of the technical schemes upon which so many projects rested. This, after all, is exactly what Desaguliers did.

It could hardly have helped any of the projects in which the Newtonians were involved to see the crisis of the South Sea Bubble emerge. In April 1720, the South Sea Bill, which engrafted part of the National Debt onto the South Sea Company, passed the House of Lords. It is instructive that the former Lord Chancellor, William, Lord Cowper, patron to both Harris and Desaguliers, was one of those to complain in the House of Lords about the Trojan Horse that was the South Sea Bill.[42] At the same moment, Swift was grumbling that he did not understand the South Sea mania although he thought that "perhaps the Frolick may go round, and every Nation . . . have its Mississippi."[43] A great fear of collapse spread across the political spectrum. As the panic reached to provincial towns like Bristol, where merchants had been heavily invested not just in South Sea but in other companies like English Copper, there were the inevitable calls for retribution that echoed those in Parliament against the South Sea Directors. The Cato letters published in the *London Journal* demanded "their Necks and their Money."[44] The entire atmosphere was one of suddenly high morals and low politics. Typical, perhaps, was the reflection of Samuel Bradford, the pluralist Bishop of Carlisle and former Boyle Lecturer, whose sermons reached their moral summit as stocks hit their nadir: "Wealth gotten by vanity shall be diminished: but he that gathereth by labour, shall increase."[45]

Insurance and bribery

The universe of projectors had a major impact on the possibilities presented to philosophers. Unfortunately, we know so little of this emerging world that we have to rely largely on the Parliamentary investigations

[42] Carswell, *South Sea Bubble*, p. 124.

[43] In Harold Williams, ed., *The Correspondence of Jonathan Swift*, II (Oxford, 1963), p. 342. Swift to Charles Ford, April 4, 1720.

[44] These were the work of the radical whigs John Trenchard and Thomas Gordon who published *The Independent Whig*, although Trenchard remarked privately that the collapse was what he had long expected. Spencer Library, MS. G.3, f. 11. Trenchard–Simpson Correspondence, John Trenchard to Simpson, n.d. See Carswell, *South Sea Bubble*, pp. 246–7.

[45] Samuel Bradford, *A Sermon Preach'd in the Parish-Church of St. Mary-le-Bow. On Sunday, November 20, 1720* (London, 1720), p. 3. On Bradford, see Clyve Jones and Geoffrey Holmes, eds., *The London Diaries of William Nicolson, Bishop of Carlisle, 1702–1718* (Oxford, 1985), p. 54.

and the brief reports of the schemes that appeared in the press. Only the major figures, like the financial wizard John Law, or investors like the Duke of Chandos, have attracted much attention.[46] Although such persons were clearly in a position to offer their patronage, it was the less well-known innovators who perhaps reflected their world more accurately. Such, I think, was Case Billingsley, Longitudinarian and arch-projector. Because he created the companies that occupied the attention of Desaguliers, Chandos, and many of their associates, the detection of his scattered traces permits a closer examination of a hidden world.

From time to time throughout the early eighteenth century, Case Billingsley surfaced in the press. Sometimes the papers happily speculated on his reluctance to tell Parliament all he knew of the sudden collapse of one scheme or other – and sometimes the papers didn't know he was in their pages with anonymous advertisements of some new plan to rid the nation of debt. He was a figure who lurked in the shadows of Exchange Alley, only emerging into the light when sought by the authorities or in desperate need of help. But he also had prominent friends, most of whom are now impossible to trace as a legacy of the secrecy that was necessary for a man who skirted the law more than once. He was, above all, a financial buccaneer who was not afraid to cross swords with Isaac Newton.

Billingsley's effort in 1714 to discover the longitude was not particularly remarkable except insofar as he had taken the trouble to consult with Flamsteed. But his proposal was symbolic, for it represented his continuous search for riches as well as his apparent knowledge of the sea. They were not unconnected. It is probable that he developed his financial expertise as a result of an early connection with the navy. Throughout the reigns of William and Anne, Billingsley was one of the clerks to Sir Edward Gregory, Commissioner of the Chatham Ship Yard near Rochester.[47] This afforded him the opportunity to become skilled in financial manipulation, although what particular expertise recommended him in the first place is unclear. There is no doubt, however, that his knowledge of victualing allowed him to be assigned as purser to a succession of

[46] Baker and Baker, *Life and Circumstances of James Brydges*. Brydges' activities have also been explored in R. S. Neale, *Bath 1680–1850. A Social History OR A Valley of Pleasure, Yet a Sink of Iniquity* (London, 1981). On Law, see H. M. Hyde, *John Law. The History of an Honest Adventurer* (Amsterdam, 1948).

[47] PRO, ADM 7/1/16, Chatham, Salary and Pension Books; ADM 7/810/16 shows Billingsley was discontinued on November 11, 1714, as was Gregory; ADM 7/809/19, Sir Edward Gregory was made Commissioner at Chatham in October 1689. See R. D. Merriman, *Queen Anne's Navy* (London, 1961), pp. 108n2, 358.

vessels from at least as early as 1697 until the reign of George I.[48] Whatever the origin of Billingsley's naval career, it was a connection that served him well, although he disappeared from the records of His Majesty's Navy very early in the reign of George I. It was to be a long, and fascinating, hiatus.

The reasons why Billingsley retired as purser are entirely unclear. It may, however, have something to do with an obviously growing family for whom he had the need to provide. In 1711, while purser of the *Vanguard*, he was in the company of two sons, Case and John, who were listed on the ship's muster as stewards.[49] Perhaps his removal was simply the result of Utrecht and the end of the war. By 1713, he had made an abrupt transition into the Brewers' Company, although such a notion might easily have occurred to a purser responsible for ships' victualing who saw the profits to be made in supply. Billingsley took his freedom by precept from the Lord Mayor probably after having purchased a redemption from the Company. And he took with him into the Brewers' Company his sons Case, junior, and John, both of whom were immediately apprenticed to their father.[50] Billingsley attempted, unsuccessfully, to operate a brewhouse at St. John's, near the Hermitage, Wapping, although this appears to have been only part of a larger venture involving leases on fourteen such houses in the vicinity of London.[51]

48 PRO, ADM 42, Chatham Yard Books, Ordinary; ADM 42/7/6, the *Northumberland* (1697); ADM 42/9/8–11, the *Windsor Castle* (1704–5); ADM 42/10/8–16, the *Marlborough* and the *Vanguard* (1709–10). Warrant Books, ADM 6/4/115; 6/7/122; 6/9/35; 6/10/31; 6/11/29.

A family connection is also suggested by his duty on board the *Vanguard* in 1711 under Captain Rupert Billingsley. PRO, ADM. 8/11, Disposal of Ships and Vessels in Sea Pay, *Vanguard*, 1711; ADM 33/538. *Vanguard* Pay Book. In 1711, the Captain's wages, interestingly enough, were paid to Case Billingsley, junior. Was the family purchasing tallies at a discount at this early date? Rupert Billingsley died in 1720. See John Charnock, *Biographia Navalis*, III (1795), pp. 194–5. ADM 6/10/31. Warrant Books, *Marlborough* (Feb. 11, 1709) indicates a Charles Billingsley, a purser as well. There was also a Thomas Billingsley, a superintendent and bosun at the Sheerness Yard, who died in 1715. ADM 7/1/57.

49 PRO, ADM 36/4386/1–2. *Vanguard* Muster Book; ADM 33/358/9, 10, 19, 28, 256. *Vanguard* Pay Book. Billingsley was not the only purser to become an entrepreneur. The South Sea Director, James Edmundson, was purser of the *Royal Anne*. Carswell, *South Sea Bubble*, pp. 228, 277.

50 Guildhall Library, London. MS. 5445/25/128, 143, 249, Brewers' Company Minutes, no. 7 (1707–24). Both sons gained their freedom of the Company in 1720 and the livery in 1725 by which time their address was the "Insurance Office, York Buildings." This was the Royal Exchange Assurance. MS. 5445/26/22, 37, Brewers' Company Minutes, no. 8 (1724–1731). Also MS. 5449a and MS. 5449, Brewers' Company, Freemen and Livery List (1665–1754). Freedom by redemption does not appear to have been unusual. See W. F. Kahl, "Apprenticeship and the Freedom of the London Livery Companies, 1690–1750," *Guildhall Miscellany*, VII (1965), 17–20.

51 In 1714, bankruptcy proceedings were undertaken against Billingsley although he seems

It was not in the nature of Case Billingsley to be concerned with small undertakings. This was a characteristic that he had in common with many projectors, although it is also evident that he seldom had the financial acumen to put his plans into practice. His entry into the Brewers' Company, however, did permit him to apply his freedom to the trading of stocks on the Royal Exchange. Billingsley was living in Silver Street about 1717, during the same period he was operating the brewhouse in Wapping.[52] He was evidently using the Silver Street location as a headquarters for his operations that he and his sons had created to exploit the wave of investment that had been gathering momentum since the Revolution.[53]

By 1717, Billingsley was caught in the enthusiasm for insurance schemes that Charles Povey had so successfully stoked in the Sun Fire.[54] Billingsley's interest in the problem of longitude and his naval background turned his attention to the insurance of maritime ventures. From 1717 to 1720, several subscriptions were established for marine insurance. Two companies were eventually formed – the Royal Exchange Assurance under the patronage of Lord Onslow, and the London Assurance under the influence of Lord Chetwynd. Incorporation, however, proved difficult because of opposition from brokers on the Royal Exchange who saw their private underwriting disappearing. An extensive lobbying exercise

to have survived. PRO, B4/2, p. 54, Bankruptcy Commission. Docket Book (1713–18). The result was a series of lawsuits in which Billingsley's Rochester connections were confirmed. Billingsley had paid about £1000 for the leases, possibly as early as 1712, although his losses in 1714 resulted in the leases being transferred to his partner John Meadows and thence to Joseph Lockwood, brewer. When Billingsley made good his debts in 1720, he and his son John bodily ejected Lockwood. See PRO, Chancery Bills and Answers, C11/460/15, Billingsley v. Lockwood (April 26, 1722); C11 465/15, Billingsley v. Lockwood (1722); C11/475/6, Billingsley v. Holland (October 22, 1723).

52 Taylor, *Hanoverian England*, p. 111; Peter John Wallis, *An Index of British Mathematicians. A Check-list. Part 2. 1701–1760* (Newcastle, 1976), p. 11; and Wallis and Wallis, *Biobibliography*, p. 62. Between 1717 and 1720, the Billingsley family was in St. Olave parish, Silver Street. Guildhall Library, MS. 11316, vols. 55, 58, 61, 64, Land tax Assessment Books.

53 Case Billingsley, junior, obtained the right to act as a Broker on the Royal Exchange in 1723. Corporation of London Record Office [CLRO], Broker's Bonds, no. 609, June 11, 1723. Significantly, his petition had been supported by some of the major figures in the numerous Billingsley enterprises such as James Bradley, John London, and Peter Hartopp. CLRO, Court of Aldermen Records, May 7, 1723; Brokers Admissions, Vol. I, f. 7. Case, junior, seems to have been unable to take much advantage of his situation, for by 1736, he was forced to petition to surrender his bond as he was a prisoner, probably for debt. CLRO, Brokers Bond, no. 609, December 1, 1736; Court of Aldermen, Minutes, Vol. 141 (1736–7), f. 55, December 7, 1736. In 1725, his brother John was excused the livery in the Brewers' Company on a plea of poverty. Guildhall Library, MS. 5445/26/37, January 14, 1725.

54 By 1708, Povey's insurance scheme was only one of several such offices insuring against fire or loss of life. See E. Hatton, *A New View of London* (London, 1708), II, p. 787.

erupted by January 1718 involving almost three hundred subscribers to a joint-stock to ensure against the loss of ships and merchandise at sea. In order to build their support, Billingsley and his partner James Bradley, now self-styled solicitors, enlisted the aid of Christian Cole, who was very well connected at Court.[55] Despite considerable political opposition, much of it probably generated out of the South Sea Company, Billingsley was more successful than he could have imagined.

Lord Onslow, who lent his patronage to Billingsley's ideas, was a member of the newly formed Mercer's Hall Marine Company. This took shape in the fall of 1717 and was the first phase of the operation.[56] Between them, Thomas Onslow and Billingsley conceived of the idea to purchase the charters of the recently united companies of the Mines Royal and Mineral and Battery Works and use them to carry out the insurance.[57] While the companies had been kept together by the eminent chemist, Dr. Moses Stringer, they had a checkered history and seemed ripe for picking.[58] Onslow was connected with the Mines Royal and the Mineral and Battery Works of which Newton had been a subgovernor since 1709. Before the new company was incorporated as the Royal Exchange Assurance, it was known as the "Governor, Assistants and Societies of the City of London of and for the Mines Royal, Mineral and Battery Works,

[55] Scottish Record Office [SRO], Gifts and Deposits, GD1/170/vol. III, Bradley and Billingsley to Cole, April 7, 1718. The *Daily Courant* reported him as a minor diplomat who had been the King's resident at Venice until 1716. It is likely they meant Charles Cole. *Daily Courant*, May 5, 1716. Christian Cole first comes to notice as an agent for John Moore, Bishop of Norwich, patron of William Whiston and Samuel Clarke. B.L., Add. MSS. 22910, fols. 482–3. Cole to Moore, June 13/23, 1699. Cole appears to have been connected with English merchants in Amsterdam. See SRO, GD1/170/vol. I, Philip Bodham to ?, February 22, 1714/15. The petition for insurance was presented to the King on January 25, 1717/18. See *A Special Report from the Committee Appointed to Inquire Into and Examine the several Subscriptions...* (London, 1720), in Shiela Lambert, ed., *House of Commons. Sessional Papers of the Early Eighteenth Century*, I (Wilmington, 1975), pp. 199–202.

[56] B.L. [Shelf Mark 8225.a.38], *A List of the Names of the Subscribers for Raising the Summe of One Million Sterling, as a Fund for Insuring Ships and Merchandise at Sea...* (London, 1718).

[57] SRO, GD1/170/vol. III, Petition to the King, February 3, 1718/19. True copy certified by Billingsley. It was presented to the King by Earl Sunderland and Earl Stanhope; Lord Onslow was received as Governor and Sir John Williams as Deputy Governor. On background from the point of view of the Mines Royal and Mineral and Battery Works, see William Rees, *Industry Before the Industrial Revolution* (Cardiff, 1968), III, pp. 658–65

[58] See John H. Appleby, "Moses Stringer (f. 1695–1713): Iatrochemist and Mineral Master General," *Ambix*, XXXIV (March 1987), esp. 39–42; Rees, *Industry*, pp. 658–660; Sidney Pollard, *The Genesis of Modern Management. A Study of the Industrial Revolution in Great Britain* (London, 1965), p. 15; and Sir George Clark, *Science and Social Welfare in the Age of Newton*, second edition (Oxford, 1970), pp. 66–7.

for assuring ships and merchandise." Its members were given preference in the purchase of shares in Billingsley's York Buildings enterprise.[59]

This was only the tip of a long-buried pyramid of bribery and political pressure. Billingsley offered one thousand pounds to both the Solicitor-General, Sir William Thompson, and to Sir Edward Northey, the Attorney-General, provided that the insurance charter was approved. After pledging absolute secrecy, the partners further reserved Thompson and Northey ten thousand pounds in the subscription.[60] Billingsley later explained that he was merely trying to recompense the officials for their losses since their normal fees for passing charters could be very large, and Billingsley did have many competitors. He told a committee of enquiry in 1719:

> Could I have once imagin'd this would be represented as Bribery, I should not have been such a Fool as to have given it under my Hand, had I not before known the Sentiments of the late Attorney and present Sollicitor General to be for such a Corporation; and had they not assisted us in obviating the Difficulties that lay in our way, to have promis'd them any thing, must have been look'd on as an Attempt of Bribery.[61]

Billingsley wasn't being entirely disingenuous as such payments were part of the expected revenues of the Augustan office holder. He was successfully able to hide more serious activities at Court. Christian Cole, who had been brought into the Mines Royal and Mineral and Battery Works by Billingsley, was given the task of trying to convince the King's German advisor, Baron Bothmer, to accept stock in the company.[62]

[59] Case Billingsley, junior, was cashier successively for the Mines Royal and Mineral and Battery Works insurance and for the Royal Exchange Assurance. Hunt. Lib., Stowe Nugent Financial MSS., No. 2, John Knight Accounts, Policies (1719–21). W. R. Scott, *Constitution and Finance*, III, 387, 396–8, 405–7, 421, 423, 480–1; and Clark, *Science and Social Welfare*, pp. 66–7. Of fifty-five men who petitioned Parliament on behalf of the life insurance venture proposed by the York Buildings Company, twenty-two were found among the original subscribers of the Mercer's Hall marine insurance scheme. Those who petitioned on behalf of the Royal Exchange Assurance were all members of the Mines Royal. Barry Supple, *The Royal Exchange Assurance. A History of British Insurance, 1720–1970* (Cambridge, 1970), pp. 24–5.

[60] A. J. G. Cummings, "The York Buildings Company. A Case Study in Eighteenth Century Corporation Mismanagement" (Ph.D. thesis, University of Strathclyde, 1980), pp. 451–4; Lambert, ed., *Sessional Papers*, I, pp. 210–14.

[61] Lambert, ed., *Sessional Papers*, I, p. 215. Case Billingsley, March 7, 1719.

[62] Cummings, "York Buildings Company," p. 454. On Cole see SRO, GD1/170/vol. I, Billingsley to Cole, January 6, 1718/19; January 29, 1718/19; on efforts to involve Bothmer see GD1/170/vol. I, Billingsley to Cole, February 21, 1718[/19?]; and Billingsley to Cole, November 16, 1719. See also Ragnhild Hatton, *George I. Elector and King* (Cambridge, Mass., 1978), pp. 147–51, 164–5, 253–5. Such practices were not uncommon as was discovered in the South Sea affair. Carswell, *South Sea Bubble*, pp. 115, 231, 240.

The Royal Exchange Assurance is of considerable interest in that it represents the early eighteenth-century entrepreneurial mentality exposed by in the South Sea Bubble. More important, for our purposes, is the strength of the connection between those who were engaged in these economic adventures and those who were the promoters and the projectors of the new science. For example, it is clear that many of Newton's early disciples were directly involved in several of the ventures associated with the South Sea period. The traffic between the instrument makers of Fleet Street or Crane Court and the coffeehouses of Birchin Lane was obviously in both directions. In the list of subscribers taken at Mercer's Hall between August 1717 and January 1718, we find not only the names of Bradley, Billingsley, and Onslow, but of James Atkinson, the mathematics teacher and publisher of works on the longitude, and John Harris, who had made a career of the propagation of the Newtonian philosophy. Moreover, in the, list we find the names of many who were also known to have directly supported by subscription the publication of Newtonian works such as Harris's *Lexicon* and Pemberton's later *View of Sir Isaac Newton's Philosophy*.[63]

As was the case with most of the enterprises, the Royal Exchange Assurance encountered severe financial strain in 1720 with the collapse of the Bubble. It appears evident, however, that there was an effort by some of those in the Royal Exchange to take control of the Sun Fire Company. The Sun Fire had originated with the idea of Charles Povey to institute the Exchange House Fire Office in 1708, eventually becoming the Company of London Insurers of the Sun Fire Office controlled by its twenty-four members. In 1720, members of the Royal Exchange Assurance had been able to gain ownership of a significant portion of the Sun shares.[64] One of those behind this maneuver was undoubtedly Thomas Watts along with his confederate Colin Campbell. Campbell appears to have been one of the Directors of the York Buildings Company.[65] Watts seems to have had early connections with the Sun Fire very

[63] *A List of the Names of the Subscribers* (1718). The list included Fellows of the Royal Society such as Robert Thompson and Maurice Emmett. Emmett was an associate of Billingsley. In the list as well were John Essington, who subscribed to Pemberton, and John Meadows of Stroud in Kent, Billingsley's former partner in the brewhouses, who subscribed to Harris. A number of Billingsley's business associates appear like John London, who, like Essington, became an M.P. Thompson, London, and James Bradley were all Whigs. See W. A. Speck and W. A. Gray, eds., *London Poll Books, 1713* (London, 1981), pp. 73, 96, 101, 113, 121.

[64] Dubois, *The English Business Company*, p. 210n342.

[65] Dickson, *Sun Fire*, pp. 41, 270–1. There were a number of individuals of the same name with an interest in mathematics. A Colin Campbell was Deputy Surveyor of the King's Works until he was dismissed with his mentor William Benson in 1719. See Colvin et

likely through a John Watts, who, in 1707, was among the "List of Trustees chosen for Povey's Proposals of Insurance." Thomas Watts, however, was also engaged with the goldsmith Sir James Hallett, one of the early members of Povey's Company of London Insurers. At the time when the Royal Exchange Assurance and the London Assurance were attempting to gain charters, the Crown also received a petition from Hallett for a joint stock to establish annuities for widows and orphans, a variant on the life insurance scheme. Possibly as early as 1716, it was Thomas Watts who undertook the calculations necessary to put this scheme into force.[66] What is significant is that these individuals, and Watts in particular, represent the deliberate effort to establish a network of interlocking directorships to control the companies. The Mines Royal and Mineral and Battery Works were the effective foundation for the Royal Exchange Assurance, which in turn came to dominate the Sun Fire.[67]

These companies were not only important because of the presence of Thomas Watts of the Academy, but also because they expose a network of opportunity for natural philosophers of the eighteenth century. The reason that many of them became engaged in these enterprises was that such schemes often required the success of some technical apparatus, however rudimentary. Desaguliers' water supply ventures were not peculiar in that regard. Before Desaguliers, the functioning of London's water supply required the employment of engineers like Sorocold or Lowthorp. This network of practical possibilities, in which an experimental philosopher might be in great demand, was far more extensive than has been appreciated. Likewise, the view that those well versed in the most recent principles of natural philosophy could actually be of value in solving technical difficulties was sufficiently widespread to indicate that Desaguliers' views had some following.[68]

al., *History of the King's Works*, pp. 59, 64, 470. There was also an astronomer of the same name who set up an observatory in Jamaica in 1731. See Taylor, *Hanoverian England*, pp. 171–2.

[66] Dickson, *Sun Fire*, pp. 268–9; L. B. Namier, "Brice Fisher, M.P.: A Mid-Eighteenth-Century Merchant and his Connexions," *English Historical Review*, XLII (October 1927), p. 520n5; Lambert, ed., *Sessional Papers*, I, p. 254. Sir James Hallett was a Director of the Mine Adventurers of England. Bodleian, MS. Rawl. D916, f. 291. General Court Minutes, December 22, 1721.

[67] For example, Billingsley's associate Christian Cole of the MRMBW was part of the directorship of the Royal Exchange Assurance. SRO, GD1/170/vol. III. Case Billingsley, junior to Cole, December 12, 1720. Despite great pressure, the Assurance companies survived by offering to pay the Crown £300,000 for a charter. Michael John Jubb, "Fiscal Policy in England in the 1720s and 1730s" (Ph.D. thesis, University of Cambridge, 1978), pp. 431–2.

[68] John S. Allen, "The 1712 and other Newcomen Engines of the Earls of Dudley,"

Steam and the joint stocks

This culture of innovation found expression not only in financial experimentation, but in technical achievement as well. In some cases, the manipulation of stock fed upon the inventions of the late seventeenth and early eighteenth centuries. The most meaningful example of the association between invention and investment was the early steam engines of Thomas Savery and Thomas Newcomen. In 1698, Savery had obtained a patent for a pump for "Raising water by the impellent force of fire" for which his rights were later extended to 1733. The device was essentially a suction pump that operated on the principle of condensation of steam and a partial vacuum. It was demonstrated before the Royal Society in 1699, and may have been witnessed by James Brydges, before Savery described it in print in his *Miner's Friend* in 1701. Very shortly thereafter, Savery was manufacturing his engine in nearby Salisbury Court, between Fleet Street and the Thames.[69]

The Savery device, effectively an atmospheric engine, proved to have limited usefulness in the mines where it was most needed. It was, therefore, soon superseded by the engine of Thomas Newcomen, an ironmonger from Dartmouth, who also now had to face the obstacle of the existing Savery patent, although the designs were quite different. The effect of the competition of the Newcomen improvement was an accommodation with those who controlled the Savery patent after Savery's death in 1715. This resulted in the marketing of the early steam engine through the increasingly ubiquitous mechanism of a joint stock.

The genesis of the division of shares in the Savery patent, now to encompass the Newcomen innovation, is as revealing as it is important. It may well be that some arrangement was made as early as 1705, although it is clear that Savery's will allowed for the establishment of an annuity in return for which his widow relinquished her rights to the patent. This scheme had been drawn up and witnessed by John Meres, of the Worshipful Society of Apothecaries, who then created a joint-stock company

Transactions of the Newcomen Society, XXXVII (1964–5), p. 65; L. T. C. Rolt and J. S. Allen, *The Steam Engine of Thomas Newcomen* (Hartington and New York, 1977), p. 42; H. W. Dickinson, *A Short History of the Steam Engine* (Cambridge, 1939), p. 36; and Musson and Robinson, *Science and Technology*, p. 38.

69 On Savery, see Rolt and Allen, *The Steam Engine*, pp. 26–8. There were numerous such inventions for drawing water proposed at about the same time, one of which may have been supplying a section of Southwark with water as early as 1705. See Scott, *Constitution and Finance*, II, pp. 480–1; III, p. 33. Savery had earlier obtained a patent for a millworks in 1696 or 1697. Bodleian Library, MS. Rawl. A.241, f. 119. Brydges' possible viewing of the engine took place on March 18, 1699. Hunt. Lib., Stowe MSS. 26, Vol. I.

to market the Newcomen engine. This partnership was in effect by 1716 and although Newcomen's name does not appear among "The Proprietors of the Invention for raising water by fire," evidently his interests were protected by Edward Wallin, pastor of the Baptist chapel at Maze Pond, Southwark.[70]

In 1716, the Proprietors set up shop in the Sword-Blade Coffee House in Birchin Lane, where those interested would be attended by Edward Elliot, one of the partners.[71] This, of course, was also the headquarters of the infamous Sword-Blade Company, whose operations in land and in finance were regarded as a threat to the Bank of England.[72] There is some evidence in the activities of the northern coal magnate, William Cotesworth, of direct connection between the steam engine group and the speculators of the Sword-Blade.[73] It was from this address that there grew an enterprise that issued licenses for the construction of the Newcomen engine all over Britain.

In the coal mines, the new engine found ample employment, especially as the Savery apparatus had proved inadequate to deal with the problem of flooding. For example, in the Whitehaven collieries of Sir James Lowther, drainage was proving to be a major obstacle to the mining of some of the most promising veins. As J. S. Allen has established, Lowther and his steward John Spedding were giving serious consideration as early as 1712 to the employment of a Newcomen engine, perhaps like the one that had recently been completed near Dudley Castle in Staffordshire.[74]

[70] Rolt and Allen, *The Steam Engine*, pp. 39–40, 58–9. Meres was the architect of the Apothecaries monopoly, the Navy Stock Company, and in 1713, the Laboratory Stock Company. On the Proprietors, see Alan Smith, "Steam and the City: The Committee of Proprietors of the Invention of Raising Water by Fire, 1715–1735," *Transactions of the Newcomen Society*, XLIX (1977–8), 5–20; John S. Allen, "Thomas Newcomen (1663/4–1729) and His Family," *Transactions of the Newcomen Society*, LI (1979–80), 11–24; and J. R. Harris, "The Employment of Steam Power in the Eighteenth Century," *History*, LII (1967), 137–8. A review of the rights of the holder of the patent confirms the agreement with Newcomen was reached in 1716. B.L., Add. MSS. 22675, fols. 27–28, Cases and Opinions of Sir Thomas Pengelly, Chief Baron of the Exchequer, July 1, 1720.

[71] Rolt and Allen, *The Steam Engine*, p. 59. On Elliott see Alan Smith, "Edward Elliott: Fire Engine Proprietor," *Transactions of the Newcomen Society*, L (1978–9), 230.

[72] Carswell, *South Sea Bubble*, pp. 30–2, 34–6; and Pat, Rogers, "Gulliver and the Engineers," *Modern Language Review*, LXX (1975), 263.

[73] Northumberland County Record Office, ZCE 10/5. William Cotesworth to Mr. Meers (of the Steam Engine Proprietors), March 4, 1720/1. A copy of this letter was kindly provided by Mr. R. M. Gard, County Archivist, NCRO.

[74] J. S. Allen, "The 1715 and other Newcomen Engines at Whitehaven, Cumberland," *Transactions of the Newcomen Society*, XLV (1972–3), 241; Allen, "The 1712 and other Newcomen Engines of the Earls of Dudley," *Transactions of the Newcomen Society*, XXXVII (1964–5), 57–84; Allen, "The Introduction of the Newcomen Engine, 1710–

By the summer of 1715, Lowther had engaged in negotiations with Newcomen and an agreement for the construction of an engine was signed in November of that year. This remained in force until Lowther was released from the annual charge upon agreement to pay a lump sum to the Proprietors in 1726.[75]

There were many such engines constructed under license, and a few independently, primarily for the purpose of rendering mines workable. A similar situation existed in the coal fields of the northeast, where, in 1717, Martin Triewald was permitted by the Proprietors to assist in the building of an engine for the Ridleys of Newcastle with the young Samuel Calley, the son of Newcomen's partner. Triewald, who ultimately became a Fellow of the Royal Society through his friendship with Desaguliers, remained in the service of the Ridleys until 1726, whereupon he returned to Sweden.[76] It was also in the northeast that Desaguliers' associate, Henry Beighton, was employed as an engineer. In 1717, three years before his election to the Royal Society, Beighton had built an engine at Oxclose Colliery, and appears to have had a share in the Heaton Colliery near Newcastle. At some point, he became involved with the steam engineer, Stonier Parrott, and he entered into a partnership that included William Cotesworth, of Gateshead, presumably with the objective of escaping what appeared to be the arbitrary rents charged by the Proprietors.[77] When Beighton returned to his birthplace in Warwickshire in 1725, he leased collieries from Sir Richard Newdigate, who, as early as 1713, had

1733," *Transactions of the Newcomen Society*, XLII (1969–70), 169–90; and Allen, "Addendum to the Introduction of the Newcomen Engine from 1710–1733," *Transactions of the Newcomen Society*, XLIII (1970–1), 199–202. On Sir James Lowther, see J. V. Beckett, *Coal and Tobacco. The Lowthers and the Economic Development of West Cumberland, 1660–1760* (Cambridge, 1981); J. E. Williams, "Whitehaven in the Eighteenth Century," *Economic History Review*, VIII (1955), 393–404; and J. V. Beckett, "A Back-Bench M.P. in the Eighteenth Century: Sir James Lowther of Whitehaven," *Parliamentary History*, I (1982), 79–98. On Carlisle Spedding, see Christine MacLeod, *Inventing the Industrial Revolution. The English Patent System, 1660–1800* (Cambridge and New York, 1988), pp. 94–5, 101; and Hugh Owen, *The Lowther Family. Eight Hundred Years of "A Family of Ancient Gentry and Worship"* (Phillimore, 1989), pp. 248–54.

[75] Allen, "The 1715 and other Newcomen Engines at Whitehaven," pp. 244–5.

[76] Rolt and Allen, *The Steam Engine*, pp. 64–7; Svante Lindqvist, "The Work of Martin Triewald in England," *Transactions of the Newcomen Society*, L (1978–9), 166.

[77] Rolt and Allen, *The Steam Engine*, pp. 61–3; A. Raistrick, "The Steam Engine on the Tyneside," *Transactions of the Newcomen Society*, XVII (1936–7), 131–63; Joyce M. Ellis, *A Study of the Business Fortunes of William Cotesworth c. 1688–1726* (New York, 1981), pp. 7–8, 61–2; J. S. Allen, "Some Early Newcomen Engines and the Legal Disputes Surrounding Them," *Transactions of the Newcomen Society*, XLI (1968–9), 198–201. Cotesworth was a cousin of Dr. James Jurin, Secretary of the Royal Society. Edward Hughes, *North Country Life in the Eighteenth Century. The North East, 1700–1750* (London, 1952), p. 99.

Figure 11. Desaguliers' view of the Newcomen engine. By courtesy of Thomas Fisher Rare Book Library, University of Toronto.

leased mines to a partnership including the engineers Stonier Parrott and George Sparrow.[78]

Improving landlords, and their agreements with the purveyors of the new steam engine, provided an atmosphere of opportunity easily mined

[78] Rolt and Allen, *The Steam Engine*, p. 63; and Allen, "Some Newcomen Engines and the Legal Disputes," pp. 181–3. The two engineers Parrott and Sparrow became embroiled in a lawsuit involving the building of engines on wasteland at Wyken and Sowe, near Hawkesbury. It was found these lands were owned by Thomas Watts of London or the Prince of Wales! Watts claimed he and his wife Anne had an entitlement from William Billingsley of London, merchant, dating from 1696. It is not known whether this Watts or Billingsley were related to our projectors. See Allen, "Some Early New-

by those well versed in natural philosophy. This was the culture of innovation that was the vehicle that made natural philosophy a topic of so much interest in the eighteenth century. It was also a culture that crossed the lines of social class and of any residual ideology that may be presumed to reinforce the distinctions between country and city.

As Desaguliers often argued, projectors had managed to build their own fortunes out of the ignorance of their patrons. He had inherited the sentiments of Bishop Sprat in the reign of Charles II and of Nehemiah Grew in Anne's reign. But projectors, through symbiosis, also provided the best rationale for the employment of natural philosophers. There proved to be many willing patrons, happy to invest enormous sums in the most remarkably etherial ventures. The difficulty, as Desaguliers and other inventors and philosophers well knew, was telling those knowledgeable in things mechanical from those with enough knowledge merely to manipulate the wealthy and the curious. Thus, it followed that the role of natural philosophers, instrument makers, and mathematical practitioners in the schemes of the early eighteenth century became a matter of moral reflection. Perhaps, there really was an ethical question to be pondered in the transformation of the social and political framework of England being fashioned by projects and financial innovation. Desaguliers knew very well that the satirists, whether anonymous complainers of Grub Street or the volatile Dean Swift, had hit the mark.

comen Engines and the Legal Disputes," pp. 186–8; Marie B. Rowlands, "Stonier Parrott and the Newcomen Engines," *Transactions of the Newcomen Society*, XLI (1968–9), 49–67: and Edward Hughes, "The First Steam Engines in the Durham Coalfields," *Archaeologia Aeliana*, XXVII (1949), 29–45.

9

The limits of projecting

Secrecy was the essential element of the eighteenth-century projectors. But time has worn away this veneer that once hid their traces. It may now be possible to expose the wary and manipulative relationship between the projector and the natural philosopher. The coexistence of engineers and experimentalists in the extensive network of opportunity followed directly from the protestations of utility of natural philosophy in the early eighteenth century. The multitude of efforts to exploit technology provided a market for natural philosophers that was happily exploited long before we would formally recognize an industrial revolution. Natural philosophers, mechanics, inventors, and projectors were often all of a piece.

The culture of innovation before 1750 made natural philosophers a valuable commodity if they showed any inclination to mechanical problems. This was what Triewald, Beighton, Desaguliers, and many lesser figures clearly recognized. For one reason, patent specifications of the early eighteenth century were so equivocal that it was virtually impossible on that basis to distinguish a genuine technical effort from a project or bubble.[1] It was precisely this that enabled natural philosophers to market themselves as experts who could avoid the disasters that were so abundantly evident amongst those who put their money into inventions that could never have succeeded or possibly never even existed.

One might argue that the careers of many natural philosophers in the early eighteenth century were infected with a wonderful irony. The very

[1] Peter Mathias, "Skills and the Diffusion of Innovation from Britain in the Eighteenth Century," *Transactions of the Royal Historical Society*, series 5, XXV (1975), 101; and Christine MacLeod, *Inventing the Industrial Revolution. The English Patent System, 1660–1800* (Cambridge and New York, 1988), pp. 41ff.

fact that mechanical innovations induced an epidemic of projectors and promoters was the best argument for the propagation of the Newtonian natural philosophy. Thus, lecturers like Desaguliers might enable others to distinguish true invention from base deception. Although natural philosophy implied the possibility of constraining projectors, manipulative minds also recognized that the language of experiment – even the occasional approbation of a Newton – could magnify the powers of persuasion. As with quacks who understood the seduction of medical language, projectors who operated on the fringes of the community of natural philosophers could hardly have failed to appreciate the capacity of coffeehouse lecturers to relieve their listeners of their coin. When philosophers pretended to become engineers, however, some schemes became so large and the investments so vast that a convincing language of conquest and attainment was an absolute necessity.[2]

Such a faith in natural philosophy was crucial – obviously to those who found a market for their lectures or their expertise, but also amongst those who were determined to exploit the resources of their estates. For example, the need for pumping engines was critical in the mines of the Leveson–Gower estates in Shropshire. It was the engineer George Sparrow who leased the Wombridge mines and employed a Newcomen engine to make them workable. Henry Beighton labored among the northern mines, and Desaguliers struggled with engines along the Thames. The same was true on the estates in North Staffordshire where coal mines and iron furnaces were worked under lease to the Foley partnerships while Sparrow took a lease on mines that he worked under royalty payments to Lord Gower.[3] It matters little that many of these enterprises encountered both technical and financial pitfalls that made it virtually impossible for them to meet the objectives that once seduced investors. We must appreciate that, in the eighteenth century, it was thought possible to overcome many of these obstacles. Such was an atmosphere of improvement that infected landlord and entrepreneur alike.

[2] Cf. Roy Porter, "The Language of Quackery in England, 1660–1800," in Peter Burke and Roy Porter, eds., *The Social History of Language* (Cambridge, 1987), pp. 73–5, 85–6.

[3] J. R. Wordie, *Estate Management in Eighteenth-Century England. The Building of the Leveson-Gower Fortune* (London, 1982), pp. 99–103. For the range of Sparrow's activities, see Colin Owen, *The Leicestershire and South Derbyshire Coalfield, 1200–1900* (Ashbourne, 1984), Chapter IV. On the Foleys, see B. L. C. Johnson, "The Foley Partnerships: The Iron Industry at the End of the Charcoal Era," *Economic History Review*, second series, IV (1952), 322–40. Cf. Francois Crouzet, *The First Industrialists. The Problem of Origins* (Cambridge and New York, 1985), Chaps. 1–2.

The York Buildings Company

This was also the case in the variety of water supply ventures, particularly those designed to meet the needs of London, where it was the lack of convenient water rather than the abundance of it that spawned joint stocks. The most well known is the New River Company conceived by the Elizabethan merchant, Sir Hugh Myddleton, which was still operating in the eighteenth century, having been built in part on the proceeds from Myddleton's silver mines.[4] The course of water supply runs in and out of the interests of the Royal Society and its Fellows. In 1664, the Society considered a proposal of Sir William Petty for "propelling with more quickness the water of the Islington river to London" by means of pumps to raise the level of the water.[5] A great deal more controversial, however, was the York Buildings Company, which we have already encountered as one of the pillars of the empire of Case Billingsley. Originally known as the Thames Water Supply Company, when it was patented in 1675 to Ralph Bucknall and Ralph Wayne, it soon emerged as a joint stock under an Act of Parliament after the Revolution. This company was sold in October 1719 for £7000 to a partnership including Case Billingsley, James Bradley, Benjamin Bradley, John Hardwar, Robert Thompson, and Edmund Watts.[6] Members of this group may hold the key to the conflict and intrigue that plagued all projectors, legitimate or not.

The original York Buildings charter contained the right to purchase land, permitted originally for the laying of pipe and the building of

[4] H. W. Dickinson, *The Water Supply of Greater London* (London, 1954), pp. 34–8. See also Charles Webster, *The Great Instauration. Science, Medicine and Reform, 1626–1660* (New York, 1975), pp. 345–8; and J. W. Gough, *The Rise of the Entrepreneur* (London, 1969), pp. 164–5.

[5] Michael Hunter, *Establishing the New Science. The Experience of the Early Royal Society* (Woodbridge and Wolfeboro, 1989), pp. 90–1, 116–17.

[6] House of Commons, *A Report from the Committee to whom the Petition of the Proprietors of the Stock of the Governor and Company for Raising the Thames Water in York-Buildings, is Referred* (London, 1733), pp. 4–8. James Bradley, who was styled as Billingsley's partner as solicitor on the Royal Exchange, was the son of Benjamin Bradley; both belonged to the Merchant Taylors Company. Guildhall Library, Index to Freemen, 1520–1929. Merchant Taylors Company. It is not known whether Edmund Watts was any relation to Thomas Watts. All five were involved in the Royal Exchange Assurance. A. J. G. Cummings, "The York Buildings Company. A Case Study in Eighteenth Century Corporation Mismanagement" (Ph.D. thesis, University of Strathclyde, 1980), p. 34. An earlier version of these events is A. J. G. Cummings and Larry Stewart, "The Case of the Eighteenth-Century Projector: Entrepreneurs, Engineers, and Legitimacy at the Hanoverian Court in Britain," in Bruce T. Moran, ed., *Patronage and Institutions* (Woodbridge and Rochester, 1991), pp. 235–61.

reservoirs. Now a much grander notion evolved to found a land bank in the manner of the Sword-Blade. And there was a great deal of land on the market, especially the forfeit estates of the Jacobite lords, which the Crown was now anxious to unload.[7] By this point, Case Billingsley had been building his network beyond the Mines Royal and Mineral and Battery Works, the Royal Exchange Assurance, and the York Buildings Company, to include the Royal Lustring Company, and the Welsh Copper Company, of which the Prince of Wales had accepted the Governorship. In 1720, however, the emerging difficulties of the South Sea Company drew Billingsley and his partners into the maelstrom and into a Parliamentary investigation largely at the instigation of John Blunt. Blunt, as Director of the Sword-Blade and South Sea Companies, did not appreciate the competition of other bubbles in Exchange Alley. The Billingsley companies, nonetheless, survived intense political scrutiny, especially because Parliament sought bigger game in the aftermath of the Bubble. Blunt was one of them.[8]

The York Buildings Company was able to emerge from this cloud of suspicion because Billingsley had taken care to cultivate influential friends. Here was a common factor in his operations, just as he had managed to convince the Prince of Wales to take on the Governorship of the Welsh Copper. The client economy explains why one of the early directors of York Buildings and Governor for a brief period from 1719 was Sir William Thompson, the Solicitor-General. Amongst the directors as well was Sir Alexander Cairnes of the Royal Exchange Assurance and the English Copper Miners Company, another Billingsley enterprise.[9] And at least as early as 1721, the Governor of the York Buildings Company was Thomas Fane, the Earl of Westmorland.[10]

Protection from people of influence was a deliberate policy of many of these enterprises, not just those originating with Case Billingsley. The York Buildings Company, however, took the strategy one step further by offering to the Crown a proposal that it found difficult to refuse.

[7] Cummings, "York Buildings Company," pp. 456–9; John Carswell, *South Sea Bubble* (London, 1960), p. 167; and Barry Supple, *The Royal Exchange Assurance* (Cambridge, 1970), pp. 24–5.

[8] Carswell, *South Sea Bubble*, pp. 167–8, 171–2, 194.

[9] It has been suggested that the first Governor of the new York Buildings Company was James Brydges, Duke of Chandos, although I have been unable to confirm this. It may have been that it was intended that Chandos should lead the company, although the impression that he was Governor may have followed from his investments in it. See Cummings, "York Buildings Company," pp. 32, 35, 461–2.

[10] SRO, GD1.170/vol. I, York Buildings Company, Court of Directors. Westmorland resigned by 1723.

York Buildings intended to purchase some of the estates of the Jacobite Lords of the 1715 rebellion, thereby relieving the Crown of the problem even of collecting rents, which the highlanders were refusing to pay. One of the Commissioners of the Forfeited Estates was Richard Steele, who lived in York Buildings, but who in 1721 was still trying to persuade the York Company to come up with the funds for the purchase of the confiscated estate of Lord Widdrington in Northumberland. For the Company, it was Billingsley's associate Christian Cole who was charged with the responsibility of managing this affair.[11]

The negotiations for this purchase presented both the Company and the Crown with major problems, especially as the Company found it impossible to pay the £57,100 that it had originally bid. Nonetheless, it did try to make the estate pay for itself mainly by exploiting mineral resources at Widdrington. It did so through the employment of agents in the north, in particular John Horsley, the Presbyterian minister and schoolmaster of the Morpeth Grammar School. Horsley evidently was engaged in the exploitation of salt pans and coal deposits. At the same time, he exhibited his interest in natural philosophy with the publication of a meteorological diary of the Widdrington Estate in the *Philosophical Transactions*. He was ultimately elected a Fellow of the Royal Society in 1730, and shortly before his death in 1732, he advertised the presentation of lectures on natural philosophy.[12] Horsley had no more magic with these ventures than did others and the Company was constantly faced with the battle of meeting the demands of the Treasury for payment. There were chronic suspicions that some of the Directors had not been operating for any benefit other than their own and the result was an investigation by the Crown of the manner in which the estates had been purchased by Cole.[13] This proved another dead alley and the York Company operated the estates for some time to come.

[11] SRO, GD1/170/vol. I. Richard Steele, Henry Houghton, Patrick Haldane, and Robert Munro (Forfeited Estate Commissioners) to Christian Cole, October 11, 1721. On Steele's role in the Commission, see Calhoun Winton, *Sir Richard Steele, M.P. The Later Career* (Baltimore and London, 1970), pp. 203–4, 241–5.

[12] SRO, GD1/170/vol. I, John Cockburne to [Christian Cole?], October 26, 1721 and November 21, 1721. On Horsley see *D.N.B.*, XVII, 382; Nicholas Hans, *New Trends in Education in the Eighteenth Century* (London, 1966), p. 59. Horsley was a friend of George Liddell, the partner of William Cotesworth, who had purchased two of Widdrington's collieries at Stella and Winlaton, co. Durham. Edward Hughes, *North Country Life in the Eighteenth Century. The North East, 1700–1750* (London and New York, 1952), pp. 22–3.

[13] SRO, GD1/170/vol. I, John Cockburne [to Cole?], September 19, 1721 and September 26, 1721[/2?]. The investigation is in PRO, Forfeited Estates Commission (FEC) 1/Y.1, Interrogatories to be Administered to the Witnesses; and PRO, Treasury Papers, T1/

The water engines

The York Buildings Company remained a central part of London commercial life, its presence announced by the tower of its steam engine on the Thames at York Buildings Stairs. Foreign visitors remarked on this apparition of English progress, and the tower dominates the paintings of Westminster Bridge from York Stairs by Canaletto. Despite its continuing financial problems, the company had remained active in the business of water supply as its 1675 charter had intended. But this was a pursuit that could be highly competitive. If any enterprise was unable to fulfill the demands of its customers, if its engines failed to work, or its reservoirs ran dry, then there were other joint stocks willing to scavenge among the broken pipes and silent pumps to make their own attempt. It was a very risky, and often controversial, enterprise.

Part of the difficulty encountered by the water utilities, as had been the case in the mines, was the technological basis from which they operated. Valuable engineers like George Sorocold had been employed in numerous efforts in provincial towns like Derby since 1692, and had rebuilt the tidal water wheel of the London Bridge Water-Work just after the turn of the century. John Lowthorp, a friend of Desaguliers, had been engaged by the New River Company and the promoter Thomas Neale, Newton's predecessor as Master of the Mint, obtained a patent for a new pump that he intended for the Shadwell Water Works, on the Thames just beyond Wapping.[14] The battle between the water works was furious, especially as the metropolis grew. The need to ensure proper operation of their pumping apparatus was so critical, as in the mines, that there was no shortage of opportunities for those with some technical expertise. This is what led, in part, to the employment of Desaguliers by the York Buildings Company in the construction of its reservoirs.

Severe competition meant that the collapse of one venture merely opened the door to others. Such was the need for water in Westminster that the Mill-Bank, a joint-stock company, was able to compete with the

238/24. Christian Cole's Accounts for the York Buildings Company, April 26, 1721. Cockburne appears to have been employed in the Admiralty Office. This may indicate his connections with Billingsley.

[14] Dickinson, *Water Supply*, pp. 24–7, 38–9, 49–50; and Guildhall, MS. 537, "A short account of the London Bridge Water Works & Memorandums relative to their Business." See also E. Hatton, *A New View of London*, Vol. II (London, 1708), pp. 791–8. On Sorocold, see F. Williamson, "George Sorocold, of Derby. A Pioneer of Water Supply," *Journal of the Derbyshire Archaeological and Natural History Society*, LVII (1937), 43–93. Neale was also one of the original proprietors of the Welsh Copper Company near Neath. See Rees, *Industry Before the Industrial Revolution*, II, p. 502.

Chelsea and York Buildings companies. In 1725, the Chelsea Company under a Major Watts bought out the Mill-Bank operation.[15] In these circumstances, especially where opposition might result in the sabotage of water mains, it was obvious that any advantage would be pressed to the full.[16] In 1708, one who had heard of the possibilities of the steam engine wrote that success in the drainage of mines would keep the inventor in a coach and six.[17] It was no less the case with harnessing steam to the provision of water.

The York Buildings Company was one of the first water supply ventures to appreciate the advantages inherent in the new steam technology. Possibly as early as 1712, it decided to construct a Savery engine to assist in the pumping of water from the Thames. It is obvious, however, that this did not achieve the results expected and the Company was looking for a buyer at the same time that Case Billingsley was looking for a charter. These were propitious circumstances and the ever energetic spirit of Billingsley was not the least dampened by the disappointing results of the Savery engine. And he had previously exhibited a willingness to tackle technical problems like the solution to the longitude.

Case Billingsley's ventures crossed the boundary between the entrepreneur and the mechanic. They are significant because they exhibit the euphoria of improvement that gripped the early eighteenth century. Difficulties with a technology could be viewed as opportunities, not merely to promote schemes unlikely ever to succeed, but also to pursue possibilities of technical advantage. This was the reason why the York Buildings Company decided on the installation of a Newcomen engine at its water works on the Thames. It was to be an enormously contentious effort with anonymous satirical comments and pamphlets swirling in the smoke from the engines that enveloped the neighborhood of York Buildings.[18] Lurking in the background was the elusive genius of Case Billingsley. As Graham Cummings has noticed, several events occurred in 1724 that tie Billingsley to the willingness to adopt this new technology. In February, the Company seized the initiative by placing an advertise-

[15] Dickinson, *Water Supply*, p. 55. It is likely that Major Watts was the Thomas Watts of the Sun Fire. See Armand B. DuBois, *The English Bubble Company after the Bubble Act, 1720–1800* (New York, 1938), p. 321n74.

[16] This was the problem of the New River Company. See *St. James's Journal* (Thursday, May 3, 1722).

[17] J.C., *Compleat Collier*, p. 29. Quoted in Sidney Pollard, *The Genesis of Modern Management. A Study of the Industrial Revolution in Great Britain* (London, 1965), p. 62.

[18] Dickinson, *Water Supply*, pp. 61–3; and Alan Smith, "Steam and the City: the Committee of Proprietors of the Invention for Raising Water by Fire, 1715–1735," *Transactions of the Newcomen Society*, XLIX (1977–8), 13.

ment in the *Daily Courant*, asking for submissions for a proper machine to meet their needs. One such proposal that remains in the York Buildings' papers suggested a machine that operated through "the ponderous quality of the water that raises itself of its own accord to any height."[19] Such a device would have been remarkable indeed, if it existed. It appears that it may have – at least, in the mind of Case Billingsley.

Why Billingsley became interested in water engines is unclear. The effort to obtain patents for water engines was obviously in fashion and virtually every year in the first half of the century spawned numerous attempts. But there is a curious association that one could speculate was part of his initiation as an inventor. We are already aware that about 1713, Billingsley had taken up residence in Wapping as a brewer. It may well be that in Wapping he encountered one of the Fire Engine Proprietors, Edward Elliott, who was also the Baptist Pastor of Broad Street until Elliot's death in 1716.[20] In any case, as early as 1720, it was thought that Billingsley was about to be granted a patent for some form of water engine. It was not, however, until 1724 that his interest in this technology could be confirmed by his application for a patent for a column of water engine "after near Thirty Years Study and having made many Chargeable Experiments." Although the report on the application by Secretary Townshend was a favorable one, the patent was not actually granted until 1727.[21]

This delay may occasion some puzzlement, but the explanation for it is also ironic. One result of the investigation of the Billingsley bubbles in 1719 and 1720 had been the passage of the Bubble Act – which had little to do with the South Sea Company. And one intent of the Bubble Act was to prevent the "illegal Use made of such Patents to raise Subscriptions." This, of course, does not prove that Billingsley's patent would have immediately been the object of such considerations. The fact is, however, that when the patent was finally issued, it was subject to these restrictions. And the circumstantial evidence is at least suggestive. During 1724, when the York Buildings Company was considering a new engine and when Billingsley actually sought his patent, there appeared an anonymous set of *Proposals for Subscriptions to a New Invention for Raising*

[19] SRO, GD1/170, *Proposals to the Worshipful and Honourable Company of York Buildings*, quoted in Cummings, "York Buildings Company," p. 19.

[20] Alan Smith, "Edward Elliott: Fire Engine Proprietor," *Transactions of the Newcomen Society*, L (1978–9), 230.

[21] PRO, SP (Dom) 35/50/5. Case Billingsley petition; B.L., Add. MSS. 36123, fols. 41–2, Townshend report to Solicitor General Yorke, July 31, 1724; Add. MSS. 36276A, Patent, May 25, 1727.

Water. There can be little doubt that this suggestion came from Billingsley. It reflects his stamp in a subscription that turns the invention over to the trustees who hold tickets in the enterprise. Besides listing the many uses to which his invention can be put, the author makes note that "This Invention performs without the Power of Men or Horse, or Wind or Fire, or Air, or Steam or Current." Here, perhaps, was Billingsley's unique column of water engine. What is equally interesting is that the proposal contained a long diatribe against the constraints placed on inventors:

> well meaning People have been deceived and cheated; this hath so exasperated them that they have fallen out with all Patents, all Projects whatsoever, and thereby have run into too great an extream, which is as Ridiculous as for a Man to swear he never more would receive any Money because he had lost by some that had been counterfeit, or because a Man had been hurt by a Quack should forswear making use of a good Physician.[22]

There were plentiful prospectors in the hunt for an effective engine to raise water. This was especially true during 1724. Possibly, the York Buildings Company advertisement had something to do with the surge in interest. The notion of subscription was not so unique to be confined to the Billingsleys, although few inventors could have easily matched their skill in financial manipulation. But it must be admitted that this anonymous suggestion might also have come from others like William Mason and Thomas Chalmflower, who, in March 1724, applied for a patent for a siphon or attracting engine that worked "partly by Attraction and partly by force."[23] It might also have come from the plumber, Valentine Flower, for an engine to raise water "by means of a Leveridge and an Horizontall Fly without the help of Fire," or perhaps even from the inventor Robert Bumpstead, who joined the parade of petitioners.[24]

There is no doubt, however, that the Billingsley family maintained an

[22] B.L., Shelf Mark 816.m.10 (110), [Anon.,] *Proposals for Subscription to a New Invention for Raising Water, and working several useful Machines* [1724]. This argument bears a striking similarity to the analogy used twenty years later by Desaguliers, although it seems unlikely that the 1724 proposal was his. Desaguliers, *Course of Experimental Philosophy* (London, 1744), II, p. 416. See also Dubois, *English Business Company*, pp. 21–2, 63n119.

[23] PRO, SP (Dom) 35/48/88 (2), March 27, 1724. Chalmflower evidently knew Case Billingsley. Thomas Chalmflower had acted on power of attorney for an Anne Holland in the purchase from Billingsley of stock in the English Copper Company. See Cummings, "York Buildings Company," pp. 462–3.

[24] PRO, SP (Dom) 35/49/34, Valentine Flower, April 21–8, 1724; SP (Dom) 35/50/10. Robert Bumpstead, May 30-June 29, 1724. On Bumpstead see below. For attempts to secure patents in 1724–5 see PRO, SP (Dom) 36/43/62–77. List of Inventions, 1724–37. The Patent warrants are in B.L., Add. MSS. 36123, Hardwicke Papers.

interest in the improvement of such engines. During 1732, when his father had been imprisoned for debt in the Netherlands, John Billingsley applied for a patent for an engine to raise water, but that also might be used in several industries, including the hammer of a forge. It is possible that John saw the application of such technology in the enterprises of the York Buildings Company, but one might equally suggest that the captivity of Case made such a moment especially propitious if funds could subsequently be obtained for the invention.[25] There was, as usual, an air of desperation in the motives of the Billingsleys. But it might be that the financial innovations and the deluge of patents that had followed the Revolution made the consequence of legal restraint of projects inevitable. In this sense, Case Billingsley became the victim of the very wave he sought to catch.

Patents and inventions appear in the careers of projectors as inevitably as the mist that shrouded York Buildings and Exchange Alley. Patents provided opportunity for commercial exploitation as had been the case with the Savery and Newcomen engines, or Case Billingsley's undeveloped column of water engine, which never produced the results he desperately hoped for. But Billingsley knew that a patent might produce a charter to permit incorporation. Alternatively, it might even be possible to purchase the charters of companies that were long established. This method had drawn him to the Thames Water and to the Mines Royal and Mineral and Battery Works. It is also what led him to entanglement in the affairs of the English Copper Company.

The Copper Miners

The Governor and Company of the Copper Miners in England was one of the many ventures in mining formed immediately after the Revolution. Based on a patent granted to John Duckett and Gabriel Wayne for a new means of refining copper ore, the company had been incorporated in 1691 at the initiative of Sir Joseph Herne and the land bank projector, John Briscoe.[26] Case Billingsley became interested in the copper enterprise

[25] PRO, SP (Dom) 36/26/94, February 21, 1731/2; SP (Dom) 36/26/99, Recommendations of Patent, March 22, 1731[/2?]; B.L., Add. MSS. 36142, fols. 269–70, Yorke recommendation, March 22, 1731[/2?]; University of London Library, Autograph Letters 5, Case Billingsley to Thomas, Earl of Westmorland, January 8, 1732. Cf. Cummings, "York Buildings Company," p. 476.

[26] W. R. Scott, *The Constitution and Finance of English, Scottish and Irish Joint-Stock Companies to 1720*, 3 vols. (Cambridge, 1912; reprint edn., Gloucester, Mass., 1968), II, pp. 430–2; Henry Hamilton, *The English Brass and Copper Industries to 1800* (London, 1926), p. 338. On Gabriel Wayne see Joan Day, "The Costers: Copper-Smelters and Manufacturers," *Transactions of the Newcomen Society*, XLVII (1974–6), 47–9.

during the height of his activity in York Buildings and Royal Exchange Assurance in 1720, very likely because of the opportunities he had perceived in English coinage. Copper coinage had caused the Mint much concern after it called for tenders in supply of copper blanks in 1717. One group involved was led by the ironmonger William Wood, who, just before the Bubble, obtained rights from the Mines Royal and Mineral and Battery Works to "raise the calamine stone for making brass and to raise, melt, smelt, etc. all ures and ores of gold and silver, quick-silver, lead, copper, tin, and all other minerals, salts and earths." Wood's efforts to supply copper were a failure.[27]

The copper puzzle still promised great wealth. Billingsley and his confederates, John Essington of the York Buildings Company and James Bradley of the Royal Exchange Assurance, put together a plan to capture the Copper Miners and to swallow some of the many smaller smelting operations. Essington was a prominent London merchant and later briefly M.P. for New Romney, and James Bradley we have already met as Billingsley's partner as solicitor and as a member of the Merchant Taylors' Company.[28] Billingsley's plans involved capturing most of the copper and brass industries. It is clear that his effort to control copper manufacturing was based on part ownership of a copper works at Wimbledon with Essington and Bradley. This mill had been in operation since 1712 with a considerable investment.[29] For Billingsley, this was merely the beginning.

During 1719, there was evidence of a major reorganization among a variety of groups concerned with copper mining. In June of that year, several individuals came to an agreement with Lord Falmouth to search and mine his lands for copper ore. This agreement was signed by John and Thomas Coster, Nehemiah Champion, Gabriel Wayne and Abraham

[27] On the background to this, see J. M. Treadwell, "William Wood and the Company of Ironmasters of Great Britain," *Business History*, XV (1974), 98–100; and Treadwell, "Swift, William Wood, and the Factual Basis of Satire," *Journal of British Studies*, XV (Spring 1976), 82–4. At about this time, an anonymous pamphlet was printed that named Wood, and Billingsley's partners Thomas Chambers, John Essington, and Henry Robinson as persons prepared to make proposals to the Treasury. Billingsley's name was notably absent although the author remarked on a particular issue of his: "through the very great Scarcity of Silver, a speedy Supply of Copper Money would have been a very great Ease to the Trading People of the Nation." B.L., Shelf Mark 712.m.1 (74), *A Letter from a Livery-Man of the City of London, to a Member of the Honourable House of Commons*. Cf. Richard S. Westfall, *Never at Rest: A Biography of Isaac Newton* (Cambridge, 1980), p. 844.

[28] Cummings, "York Buildings Company," p. 459; Romney Sedgwick, *The House of Commons, 1715–1754* (New York, 1970), II, p. 17; Guildhall Library, Index to Freemen, Merchant Taylors Company.

[29] Hamilton, *English Brass and Copper*, p. 148.

Elton of the Conham works of Bristol, Thomas Chambers, and John Essington. Essington and Chambers were already engaged in the English Copper Company affairs at Lower Redbrook. John Coster's family had been engaged in copper smelting in the Bristol area before the Revolution. In 1714, John had secured a patent for an engine to drain mines that appears to have been powered by a water wheel. Nehemiah Champion, a brass founder at Bristol, is even more interesting, for it was he who was Newcomen's Bristol agent and who was probably supplying brass boilers and steam cylinders for the engines.[30]

By 1720, Billingsley was ready to put the pieces of another project into place. In July, he entered into agreement with Thomas Chambers, junior, of Lower Redbrooke and India House, as well as Bradley and Essington to control the copper operations of the English Copper Company. It was to be amalgamated with the mill at Wimbledon and with the works at Lower Redbrook. Soon joining the concern were the apothecary, William Lilly, and Sir Alexander Cairnes of the Royal Exchange Assurance.[31] Both were confederates of Billingsley. So, too, was Christian Cole, who was approached in 1721 for his vote for a slate of directors approved by Billingsley, Chambers, and Essington.[32]

On the surface, the affairs of the English Copper Company hardly appear unusual among the Billingsley enterprises. But this adventure was much more complicated than the York Buildings charter purchase or the Royal Exchange subscription. The transfers of shares that built the strength of English Copper meant the trading of rights to leases held in several locations and from several different sources, including Essington's mills at Wimbledon and at Wandsworth as well as leases purchased from the Mines Royal and Mineral and Battery Works. All of this came out in the inevitable lawsuits when the partners had a falling out, as was soon to be the case. Probably the immediate cause of the difficulties was the

[30] Hamilton, *English Brass and Copper*, pp. 146–7; Day, "The Costers," pp. 51–3; Joan Day, *Bristol Brass: A History of an Industry* (Newton Abbot, 1973), pp. 49–53; Cyril Hart, *The Industrial History of Dean* (Newton Abbot, 1971), pp. 104–11. On Champion, see Rolt and Allen, *The Steam Engine*, p. 67; John S. Allen, "Thomas Newcomen (1663/4–1729) and His Family," p. 18. With John and Thomas Coster, Champion had been one of the originators in 1702 of the Bristol Brass Wire Company at Baptist Mills. See Arthur Raistrick, *Quakers in Science and Industry* (Newton Abbot, 1968), pp. 190–2; and Allen, "The 1715 and other Newcomen Engines at Whitehaven, Cumberland," *Transactions of the Newcomen Society*, XLV (1972–3), 250.

[31] B.L., Shelf Mark 522.m.12 (3). *Articles of Agreement Made, Concluded, and Agreed upon the twentieth Day of July, in the Year of our Lord One thousand, seven hundred and twenty...* (London, 1725); Cummings, "York Buildings Company," pp. 459–62; Hamilton, *English Brass and Copper*, p. 148.

[32] SRO, GD1/170/vol. III, Essington, Billingsley, and Chambers to Cole, March 31, 1721.

fall in stock in the aftermath of the South Sea crisis, but John Essington traced the difficulties to the arrangements made over the copper works at Neath in 1719.

Without entering into the terribly tangled traces of stocks, shares, leases, and lost receipts, it is clear that Billingsley was at his wits end by 1723 and had initiated a suit against his former partner, Essington. It was revealed that Bradley had become involved over the Neath works, and that also at stake were leases held by William Lilly, William Dale, and Daniel Peck from the Mines Royal and Mineral and Battery Works. One of those named in the lawsuit as a subscriber to the English Copper maneuver was Henry Robinson, one of the Proprietors of the Fire Engine, who amassed a large fortune by the time of his death in 1762.[33]

The English Copper Company had promised the control of most of the copper smelting in England, but it had outstripped Billingsley's financial power to manage it. In 1723, its nominal Governor was the Prince of Wales, although it was presumably being run by John Essington as Deputy Governor.[34] It would seem that the Company was having difficulties over the water supply for its stamping engines on the Blackwater River. This suggests that the proposals by Billingsley for an engine may

[33] Among Robinson's estate, curiously enough, were Wimbleton Mills and houses in Wapping. He also left £100 to Case Billingsley although Case, senior, had died some fifteen years earlier. PRO, C11/1455/8, Answers to John Essington to the Bill of Complaint of Case Billingsley, April 29, 1725. See also Alan Smith, "Steam and the City," p. 11. These complex transactions are explored in Cummings, "York Buildings Company," pp. 460–76.

Daniel Peck was undoubtedly a relative of Philip Peck, who also had an interest in the same leases as well as being secretary to the Welsh Copper Miners, which took over the Neath works. The entire affair became a legal nightmare beyond the lifetime of the original participants. PRO, C11/1455/8, Billingsley Bill of Complaint; The Further Answer of John Essington; C38/380/1726, A.C. Billingsley v. Essington, Queries, July 19, 1726; Cummings, "York Buildings Company," p. 470; PRO C1111863/12, Billingsley v. Lilly, Bill of Complaint, May 28, 1742; The Answer of Mary Lilly, Widow, September 10, 1742.

On the Pecks' connection with the Duke of Chandos and the York Buildings Company, see Hunt. Lib., Stowe MSS. 57, XXIV, f. 262, Chandos to John Payne, September 29, 1724; and XV, f. 69, Chandos to Mr. Peck, December 13, 1724; XXVIII, f. 255, Chandos to Daniel? Peck, February 3, 1725/6. Daniel Peck was bankrupt in 1724. On Daniel Peck's involvement in copper mining in Scotland with Sir John Erskine, see T. C. Smout, ed., "Journal of Henry Kalmeter's Travels in Scotland 1719–1720," in *Scottish Industrial History*. Scottish History Society, series IV, vol. XIV (1978), 21, 37–8. On Chandos's relations with Philip Peck, see Hunt. Lib., Stowe MSS. 57, XXIV, f. 249, Chandos to Peck, September 22, 1724; XXIV, f. 269, Chandos to Peck, October 5, 1724; XXVIII, f. 28, Chandos to Peck, March 28, 1726; XXVIII, f. 34, Chandos to Peck, March 30, 1726; XXX, f. 247, Chandos to Philip Peck, August 30, 1727.

[34] *St. James's Journal* (March 30, 1723). The Court of Assistants included Henry Robinson, and Benjamin Bradley, the father of James Bradley, Billingsley's partner.

have been adapted in 1724 to meet this contingency.[35] By 1725, however, the Prince thought better of his connection and it was the senior Thomas Chambers who was elected Governor with the assistance of Essington. At this point, the Company was still trying to expand its interests and it entered into negotiations with the "Bull Wharf Copartnership for refining Copper, Lead, Ashes & making Brass and Compounding Metalls at Lambeth." This operation was controlled by Sir Thomas Mackworth. The projected union failed to succeed because of a dispute over profits and shares.[36] Essington emerged as Governor in 1727, but his reign was brief. Of the original partners, not only was Daniel Peck bankrupt, but by 1728, so, too, was Essington. Although Essington was also elected M.P. for New Romney in 1727, he was unseated for bribery in 1728 and died the following year in Newgate, where he had been imprisoned for debt.[37] He was not the last of Billingsley's associates to see the inside of debtor's goal.

Lotteries and coinage

It was Billingsley's connections with the Hanoverian Court of George I that allowed his ventures to fly as far as they did. One of the reasons why his concerns in York Buildings or English Copper ran into difficulties after the Bubble was that he often exploited his connections more than was obviously wise. It is not difficult to see why, however; he simply did not have the personal economic resources to carry out his innovations. Indeed, the plan was to use money obtained by subscription or by lottery where possible. The most revealing example of these speculations was the Harburgh Company, in which he became entangled in 1721. The city of Harburg on the Elbe had come under the control of George I, as Elector of Hanover, in 1705, and efforts to develop the port for the exploitation of trade with England had gathered momentum since the Succession in 1714. By 1720, a group of projectors had received a charter from the King, the plan by then including the formation of a bank and a manufacturing enterprise. The charter did not permit the entrepreneurs to operate directly in England, especially given the power of the Bubble

35 PRO, C38/352/Part I, Chancery Order Books (1728), pp. 157–9. Queries, ca. 1723.

36 PRO, C11/1483/14, Copper Miners v. Benjamin Bradley, June 6, 1730, Bradley answer. Thomas Mackworth was a Director of the Mine Adventurers.

37 PRO, C11/1456/7, Billingsley v. Essington. The Further Answer of John Essington, October 10, 1726; PRO, B4/6/134, no. 1358, Bankruptcy Commission. Docket Books, May 2, 1728; Sedgwick, *House of Commons*, II, p. 17; and Cummings, "York Buildings Company," p. 462.

Act, so Case Billingsley was brought in to devise a lottery to finance the project.[38] The result, once more, was a Parliamentary uproar.

The Billingsley plan for the Harburgh Company brought into play his allies at Court and in the city. The Governor was the young Prince Frederick, the grandson of George I, although he was obviously a mere figurehead. The City merchant John Eyles, subgovernor of the South Sea Company, and Whig M.P. for Chippenham, was named subgovernor of the Harburgh Company as was John Shute Barrington, M.P. for Berwick-on-Tweed, an active campaigner for the rights of Dissenters.[39] This was a great Whig enterprise. The list of the Harburgh operation included Lord Nicholas Lechmere, the friend of Steele and defender of Whiston, Sir Charles Wager, M.P. and later Admiral, who was at the time employed in the Admiralty Office, Sir Thomas Webster, M.P, as well as Charles Delafaye, Under-Secretary of State to the Earl of Sunderland.[40] As we might expect, Billingsley drew into the scheme a large number of his confederates from the York Buildings Company, the Royal Exchange Assurance, and the English Copper Company such as Sir Alexander Cairnes, William Squire, and John London. This included Billingsley's entire family, along with Benjamin and James Bradley, several members of the Lilly family of apothecaries, along with numerous merchants with whom he was connected like John Dale, Benjamin Burroughs, Peter Hartopp, and John Thompson. What is especially interesting about his merchant friends is that they too tended to exhibit strong, if occasionally independent, Whig links.[41]

[38] A. J. G. Cummings, "The Harburgh Company and its Lottery, 1716–1723," *Business History*, XXVII (July 1986), 1–18.

[39] Cummings, "The Harburgh Company," pp. 3–5.

[40] B.L., Add. MSS. 6406, "Diploma Societatis Harburgensis," May 18, 1722; Sedgwick, *House of Commons*, II, pp. 202–4, 502–4, 527. Billingsley later claimed to have been introduced to Sir Charles Wager over his longitude proposal. U.L.L., AL.5, Billingsley to Earl of Westmorland, April 17, 1732. On Delafaye, see Michael Harris, "Newspaper distribution during Queen Anne's reign. Charles Delafaye and the Secretary of State's Office," in *Studies in the Book Trade. In honour of Graham Pollard* (Oxford, 1975), pp. 139–51; and Margaret C. Jacob, *The Radical Enlightenment. Pantheists, Freemasons and Republicans* (London, 1981), pp. 133–6. Delafaye was an entrepreneur of sorts who formed a salvage company with John Rowe, Jacob Rowe, and John Phillips to recover ships' cargo, especially bullion. PRO, SP (Dom) 35/29/48, December 15, 1721. On Delafaye's connections with the projector William Wood, see J. M. Treadwell, "Swift, William Wood, and the Factual Basis of Satire," *Journal of British Studies*, XV (Spring 1976), 56–7.

[41] Several of those invested in the Harburgh fiasco also lent their support to the application of Case, junior, to access to the Royal Exchange as a broker. B.L., Add. MSS. 6406, "Diploma Societatis Harburgensis," May 18, 1722; Corporation of London Record Office, Court of Aldermen Records, May 7, 1723; W. A. Speck and W. A. Gray, eds., *London Poll Books, 1713* (London, 1981), pp. 101, 113, 121; House of Commons, *The*

The Parliamentary committee that was probing irregular practices in the lottery soon found the traces of Case Billingsley. It was discovered that the tickets for the lottery had been printed on Tower-Hill and distributed from York Buildings House. One of the subscribers to the plan was Richard Mount, the chartmaker and printer at Postern Gate on Tower Hill, who had published Billingsley's proposal on the longitude in 1714.[42] But the committee felt that it was necessary to interrogate Billingsley and when it was rumored that he had fled, it was ordered he be taken into custody. He appears to have found the scrutiny a bit too close and the London papers reported his flight to Holland to escape the wrath of the Commons.[43] If this were true, then perhaps his absence might explain the reason Case, junior, had need to act as a broker on the Exchange to manage his father's various affairs. In any event, no one appears to have doubted the senior Billingsley's involvement and the failure to account for the funds committed by ticket purchasers raised serious problems for the committee.[44] The result was that the Harburgh Lottery was prohibited by Parliament. Billingsley seems to have escaped any significant consequences, perhaps shielded by his nervous Whig connections. The real victim was the indignant Lord Barrington who was expelled from the House.[45]

Given Billingsley's fascination with lotteries, this was not the last we were to hear of him. When an Act was finally passed in 1736 for the construction of a new bridge to cross the Thames at Westminster, after decades of talk and partly after the reassurances of Desaguliers, Billingsley surged into action. Here, he hoped, was a golden opportunity

Report of the Committee, Appointed to Examine into the Project, Commonly Called the Harburg-Lottery (London, 1723), pp 5–15; and *The Supplement to the Freeholder's Journal of Wednesday April 25* (Friday, April 27, 1722). On William Squire and Alexander Cairnes, see Alfred P. Wadsworth and Julia De Lacy Mann, *The Cotton Trade and Industrial Lancashire, 1600–1780* (reprint edn., New York, 1968), pp. 214–15; and T. S. Willan, *River Navigation in England, 1600–1750* (London, 1964), p. 70. The Harburgh List also included a John Meres, who may have been the apothecary John Meres, F.R.S., who was one of the Proprietors of the steam engine patent. It is also possible this may have been Sir John Meres, who was to become Governor of the York Buildings Company. See Alan Smith, "Steam and the City," pp. 5–20.

42 *Committee Appointed to Examine into . . . the Harburg-Lottery*, pp. 5–7; Sarah Tyacke, *London Map-Sellers, 1660–1720* (Tring, 1978), pp. 125–6; and E. G. R. Taylor, *The Mathematical Practitioners of Tudor and Stuart England* (Cambridge, 1954), p. 280. Mount died accidentally in 1722.

43 Cummings, "The Harburgh Company," p. 11; *St. James's Journal* (Saturday, March 9, 1723).

44 B.L., Shelf Mark, 376.b.7 (48), [Anon.,] *A Letter from a Citizen to J________ B________, Esq; Member of Parliament Relating to the York Buildings Company's Affairs* (London, 1724); *Committee Appointed to Examine into . . . the Harburg-Lottery*, p. 6.

45 Sedgwick, *House of Commons*, I, p. 438; Cummings, "The Harburgh Company," p. 13.

in a life that had seen more than its share of setbacks. The purchase of land and the costs of construction were to be offset by a lottery that, by the spring of 1737, was already proving a dismal disappointment. It was at this juncture that Billingsley's daughter Priscilla attempted to interest the Duke of Chandos in promoting a scheme that the inventor had devised. Chandos declined any part in it, conceivably on the advice of Desaguliers.[46]

Case Billingsley was, all the while, an opportunist circling the dockyards and coffeehouses waiting for hints of movement in the flow of money that revealed the trail he had to follow. This was natural to someone who had learned his trade as a purser, where gains were to be had in the judicious handling of ships accounts or the trading of sailors' tallies at a discount. But we should not see him merely as a predator, although he was surely that. He was an innovator as well for whom the climate of the bubbles proved as natural as a ship's bunk. The innovations in finance that had been unleashed by the Revolution and England's foreign wars made rich men of those, like James Brydges, who had been fortunate enough to obtain positions ripe with advantage. For Billingsley, who could not so easily seek office or place in Parliament, his horizons were what his imagination could construct. This meant the focus of his attention on the millions of pounds tied to the emerging joint-stock companies and, over the length of his career, a compelling interest in monetary theory.

Such a concern was hardly unique in the early eighteenth century, where coinage was a matter of public dispute. And so it had been for decades. As Charles Webster has shown, the policy of coinage had been hotly debated during the Interregnum as it was to be at the end of the century. So, too, was it to be a matter of direct concern to Newton at the Mint.[47] Billingsley's personal fascination with the subject, therefore,

[46] Hunt. Lib., Stowe MSS. 57, XLVIII, f. 199, J. Pudsey (Chandos's Secretary) to Mrs. Grahme, February 6, 1736/7. Mrs. Grahme was Billingsley's daughter, who had married the son of William Grahme, Dean of Carlisle. Billingsley laid claim to a Scotch peerage on behalf of his grandson, William. B.L., Add. MSS. 24120, f. 392, "Letters Addressed to Thomas Wotton, Relative to his Baronetage, 1725–1746," Billingsley to Wotton, November 23, 1739. Billingsley may have had some knowledge of the Bridge design because he appears to have been involved with the engineer, Batty Langley, nemesis of Desaguliers. B.L., Add. MSS. 36030, Hardwicke Papers, Notes of Cases, fols. 132–3, Edward Billingsley v. William Barnes, November 8, 1735. These notes are virtually indecipherable. Patricia Carson, "The Buildings of the First Bridge at Westminster, 1736–1750," *Journal of Transport History*, III (November 1957), 111–22, esp. 115–16. On Langley and Desaguliers, see R. J. B. Walker, *Old Westminster Bridge. The Bridge of Fools* (London, 1979), *passim*, esp. p. 69.

[47] Charles Webster, *The Great Instauration*, pp. 403–20. See also John Craig, *Newton at the Mint* (Cambridge, 1946).

places him among those mathematical practitioners and projectors who haunted Parliament and the Mint with incessant schemes to restore a depleted coinage.

Toward the end of his life, Billingsley wrote to the Duke of Newcastle with a "Memorial" on coinage that he had addressed to the King. In his letter of 1744, in which he requested the Privy Council examine his proposals to prevent the export of silver coin, Billingsley had the temerity as well to recount his control of several ventures in 1719 when he set up insurances and lotteries that had raised money for the crown – or so he claimed. This was exceptionally audacious, even for Billingsley. But he also recalled that his interest in the problems of coinage had dated from the reign of William, and that under George I, he had privately printed his own suggestions on how to counteract the export of silver and gold coin.[48] These pamphlets, printed anonymously, can now be identified.

Throughout the late seventeenth century and well into the eighteenth, England experienced great difficulties in the shortage of coin. This was especially the case because of the export of silver, which, because it was undervalued in England, found higher prices on the Continent. It was such a severe problem that Newton was involved in an attempt to revaluate the coinage to reduce the temptation to export bullion rather than bring it to the Mint to be coined.[49] It was in these circumstances that the public began to hear from several commentators, but especially from an anonymous pamphleteer who signed his pamphlets merely "B" or, occasionally, "B/B" – the latter undoubtedly signifying the insurance promoter Thomas Burgess and Billingsley.[50] These pamphlets were concerned about the supply of coin for trade and with the impact on the National Debt, the reduction of which, one may recall, was one of the selling points ostensibly behind the charters for the new insurance schemes in 1717.

48 B.L., Add. MSS. 32702, fols. 31–2, Billingsley to Newcastle, January 28–9, 1743/4.

49 On Newton's role in this issue, see Craig, *Newton at the Mint*, pp. 102–10; Craig, *The Mint. A History of the London Mint From A.D. 287 to 1948* (Cambridge, 1953), pp. 213–22; Westfall, *Never at Rest*, pp. 836–8; and C. R. Fray, "Newton and the Gold Standard," *Cambridge Historical Journal*, V (1935), 109–17. On the politics of Newton's place at the Mint, see Peter Laslett, "John Locke, the great recoinage, and the origins of the Board of Trade: 1695–1698," in John W. Yolton, ed., *John Locke: Problems and Perspectives* (Cambridge, 1969), pp. 137–64.

50 The identification of Thomas Burgess as Billingsley's confederate is confirmed by a petition from them to the King and Lord Sunderland in 1716 in which they refer to the scarcity of silver money and a notion to produce copper coinage. B.L., Add. MSS. 61605, fols. 21–4, Case Billingsley and Thomas Burgess to the King, June 21, 1716; and Add. MSS. 61640, f. 110, Case Billingsley and Thomas Burgess to Lord Sunderland, June 21, 1716. Burgess was a druggist and promoter associated with Sir James Hallett. See Scott, *Constitution and Finance*, III, p. 446; and P. G. M. Dickson, *The Sun Insurance Office, 1710–1960* (London, 1960), p. 269.

From 1716 to 1718, a period of intense activity on the part of Billingsley, there was a series of six printed pamphlets that appear to come from the same person. These were likely the very proposals to which Billingsley referred in his letter to Newcastle almost thirty years later. It is impossible to be certain of the order in which they were all published, and several of them were really new editions of ones that had been previously printed. In any case, they all dealt with the problem of loss of English coin as silver was exported to the Continent. In the spring of 1716, two pamphlets appeared in quick succession in which the author, signed merely B, argued that the experience of his family and a recent tour through England had shown him the general decline of trade caused by the lack of specie and its melting down into bullion by goldsmiths who had recognized the profit to be made in its export. The author, therefore, argued that the value of coin should be raised in order to eliminate this profit, which operated to the general detriment of the traders of the nation.[51]

During 1717, there was alarm at the Treasury over the inadequacy of the amounts of coin in circulation. Several proposals were circulated and considered by Newton for the establishment of a copper coinage that might, in part, overcome the shortage of silver money. The experiment proved to be problematic, although it did continue until 1725 after several false starts.[52] The notion of copper coinage, of course, would explain the focus of attention that Billingsley had given to this issue and his enthusiasm for the plans of the English Copper Company. More serious, however, was the fundamental difficulty of the shortage of silver coin that had scarcely been minted at all in recent years. The reason, as Newton and Billingsley both understood, was the high price it could fetch on the Continent. In August, Newton was instructed by the Treasury to prepare a report on the relative values of silver and gold. In his Report of September 21, Newton proposed to deal with the imbalance between gold and silver by reducing the price of gold guineas by at least 6d., hoping, thereby, to erode the profit to be made by the export of silver and the purchase of gold on the continental markets.[53]

[51] B., *Some Proposals Humbly offer'd to the Parliament of Great Britain, For Easing the Nation of the Old Navy-Debt, (Now the South-Sea-Debt) of Ten Millions Sterling, Principal Money*... (London, 1716), pp. 4, 18, 15. Dated March 31, 1716, this appeared on April 7, 1716. *Daily Courant* (April 6, 1716). Also B., *Further Reasons Most humbly offer'd, for Increasing our Money; putting an immediate Stop to its being Melted Down, and sent Abroad* (London, 1716), pp. 2–6. This is dated May 16, 1716. Both pamphlets were republished on July 28, 1716, and the first appeared again before February 26, 1717. See *Daily Courant* for those dates.

[52] Craig, *Newton at the Mint*, pp. 98–101.

[53] Westfall, *Never at Rest*, pp. 837–9; Craig, *Newton at the Mint*, pp. 106–9.

Later reprinted in the London press in December 1717, Newton's report to the Treasury encouraged further dispute, although the Government chose to accept his relatively mild recommendations. There were those who were not the least convinced that lowering guineas would achieve anything. It was very likely Burgess and Billingsley who published a pamphlet in January 1718 accepting "That the way to preserve our Silver, is to raise it in it's Value" as Newton had recognized. What they suggested, however, was that an Act of Parliament be passed that would prevent the export of coin without which the change of value, certainly of the extent proposed by Newton, would achieve nothing.[54] The pair proved to be correct. As John Conduitt acknowledged several years later, "the reduction of 6d. upon the guinea had no lasting effect upon the exchange; did not apparently lower the price of silver; nor bring an ounce of silver to the Mint."[55] In a pamphlet that may have appeared during 1717, Burgess and Billingsley pointed out that the purpose should have been to encourage both gold and silver to be minted and that this could best be achieved by the increase in the value of gold and silver coin. As they put it, "the lowering Guineas, will never bring one Ounce of Silver to our Mints, while the Price of Bullion remains higher elsewhere" and that "The Depreciating of Guineas would prove a great Discouragement (if not put a full stop) to the Coinage of Gold, Will be an Injury to the Nation in general, And to every one that has Guineas in particular."[56] But it was to Newton the Treasury chose to pay heed.

A Dutch prison and influential friends

Billingsley's prospects during the next decade proved far more dismal than the excitement of the Bubble had first promised. Although some of his ventures endured, he had engaged in so many squabbles with his former partners, it is difficult to see how he might have managed to

[54] B./B., *Reasons Humbly Submitted to the Consideration of the Honourable House of Commons; against the further lowering of Guineas; And against giving a Liberty to Export our Coin. With some Remarks upon Sir Isaac Newton's Representation of the 21st September last, to the Lords of the Treasury* (London, January 13, 1717–18), p. 2. Newton's report was printed in the *Daily Courant* (December 30, 1717). It is reprinted in J. R. McCulloch, ed., *Old and Scarce Tracts on Money* (London, 1933), pp. 274–9.

[55] Quoted in Craig, *Newton at the Mint*, p. 109.

[56] B./B., *Some Proposals, Humbly Offer'd, For the Increase and Preservation of our Currant-Coin, Particularly Silver, which is now become very scarce, And (if a speedy Remedy be not apply'd) will all be gone in a little time; For the further Payment of the Publick-Debts, So as to give every one their full Due, And instead of laying on any New Tax to accomplish this Work, To make the doing it an Advantage to all concern'd, And an Encouragement to Trade* (London, n.d.), p. 2.

survive. But survive he did, and possibly this may been the result of evident contacts he had in the merchant community in Holland. Whenever he bemoaned the dismal state of the coinage, he always referred to the trade in bullion to Holland. Billingsley was obviously well connected, and this too may explain why he was able to escape the bankruptcies that too frequently visited his former partners. In the spring of 1730, he reverted to his familiar form of printing pamphlets on his ideas to reform the coinage, utilizing the influence of the Rochester M.P., David Polhill, to bring his notions to the attention of Walpole. Billingsley's objective was to reduce the National Debt, and therefore taxes, by increasing both the Sinking Fund and the circulation of coinage. His view, which was not unique, was of a wide gulf between those who benefitted from investment in public funds and those who were forced to carry the burden, so that "these Gentlemen do not consider how their glorious and easy Estates are (immediately) produced, Neither by the prolifick Virtue of the Surface or Bowels of the Earth or Sea, But by the care, sweat and labour, and (too evidently) by the ruin of the most industrious and useful Part of this MAJESTY'S Subjects, I mean, the Trading and Midling Part of the Nation, upon whom Taxes have and will always be heavyest." Billingsley went on to draw the conclusion that the inability to recover debts was "one of the chief occasions of the unaccountable Number of Bankrupts, who have for many Years past filled our Gazets, Prisons, and Foreign Citys and Towns with the Persons and Effects of these Insolvents, To the impoverishing our Country."[57] Billingsley's analysis was based on increasingly bitter experience and it also proved prophetic.

Further communication with Walpole came from Amsterdam in September 1730. This time it was a suggestion for yet another lottery, the proceeds of which would be used to pay down the National Debt. Billingsley claimed to have been examining Dutch efforts to achieve the same thing and promised to send further details to Walpole.[58] What Billingsley did find in the Netherlands was more directly revealed in a simultaneous correspondence with the Earl of Westmorland, the former Governor of

[57] Cambridge University Library, Cholmondely-Houghton MSS. 1701, Billingsley to Walpole, March 25, 1730; with enclosure, *Reasons Humbly offered to the Consideration of the Parliament of Great Britain. For lessening the present heavy National-Debts and Taxes, Inlarging the Sinking-Fund, Remedying the ill State of the Coin, And thereby reviving Trade and Credit*. See also Michael John Jubb, "Fiscal Policy in England in the 1720s and 1730s" (Ph.D. thesis, University of Cambridge, 1978), p. 408; and P. G. M. Dickson, *The Financial Revolution in England, 1688–1756* (London and Melbourne, 1967), Chapter 9.

[58] C.U.L., Ch(H) MSS. 1771, Billingsley to Walpole, September 15, 1730; Ch.(H) MSS. 1772. Billingsley to Walpole, September 26, 1730.

the York Buildings Company, in which Billingsley's son John was still employed. Case claimed that the mints in the Netherlands were kept busy melting down English coin, and if this practice continued, the shortage would soon bring a stop to English trade. He also complained that he had asked Walpole for assistance in carrying out his investigation. When pressed, Walpole evidently told John Billingsley that he had not given much consideration to the matter. Case sought Westmorland's intervention with the Minister.[59]

By January 1731, Case Billingsley was clearly desperate. He claimed to have spent hundreds of pounds getting to the bottom of the trade in bullion and that without assistance he would be unable to discharge the debts he had contracted. He merely requested two hundred pounds to clear away his affairs in a matter he was encouraged to examine by David Polhill, an adherent of both Walpole and Newcastle. There was an urgency in Billingsley's tone that required more willingness to act than Walpole cared to demonstrate.[60] Without assistance from London, which even his sons were incapable of supplying, Billingsley was in great difficulty. In the spring of 1731, Billingsley was arrested for debt and imprisoned at Vianen, near Utrecht, possibly with Case, junior, or a younger son, Edward. He wrote in desperation to Westmorland in June, indicating that Walpole was awaiting the judgment of John Conduitt at the Mint on the value of Billingsley's scheme for the coinage, although Billingsley claimed that he had already informed Conduitt of all the steps he had taken to ascertain the extent of the bullion trade in Holland.[61] Conduitt agreed that the proposals of Newton had done nothing to slow the export of silver coin, just as Billingsley had predicted.

But there was no immediate rescue. In January 1732, Billingsley was still in Vianen in dire need of funds. In a plaintive letter, Billingsley begged Westmorland to speak to the King about his earlier efforts to solve the riddle of the longitude, and to explain how the Protestant interest was being destroyed by the failure to augment the sinking fund and reform the coinage. He reminded Westmorland that the money he had spent had been in the service of the King for which his reward was to lie "in the Most nasty cold & dangerous Prison here now near nine Months to the ruin & heart breaking of my self, a beloved wife, & thirteen beloved Sons-Grandsons & Daughters." What angered Billingsley was that he

[59] University of London Library, Autograph Letters 5/1, Case Billingsley to the Earl of Westmorland, September 12, 1730; also AL 5/2, John Billingsley to Westmorland, October 6, 1730. John requested a letter be left at Coles Coffee House in Birchin Lane.

[60] U.L.L., AL5/3, Case Billingsley to Westmorland, January 5 and 9, 1731.

[61] U.L.L., AL5/4, Case Billingsley to Westmorland, June 13, 1731.

could not pay what was demanded of him in Holland, but in England there were debtors "Who justly owe me more than ten thousand pounds sterling."[62] Billingsley finally did secure his release, although to what extent his influential friends were responsible is not clear. By April 1732, he was up to his old games of attempting to secure a patent for his longitude proposal for which he claimed the approval of Newton and the Lords of the Admiralty fifteen years earlier.[63]

Billingsley's interest in practical problems like the longitude and the coinage was established in the emerging capitalist framework of the eighteenth century. We should not succumb to the temptation to condemn his manipulations as merely efforts to secure his own advantage. This opportunism and use of connection were little more than Newton and many of the philosopher's disciples had already attempted even if Billingsley seems more unscrupulous. Moreover, he reflects the effort to control and manipulate nature through trade that, it was widely held, would establish the power and prestige of Hanoverian and Protestant interests.

What became of Billingsley upon his release from his Dutch dungeon is difficult to determine. But it comes as little surprise to find him trying to obtain the support of the Duke of Chandos early in 1735. The Duke's secretary was instructed to tell the promoter that he had "a very good claim to relief from the Administration." Chandos had suffered his own losses, ironically in York Buildings, so that "it is as much as he can do to find money to support Himself."[64] By 1736, however, Billingsley was still hounding Walpole with his lotteries, which he had amended in conjunction with a scheme to fund the new Westminster Bridge. By this time, apart from a stay in Newgate for tangling with the postal privileges of the House of Commons, his address was in Killigrew Court, Whitehall, near the Admiralty Office, perhaps indicating the reestablishment of his contact with the Navy.[65] If he planned to use his prominent friends to enable him to resume his activities in the Navy, Billingsley succeeded. The operation, however, was ultimately an effort to rescue his family from the straightened circumstances in which it found itself. Far too often it had experienced the debtor's prison. War and active service, as in William's reign, proved its salvation. By 1744, when Billingsley wrote to

[62] U.L.L., AL5/5, Case Billingsley to Westmorland, January 8, 1732.

[63] U.L.L., AL5/6, Case Billingsley to Westmorland, April 17, 1732.

[64] Hunt. Lib., Stowe MSS. 57, XLV, f. 288, J. Pudsey to Billingsley, February 8, 1734/5.

[65] C.U.L., Ch(H) MSS. 2648, Billingsley to Walpole, December 23, 1736; and B.L., Add. MSS. 24120, f. 392, Billingsley to Thomas Wotton, November 23, 1739.

Newcastle about the coinage, he was signing himself as purser to the Royal George, which was at Chatham fitting out. This ship, with a complement of about eight hundred men, would have provided ample opportunity for the sharp pen of Billingsley. Amongst the ship's company as well was his son, Edward, who had already gone bankrupt in 1740.[66] Both were discharged to the Prince George in April 1747, after which very little trace can be found although Edward appears to have been paid out in 1749.[67]

Case Billingsley died in 1747 in the parish of St. Bennet, Gracechurch, just a stone's throw from Birchin Lane and Exchange Alley, where he had invested so much of his life and imagination. From the inventory of his possessions, it can be determined that to the end, he claimed shares in the Mines Royal and Mineral and Battery Works and listed those debts that remained unsatisfied from William Lilly, John Essington, James Bradley, and William Thompson. His survivors carried on, true to form. The sons fell to squabbling over the remains, with Case, junior, accusing Edward of failing to account for £1000. For his part, Edward Billingsley died in King's Bench Prison in 1750, where he had been imprisoned for debt. His executors had to sue Case, junior, to settle Edward's will. It may have been an ignominious end to such exalted dreams, but lawsuits were typical of the adventures of the Billingsleys and thousands of other projectors.[68]

The importance of projectors

The career of Case Billingsley was a remarkable quest through the manipulation of millions of pounds, to foreign imprisonment, to an old

[66] B.L., Add. MSS. 32702, f. 32, Billingsley to Newcastle, January 29, 1743/4; PRO., ADM. 36/3041, "Royal George" Muster Book, 1744–5; PRO B4/10/22, Bankruptcy Commission, Docket Books, October 28, 1740.

[67] PRO., ADM. 33/411/ nos. 9 and 42, "Royal George" Pay Book; ADM. 6/17/216. Warrant Book, 1745–51, April 4, 1747.

[68] PRO., PROB. 3/48/40, Case Billingsley Inventory; PROB. 31/320/375, Edward Billingsley, Will, August 12, 1748; Affidavit of Richard Rooke, April 24, 1750; PROB. 31/376/24, Edward Billingsley, Inventory, January 20, 1755; PROB. 31/377/167; Edward Billingsley, Inventory, February 1, 1755. In his will of 1748, Edward left his possessions to his nephew, William Graham, Lord Viscount Prestoun, the eldest son of his sister, who had remarried a merchant, Joseph Bezeley. Case Billingsley may also have been survived by a brother Samuel, who seems to have been one of the products of Christ's Hospital School in 1692. He appears in the Navy accounts as an instructor of navigation from 1734 to 1740. Spencer Library, MSS. Q8:29, f. 16; Q8:30, f. 12. Accounts of Arthur Onslow, Treasurer of the Navy; Guildhall Library, MS. 12876, vol. II, Christ's Hospital Discharged and Apprenticed (1692), no. 52; SRO, GD1/170/vol. III, Case Billingsley to Christian Cole, March 25, 1720.

man's berth in the navy. He had followed the wider tradition of hundreds of innovators in his efforts to secure the longitude prize, and with swarms of others, he sought a patent for a new engine for pumping water. His explorations were overreaching, venturing further than his influential allies would wish, skirting shoals of accusations of bribery and misappropriation of funds. Yet Billingsley reflects the mentality of improvement that sometimes got out of control, often beyond the reach of Government even after the Bubble Act was passed. It was easy for men of the age to style themselves solicitor or engineer. As with his water engine, his creation of new joint stocks took advantage of practical demands, whether in water supply or in the mining of estates. His companies were unique only in the degree to which they were controlled by interlocking directorships adept at the manipulation of stock. And his companies survived the Bubble in ways that others did not.

Far more significant, especially in the context of Newtonian England, was the fact that these companies became exceedingly important vehicles for the adoption of technical innovation. Perhaps their mechanical interests have meant they have been dismissed in the accounts of natural philosophy. But Case Billingsley's companies were central to a community that had issued out of the late seventeenth century, and that was to provide opportunity for the patronage of natural philosophers. Private support for natural philosophy now gave way to public employment. The popular lectures of the early eighteenth century had only been a small hint of what was to come.

In March 1748, a fire struck Exchange Alley, erasing the scene of generations of intrigue. For anxious minds, fears of the Great Fire of 1666 were reborn. Even without great winds, the London Assurance Office was swiftly gutted as well as the Sword Blade Coffee-House, which had been one of the locations, ironically, from which the Sun Fire had transacted its business of insurance against this very catastrophe. Nature's force still ruled among the cinders of Birchin Lane. There was little trace left of the Marine Coffee-House, which for decades had heard the lecturers of natural philosophy and witnessed the auctions and the bargaining that had launched more than one invention into the netherworld of stocks.[69] Gone, too, by then was a generation that had watched these warrens and alleys grow into an image of infamy. Billingsley had died a year earlier. He may not have lamented the final conflagration.

For Desaguliers and the Duke of Chandos, had they survived beyond 1744, the reaction might have been different. They may have felt some

[69] Lillywhite, *London Coffee Houses*, pp. 342, 361, 565.

regret for all the dreams that had formed in the embryo of Exchange Alley. But even that would have been a passing moment. Too often nightmares had been the result. Yet there would have been some reason for nostalgia. They had both fashioned a career out of the rough and tumble of stocks and the company of coffeehouses that had brought Britain into the brutality of commerce and early industry. What had emerged was not just the myriad of companies and ventures we have encountered, but also a science that could not afford to be oblivious to the opportunities a flood of projects would provide.

Public lectures had transformed science into practical employment. In these circumstances, the Royal Society was eventually relegated to the periphery of technical possibility. Increasingly, the Royal Society seemed an anachronism that had failed to appreciate the forces of commerce and finance unleashed by the Revolution. But there were those, still part of the Royal Society to be sure, like Desaguliers and Chandos who played a significant role in the transmutation of natural philosophy into a scientific culture that had as much in common with Bacon as with Newton.[70]

[70] Cf. Michael Fores, "Newton on a Horse: A Critique of the Historiographies of 'Technology' and 'Modernity,'" *History of Science*, XXIII (1985), 351–78.

10

The Chandos connection

The business of natural philosophy had become the business of projectors. We have seen the lure of the patronage of companies and their courts of directors. We have seen the efforts – for example, of New River or of Sir Richard Newdigate – to wed improvement with the knowledge of engineers and philosophers. This was the product of deliberate policy. The Duke of Chandos was its greatest practitioner – not because he was a Fellow of the Royal Society, but because he was, above all, looking for a practical advantage that he and his companies might exploit. Natural philosophers were potentially as much a resource as nature itself.

By 1744, this had become a commonplace born of practice. Desaguliers effectively asserted this in print, although only late in his life. This was also a sentiment that the intellectualist tradition of the history of science would necessarily fail to discover. There was more to the life of the eighteenth-century natural philosophers than they committed to print. And there was more to the power of natural philosophy than treatises alone could convey. The written tome had once been a monument to private patronage. Subscription publication changed all that. The public interest in natural philosophy was reflected in the long lists appended to the works of science. And the culture of the entrepreneurs meant that public subscription to joint-stock companies would provide yet another vehicle for the support of philosophers. The full meaning of this cannot be appreciated until it is understood that philosophers might seriously respond to the demands of projectors – either by revealing dangerous frauds or by assisting in the development of a technology by which nature was to be improved. This was an ideology of investment and industry of which the Newtonians were acutely aware.

The building of James Brydges

As with the Restoration, the Revolution had transformed ambition. The financial innovations following 1688 had created new opportunities in the world of commerce, unleashing a wave of patents and projects that was soon out of control. Likewise, the ground of politics had suddenly shifted, making necessary swift recalculation of the avenues of advancement that only nonjurors stubbornly renounced. But they were a minority. More common were ambitious young men willing to accept new rules of political conduct that had less to do with principle than with the pragmatism of survival while the impact of the Stuart exile could be assessed.

One of this generation, born in the last years of the reign of Charles II, was James Brydges of Herefordshire, son of the union of landed gentry and the daughter of a merchant in the Turkey Company. He symbolized the new men who would be as much at home among the negotiations of traders and parliamentarians as in the management of country estates. And he was ambitious. After a tour of the continent, having met Leibniz in Hanover, he was in London in 1694, where he began his seige of the influential to secure himself a place in the Excise. In 1694, he was also elected a Fellow of the Royal Society in which he took great interest perhaps as much for the possibilities of preferment offered by such prestigious company as for natural philosophy. Here he was in the sight of influential men like Sir John Hoskins, former President of the Society and M.P. for Herefordshire, Sir Robert Southwell, the diplomat and President of the Society from 1690–5, and Newton's benefactor Charles Montagu, the Chancellor of the Exchequer, who succeeded Southwell as President.[1] But as the young Brydges was to discover, seeking preferment was one thing and finding it was quite another – especially for a young man without strong connections and with a father who had no use for the ascendant Whigs.

While Brydges learned the rituals of place, his initial overtures bore little tangible fruit except to extend his list of impressive contacts. In these years, he was often seen with Sir Godfrey Copley at the Royal Society and Sir Josiah Child, with whom he explored the world of trade. He sought the assistance in his quest for office of Archbishop Tenison, with whom he discussed plantations and new discoveries yet to come.[2]

[1] C. H. Collins Baker and Muriel I. Baker, *The Life and Circumstances of James Brydges, First Duke of Chandos* (Oxford, 1949), pp. xiv, 7–23; Hunt. Lib., Stowe MS. 26, "A Journal of My Daily Actions," 1697–1702, Vol. I, March 13, 1697; June 29, 1698; Nov. 23, 1698.

[2] Hunt. Lib., Stowe MS. 26, Vol. I, January 21, 1696/7; February 6, 1696/7. He appears

But still there was no advancement until he recognized the futility of the exercise without a political base of his own. This was the reason for his decision in 1698 to contest Hereford in the summer elections. It too proved more difficult than he had first thought. His search for preferment in the new regime left him vulnerable to his political enemies, who spread rumors that he was a beholden to the favors of the Court. Although ultimately successful in the election, Brydges was stung by the accusation that he was "so entirely of ye Court party as even now to receive a pension from them." The notion, he claimed, was an abomination. He was still trying to find his feet in a world where the conflicts of Court and Country were in a flux that would meld into the contests of Whigs and Tories.[3]

It took time before the Court decided it could trust Brydges. In 1703, he was appointed one of the Commissioners of the Admiralty, and not until 1705 was he finally to receive the lucrative post of Paymaster of the Forces Abroad which gave him marvelous opportunities to apply the funds at his disposal for investment and speculation. Brydges seized the occasions his post presented and, while he ran the risk of political attacks on his conduct, he did not engage in a practice unexpected of placemen.[4] There remained questions of Brydges' political loyalties primarily, one suspects, because he was a trimmer who watched the political horizon carefully for any sign of heavy weather or for a shift in the winds. During Anne's reign, he was haunted by the specter of both Whigs and Tories. He attempted frequently to maintain his distance, complaining of both extreme Whigs and Tories to Marlborough's aide Cadogan, and of the "Violence of Parties" to the Lord Treasurer Godolphin.[5]

This is not to say that Brydges was entirely without principles. He simply sought to escape the extremes of political passions that might render regimes and their favors uncertain. Reasonableness was the other

to have met Copley, if he did not already know him, at the Grecian Coffee-House in the company of John Toland and John Trenchard. See Vol. I, February 12, 1697/8.

[3] Baker and Baker, *Life and Circumstances of James Brydges*, pp. 19–20; Hunt. Lib., Stowe MSS. 57, I, fols. 26–7, Brydges to Mr. Powell at Hegdon, n.d., probably 1698. On this problem, see David Hayton, "The 'Country' interest and the party system, 1689-c. 1720," in Clyve Jones, ed., *Party and Management in Parliament, 1660–1784* (New York, 1984), pp. 37–42.

[4] Baker and Baker, *James Brydges, First Duke of Chandos*, p. 47.

[5] Hunt. Lib., Stowe MSS. 57, II, f. 6, Brydges to Cadogan, December 24, 1707; II, f. 29, Brydges to Godolphin, March 31, 1708. See Clyve Jones, "James Brydges, Earl of Carnarvon, and the 1717 Hereford By-election: A Case Study in Aristocratic Management," *Huntington Library Quarterly*, XLVI (Autumn 1983), 311. On frequent elections and political passions, see Geoffrey Holmes, *British Politics in the Age of Anne* (London and New York, 1967), p. 218.

side of opportunism. His comments on the Sacheverell affair reveal a man who could feel the political winds shifting and who was trying to find a haven from the worst violence. In the spring of 1710, when the electoral force of Church and King was multiplying, he complained that "ye Tories are very free & open in fixing the cause & warmth of this Impeachment on our Friends, & ye Wiggs on ye other side are not very forward in taking ther share, they ought of it, to themselves."[6] The tide was even more worrisome, as he explained to his banker John Drummond in Amsterdam, because it followed "such a Billingsgate nonsensical Sermon, so trifling in it self, if compear'd to ye consequences that may attend it."[7] Brydges had been right. The Godolphin ministry to which he was attached, partly through his alliance with Marlborough, fell in the late summer of 1710 with the attendant rise of the Tories as the chief beneficiaries of Sacheverell's bilious trail through the countryside.

But Brydges was not fool enough to think he should challenge the Earl of Oxford's new ministry. On the contrary, he was soon hard at work trying to convince Oxford of his usefulness. In 1711, he told Oxford that "by reason of my acquaintance in ye City & among ye monied men" he might be useful amongst the Directors of the new South Sea Company.[8] Brydges had become adept at trading in stock and appears to have taken a large position in the new company. But his aim, one suspects, was not improved management of the South Sea Company, but political connection among the dominant Tories. In 1712, he paid his respects to the Oxford administration having explained that "The Obligations I had to My Ld Godolphin & ye Duke of Marlborough were wholly personal, & as the death of ye one & ye retreat of ye other from Publick affairs have left me perfectly at liberty So nothing shall hinder me from following ye dictates of my own mind."[9] But Brydges was unsure of where his obligations truly lay, having been tied to the Marlboroughs as they fell from grace. He resigned his position as paymaster in 1713, although he took time to comfort the old soldier, arguing that "when ye anger of Parties against great Generals & Ministers has no real foundation in truth, it softens by degrees, & ye eyes of ye people begin to be open'd."[10]

Brydges was the quintessential trimmer, especially where allegiances

[6] Hunt. Lib., Stowe MSS. 57, III, f. 204, Brydges to Cadogan, April 7, 1710.
[7] Hunt. Lib., Stowe MSS. 57, IV, f. 56, Brydges to John Drummond, July 5, 1710.
[8] Hunt. Lib., Stowe MSS. 57, V, f. 95, Brydges to Oxford, June 21, 1711; Baker and Baker, *Life and Circumstances of James Brydges*, p. 51.
[9] Hunt. Lib., Stowe MSS. 57, VIII, fols. 111–12, Brydges to Oxford, November 1712.
[10] Hunt. Lib., Stowe MSS. 57, IX, fols. 123–4, Brydges to Duke of Marlborough, July 8, 1713.

might obstruct preferment. His constant refrain was the difficulties presented by the warfare between the Tories and the Whigs. This enabled him to maintain contacts in both camps, from Robert Walpole to Viscount Bolingbroke. In the summer of 1714, when it was apparent that the Queen's days were few, he was prepared to profess his allegiance to the Hanoverians.[11] His reward was his creation as Earl of Carnarvon by the new King, but in the short term, this was to prove a hindrance to his political fortunes. In the election of 1715, he was unable to secure the election for the city of Hereford of his candidates, Nicholas Philpott and Herbert Rudhale Westfaling, because of their Whiggish tendencies. This result merely confirmed the conclusions he had earlier reached to stay out of active pursuit of a seat for himself. As he explained to Philpott soon after the Succession, he had retired from politics because it was futile "to engage afresh in a life of hurry & envy & in a Nation so divided into Parties, that no one is allow'd any good quality, by ye opposite side."[12]

Brydges and the Tories

Brydges' effort to maintain his independence, or perhaps to rationalize his reluctance to risk the electorate, may explain his consistent links with prominent Tories. His efforts on behalf of Viscount Bolingbroke, whose political fortunes were shattered by the Succession and accusations of Jacobitism, could not have been widely known. It was Brydges who warned Bolingbroke of the possibility of action by the new Whig ministry in the spring of 1715 "when no body knows who is Safe, & who is not will be very prudent."[13] Slightly over a month later, Bolingbroke fled to France after having borrowed £20,000 from Brydges. The connection between the two was furtively maintained with Brydges managing Bolingbroke's affairs in England and arranging funds to be drawn on the banker Richard Cantillon in Paris for use of the exile. Bolingbroke's decision in July 1715 to join the Pretender was the last straw and Brydges had to contend with the fact that Bolingbroke had been the object of a Act of Attainder passed by Parliament in September. From that point,

[11] Hunt. Lib., Stowe MSS. 57, X, f. 34, Brydges to Marlborough, January 9, 1713/14; X, fols. 169–70, Brydges to George, Elector of Brunswick, July 31, 1714.

[12] Jones, "James Brydges, Earl of Carnarvon," pp. 311, 314; Hunt. Lib., Stowe MSS. 57, XI, f. 10, Brydges to Philpott, September 19, 1714.

[13] Hunt. Lib., Stowe MSS. 57, XI, f. 251, Brydges to Bolingbroke, February 7, 1714/15.

Figure 12. James Brydges and family, by Godfrey Kneller, ca. 1713.
By courtesy of National Gallery of Canada, Ottawa.

Brydges had to be very circumspect and he relied on the information provided through Cantillon in Paris for news of the fugitive.[14]

He had reason to be afraid. The Jacobite rebellion had been crushed, Habeas Corpus suspended, and the hunt was on for sympathizers of the Stuarts. Constables searched the house of the nonjuring bishop Nathaniel Spinckes for links to the Pretender. His mentor George Hickes, whom Brydges had patronized, pleaded with Hans Sloane to intervene with the Justices to prevent Hickes' imprisonment in view of his age and infirmity.[15] While the ferrets furiously hunted Jacobites, Brydges was

[14] H. T. Dickinson, *Bolingbroke* (London, 1970), pp. 134–43; Hunt. Lib., Stowe MSS. 57, XI, fols. 315–19. Brydges to Bolingbroke, April 14, 1715; XI, fols. 331–2, Brydges to Bolingbroke, April 17, 1715; XII, fols. 11–15. April 29, 1715; XII, f. 60, Brydges to Cantillon, July 11, 1715; XII, f. 61, Brydges to Bolingbroke, July 11, 1715; XII, f. 307, Brydges to Cantillon, March 14, 1715/16.

[15] Baker and Baker, *Life and Circumstances of James Brydges*, p. 67; B.L., Sloane MSS. 4059, f. 163, George Hickes to Hans Sloane, n.d.

warming to Walpole in anticipation of the resignation of the Duke of Somerset from the Court following the arrest of Somerset's son-in-law, Sir William Wyndham, on suspicion of Jacobitism.[16] The new Earl of Carnarvon was cautiously circling the chaos for spoils.

Brydges maintained his ties with a number of Tories after the Succession. None was quite as dangerous as Bolingbroke, but they were a prominent force in the intellectual world. A long-term confidant of Brydges was Dr. John Arbuthnot, formerly physician to Queen Anne and close both to Harley and Swift. On the death of the Queen, Arbuthnot was removed from Court by the new ministry. Brydges attempted to intercede with Walpole on Arbuthnot's behalf but was quite unsuccessful. Once again, he was convinced that "where violence in Party-matters Supplies ye defect of men, & is allow'd to bear Sufficient cause for doing unjustifiable hardships."[17] But Brydges' association with Tory intellectuals was not restricted to Arbuthnot whom he undoubtedly knew through the Royal Society.

It is also the case that Brydges patronized a group of Tory Newtonians, among them Desaguliers' Oxford teacher, John Keill. Desaguliers, certainly no Tory, had entered into the pay of Brydges in 1716. Shortly thereafter, John Keill, who had earlier emerged as Newton's champion in the dispute with the Hanoverian Leibniz, established a close relationship with Brydges. Keill was a frequent visitor to the estate at Cannons which Brydges was in the process of developing, and in 1718, it was Brydges who became godfather to Keill's son. Keill dedicated his *Introduction to the True Astronomy* to Brydges, possibly because the subject was one in which Brydges had evinced an interest through the construction of an observatory on the estate.[18]

The issue of science and political ideology is a difficult one. There can be little doubt that the most prominent early Newtonians had established Whig political connections. But were they any different from Brydges, who eventually caught the Hanoverian wave? Or was there some fun-

[16] Hunt. Lib., Stowe MSS. 57, XII, Brydges to Walpole, October 3, 1715.

[17] This was a chronic refrain. On the issue of sectarianism and authority see S. J. Schaffer, "Anglo-Dutch Medicine and the Glorious Revolution," *Notes and Records of the Royal Society*, XLIII (July 1989), 167–90; Hunt. Lib., Stowe MSS. 57, XV, fols. 52–3, Brydges to Arbuthnot, September 25, 1717. See also Harold Williams, ed., *The Correspondence of Jonathan Swift* (Oxford, 1963), II, 121–3. Arbuthnot to Swift, August 12, 1714; and Lester M. Beattie, *John Arbuthnot: Mathematician and Satirist* (Cambridge, 1935), pp. 5, 229.

[18] Hunt. Lib., Stowe MSS. 57, XI, f. 94, Brydges to Dr. Kehl [sic], November 16, 1717; XVI, fols. 51–2, Brydges to Dr. Keil, November 20, 1718; Baker and Baker, *Life and Circumstances of James Brydges*, pp. 67–8.

damental Whig ideology to Newtonianism as the Jacobs have proposed? It would seem to me when the success of the Whigs was still an open question – in the reign of Anne, for example – then the ideological basis of Newtonianism was less obvious and less successful than would be the case under the Robinocracy. After all, as Anita Guerrini has suggested, it is arguably the case that the commitment to Newton of Tories like John Freind or John and James Keill "had little or nothing to do with political or religious ideology." John Keill had been patronized by Robert Harley up to the Tory implosion in 1714, and it is evident that Keill's support of Newton neatly coincided with an anti-Hanoverian, and hence his anti-Leibnizian, position when it looked as though Leibniz might descend upon England in the baggage train of a new king.[19] It was clearly possible for the Tories to see the concordance between their own political sympathies and the Newtonian polemic issuing from the Royal Society in the last years of Anne's reign. Perhaps Keill or Freind sensed that the Society might be cautious in its allegiance to a Hanoverian. Thus, it is less an issue of what is inherent as what is coincident. Self-interest and ideology are really separate currents – and, of course, they could intersect. It is equally the case that even after the South Sea Bubble threatened the new reign, Brydges would not turn his back on the impoverished Tory Brook Taylor, the Newtonian mathematician and friend of Bolingbroke.[20]

The question, therefore, of the inherent ideological position of the Newtonians cannot be regarded as a simple one. Certainly proponents of the Latitudinarian Church, like Whiston and Clarke, who were hated enemies of Tories like Atterbury and Sacheverell, were heirs to Newton's metaphysics or to his religious sympathies. But the foundation of connection also confuses the role of ideology. Ideology is at least as complex an issue as the evolution of Brydges' own political alliances. Success could

[19] Anita Guerrini, "The Tory Newtonians: Gregory, Pitcairne and their Circle" (unpublished paper presented to the William Andrews Clarke Library seminar, July 1985), p. 29; see also Guerrini, "Newtonian Matter Theory, Chemistry, and Medicine, 1690–1713" (Ph.D. thesis, Indiana University, 1983), Chapter 2; Guerrini, "James Keill, George Cheyne, and Newtonian Physiology, 1690–1740," *Journal of the History of Biology*, XVIII (Summer 1985), 247–66; and John Gascoigne, *Cambridge in the Age of the Enlightenment. Science, Religion and Politics from the Restoration to the French Revolution* (Cambridge, 1989), pp. 145ff. Cf. Arnold Thackray, "'The Business of Experimental Philosophy, The Early Newtonian Group at the Royal Society," *XIIe congres international d'histoire des sciences* (1968), *Actes*, IIIB (1971), 155–9.

[20] Private Collection, Stirling of Garden MSS., James Stirling Correspondence (1715–62), Bundle 143, James Stirling to John Stirling, June 5, 1725; Hunt. Lib., Stowe MSS. 57, XX (MS. film 417), f. 71, Brydges to Brook Taylor, December 5, 1721; Gascoigne, *Cambridge in the age of the Enlightenment*, p. 154. For Desaguliers' own experiments with Taylor, see Desaguliers, *Course of Experimental Philosophy*, II, pp. 296–8.

breed its own allegiances, no less with Newton than with Walpole and the Whigs.

The political landscape was rapidly transformed by the Hanoverian Succession and by the abysmal debacle of the Jacobite risings of 1715 in which Bolingbroke had so foolishly involved himself. From that point, James Brydges cautiously went about repairing his political fortunes. When he had resigned the paymastership in 1713, this seemed to presage his seeking of a seat in Parliament. After the Succession, he evidently found it difficult to reconcile his Whig connections with the Tory establishment of Hereford. The death in December 1716 of the Herefordshire Tory Lord Scudamore changed all that, and in a carefully orchestrated campaign in 1717, Brydges was able to engineer the election of the inveterate Whig Herbert Westfaling, who remained in the seat until 1727. Westfaling would prove most helpful. Like many Brydges' associates, he was a cipher in the courts of directors of many of the companies in which Brydges invested.[21] This was also the case with Humphry Walcot, the Tory M.P. for Ludlow, a protege of Brydges and his deputy as paymaster of the forces, who was part of the Court of the African Company that was virtually manipulated by Brydges.[22] By 1720, Brydges, now invested as Duke of Chandos by George I, was able to recruit to his patronage men of ability of various political stripes. The voracious packs of Whigs and Tories with which he was not easily comfortable required on his part some adaptation from the Court-Country conflicts that had marked his origins. But Brydges had become a Duke not through adherence to old principles, rather by malleability to new realities. Whiggism was obviously ascendant and a Hanoverian loyalist he would increasingly be.

The Duke of Chandos and his agents

Brydges had been alert to Baconian political and philosophical ideals.[23] And the efficacy of natural philosophy was a faith he was willing to follow. As we have already seen, James Brydges was prepared to extend his patronage among the natural philosophers – especially to those who might provide beneficial advice concerning the many projects in which

[21] Jones, "James Brydges, Earl of Carnarvon," pp. 311–17; Hunt. Lib., Stowe MSS. 57, XXV, f. 223, Brydges to Sir John Mears, February 12, 1724/5.

[22] Sedgwick, *The House of Commons*, II, p. 504; Walcot was something of an entrepreneur having formed a company in 1701 to promote an invention of his uncle for desalinating sea water. I wish to thank Dr. Eveline Cruickshank of the History of Parliament Trust for the opportunity to examine their files.

[23] Hunt. Lib., Stowe MS. 13. Observations out of ye Old & New Testament, also out of Lord Bacon's Essays & Fuller's Holy State. [Begun] Jan. 16, 1696/7.

he was engaged. This was the reason that he employed the luckless Richard Bradley in the construction of a physic garden and a hothouse at Cannons.[24] The same kind of motives was undoubtedly behind his provision of a clerical living to Desaguliers to whom was given the task of the management of the improvement of the grounds, including overseeing Bradley's efforts, irrigation of the estate, as well as heating of the Duke's house. Chandos's characterization of Desaguliers as "the best Mechanick in Europe" leaves little doubt of the Duke's view of the philosopher.[25] As Chandos became more concerned directly in the fancies of the South Sea Bubble, he increasingly relied upon those who claimed expertise in the practical aspect of natural philosophy.

Chandos was, above all, an entrepreneur. He was always on the lookout for a venture that had some chance of succeeding. That, at least, was part of the reason for his support of Richard Steele's Fish Pool. But as the rage of the Bubble became difficult even for the steadiest of nerves, Chandos spent an increasing amount of time attempting to revive the sagging fortunes of the African Company in which he had heavily invested. Chandos virtually controlled the Company from 1720 and he tried to interest Walpole in investing in a subscription in the African trade. When Walpole was rushing to reestablish public confidence and rescue the South Sea Company victims, Chandos was trying to get Walpole to turn over part of the Company's trade to the African Company.[26]

Chandos's dealing on behalf of the entire African Company may seem slightly presumptuous, but he had every reason to be confident of his authority. The Court of Assistants was packed with his men, including his banker John Drummond, Humphry Walcot, who was chosen subgovernor in 1721, and Charles Hayes, the friend of John Harris and author of works on the fluxions and the longitude.[27] By 1721, active among the Court of Assistants were Henry Levett, Chandos's personal physician, and the Duke's confidant, John Arbuthnot.[28] They acted in concert as

[24] Hunt. Lib., Stowe MSS. 57, XV, f. 275, Brydges to Bradley, July 9, 1718; XV, f. 195, Brydges to Bradley, August 8, 1718; XVI, fols. 407–10, Brydges to Bradley, December 16, 1719.

[25] Hunt. Lib., Stowe MSS. 57, XV, f. 252, Brydges to Mr. William Mead, June 16, 1718; XVI, f. 410, Brydges to Desaguliers, December 16, 1719.

[26] Hunt. Lib., Stowe MSS. XVIII, f. 2, Chandos to Walpole, February 27, 1719/20; XVIII, f. 200, Chandos to Walpole, December 1, 1720.

[27] PRO, T70/53/56–58. Royal African Company Letters, March 16, 1720; May 16, 1721. On Hayes see *Gentleman's Magazine*, XXI (December 1761), 543–6; and R. V. Wallis, and Peter Wallis, *Biobibliography of British Mathematics and its Applications*, Part II, 1701–60 (Newcastle, 1988), p. 14.

[28] PRO, T70/91. African Company, Minutes of the Court of Assistants; p. 11. August 24, 1721; p. 13. August 29, 1721; p. 19. September 12, 1721; p. 29. September 28, 1721.

the Duke's agents and they represented an effective method to control the direction of a company that Chandos employed more than once.

The role of the Duke of Chandos in the affairs of the African Company reinforced his tendency to rely on the expertise of those well versed in natural philosophy. The dismal state of the African Company afforded ample opportunity for revival once he recognized that the slave trade was a losing proposition.[29] He was convinced that slaves and ivory were not likely to produce much profit, but more might be said of the trade in gold, cotton, potash, and various medicines. It is highly probable that the increasing fears of smallpox that stalked London in 1721 made him realize that medicine and money were hardly unrelated. The African Company, while it certainly had its own special experience with survival in tropical climates, was nonetheless also well placed to take advantage of the importation of medicines. In 1721, he ordered James Phipps, the Company's Captain General at Cape Coast Castle, "to enquire into the Physick the Natives use for curing distemper, & if they have any Specificks to find out, of what they are made, & to send over some of their drugs with an Account of their Nature, & how used & in what quantities, viz. for Fevers, fluxs, &c."[30]

One reason that Chandos sought out medicines was that the African Company had itself been struck by a series of illnesses not all of which originated in the tropics. The fortunes of the Company were undermined by the increasingly ubiquitous smallpox, which killed both slaves and their keepers in the trade. By the end of 1721, the Company became aware of the extent of this problem and Chandos took some initiative to try to secure for the African forts a surgeon who understood the techniques of inoculation. The result was the employment of a Scottish physician, Dr. James Houstoun, who had applied for a position as a surgeon with the Company's African settlements. It is now clear that his petition for employment was drawn up by John Drummond of the Court of Assistants, who was also Chandos's personal banker. Recruitment of Houstoun had been through John Arbuthnot, who made the initial recommendation. This reveals not only the extent to which Chandos effectively ran the Company but also the degree to which he was dependent on his associates in doing so. These included not just Arbuthnot, but also Henry Levett, and Hans Sloane of the Royal Society.[31]

[29] Hunt. Lib., Stowe MSS. 57, XIX, f. 13, Chandos to Ambrose Baldwin at Wydah, July 7, 1722.

[30] Hunt. Lib., Stowe MSS. 57, XIX, f. 222, Chandos to James Phipps, October 5, 1721.

[31] PRO, T70/91/106, February 22, 1721[/2]; James Houstoun, *Memoirs of the life and times of James Houstoun, M.D. (formerly physician and Surgeon-General to the Royal*

The search for new remedies to disease included not only the application of inoculation, but also the attempt to examine botanical specimens for medicinal uses. Chandos's letters in 1721 and 1722 abound with instructions to the African Company's factors on the slave coast to seek out plants that might prove beneficial. They were especially instructed to try to find what Chandos knew as "quinaquina" or Jesuit's Bark, balsom for use in rheumatism, as well as a dried gum that produced an indigo dye. The evaluation of botanical specimens was largely to be left to Hans Sloane. On his orders, factors were warned that the specimens that were sent should be in sufficient quantity and, where possible, bearing fruit and flower, so that Sloane might be able to examine them fully.[32]

As of 1720, there was an air of desperation in the pronouncements of the Duke of Chandos. He watched with dread as the stock-exchange furies played out their roles in the chaos and collapse of stocks from the heights they had reached in early summer. But he had, I think, acknowledged the dangers in the rampant enthusiasm of investors. In these circumstances, Chandos realized, of course, that the future of the African Company could not rest entirely upon new drugs or dyes. He was virtually obsessed with discovering the source of the small amounts of gold the traders sometimes brought out of the African coast. James Phipps of Cape Coast Castle was instructed to find the gold mines so that the Company could send miners for their exploitation. Perhaps, this is what would enable the African adventurers to escape the collapse that Chandos described to Bolingbroke as "so universal a scene of Misery" as to be "inconceivable."[33]

If Chandos were to put any hope in African gold mines, he had first to encourage exploration for them. But often the agents on the coast were reluctant to venture into the interior – sometimes because of the hostile reception but largely because the health of Europeans deteriorated rapidly in the jungles of the Gambia. He obtained from William Derham, the rector of Upminster, a transcription of a journal of exploration up the river Gambia that had found its way into the papers of Robert Hooke.

African Company's settlements in Africa, and late surgeon to the Royal Asiento Company's factories in America) (London, 1747), pp. 77, 127–8. For background, see L. Stewart, "The Edge of Utility: Slaves and Smallpox in the Early Eighteenth Century," *Medical History*, XXIX (1985), 54–70.

32 Hunt. Lib., Stowe MSS. 57, XVIII, f. 387, Chandos to Baldwin, chief factor at Wydah, March 3, 1721/2; XVIII, f. 435, Chandos to Baldwin, May 11, 1722. See Stowe MSS. 28, James Brydges, "A Collection of Some Remarkable Papers," A Receipt for drawing Dyes out of Wood or Plants given me by Coll. [Samuel] Horsey. 1723.

33 Hunt. Lib., Stowe MSS. 57, XVII, f. 219, Chandos to Phipps, July 20, 1720; XVII, fols. 206–7, Chandos to Bolingbroke, October 5, 1720.

Derham, who edited Hooke's works, presented this to Chandos who made certain that it was immediately passed on to Henry Glynn, the chief merchant at Gambia, as proper encouragement.[34]

Chandos was prepared to consider the ends of the earth to discover his gold. His attention was momentarily diverted from Africa to the West Indies terminus of the African Company in Jamaica. In the summer of 1720, when the projects bubbled furiously, a group of West Indies merchants and landowners received a patent from the Crown to mine for gold and silver in Jamaica. Chandos was immediately interested, although he couched this, as he was later often to do, in a note of false caution:

> The humour of the Age hath made people of all degrees so fond of Projects wch: we call here by the Name of Bubbles that for these 8 months past nothing could be propos'd tho' ever so extravagant & whimsical for the promoting of which a Subscription of as many Hundred Thousand pounds was desir'd was not greedily entered into amongst the rest. . . .[35]

One West Indian landowner curious about Chandos's schemes for a gold trade was Martin Bladen. Bladen had married wealth, but his connections with Chandos's business affairs dated from paymaster days in Anne's reign. Bladen was one of those drawn to the African Company. During 1721, when Chandos was trying to get projects off the ground at both ends of the slave trade, he reported to Bladen of the great wealth that seemed about to be generated by the refining of ores in Africa. On one voyage alone, £4000 of gold and ivory tusks had been shipped to London. The tests performed on ore sent by James Phipps revealed enough lead and copper to whet the Duke's appetite and he was optimistic that miners sent by the African Company would make important discoveries. Within a few months, most of the miners were dead from disease.[36] This could be one reason why Chandos was so enamored with

[34] Hunt. Lib., Stowe MSS. 57, XIX, fols. 21–2, Chandos to Revd. William Derham, March 18, 1720/21; XIX, fols. 75–6, Chandos to Glynn, May 23, 1721. There are several copies of this journal in the Duke's papers. Hunt. Lib., Stowe MS. 9, "Papers Relating to African Affairs," pp. 1–7, March 14, 1720/21; Stowe MSS. 28, "A Collection of Remarkable Papers," pp. 34ff. Another version is in Bodleian Library, MS. Gough (Somersetshire) 7, no. 23. See Thora G. Stone, "The Journal of Cornelius Hodges in Senegambia, 1689–90," *English Historical Review*, XXXIX (1924), 89–95. Chandos's search for charts of the coast was aided by Hans Sloane. Hunt. Lib., Stowe MSS. 57, XX [MS. film 417], f. 48, Chandos to Sloane, December 19, 1721.

[35] Hunt. Lib., Stowe MSS. 57, XVII, f. 181, Chandos to Sir Richard Lawes, September 23, 1720. Lawes was probably the brother of Sir Nicholas Lawes, whose son-in-law Charles Long was an absentee Jamaica landowner and a central director of the Company.

[36] Hunt. Lib., Stowe MSS. 57, XVIII, f. 295, Chandos to Col. Bladen, September 21, 1721; XIX, fols. 220–2, Chandos to Phipps, October 5, 1721. On Bladen, see Sedgwick, ed., *House of Commons*, I, pp. 465–6. On the fate of the miners, PRO, T70/7/44–45, Abstracts of Letters Received from the Coast of Africa, July 2, 1722.

the introduction of inoculation amongst the slave trade. In any case, the Directors seem to have taken these disasters in stride. The Company continued to send miners and apparatus, including a Savery's Bellows, to the African coast for the purpose of smelting and making assays of the ores. Yet all calamities in these ventures were not easily shrugged off, as when a band of miners murdered the captain of a company ship and turned pirates.[37]

But Chandos persisted. He remained encouraged by the advice of Captain Trengrove, probably Ezekial Trengrove of the Proprietors of the Newcomen engine, on methods relating to the "buddling" of earth in search of gold. Trengrove had a special expertise in mining. This likely followed some connection with Devon and Cornwall, or perhaps from his association with the Apothecaries Company which figured prominently in the history of the Newcomen engine.[38] Although mining proved discouraging, Chandos used Desaguliers to explore the likelihood of success of a soap manufacturing venture with Colonel Samuel Horsey, a future Governor of the York Buildings Company. The African Company was prepared to contract with Horsey for the supply of potash from Africa. Chandos was interested not only in the advantage to the African Company, but Horsey was thinking of a joint-stock venture in which the Duke was seemingly prepared to invest. This proposal eventually involved the York Buildings Company even more directly.[39] But the

[37] Hunt. Lib., Stowe MSS. 57, XXI, f. 11, Chandos to Mr. [Henry] Neale, July 7, 1722. This was probably Neale of the Mine Adventurers Company. Stowe MSS. 57, XXII, fols. 275–6, Chandos to Mr. Payne, July 9, 1723.; PRO, T70/91/235–236. African Company, Minutes of the Court of Assistants, July 25, 1723. This was John Payne, a timber merchant who acted as Chandos's agent in Bridgwater. Payne was an entrepreneur and inventor in his own right. Baker and Baker, *Life and Circumstances of James Brydges*, pp. 222–6, 353–4.

[38] Hunt. Lib., Stowe MSS. 57, XIX, fols. 183–4, Chandos to Phipps, September 11, 1721; XXII, fols. 132–5, Chandos to Mr. Dodson, December 5, 1722; Hunt. Lib., Stowe MSS. 28, "A Collection of Remarkable Papers," p. 14. Capt. Trengrove's Computation of the Number of Men and Boys Necessary for the Management of the Buddle and the Quantity of Sand to be washt by one Buddle in a Day. June, 1723; also Bodleian Library, MS. Gough (Somersetshire) 7, f. 262. June 1723. On Trengrove, see Alan Smith, "Steam and the City: the Committee of Proprietors of the Invention for Raising Water by Fire, 1715–1735," *Transactions of the Newcomen Society*, XLIX (1977–8), 14; John S. Allen, "Thomas Newcomen (1663/4–1729) and His Family," *Transactions of the Newcomen Society*, LI (1979–80), 18; L. T. C. Rolt and J. S. Allen, *The Steam Engine of Thomas Newcomen* (New York, 1977), p. 60; and J. S. Allen, "Bromsgrove and the Newcomen Engine," *Transactions of the Newcomen Society*, XLIII (1970–1), 190–1.

[39] Hunt. Lib., Stowe MSS. 57, XXIII, fols. 33–4, Chandos to Desaguliers, October 24, 1723; XXIII, f. 71, Chandos to Desaguliers, November 13, 1723; XXIV, f. 219, Chandos to Col. Samuel Horsey, August 27, 1724. See also Stowe MSS. 57, XXIV, fols. 128–9, Chandos to Capt. Genl. Tinker at Cape Coast Castle, June 27, 1724; Stowe MSS. 28, "A Collection of Remarkable Papers," Coll. Horsey's Method of extracting Salts from

persistence of the African Company failed as forts were eventually abandoned and jungle and climate claimed European lives. Fear eventually conquered greed. But the effort did suggest that Chandos's intention to exploit all avenues available was not unique to him even if it meant unexplored territory far from a country gentleman's estate.

If the rage of investment in 1719 and 1720 is any guide, there were many who shared the Duke's enthusiasm for projects. The degree to which many of these ventures interlocked is truly remarkable. The possibilities that Chandos felt in the African trade, and which he thought could be improved by the philosophers, were also present in other companies. This was, as we will see, a prevalent feeling in the York Buildings Company, to which some of the Newtonians were ultimately tied. What is notable is the degree to which Chandos's African Company was linked to York Buildings and some of the other ventures of Case Billingsley. Besides Charles Hayes, who was one of Billingsley's rivals for the longitude prize, the Court of Assistants of the African Company included John Essington of the York Buildings Company and of the English Copper Company. The African Company Court also involved Christian Cole, the Billingsley operative, who was behind many of the maneuvers of the Mines Royal and Mineral and Battery Works, the Royal Exchange Assurance and the English Copper Company.[40]

The network of new companies formed by joint stock after the Revolution provided mechanics and philosophers alike with an abundance of opportunity. The Royal Exchange became as important as the Royal Society in establishing links between the world of commerce and the universe of philosophers. Increasingly, natural philosophers acknowledged not only the opportunities, but also their own utility in a time when so many projectors deceived the unwary. That is not to say that

Wood Ashes; and PRO, T70/92/128, African Company. Minute Book of the Court of Assistants, November 26, 1724. On Horsey, see Baker and Baker, *Life and Circumstances of James Brydges*, p. 230.

40 PRO, T70/91/198, African Company, Court of Assistants, Minutes, 1723; T70/53, African Company Letters, p. 191, February 28, 1722/3; p. 240, February 21, 1723/4; p. 268, August 11, 1726. Among the Court of Assistants was John Torriano, one of those involved in the Charitable Corporation that had dealings with Chandos and the York Buildings Company. PRO, T70/53/282, July 6, 1727. See also Cummings, "York Buildings Company," p. 416; PRO, T70/116/24, African Company. Committee Book of Accounts, Payments to Directors. It is unclear that the Mr. Peck with whom Chandos was dealing in 1724 was Philip Peck; he may have been Daniel Peck of the Welsh Copper Miners and a former partner of Billingsley. Hunt. Lib., Stowe MSS. 57, XXIV, f. 262, Chandos to John Payne, September 29, 1724; XXV, f. 71, Chandos to Mr. Peck, December 13, 1724. Note also A. Raistrick, "The London Lead Company, 1692–1905," *Transactions of the Newcomen Society*, XIV (1933–4), 128–9.

all of the joint-stock ventures were intended to defraud. Chandos recognized, above all else, that in the rage of investment, it was the advice of natural philosophers that might permit his projects to survive. He was soon to learn that even Desaguliers offered no preservative against disputes. And even legitimate endeavors like the Savery engine patent, with the new life breathed into it in 1715 by the Proprietors of the Invention for Raising Water by Fire, was to have its share of controversy.[41]

Desaguliers and water supply

Some of the most remarkable sights in London in the early part of the century were the machines that been built to supply the rapidly expanding city with a sure supply of water. These consisted of reservoirs, deep trenches, or pipes made usually of wood that spanned streams and sprayed leaks, as well as the boilers and hulking beams of steam engines with large towers belching smoke and cinders so that visitors marveled as often as the city inhabitants grumbled.[42]

One of the most important of the enterprises at work in the early eighteenth century was the New River Company, which had its origins in Islington in the reign of James I. From its offices in Puddle Dock, on the Thames not far from Fleet Ditch, the New River ran a joint-stock company that drew water from Amwell and Chadwell in Hertfordshire. The water was brought from nearby Islington in a canal from which it was then conveyed to Westminster and London by lined trenches running, some would have no doubt noted, by the mad house on Islington Road.[43] By the turn of the century, the New River group had expanded to the point of challenging the interests of the London Bridge Company, which, under the control of the goldsmith Richard Soame, had harnessed the

[41] J. R. Harris, "The Employment of Steam Power in the Eighteenth Century," *History*, LII (1967), 136–7; Alan Smith, "Steam and the City," pp. 5–19. These arrangements did not stop other inventors from trying to establish their own claims to a patent even before the Savery term expired in 1733. For these controversies, see, for example, J. S. Allen, "The 1715 and the other Newcomen Engines at Whitehaven, Cumberland," *Transactions of the Newcomen Society*, XLV (1972–3), 237–68.

[42] Nicholas J. Barton, *The Lost Rivers of London* (London, 1962), pp. 69ff.; A. de la Mottraye, *The Voyages and Travels...*, translated from the French. Vol. III (London, 1732), p. 226; E. Hatton, *A New View of London*, Vol. II (London, 1708), Section VII, pp. 784–5, 791–8; *Survey of London*. XVIII. The Strand. The Parish of St. Martin-in-the-Fields, Part II (London, 1937), pp. 48–51

[43] Hatton, *A New View of London*, III, pp. 792–3; H. W. Dickinson, *The Water Supply of Greater London* (London, 1954), pp. 34–45; and W. R. Scott, *The Constitution and Finance of English, Scottish and Irish Joint-Stock Companies to 1720* (repr. Gloucester, Mass., 1968), I, *passim*.

hydraulic engineer George Sorocold to rebuild their waterworks in 1701 or 1702. The competition may be one reason that Desaguliers' associate, John Lowthorp, was employed in the examination of engines for the New River Company shortly thereafter.[44] It became rapidly obvious that engineers well versed in water supply technology could provide an advantage. These ventures became as natural a source of opportunity as had the creation of private water supplies for the estates and households of the landed gentry in which Desaguliers and Lowthorp had pooled their talents in 1718. Immediately thereafter, Desaguliers himself was employed in a project, possibly orchestrated by the New River Company, to expand the flow of water to the City.

All of these water supply ventures were plagued by controversy. The problems were usually financial, although they could just as easily be complicated by technical or political difficulties as well. Desaguliers' scheme was no exception. Troubles seemed to haunt his every step. In 1719, an effort had been made to obtain a charter to bring water from the river Colne into London and Westminster. Despite the testimony and support of Edmund Halley, a Bill presented to the House of Commons soon ran into difficulties. Opposition evidently came from those who suspected that this was actually directed by the New River Company, suspicions that were undoubtedly heightened by the stock manipulations of the South Sea Bubble the previous year. By the early eighteenth century, New River had swelled to the extent of eight hundred bridges and aqueducts from Ware in Herefordshire to the rapidly growing metropolis. The fallout over South Sea obstructed any further response to demand for water from London. During 1721, in debates over the Bill, specific amendments were made to prevent any transfer of stock until after the water had been brought into the new reservoir in Marylebone Fields, and it was likewise suggested that "No Person concerned in the New River Company be capable of being Governor or Assistant of this Company or to vote."[45] In the aftermath of the South Sea Bubble, the timing of this project could not have been worse and the bill died before Parliament.

[44] Scott, *Constitution and Finance*, I, pp. 294, 332–3; Dickinson, *Water Supply of Greater London*, pp. 24–7; Hatton, *New View of London*, II, p. 791; and F. Williamson, "George Sorocold, of Derby. A Pioneer of Water Supply," *Journal of the Derbyshire Archaeological and Natural History Society*, LVII (1937), 81ff. One engineer employed by the New River after 1720 was Henry Mill, who built the water supply for Robert Walpole at Houghton. William Matthews, *Hydraulia; An Historical and Descriptive Account of The Water Works of London* (London, 1835), *passim*, esp. Chapter 2; Dickinson, *Water Supply of Greater London*, p. 40; *D.N.B.*, XXVII, 381–2; Wallis and Wallis, *Biobibliography*, p. 26; and Stephen Switzer, *An Introduction to a General System of Hydrostaticks and Hydraulicks* (London, 1729), pp. xxx, 103.

[45] *A Bill for Incorporating several Undertakers, for the Better Accommodating the Inhab-*

From the beginning, Desaguliers was involved in the scheme to expand the water resources of the rapidly expanding metropolis. In 1719, he reported he was engaged in calculations of the capacity of the Cowley stream, one of the branches of the Uxbridge River. He argued that the increase of new building required additional sources of water.[46] A considerable part of this new development was in the hands of the Duke of Chandos. How much influence Desaguliers actually had is difficult to tell, although he appeared before a committee of the House of Commons to testify on behalf of the proposal in 1721. There is no question, however, that Chandos was keeping himself well informed of developments as they would have some effect on his own plans to develop the Cavendish Square properties as well as on his negotiations with the rival York Buildings Company for a reservoir. And both Desaguliers and Captain Robinson, who provided the analysis of water flow for the Uxbridge project and supported it before Parliament, were in his employ.[47]

The plan faltered in the House because of enormous opposition. During the summer of 1721, powerful political forces were obviously marshalled against the scheme. The House was inundated with petitions from the surrounding counties. One group that petitioned in support of the new water venture was John Harris's old company, the Sun Insurance Office, as it was in the business of insuring houses, and hence, lamented the fact that several had been destroyed for want of a more ready supply of water.[48] Sun Insurance was another of Chandos's many interests. Their proposal further suggested the erection of new engines, as the result of which one petitioner asked Parliament for a clause in the proposed Act that would prevent his exclusion from supplying the same places with water.[49] It also appears to have unleashed other efforts, such as the Chelsea

itants in and about the Cities of London and Westminster, and the Parts adjacent, with Water... in the Name of the Governor and Company of St. Albon's Waters, n.d. [B.L., Shelf Mark 816.m.14 (32)]. See *Journals of the House of Commons*, XIX (1718–21), p. 297 (March 9, 1719); Dickinson, *Water Supply of Greater London*, p. 45; see also A. B. DuBois, *The English Business Company after the Bubble Act, 1720–1800* (New York, 1938), pp. 55n67, 179n172, 180n184. Chandos was deeply involved in the development in Marylebone and negotiated with several water companies over the site. See Baker and Baker, *Life and Circumstances of James Brydges*, Chapter 12.

[46] Desaguliers, *A Course of Experimental Philosophy*, II, pp. 419ff.

[47] Hunt. Lib., Stowe MSS. 57, XIX, f. 30, Chandos to his son, Lord Carnarvon, April 3, 1721; XXV, f. 71, Chandos to James Marye, December 11, 1724; on Desaguliers and Robinson, see *Journal of the House of Commons*, XIX, 526–7, April 24, 1721.

[48] *Journal of the House of Commons*, XIX, 324–604. The petition of the Sun of June 10, 1721, followed by three days an identical one from the Hand-in-Hand Office.

[49] *The Humble Petition of Robert Bumpstead, In Behalf of himself and several others. To the House of Commons*, n.d. [BL, Shelf Mark 357.b.6 (93)]. Bumpstead was an inveterate inventor. In 1721, he applied for a patent for a new glass light for streets and ships, and

Water, which was incorporated in 1722 by Act of Parliament to supply reservoirs in Hyde Park and St. James's Park with water from the Thames.[50]

The increase in building mentioned by Desaguliers and the arguments of the fire insurance companies had not carried the day in the summer of 1721. Desaguliers had tried, however, to answer the complaints of those millers who feared the further transfer of water away from the streams in the vicinity of Uxbridge. He seemed to have allayed any concerns the Crown might have had about the water works at Hampton Court, and pointed out that the fears of the Duke of Somerset that his mills on the Thistleworth River might be ruined were groundless because both were supplied by an old cut. Either Desaguliers or one of his confederates in the spring of 1721 indicated the benefits of a further supply of water in the control of pestilential diseases in the populated areas of London. This rather telling argument, at a time when London was experiencing one of its worst outbreaks of disease in decades, still did not turn the argument in favor of the plan. In the short term, it may have been one of the victims in the acrimonious debate over the Quarantine Bill.[51]

In the longer run, however, the project was not permitted to die. Although it was not very popular in the area about Uxbridge, which was a market town, and especially amongst the millers thereabouts, the projectors kept up their efforts to convince the opposition of benefits of their proposal. Desaguliers was in the thick of the action. But those who opposed the Bill in the first place were not easily convinced. They too marshaled their resources and the owners of land, like Sir George Cooke, continued to solicit the Members of the House to prevent its revival. Each side recruited surveyors to demonstrate the validity of its position relating to the amount of water in the Uxbridge system. When it appeared that the projectors had engaged John Rowley, Master of Mechanics to the King, to undertake a survey in 1724, Sir George Cooke tried to have an associate intercept him to show him the numerous faults in the scheme. Not to mention the opposition of the landowners, Cooke argued that it

in 1724, for a new water engine. See Public Record Office, State Papers (Dom.) 35/28/8, August 12, 1721; and SP (Dom.) 35/50/19, May 30–June 29, 1724.

[50] Dickinson, *Water Supply of Greater London*, pp. 55–6; and Dubois, *English Business Company*, p. 110.

[51] PRO, MPF 259. A Plan for the Design for bringing Water from the Village of Drayton for the better Supplying the Cities of London and Westminster with Water [marginalia, March 1720/21]. The cross-sections and calculations are in PRO, SP (Dom.) 35/25/104. March 4, 1720/21.

Figure 13. The Chelsea Water Works, according to Stephen Switzer.
By courtesy of Thomas Fisher Rare Book Library, University of Toronto.

would require cutting a canal through porous chalk for a mile and a half that, unless it were lined at great expense, would render the plan useless.[52]

The opposition, of course, had its own expert opinion in a project that seems to have divided not only landowners and entrepreneurs, but the

[52] B.L., Lansdowne MS. 841, fols. 120–1, George Cooke to ?, February 8, 1723/4; *Journals of the House of Commons*, XX, 284. March 5, 1724. Rowley was probably John Rowley,

mathematical community as well. Some of the criticism was directed personally at Desaguliers in a fashion that exposed a bitter rivalry. Within a year of Sir George Cooke's efforts, the furor burst into public prominence when Desaguliers printed a letter directed at a William Bower of Denham, near Uxbridge in January 1725. Bower had been part of the obstruction, and Desaguliers had attempted to show him how the latest proposal could be of benefit to the millers below Uxbridge. A meeting was arranged but failed to take place.

Desaguliers, who had learned of an attack upon his reputation, decided to make the new proposals public. He revealed that the projectors had suggested to the millers that their mills be converted from undershot to overshot, which he estimated would improve their capacities by about one-fifth. To some, this had already proved an attractive option. It is obvious that those engaged in the project, where Desaguliers had become the principal actor, were very confident. The philosopher attempted to divide the opposition by a proposal that he hoped would prove difficult to reject: "if I can't make good All that I have here asserted, I have Authority from the Gentlemen principally concern'd to say, that they will drop their Design – Provided the Millers will agree, on the other Hand, that if I do make it all clearly out to be True in Fact, they will drop their Opposition."[53] But oppose it they did. This must have been a remarkably frustrating situation for Desaguliers. He had steadfastly maintained the role philosophers could play in preventing the many fraudulent and ill-conceived plans that he had seen spawned since the Revolution. But here he was, in the aftermath of the South Sea Bubble, the victim of the very suspicions that he knew too well were often warranted.

In his account of the controversy, which he published twenty years later, Desaguliers conflated the original proposal of 1720 with the events of 1725. The project was the same, but the animosity had blossomed. The philosopher later explained how the dispute had been fostered by

a maker of scales and instruments who had sold tickets to the lectures of Francis Hauksbee and James Hodgson. Desaguliers complained that because of an error by Richard Steele, Rowley got much of the praise due George Graham for the invention and manufacture of the orrery. See Desaguliers, *Course of Experimental Philosophy*, second edition, I (London, 1745), pp. 448–9; E. G. R. Taylor, *The Mathematical Practitioners of Tudor and Stuart England* (Cambridge, 1954), p. 294; cf. Peter Eden, ed., *Dictionary of Land Surveyors and Local Cartographers of Great Britain and Ireland, 1550–1850*, Part II (Folkestone, 1976), entry R. 272.

53 B.L., Lansdowne MS. 1056, f. 47, J. T. D ________rs to W. Bow___r, Esq., January 9, 1724/5. See also Lansdowne MS. 841, fols. 116–18, Desaguliers to William Bower, January 12, 1724/5. Such a proposal was placed before the House of Commons by the Petitioners. *House of Commons Journal*, XX, 384–5 (January 28, 1724/5).

the failures of the opposition to properly calculate the flow of the Cowley stream. "A Person who call'd himself a Mathematician" had come to an estimate less than half than what Desaguliers had concluded.[54] Desaguliers was referring to Jonathan Warner, who had been asked for his "advice in opposing the Machinations and evil designs of some Projectors." In some nicely considered prose, in a letter in which Warner described his solitary day of examination of the flow of the streams, he remarked of Desaguliers, "I am apt to think he had more regard to the Currency of his Project than to the Quantity of the Currents." The Doctor had been right, he was going to be abused. Warner described defects in the proposals that might be rectified in other ways that "the Doctr: as a projector, a profess'd Mathematician, and Experienced Philosopher cannot or at leastwise should not, be a stranger." Warner, typically, concluded with a reference to a friend with a newly invented engine that might supply the gardens of Hampton Court with water from the Thames without diverting streams and ruining millers.[55]

What is remarkable about this entire affair is the degree of hostility it provoked in the community of mathematical practitioners whose expertise was part of the arsenal of the opposing camps. Warner's remarks were doubly revealing. This dispute demonstrated that technical projects, as with popular experiments, increasingly relied upon natural philosophers in the achievement of public assent. Natural philosophers could confer the status of authority. There was, therefore, a much broader concern that cut deeply into the foundation of British society, especially the part of it that had cause to try to understand the havoc created by the projects that helped inflate the South Sea Bubble. This is why the satires of the period were not misplaced complaints; rather, they were also part of the weaponry of the political battlefields of Hanoverian England. When Jonathan Swift made allusions in 1726 to an "Academy of PROJECTORS in Lagado," his comments were not without a resonance in the Royal Society, or amongst those sympathetic with the kind of Society that Denis Papin had envisaged. In *Gulliver's Travels*, the former governor of Lagado, Lord Munodi, described a scheme that had, of course, gone awry:

[54] Desaguliers, *Course of Experimental Philosophy*, II, pp. 419–20.

[55] B.L., Lansdowne MS. 841, fols. 113–14, Jno. Warner to [Sir George Cooke?], January 9, 1724/5. It is possible the surveyor is meant to be John Warner (1695–1725). See Peter Eden, ed., *Dictionary of Land Surveyors*, Part III (Folkestone, 1976), entry W. 113; and Joyce Brown, "Guild Organization and the Instrument-Making Trades, 1550–1830: The Grocers' and Clockmakers' Companies," *Annals of Science*, XXXVI (1979), 24. There was a land surveyor named Samuel Warner, who taught mathematics in Hatton Garden where Whiston lived. Whether Desaguliers' antagonist was connected with this practitioner is uncertain. E. G. R. Taylor, *The Mathematical Practitioners of Hanoverian England* (Cambridge, 1966), p. 194; and Eden, *Dictionary*, entries W.112 and W.116.

he had a very convenient Mill within Half a Mile of his House, turned by a Current from a large River, and sufficient for his own Family as well as a great Number of his Tenants. That, about seven Years ago, a Club of those Projectors came to him with Proposals to destroy this Mill, and build another on the Side of that Mountain, on the long Ridge whereof a long Canal must be cut for a Repository of Water, to be conveyed up by Pipes and Engines to supply the Mill.[56]

We cannot be certain that Swift had Desaguliers' project in mind. There was a number he might have chosen. But there remained the question of what it was that was being destroyed by the proposals of projectors. Was it the rights of those landowners who wanted nothing to do with the inventions and schemes of joint stocks; or was it the view of a political and economic structure that had come close to collapse because of innovations in finance – as a result of which the South Sea Bubble presented an urgent opportunity to rectify wrongs perpetrated since the Revolution?

The importance of philosophers

It was not Swift but Desaguliers who had lived through many of the attempts to utilize the steam engines both in mines and in water supply. It was precisely for this reason that Desaguliers could claim in the final volume of his *Course of Experimental Philosophy* that he had tried to prevent some of the ruin that otherwise have resulted:

I hope at last I have one Merit to plead; which is, that I have put it out of the power of any Person to be impos'd upon for the future, by those that pretend to great Performances in Water-Works, by shewing the utmost that can be done that way. . . . And it is this: No Man must hope that with any Engine in the World, a Man shall raise more than one Hogshead of Water ten Feet high in a Minute, with moderate Work, or such as he may hold all Day: nor an Horse above one Hogshead 50 Feet high in the same time. This is a very necessary Caution; for there are several Persons who have Money, that are ready to supply boasting Engineers with it, in hopes of great Returns; and especially if the Project has the Sanction of an Act of Parliament to support it – and then the Bubble becomes compleat, and ends in Ruin.[57]

There were any number of projects that Desaguliers might have had in mind. He might have meant the New River Company, or the schemes at Islington, or the Colne River plans, for which he had encountered so

[56] Quoted in Pat Rogers, "Gulliver and the Engineers," *Modern Language Review*, LXX (1975), 261.
[57] Desaguliers, *Course of Experimental Philosophy*, II, pp. vii–viii.

many rivals.[58] Or he might have meant some of those who tried to involve themselves in the affairs of the York Buildings Company by suggesting improvements to its engine on the Thames. He may have been referring to the proposal of Edward Newton that he erect for the York Buildings Company a new engine to raise water ninety feet above the Thames, which would outdistance the ability of Marchants Water Works, the New River Company, or the Hampstead Water to compete. Newton offered a trial which, if successful, he would then sell the engine to the Company for a mere £5,000.[59] But Desaguliers had no lack of candidates for his reflections. Even Case Billingsley's proposed engine could easily have qualified.

When Desaguliers saw the benefit arising from natural philosophy and from his public demonstrations of mechanics and hydraulics, there is a high likelihood that he was thinking of the York Buildings Company. This Billingsley enterprise, at least in the creation that Desaguliers best knew it, was a company with which the philosopher had significant contact once he had come into the employ of the Duke of Chandos. The reason for this is simply that Chandos was a major player in the various phases of the York Buildings' metamorphosis. What began as a water supply venture was transformed into insurance scheme, mining company, major landowner, as well as water utility in London and Westminster. Chandos and Desaguliers were to be important figures in those transitions. Even the Earl of Westmorland, Governor of the York Buildings Company until 1723, and savior of Case Billingsley in the 1730s, was one of Desaguliers' patrons. Like many major landowners, Westmorland was involved in mining. It was Desaguliers who, in 1727, designed for Westmorland a machine that would clear foul air from mines. It was this engine that Desaguliers demonstrated to the Royal Society in the same year.[60]

As had been the case with the African Company, the Duke of Chandos tried to exercise extensive control over enterprises in which he was heavily

[58] As late as August 1735, Desaguliers had again been called to measure the flow of water for the New River Company in a plan for the Lea navigation, and in 1738 to testify before Parliament. *Journals of the House of Commons*, XXIII, (March 21, 1738) 294–5.

[59] SRO, GD1/70/vol. III. The Proposal of Edwd. Newton, Gent. for Raising Very Great Quantitys of Water by an Engine of a New Invention at less than half the Charge of any Horse-work, n.d.; see also Dickinson, *The Water Supply of Greater London*, pp. 38–40, 45, 51–2.

[60] Desaguliers, *Course of Experimental Philosophy*, II, p. 561; Desaguliers, "An Attempt made before the Royal Society, to shew how Damps, or foul Air, may be drawn out of any Sort of Mines, &c. by an Engine contriv'd by the Reverend J. T. Desaguliers, LL.D. and F.R.S.," *Philosophical Transactions*, XXV, no. 399 (July–Sept., 1727), 353–6.

invested. As a result, especially after the Earl of Westmorland resigned in 1723, Chandos took steps to ensure that he was effectively represented among the Directors of the York Buildings Company. Although it took until the end of 1724, Chandos brought sufficient pressure to bear that he was able to force the Company into accepting Samuel Horsey, with whom he had dealt through the African Company, as well as Herbert Westfaling, the Whig M.P. for Hereford. The qualifications of stock to enable their election was achieved simply by transferring stock to them in trust for Chandos. By 1725, Chandos was a considerable force in York Buildings at a time that was to prove important to its history and to that of Desaguliers as well.[61]

Philosophers like Desaguliers found it virtually impossible to avoid such problematic enterprises. Perhaps that was why he felt so strongly that philosophers were the best preventative against frauds. If he could establish his case for a fundamental, and necessary, link between philosophers and mechanics, then the chances of disaster might be reduced. Because he was so deeply engaged in a wide variety of ventures of a technical and commercial nature, Desaguliers, perhaps more than any of his contemporaries, understood the root of the problem. In the end, the force of satire against philosophers would not only be blunted, but it would be turned on its head. Natural philosophy would be one path to prevent investors from being mangled in the wreckage that was so often all that remained of projects. And, in the end, philosophers would be employed in serving the stability of Hanoverian England. But what Swift and his friends saw in 1720 was the crushing wreckage of Whig experiments.[62]

[61] Hunt. Lib., Stowe MSS. 57, XXV, f.77, Chandos to Col. Horsey, December 19, 1724; fols. 76–7, Chandos to Mr. [Thomas] Haley (of the York Buildings Company), December 18, 1724; fols. 79–80, Chandos to Mr. [James] Marye (of York Buildings), December 19, 1724; f. 80, Chandos to Sir John Mears, December 20, 1724; f. 84, Chandos to Westfaling, December 20, 1724. Samuel Horsey had been Lieutenant Colonel of the Fourth Troop of Foot and Horse Guards from which he resigned after serving for twenty-one years. Horsey, in pursuit of a governorship of South Carolina, then came into contact with the Earl of Westmorland, First Lord Commissioner of Trade. PRO, SP (Dom.) 36/2/59–60. July 20, 1727. Petition of Samuel Horsey.

[62] *The Correspondence of Jonathan Swift*, III, ed. Harold Williams (Oxford, 1963), pp. 344–5. Duchess of Ormonde to Swift, April 18, 1720; p. 361. Swift to Mrs. Esther Vanhomrigh, October 15, 1720.

11

The dragons on the Thames

As though following the doctrine of Thomas Sprat, the promise of the public lecturers was the transformation of nature. For landowners and entrepreneurs, the engine of improvement stoked the demand for natural philosophy. And for philosophers, projects increasingly provided the focus for their lives. Hence, Desaguliers' concern in the commercial and mechanical fancies of the York Buildings Company had deepened as Chandos sunk more money into its affairs. Many of the technical issues of hydraulics that he later addressed in his *Course of Experimental Philosophy* arose directly out of these ventures.

Desaguliers was to be given frequent opportunity to display his expertise in hydraulics. For example, most water supply ventures suffered chronic constrictions from vapor locks and rotting pipes springing leaks. The constantly crumbling and decaying pipes were obviously unable to cope with the water pressure a series of reservoirs or the occasional engine might pose. Experiments were tried both at Cannons and in Cavendish Square on various types of pipe, composed variously of elm, lead, iron, or copper.[1] Chandos intended the stronger copper pipe for use at Cannons. This was to be produced by Daniel Niblet, the coppersmith with whom Desaguliers and Vream had received a patent for a steam boiler in 1720.[2] Desaguliers' role in the irrigation of Cannons recommended him to the York Buildings Company as its network of pipes expanded with

[1] See Royal Society MSS., Journal Book (C), XI (1714–20), pp. 456ff. March 3, 1719/20; Hunt. Lib., Stowe MSS. 57, XXVIII, f. 79, Chandos to Desaguliers, April 25, 1726; Stowe MS. 28, p. 32. "Colonel Horsey's computation of the Price of Iron Pipe."

[2] Hunt. Lib., Stowe MSS. 57, XXVIII, f. 33, Chandos to Desaguliers, March 30, 1726; Royal Society MSS., Classified Papers, XVIII (2), f. 40, "Steam Confined," Read April 6, 1721.

its system of reservoirs. Although there was some fear of breakage in places where traffic was heavy, in 1728, the York Buildings Company tried a type of clay pipe for which William Edwards of Monmouth held a patent.[3] Desaguliers was likewise consulted about the best route for pipes to the Duke's projected development in Cavendish Square. All hinged upon negotiations with the York Buildings Company for land upon which to construct a reservoir.[4] Desaguliers would be involved in this as well.

By the 1740s, Desaguliers could draw upon a wealth of practical experience. This was especially true in relation to the steam engine employed by the York Buildings Company and, thus, forms the background to his lengthy disquisition on the Newcomen engine in his *Course* of 1744. But like many entangled in the strife over the early steam engines, Desaguliers' experience was not entirely a happy one and his later reflections must be seen in that light.

The prestige that had been accorded natural philosophy failed to assuage the rivalries spawned by many of these ventures. None of the claims made by even the most knowledgeable natural philosophers could stem the tide of innovation and the internicene conflict among companies of promoters and their engineers who built the monsters that spit cinders and soot over the great houses of the Thames. The winds of innovation promised the mastery of nature – and the control of the fortunes of companies of investors, from the old Jacobite estates to the markets of London.

The technology of water supply

In the early eighteenth century, the application of new technology and the employment of those practitioners of natural philosophy never became so entwined as in the development of the Newcomen engine. Desaguliers

[3] PRO, SP (Dom.) 35/76/21. Patent of William Edwards, February 26, 1724/5. There is an extensive discussion of these pipes and their problems in Stephen Switzer, *An Introduction To a General System of Hydrostaticks and Hydraulicks, Philosophical and Practical.... Containing in General A Physico-mechanical Enquiry into the Original and Rise of Springs, and of all the Hypotheses relating thereto; as also the Principles of Water-works,... Deduc'd from the Theory of Archimedes, Gallileo, Torricelli, Boyle, Wallis, Plot, Hook, Marriotte, Desaguliers, Derham, Hawksbee, and others* (London, 1729), vol. I, pp. 116ff., and notes. See also H. W. Dickinson, *The Water Supply of Greater London* (London, 1954), p. 33.

[4] Hunt. Lib., Stowe MSS. 57, XXVI, fols. 158–9, Chandos to Desaguliers, July 6, 1725; XXVIII, f. 106, Chandos to Coll. [George] Watkins, May 13, 1726; XXVIII, f. 177, Chandos to Horsey, June 28, 1726; XXVIII, f. 81, Chandos to Watkins, April 29, 1726. On the background to the Cavendish Square project, see Baker and Baker, *Life and Circumstances of James Brydges*, pp. 289ff.

had quickly become a prominent advisor regarding the application of these engines. The York Buildings Company had experienced great difficulty with its Savery engine on the Thames. Chandos frequently enquired of Desaguliers for the details of such engines, more especially in 1724, when it became apparent that the York Buildings Company was considering the erection of a Newcomen engine at its works at the lower end of Villiers Street, off the Strand.[5] One reason Chandos was so enamored of the efforts of the Company to raise water was the fact that its debt on the purchase of the estates of the Jacobite Lords was so high that they failed to satisfy the demands of the Government for payment. While Billingsley had designed the use of the York Buildings' charter to be applied directly to the purchase of these lands, with a view to their possible exploitation, the Company soon realized that if the utility could be made to operate effectively, then its original purpose of providing water might also subsidize their other schemes.

The York Buildings Company was obviously the hub of a vast network of ventures. Like the Royal Exchange Assurance and English Copper, it was also capable of establishing a firm base until other plans could prove effective. By the end of 1724, Chandos had understood this, and in those circumstances, the efficient operation of an engine to raise water seemed essential to the health of the Company. The lure of steam engines would soon inflate the promises of philosophers. He even wrote to Ralph Bucknall, the recipient of the original patent under the Great Seal in 1675 to raise water at York House, enquiring "whether you did not make an attempt to save the Charge of Horses by trying to raise [water] by the Fire Engine, how large an Engine you had for that purpose, how many Tons of Water it flung up in an hour, where your Great Reservoir was & how it came to fail of Success."[6] For more current information on the technical possibilities of this, and the rival London Bridge Water Works, he relied on the engineer John Lowthorp.[7]

[5] Hunt. Lib., Stowe MSS. 57, XXIII, f. 280, Chandos to Desaguliers, February 11, 1723/24; XXV, f. 69, Chandos to Desaguliers, December 13, 1724; see also L. T. C. Rolt and J. S. Allen, *The Steam Engine of Thomas Newcomen* (New York, 1977), p. 81; Dickinson, *Water Supply*, pp. 60–1. The York Buildings Company had already erected one such engine at the Tranent Colliery in East Lothian in 1719.

[6] Hunt. Lib., Stowe MSS. 57, XXV, f. 53, Chandos to Mr. [Ralph] Bucknall, December 8, 1724; XXV, fols. 52–3, Chandos to Mr [James] Marye [of the York Buildings Company], December 6, 1724. On the background to Bucknall, see Dickinson, *Water Supply*, p. 48. On the problems arising from the purchase of the estates, see PRO, T1/258, York Buildings Company Minutes; PRO, FEC1/1721, "Interrogatories to be Administered to Witnesses."

[7] Hunt. Lib., Stowe MSS. 57, XXIX, f. 133, Chandos to Col. Samuel Horsey, December 24, 1726; f. 133, Chandos to Lowthorp, December 24, 1726; XXX, f. 5, Chandos to Horsey, May 10, 1727.

But the Duke of Chandos was also anxious about the competition of other water supply companies. Desaguliers had been enveloped in the plans to transform the Uxbridge River into a reliable source of water. Interested in this development was the rival New River Company on whose behalf Desaguliers was apparently engaged. Despite enormous opposition to its plan, it had been resurrected in 1724 and Chandos entertained the possibility of the sale of land for a reservoir with its proponents although his large interests in the York Buildings Company inclined him to make it the first offer.

Challenging the steam engine proprietors

Chandos was scheming at several levels. In the spring of 1725, shortly after Desaguliers had been publicly abused by the opposition to the Uxbridge plan, Chandos instructed Herbert Westfaling to contact James Lowther, then M.P. for Appleby. Competing interests ensured the matter of water works was much disputed in the House of Commons. Since the York Buildings Company was subject to special scrutiny because of its difficulties over the Jacobite estates, Chandos hoped to convince Lowther to make a statement in the House regarding the capabilities of steam engines in the drawing of water. As York Buildings was known to be erecting a new engine, such a statement from Lowther could only help. Lowther, who was also a Fellow of the Royal Society, was regarded as something of an authority on the Newcomen engine, having employed one to drain his colliery at Whitehaven since 1717.[8]

In order to understand the gyrations over the Newcomen engines and the role of the engineers, it is important to recognize that the frustrations over the engines also represented opportunities. By the middle of the 1720s, this was even more the case because the continued disputes over the erection of engines in the North were going to lead to plans to challenge the patent originally held by Savery but now being exercised

[8] Hunt. Lib., Stowe MSS. 57, XXV, f. 71, Chandos to James Marye, December 11, 1724; XXVI, f. 61, Chandos to Westfaling, May 27, 1725. Lowther's experience with mining was extensive. He is reported, on one occasion, to have demonstrated to the Royal Society the combustible qualities of mine damp, which he brought from Whitehaven in a bladder and ignited with a candle. See Royal Society MSS., Letters and Papers, Decade I, vol. I, no. 78. Carlisle Spedding to Sir James Lowther, November 14, 1740; Romney Sedgwick, ed., *The House of Commons, 1715–1754*, 3 vols. (New York, 1970), II, pp. 226–7; and J. S. Allen, "The 1715 and other Newcomen Engines at Whitehaven, Cumberland," *Transactions of the Newcomen Society*, XLV (1972–3), 237–68. On background to the York Buildings problem, see Baker and Baker, *Life and Circumstances of James Brydges*, pp. 289–90.

by the Committee of Proprietors of the Invention for Raising Water by Fire.[9] By odd and confusing coincidence, one of the major figures in this joint stock was not Sir John Meres of York Buildings, but a John Meres of the Apothecaries Company, who was nominated a Fellow of the Royal Society in 1724 by Sir Hans Sloane.[10] In the summer of 1725, John Meres, who was effectively the manager of the fire engine patent, had gone to Paris in an effort to construct an engine at Passy and to obtain a patent from the French Crown.

By 1725, the introduction of steam engines in the North East coal fields had not produced expected benefits. One of the reasons was the inability to get accomplished engineers like Stonier Parrott to complete their work. The mine owners saw few advantages and continual conflict. The frustration produced a two-phase agreement consummated in the spring of 1725 between the coal magnates George Liddell, William Cotesworth, and their rival Richard Ridley to erect engines at Heaton colliery and "for the preventing all Strife and controversy for the future and to putt an End to Suit, Suits and Quarrells as have already happened and are now depending."[11] By 1726, the new tone surfaced in the "Grand Allies Agreement" through which the North East coal trade was to be controlled by an understanding among the major players including Cotesworth, Liddell, and Ridley.[12]

Such arrangements made it easier to put an end to the costly and bitter feuds that had preoccupied the entrepreneurs. But it did little to make more effectual the construction of Newcomen engines then commanded by the Proprietors in London. It was this realization that produced a scheme toward the end of 1725 to break the monopoly that some felt the Proprietors had improperly expanded from the original Savery patent. The attraction was engines that would be rent-free. Stonier Parrott proposed that a petition and a Bill should be brought before the House of Commons to annul the Patent Act originally establishing Savery's claim and that had proved permissive enough to include the Newcomen engine. Parrott claimed considerable support of "1000 hands from this Corpo-

[9] See, for example, the dispute in 1724 between John Potter of Chester le Street and Samuel Calley over the rights to construct fire engines by commission from the Proprietors in London. *Newcastle Courant* (Saturday, July 18, 1724, and Saturday, November 21, 1724).

[10] Alan Smith, "Steam and the City: The Committee of Proprietors of the Invention for Raising Water by Fire, 1715–1735," *Transactions of the Newcomen Society*, XLIX (1977–8), 5–6.

[11] NCRO, Ridley (Blagdon) MSS, ZR1/11/1. Articles of Agreement, March 13, 1725; see also, Marie B. Rowlands, "Stonier Parrott and the Newcomen Engines," *Transactions of the Newcomen Society*, XLI (1968–9), 63–4.

[12] NEIME, Shelf 18. Grand Allies Agreement, June 27, 1726.

ration and City [London?] as well as from Newcastle under Lyme, Woolverhamton and Birmingham among other places" and stated a willingness to "bring in a Bill at the same time in behalf of Sr Richd Newdigate & others in those parts." He thought it wise, however, not to mention the plan to those involved in Newcomen engines around Newcastle lest the Proprietors in London learn of it.[13]

Parrott's challenge was based on the argument that the Savery patent could not be legitimately extended to the Newcomen engine, which was an entirely separate invention. In a memorandum that he prepared for Cotesworth's benefit, Parrott argued that the Savery engine was in fact based on principles described by the Marquis of Worcester in 1663. Thus, he argued,

> This Engine of Mr. Savory's appears to be an engine of a quite different form and working by quite different powers to that now in use, Mr. Savory's engine working only by the impellant force of fire and the present engine no wise impelled by fire but wrought by the impellant force of the atmosphere and no otherwise impelled, the fire being only made useful to raise steam which being condensed makes a vacuum for the atmosphere to take pressure upon.[14]

The view that there was a great difference between Savery's siphon and Newcomen's piston was later adopted by Desaguliers. Parrott's objections to the application of the patent could hardly be considered to be especially unique.[15] In 1725, Parrott's immediate objective was to break the monopoly held by the London Proprietors. At a time when the Newcomen engines were engendering so much controversy, an end to the Patent was widely desired. An alliance between entrepreneurs and engineers would have been a natural result. Parrott further alleged that the Patent had been an impediment to improvement:

> By misuseing the Patent and making it Prejudicial to the Subjects in insisting on exorbitant rents of £400 per ann. of some, very small rents of £20 per ann. from others who have as great a quantity of water as good collierys, in letting some engines without any rent at all and in refusing others the use of Engines on any terms whatsoever, and in taking considerations from others to grant exclusive covenants not to suffer engines to be erected within certain districts of the kingdom

[13] Gateshead Public Library, Cotesworth MSS, CK4/45. Stonier Parrott to Cotesworth, December 16, 1725. On the background to the conflict, see Edward Hughes, "The First steam engines in the Durham coalfields," *Archeologia Aeliana*, XXVII (1949), 29–45.

[14] Gateshead Public Library, CK4/46. "A memorandum giving the history of the fire engine patents." December 16, 1725.

[15] Desaguliers, *Lectures on Experimental Philosophy*, II, pp. 464–7.

and particularly not within the City or suburbs of London or Westminster except at York Building.[16]

Parrott's scheme failed to gain much public support. But his complaints were very well-founded. Desaguliers, for example, later agreed with Parrott and certainly regarded Parrott as an important improver on the expensive and unreliable copper boilers for these engines.[17] The campaign does, however, form part of the extensive background to the troubles then in full fury at York Buildings. Parrott's plot was being put in place in December 1725, precisely when an anonymous satire on the York Buildings follies materialized. These struggles suggest that the control of the Proprietors of the Patent was beginning to slip away. Demand had reached even to an international market.

The increasing difficulty of managing the English competition is confirmed in the escalating search for opportunities on the Continent. This had been understood by John O'Kelly at Liege, who, in 1725, was preparing to go to Spain to erect another engine. In March 1725, he had paid a visit to the Swedish Ambassador at The Hague with a view to offering his services in the Swedish mines. O'Kelly's workman, one Saunders, had been in Sweden late in 1724. By 1725, both O'Kelly and Saunders were in London when storms were brewing over the Savery Patent and as Martin Triewald was giving thought to returning to Sweden to take advantage of the overtures of the Government.[18] Neither O'Kelly nor Triewald were to make much more of a success in the transfer of technology than was John Meres of the London Proprietors.

But it does seem probable that the Proprietors were trying to position themselves for the expanding demand across the Channel. John Meres' venture at Passy encountered its own share of opposition, although not all of it was as ill-informed as the Lord Mayor and Aldermen of Paris, who, Lord Percival wrote, "fined Meers a thousand Livres for designing to drain the River dry and so to Starve Paris by water, and would not comprehend that the water must have its discharge into the River again from his Engine."[19] Of course, there were French engineers who might have resolved the conflict, but there is no evidence they were

[16] Gateshead Public Library, CK4/46. December 16, 1725. This memorandum is reproduced in Rowlands, "Stonier Parrott and the Newcomen Engine," Appendix II, pp. 61–2, and in Hughes, "The First Steam Engines in the Durham Coalfields," pp. 41–3. See also Christine MacLeod, *Inventing the Industrial Revolution. The English Patent System, 1660–1800* (Cambridge and New York, 1988), p. 91

[17] Desaguliers, *Course of Experimental Philosophy*, II, pp. 480–1.

[18] Svante Lindqvist, *Technology on Trial. The Introduction of Steam Power Technology into Sweden, 1715–1736* (Stockholm, 1984), pp. 141–55.

[19] BL, Add. MSS. 47031, ff. 164–5. Lord Percival to Daniel Deering, May 4, 1726.

ever consulted. Such might have been Belidor, who later corresponded with Desaguliers on steam engines, but especially useful could have been Philippe Vayringe, a locksmith and watchmaker, who had studied with Desaguliers in 1721–2.[20] In any case, the human obstacles proved too much for Meres. He was dead within a month of his encounter with Parisian officialdom.[21]

The tide of innovation along the Thames proved far more compelling than the limits of the Seine. Getting the Newcomen device to work proved to be exceedingly difficult and, as a result, Desaguliers was urgently consulted about the London trials.[22] Meanwhile, Chandos continued his negotiations with the York Buildings Company for land for a reservoir near Cavendish Square in consideration for which the Duke's house was to be continually supplied with water and two fire engines would be put in place in the event of a blaze. But there were a series of reservoirs being planned and possibly even notions of several engines to fill them. The grandeur of their plans proved their undoing. The intransigence of the Duke of Somerset was magnified by the possibility that smoke from an engine might be a nuisance to his house.[23] This fear proved the most damaging of all.

York Buildings Dragons

By the winter of 1725–6, the engine still had not been completed. Desaguliers was responsible for the construction of reservoirs for the Company, but the extent of his involvement with the engine is not so clear. Apparently, there was much dissatisfaction over the pace of the reservoir construction, and by September 1726, the York Buildings Court refused further payments to Desaguliers until the work for which he had contracted was complete.[24] The fact that there was so much difficulty over this matter may provide some of the background to the appearance of a mysterious pamphlet that was announced in the Whig *Daily Courant* in

[20] Alan Smith, "The Newcomen Engine at Passy, France, in 1725: A Transfer of Technology Which Did Not Take Place," unpublished paper, November, 1979, pp. 2, 15–16.

[21] Smith, "Steam and the City," p. 13. There are hints that Meres' health may not have been the best during the 1720s. Gateshead Public Library, CK4/38. Meres to Cotesworth, July 1, 1721.

[22] Hunt. Lib., Stowe MSS. 57, XXV, f. 32, Chandos to Desaguliers, April 7, 1725.

[23] Hunt. Lib., Stowe MSS. 57, XXVI, fols. 188–9, Chandos to Mr. Peere Williams (his attorney), July 17, 1725; XXVI, fols. 275–6, Chandos to Sir John Meres (of York Buildings Company), September 1, 1725.

[24] Cummings, "York Buildings Company," p. 18.

Figure 14. View on the Thames from York Stairs, by Canaletto, ca. 1747. By permission of Yale Center for British Art, Paul Mellon Collection. The York Buildings engine was housed in the tower at the right.

December 1725 and subsequently went through another edition the following year.[25] The anonymous tract entitled *The York-Buildings Dragons*, Chandos suggested, may well have come from Desaguliers.

But its authorship is by no means certain. The printing of this pamphlet has been the source of some puzzlement, without even considering that it went quickly through two peculiar editions. As a satire, it may well have backfired. Yet, it does reveal a deep conflict in London over these ventures. In order to understand this, we must look more closely at the tract itself. Its full title was *The York-Buildings Dragons; OR, A Full and true Account of a most Horrid and Barbarous Murder, (Whose Fatal Consequences Will not only Distract The Present Age, But Affect the Three Next Generations) Intended to be Committed On Monday the 20th of December Next, On The Bodies, Goods, and Name of the greatest Part of his Majesty's Liege Subjects, dwelling and inhabiting between Temple-Bar in the East, and St. James's in the West; and between Hungerford-*

[25] It was in the *Daily Courant* that Desaguliers was accustomed to advertising his lectures. See Dickinson, *Water Supply*, p. 62; *Daily Courant*, December 14, 1725.

Market in the South, and St. Mary la Bonne in the North, by a Sett of Evil-minded Persons, who (by the Instigation of PLUTUS, and not having the Fear of several Lords, Knights, and Gentlemen, before their Eyes) do assemble Twice A-week; to carry on their wicked Purposes, in a private Room over a Stable, by the Thames Side, in a remote Corner of the Town. Scavanged parts made of the engine a great monster that "on the sudden, will . . . clap his Wings several Times successively, with prodigious Force and incredible Velocity; and so terrible will be the Noise thereof, that it will be heard as far as CALAIS, and farther too if the Wind fit right." The leviathan (possibly the first of two) was to live "upon Newcastle and Scotch Coal," in which, of course, the York Buildings Company was otherwise concerned, thereby inducing by immense consumption a calamitous shortage of coal. Thus would be produced from its nostrils "perpendicularly up to the Skies two such vast dense and opaque Columns of Smoak, that those who live in the Burrough will hardly see the Sun at Noon-day." Disaster would follow disaster, with remarkable echoes of Meres' problems at Passy and of Desaguliers' encounters with the alarms of bargemen who feared the fall of water levels. It was foretold that a Welshman (probably William Edwards) "would make the dragon drink, who immediately dipping his two heads into the Thames, will suck out thence such a prodigious quantity of water, that barges will never after be able to go through bridges; the wharfs will become useless from the Steel Yard to Millbank; and the tide will not rise high enough to fill the basin of a set of good-natured gentlemen who have been at immense pains to serve the new buildings with water." The anonymous author concluded, "I must let the World know, that these Conspirators are Enemies to the Souls as well as the Bodies of all Persons they can have any Influence over, by setting up a New Kind of Popery, and have already persuaded several Families to worship these Dragons."[26]

While this tract was being published, Desaguliers was racked with one of his chronic attacks of gout. Chandos fumed that his philosopher's illness had "retarded so long the finishing of the fire Engine." Desaguliers' indisposition of six weeks appeared to be having ruinous consequences amongst investors in the City. Of the engine, Chandos wrote, "They now speak of its being ready to play in the matter of three Weeks time but the frequent hindrances they have met with have much dispirited the Town, and given people a very indifferent impression of what they are

[26] *The York-Buildings Dragons* (London, 1726), pp. 5–6, 9. This pamphlet is now very rare. The first edition is in the Spencer Research Library, University of Kansas. A second, revised, edition is in the Bodleian Library, Oxford.

to expect from it." Now the fortunes of the Company were so much worse, he thought, because of the publication of the pamphlet. He complained to Samuel Horsey that "There has been published a very stupid performance called the York Buildings Dragons; they father it upon the Doctor, but I hope without reason; for surely he could never be the Author of such a piece, unless it was whilst the Gout was in his head; & if so, I wish it had at the same time been in his right hand."[27]

Chandos was surely suspicious of Desaguliers. But was there any evidence that the philosopher was the culprit? Some have recently suggested that the author was likely involved with one of the rival water supply schemes such as the New River Company.[28] As we know, Desaguliers' role throughout the early twenties in the Uxbridge River scheme and the effort to divert the flow of the Colne would place him among the chief suspects. But there might be other possibilities as well. There was, as Parrott revealed, great conflict over the Savery patent. There was also competition in the spring of 1726, from the recently formed Chelsea Water Works under John Fane, brother of the former Governor of the York Buildings Company, the Earl of Westmorland. The Chelsea company sought to provide new supplies of water to Westminster from reservoirs in Hyde Park and St. James's Park.[29] Obviously, the evidence pointing at Desaguliers was circumstantial at best. Perhaps Desaguliers' dissatisfaction over payment for his services might have been an issue sufficient for him to want to reflect on the power of "PLUTUS" but this merely suggests a possible motive.

Slightly more meaningful may be the fact that the creator was clearly conversant with the current disputes and debates in philosophy and mechanics. It was revealed that the parts of the monster were imported by sea by some virtuosi and passed thorugh the Custom House as a curiosity. "And you know natural knowledge is so much encouraged, that such things never pay any duty, but pass unexamined; – witness Villette's great burning glass" with which Desaguliers had done experiments. Toward the end of the original assault on York Buildings was an acknowledgment that it was written by "A Club of ingenious Gentlemen, Who meet once a Week, at B______'s Coffee-House, to settle the Mechanical, Physical, and Moral World." Button's Coffee-House, to which the author was likely referring, had been the site of lectures on natural philosophy for

27 Hunt. Lib., Stowe MSS. 57, XXVII, f. 207. Chandos to Horsey, January 15, 1725/6.

28 Cummings, "York Buildings Company," p. 21; and Rolt and Allen, *The Steam Engine of Thomas Newcomen*, p. 84.

29 Dickinson, *Water Supply*, p. 56; Guildhall Library, MS. 173/1, Papers Relating to Water Companies, 1719–1819. "The Proprietors of the Chelsea Water Works."

years. More to the point, however, was the fact that the author identified some of the members of the secret club as S. Cathedral, Wm______ Bump______d, B______ Lang______y, Matt______ Pal______r, Fr______ C______de Hatz______d, G______ Gord______n, J______ Osm______d, Golden Quill, and the Reviver of ancient Flocution. The Secretary was called Anodyne Necklace, which was a name given to a popular remedy for teething amongst children and was intended to represent no more than a fraud.[30] I cannot identify Golden Quill, nor the Reviver who was second Chairman of the group.

The rest, however, are more revealing. Samuel Cathedral (or Catheral), M.A., was Chaplain to Lord Cholmondeley, and in 1720, had written a minor *Essay on the Conflagration*. The anonymous pamphleteer clearly had some familiarity with the debates concerning Thomas Burnet, John Woodward, and William Whiston. Why Catheral should have required special mention remains unclear. George Gordon, however, is no stranger. He was not only one of those who had chased the longitude prize in a tract that appeared in 1724, but in 1719, he had published *Remarks upon the Newtonian Philosophy: wherein it is proved to be false and absurd*. It would not have been peculiar for Desaguliers to think of Gordon, an instrument maker, as badly misguided for his criticism of attraction as an occult quality or as some force of "extravagant" power.[31]

Amongst the members of the mysterious club were four Chief Engineers. Of the four, William Bumpstead is the most difficult to identify. It is likely that the author meant Robert Bumpstead, who had petitioned Parliament to prevent the proprietors of the Uxbridge scheme from excluding him from supplying London and Westminster by means of an engine he had invented.[32] Johann Conrad Francis de Hatzfeld was a minor critic of natural philosophy, having published an attack on the Royal Society in 1724.[33] Hatzfeld was greatly irritated because, as one of Desaguliers' perpetual motion men, he got short shrift when the Society

[30] On the necklace and its sale in the vicinity of the Royal Exchange, see Francis Doherty, "The Anodyne Necklace: A Quack Remedy and Its Promotion," *Medical History*, XXXIV (July 1990), 268–93.

[31] Geoffrey Bowles, "The Place of Newtonian Explanation in English Popular Thought" (Ph.D. thesis, Oxford University, 1977), pp. 222–3; E. W. Strong, "Newtonian Explications of Natural Philosophy," *Journal of the History of Ideas*, XVIII (1957), 81–2; and Taylor, *Hanoverian England*, pp. 119–20. Apparently, in 1726, Gordon secured from Desaguliers the recommendation that his *Introduction to Astronomy, Geography and Dialling* was "very useful to beginners."

[32] PRO, SP (Dom.) 35/50/19, May 30-June 24, 1724.

[33] *The Case of the Learned represented, according to the merit of the ill progress hitherto made in Arts and Sciences, shewing I. The cause of gravity and attraction. II. What Nature is and the effects if is capable of* (London, 1724).

ignored his ideas in the summer of 1725. He apparently sent copies of his proposal to both universities and to the Royal Academy in Paris as well as having demonstrated it to an audience that, he claimed, included Fellows of the Royal Society. Not to be ignored, Hatzfeld suggested the Society's silence was "an evident proof, that neither the Royal Title of your Society, nor the Divinity Mr. Wiston [sic] ascribes to your President, are able to give you more light than you naturally are endued with." The Society, he alleged, had dismissed his experiments because they were "sufficient to dash your royal Splendour, and to make you look as so many hogs in Armor." There was a great sense of exclusion behind his anger – "because you won't come into ye truth of things, when it is discovered to you by others."[34] Such were the burdens philosophers had to bear.

Desaguliers never tired of pointing out how such petitioners made natural philosophers indispensable. Of the remaining engineers, Matthew Palmer was one of the legion of obscure inventors who offered an invention to raise water. In 1722, Palmer claimed to have a machine that "works of it selfe without Intermission by the Pondorous quality of ye Water, as well as by a Second Cause when required."[35] It was these kinds of notions that particularly vexed Desaguliers. Much more interesting was Batty Langley who did lay claim to being an engineer, and, judging from works that he published throughout the 1720s, he appears to have been one of that very large group of estate surveyors who looked to aristocratic patrons for employment. It is well established that there was no love lost between Langley and Desaguliers later on. This conflict involved Charles Labelye, a Swiss Huguenot whom Desaguliers described as "my Disciple and my Assistant." Labelye was later appointed the engineer for the New Westminster Bridge in 1737 over applications from Langley and others. Labelye's rise to prominence had a great deal to do with the support of his Huguenot and Masonic brother, Desaguliers.[36] None of this Club of Engineers was of much use to Desaguliers.

[34] Royal Society MSS., Letter Book (C), XIX, 241–5. H. F. Hatzfeld to ye Royal Society, July 28, 1725, and October 17, 1725.

[35] PRO, SP (Dom.) 35/31/151.

[36] R. J. B. Walker, *Old Westminster Bridge. The Bridge of Fools* (London, 1979), pp. 69, 83–5; Blanche Henrey, *British Botanical and Horticultural Literature before 1800*, 3 vols. (London and New York, 1975), II, pp. 470–2, 498–500. In 1726, Langley published *Practical Geometry, applied to the Arts of Building, Surveying, Gardening and Mensuration* and *Sure Guide to Builders*. See also *D.N.B.*, XXXII, 108; and Patricia Carson, "The Building of the First Bridge at Westminster, 1736–1750," *Journal of Transport History*, III (November 1957), 116–19 on Labelye. Labelye's challenge to Leibniz's views on impact in the *vis-viva* controversy was published by Desaguliers. *Course of Experimental Philosophy*, II, pp. 76–93, 506.

All in all, the author had intimate knowledge of the disputes surrounding the mechanics and philosophers who were drawn to such schemes as the York Buildings engine. Undoubtedly, those who supposedly attacked the engine were enemies of Desaguliers. One of the exposed schemers was "J——— Osm———d, Plumber, great Pencil-Bearer, Describer of the Velocities." In the 1726 edition of the pamphlet, his initial had been corrected to "G". The change was significant and gives us a substantial clue to the intentions of the anonymous author. George Osmond was the surveyor who had described the velocity of the Cowley Stream before a Commons committee enquiring into the Uxbridge plan in January 1725. For Osmond to have become a target, especially because a year earlier Desaguliers had been a promoter of the Cowley scheme, there had to have been a falling out.[37] It is clear, however, that the pamphlet actually pilloried those "Ingenious Gentlemen" who opposed the new engine and who even had their own to sell. Desaguliers was evidently settling several scores. But the tract also severely embarrassed York Buildings.

Desaguliers as engineer

Desaguliers was in a conflict of interest with the Duke's plans. Perhaps he was determined to harm the York Buildings effort, but that seems less likely than an effort to discredit opponents of the engine – as appears to be the point of the cast of characters. But the intent seems to have done the York Buildings group more harm than good. In November 1724, a year before the *Dragons* appeared, the Uxbridge proposal had been revived before the House of Commons. Within a month, the adventurers had approached the Duke of Chandos to lease some of his land for the construction of a reservoir. However, Chandos was heavily invested in the rival York Buildings Company and this persuaded him to offer the land first to the directors of York Buildings.[38] This complicated matters for Desaguliers for he was, at this point, well disposed to the New River Company's proposal for the rivers and streams about Uxbridge.

By Christmas, Chandos had received an offer from the New River group for his land in order that they might build a reservoir near Cavendish Square. Yet, he expressed his willingness to assist York Buildings even if it meant disposing of the land at a lower price than that offered by its rivals. At this point, Chandos was busily purchasing York Buildings

[37] *House of Commons Journal*, XIX, 526, April 24, 1721; XX, 381–2, January 26, 1724/5.

[38] Hunt. Lib., Stowe MSS. 57, XXV, f. 71, Chandos to James Marye, December 11, 1724.

stock. His main fear was that Edward Harley, second Earl of Oxford, from whom he had first obtained the land might make trouble if he learned of Chandos's plans. Oxford was one of those behind the New River project.[39] The fact that Desaguliers was also, at this point, lobbying for the New River proposal may have planted the seeds of suspicion in the Duke's mind.

There were significant advantages to Chandos if he could secure an agreement to develop a reservoir in the Cavendish Square area. Not only was this a desirable use for the land he held near Marylebone, but, as he pointed out to Sir John Meres, the York Buildings Company might be put in a flourishing condition if it could secure contracts to supply the estimated six thousand houses that had been constructed over the previous four or five years.[40] But Chandos was not prepared to leave this to chance. He pressured Meres to accept into the York Buildings' directors his nominees Westfaling and Colonel Samuel Horsey.[41] Horsey especially proved invaluable in the Duke's dealings with the company and he became the major instrument of Chandos's plans to revive the water supply as a source of revenue. It was Horsey who first learned from Chandos that the Duke of Devonshire was "mightily dissatisfied with the N. River Comp.," and, as trustee of the Duke of Bedford, Devonshire was prepared to offer York Buildings land in Bloomsbury Square in which to construct what would have been a strategically located reservoir.[42] By July 1725, Chandos was becoming frustrated with the dithering of the York Buildings' directors and told Horsey that he was willing to go to the expense of constructing the reservoirs himself at easy terms to the Company.[43]

It was at this very moment that Desaguliers was once again invited by Chandos to enter into the plans for the York Buildings Company. Presumably, it was intended that the Doctor should provide the Company with advice on how to conquer difficulties it was having with getting its engines running and with the pipes that were giving a great deal of trouble by bursting under the pressure.[44] Chandos at least recognized the im-

[39] *Ibid.*

[40] Hunt. Lib., Stowe MSS. 57, XXV, f. 212, Chandos to Sir John Meres, February 7, 1724[/5].

[41] Hunt. Lib., Stowe MSS. 57, XXV, f. 223, Chandos to Sir John Meres, February 12, 1724/5.

[42] Hunt. Lib., Stowe MSS. 57, XXVI, f. 135, Chandos to Samuel Horsey, June 25, 1725; Baker and Baker, *Life and Circumstances of James Brydges*, p. 291.

[43] Hunt. Lib., Stowe MSS. 57, XXVI, f. 193, Chandos to Samuel Horsey, July 20, 1725.

[44] Hunt. Lib., Stowe MSS. 57, XXVI, f. 192, Chandos to Sir John Meres, July 20, 1725; f. 191, Chandos to Mr. Wilkinson, July 19, 1725.

portance for the company of success in its attempts to secure leases to supply the growing number of houses. For this reason, in the summer of 1725, Desaguliers was consulted in the intended construction of reservoirs in Cavendish Square and later in St. James's Square in order to defeat the intentions of the New River Company. Desaguliers had developed into an entrepreneur quite capable of playing off one venture against another. Around September, he had insinuated himself into negotiations for land for the York Buildings' reservoirs and claimed to act on behalf of the Company in its dealings with Chandos. By November, the deal still had not been brought to a close.[45] Frustration was followed by the birth of the *Dragons*.

More complicated yet was the provenance of a second enlarged version of the *Dragons*, which appeared in the spring of 1726. Further delays of completing the engine because of "excessive Rains and Frost" induced a bitter debate in the York Buildings Company and complaints of abuses by the management.[46] The new *Dragons*, penned by a classical scholar capable of citing in Greek from Milton's *Paradise Lost* about "Gorgons and Hydra's, and Chimera's dire," contained a speech by an opponent of the Company, Orlando Furioso, who was answered by a more sensible Prudentio. *Orlando Furioso*, as the title of an epic Renaissance satire by Ariosto that was commonly available, would have been instantly recognizable to many in the eighteenth century. Three copies alone were in Chandos's library.[47] It is revealed in a postscript that "About five Weeks ago, an ill-set Ladder gave a plaguy Fall to another of their Conjurers, who (as they pretend) has only a Fit of the Gout."[48] This might suggest that the pamphlet was directed at Desaguliers rather than written by him, although gout was by no means restricted to philosophers. But the answer by Prudentio, to the accusations of the agitated Furioso, was clearly by a defender of the York Buildings Company. The addition therefore coincides with reflections upon the Club of Ingenious Gentlemen launched in the earlier pamphlet. The Bodleian copy of this rare tract has a marginal

[45] Hunt. Lib., Stowe MSS. 57, XXVII, f. 92, Chandos to Capt. Burroughs, November 6, 1725; Baker and Baker, *Life and Circumstances of James Brydges*, pp. 292–3.

[46] *Daily Post* (Tuesday, February 15, 1726).

[47] Hunt. Lib., A Catalogue of the Large and Valuable Library of the most Noble and Learned James Duke of Channdos . . . which will be sold at Auction, by Mr. Cock, At the Great Room in Exeter-Exchange in the Strand, On Thursday, the 12th of March 1746/7, and The Twenty-nine following Evenings (Sundays excepted) [copy of British Library K.r.2a]; and Stowe MSS. 329. Catalogue of Library, ca. 1761.

[48] *The York Buildings Dragons . . . The second edition, Augmented by almost Half, Containing Several more wonderful Discoveries, and concluding with Orlando Furioso's Speech, and Prudentio's Answer* (London, 1726), p. 12.

annotation identifying Desaguliers as the author at least of Prudentio's reply. It now seems possible that the decline of Desaguliers in the New River coincided with the clarification of the second version of the *Dragons*. The new edition was intended as a defense of York Buildings to erase the clumsy embarassment of the first.

The circumstantial evidence substantiates Chandos's suspicions of Desaguliers as the author. Could Desaguliers really have been so provoked as to pen reflections on the motives of the York Buildings Company? If that was the intention, Desaguliers may well have had sufficient motive in his difficulties in obtaining money to finish the reservoirs that he had contracted to complete. *York Buildings Dragons* simply reveals the great distrust of these engines and the companies that tried to promote them. Despite these problems, in the spring of 1726, Chandos was convinced that "the Company had paid to Dr. Desaguliers & other Charges of the Waterworks the greatest part of what was due."[49] The crisis passed as the engine rumbled to life. Moreover, if Chandos did suspect Desaguliers had conceived the *Dragons*, this did not stop him from continuing to employ the philosopher. In fact, because Desaguliers was engaged by both Chandos and York Buildings, the Duke had some control over the terms and conditions under which the reservoirs would be constructed on land he leased to the Company. In the end, one of the reasons why Desaguliers was denied further payment by the York Buildings Court late in 1726 was that Chandos was dissatisfied with the pace of negotiations between himself and the Company. By June, he ordered Desaguliers to cease laying in York Buildings' pipe to his house in St. James's Square until the matter was resolved.[50]

Desaguliers' intense relationship with the affairs of the Duke of Chandos was a remarkable example of the way in which natural philosophers were regarded. Far from being mere commentators on etherial metaphysics or on abstruse mathematical forces, they were thought to be the holders of a practical grail. For a Newtonian to be engaged in the projects of water supply and the construction of reservoirs was perceived as an advantage to the companies that employed him. Thus, when Desaguliers became involved in the machinations of the competition between the New River Company and the York Buildings group, he was a resource upon which much hinged. The schemes of the companies revolved also around

49 Hunt. Lib., Stowe MSS. 57, XXVIII, fols. 99–101, Chandos to Horsey, May 6, 1726.

50 At the end of May 1726, Chandos was involved with Desaguliers as trustees for a turnpike road. Hunt. Lib., Stowe MSS. 57, XXVIII, f. 125, Chandos to Bucknall, May 30, 1726; XXVIII, f. 176, Chandos to Mr. Peters (his attorney), June 27, 1726; f. 176, Chandos to Desaguliers, June 28, 1726.

their dealings with the Duke of Chandos and, thus, Desaguliers was doubly entangled. This may go some way to explaining the genesis of the *York Buildings Dragons*.

In any case, by the spring of 1726, the furor over the *Dragons* had begun to dissolve and a draught of the engine was announced to be published to coincide with its operation.[51] Desaguliers had then been paid what appears to have been several thousand pounds for his efforts in the York Buildings water works.[52] Presumably, this work included both reservoirs and his attempts to get the York Buildings engines in practical working order. That indeed was the reason Desaguliers had been consulted in the summer of 1725, when the engines were much delayed in their construction. Natural philosophy may have been Desaguliers' first passion; his entrepreneurial instincts emerged from the competition experienced in the water supply wars of Hanoverian London.

Patents and John Payne

If the Duke of Somerset had envisaged disaster in the entrails of the York Buildings dragons, Chandos had cultivated another view. Such engines might prove able to fill the reservoirs around Cavendish Square and thus could water the ground to keep down the dust and the smoke that emanated from the construction and from the bricklayers' works thereabouts.[53] The Duke's correspondence with members of the York Buildings Company suggests that the proposal to construct a Newcomen engine probably came in 1724 – possibly as much as a year earlier than has been recognized. Chandos's queries of Ralph Bucknall followed immediately upon his declaration of the water supply apparatus that "it is the Success of this undertaking which must for some time, at least, support the Credit of the Company."[54] The Company did have grave problems in attempting to raise sufficient funds to meet its obligations in the purchase of the Jacobite estates. Until those ventures could be made to pay, the water supply must support the rest. Throughout 1724, intense debate confounded the York Buildings Court. It is conceivable that Case Billingsley's

[51] *Daily Courant* (Monday, March 28, 1726).

[52] Hunt. Lib., Stowe MSS. 57, XXVIII, fols. 99–101, Chandos to Horsey, May 6, 1726.

[53] Hunt. Lib., Stowe MSS. 57, XXVI, fols. 188–9, Chandos to Peere Williams, July 17, 1725; also Baker and Baker, *Life and Circumstances of James Brydges*, pp. 288–9.

[54] Hunt. Lib., St. 57, XXV, f. 53, Chandos to [James] Marye [of the York Buildings Company], December 6, 1724. Note also his comment shortly after that: "I see that the Stock rises but I shall never think 'em in a sure good Way till they have discharged their Debts to the Government & began to raise the Thames Water, with their Fire Engine." St. 57, XXV, f. 189, Chandos to Col. Horsey, January 29, 1724/5.

column of water engine, for which he applied for a patent that summer, was designed with York Buildings in mind. This would not have been unusual given Case Billingsley's concern in the Company. That he would have been well-informed is made all the more likely by the fact that John Billingsley was still Secretary to the Company in the difficult days of 1724 and 1725.[55]

The interest of the Duke of Chandos in the Newcomen engines was typical of his efforts to protect his investments by the application of technical expertise. This was the main reason he continued to employ Desaguliers even when the Duke was suspicious of his conduct in the affair of the *Dragons*. But Desaguliers was far from alone. Among the agents of the magnate was John Payne, a timber merchant of Bridgwater who oversaw Chandos's development of the town. Payne, who was one of those obscure men who follow in the wake of more prominent figures like Chandos, was quite remarkable as an industrial experimentalist.

John Payne first came to attention in his effort to recruit buddling boys and miners for the African Company when Chandos was determined to locate African gold.[56] As we know, this proved a disaster for those who actually went to explore the Gambia, most dying in a very short time. But Chandos continued to involve Payne in other objectives he had for the African Company. With Colonel Horsey, Chandos had the idea to create a soap manufactory based on potash from African sources. Such notions involved the York Buildings Company and efforts to exploit Scottish forests for the same purpose. Horsey and Chandos were convinced there was a market and the Duke employed Payne to establish a factory at Bridgwater. To this end, they appointed an experienced manufacturer, Charles Bellair, as soap boiler and the project was set underway early in 1725 just as York Buildings was struggling for survival. Bellair was eventually forced to admit defeat as the potash created from the burning of fern produced a soap too soft for merchants to be interested.[57]

It is not the failure of these projects that makes them important, but rather that they were attempted at all. Thus, the Duke of Chandos expressly recruited to his service those he thought might be able establish his projects on a firm technical foundation and thereby secure him an

[55] PRO, SP (Dom.) 36/43/62–66,. "List of Inventions"; PRO, T1/258/98–107. "Narrative and Account of several Proceedings relating to the Affairs of the York Buildings Company," 1723–5.

[56] Hunt. Lib., Stowe MSS. 57, XXII, fols. 203–4, Chandos to Payne, March 18, 1722/23; fols. 275–276, Chandos to Payne, July 9, 1723; PRO, T70/91/ 235–236. African Company Papers, Minutes of the Court of Assistants, July 25, 1723.

[57] Hunt. Lib., Stowe MSS. 57, XXVI, f. 43 Chandos to Payne, May 17, 1725; Baker and Baker, *Life and Circumstances of James Brydges*, pp. 230–1.

advantage over his rivals. The soap works was merely one example. Another was the effort to set up a glass factory at Bridgwater during 1725. Once again, this followed the York Buildings' lead. It had invested in glass ventures in Scotland with Robert Hackett, but these had to be stopped in 1723. Thus, when Payne proposed Bridgwater as a site for a glass partnership, both Samuel Horsey and Chandos were eager. Like much else, the glass house never met expectations, although Chandos played a part in keeping it operating until 1733 when the fires were extinguished for the last time.[58]

Despite his enthusiasm for industrial ventures backed by the wealthy Chandos, John Payne was no mere projector. He established a reputation as an inventor just as it appeared the York Buildings plans for its mines were going to follow the way of its soap prospects. Another chimera diverted attention from the dragons. Late in 1725 Payne requested the support of Chandos for an invention that he claimed would enable the smelting of iron with pit coal rather than with charcoal. The Duke was clearly intrigued as the search for such a method had occupied a great deal of recent attention. Chandos wanted to see the results of some experiments and offered to speak to the Lord Chancellor on Payne's behalf, although he pointed out to Payne that there had been a patent recently issued for such a claim. The results of Payne's process were dismal and the sample produced proved far too brittle for making into malleable bar iron. In the end, Chandos did not close the door, arguing simply that "it will require a good deal more of time and skill to perfect the Invention."[59] The Duke's attitude was a sound one. As William Wood discovered, results were far too often elusive, and during the 1720s, there were rivals enough. But a patent was an obvious possibility for the person who succeeded – especially one so well connected as Payne.

As early as 1725, the Duke of Chandos was sufficiently interested in the efforts to smelt iron with coal that he encouraged his own agent to persist. But it was into this unresolved technical problem that there entered the infamous William Wood, whose project for a copper coinage in Ireland had recently miscarried. William Wood successfully convinced

[58] Baker and Baker, *Life and Circumstances of James Brydges*, pp. 232–4; Hunt. Lib., Stowe MSS. 57, XXV, f. 229, Chandos to Col. Horsey, n.d.; PRO, T1/258/99–100, York Buildings Company Minutes, October 26, 1723, May 16, 1724. See also Hunt.Lib., Stowe MSS. 28, f. 19, "A Computation of the Profit to be Made on Mr. Salters Glass House at Flint in Flintshire, December 1724."

[59] Hunt. Lib., Stowe MSS. 57, XXVI, f. 294, Chandos to Payne, September 6, 1725; XXVII, f. 45. Chandos to Payne, October 18, 1725; f. 59, Chandos to Payne, October 26, 1725. The details of Payne, "Method of extracting Iron with Pit Coal" are in Stowe MSS. 28, fols. 25–7, September 18, 1725.

the Crown of the probable benefit from the intense heat of a reverberatory furnace in the process of smelting with coal and he was granted a patent in 1727. This seems to be precisely the method envisaged by John Payne although Payne does not appear to have directly challenged Wood's claims.[60]

What is particularly striking about the project of Wood and his rivals is the way that it seduced several individuals with whom Chandos was also intimately concerned. Without going into the intricacies of the investments surrounding this patent, it is notable that by February 1728, Wood approached Sir James Lowther for financial backing and shortly thereafter entered into an agreement to supply the Mines Royal and Mineral and Battery Works with malleable bar iron. It soon became well known that Wood's method had run into difficulties. This is likely what encouraged John Payne in the summer of 1728 to lay another claim to a patent to render pig iron malleable. In the conflict over patents, Payne was lost in the storm. At this point, Payne had approached the Royal Society with reports of his earlier experiments on iron ores and magnets.[61] Chandos looked on with well-worn cynicism.

Others were positioning themselves to take advantage of any successful invention. By November, Lowther had become Deputy Governor of the Mines Royal and Mineral and Battery Works under Sir John Meres, the former Governor of the York Buildings Company. In addition, there was a rival patent held by Thomas Tomkyns in which were engaged a group of partners with whom Chandos had dealings, including Sir Archibald Grant and Sir Thomas Mackworth of the ill-fated Charitable Corporation.[62] What is clearly of consequence here is that inventions and their ensuing patents were natural prey for the projectors like William Wood, whom Swift had vilified.

[60] Hunt. Lib., Stowe MS. 28, f. 25, Payne's "Method," September 18, 1725; PRO, PC1/4/106, papers relating to Wood's patent and subsequent disputes. On Payne and patents, see MacLeod, *Inventing the Industrial Revolution*, pp. 30, 52–3, 88.

[61] Royal Society MSS., Register Book (C), XIII, 144–5. "Experiments on Iron and Its Oars by the Magnet in the Year 1724 and 1725" (read May 30, 1728).

[62] Michael W. Flinn, "William Wood and the Coke-Smelting Process," *Transactions of the Newcomen Society*, XXXIV (1961–2), 55–71; J. M. Treadwell, "William Wood and the Company of Iron Masters of Great Britain," *Business History*, XVI (1974), 93–112; PRO, SP (Dom) 36/7/95, Payne's claim to a patent, June, 1728; Hunt. Lib., Stowe MSS. 57, XXX, f. 152, Chandos to Horsey, June 18, 1729; PRO, SP (Dom) 36/43/68, Wood's patent, January 9, 1726/7; B.L., Add. MSS. 36142, fols. 91–100, report on the petitions of Wood and Tompkyns, June 23, 1730. The York Buildings Company and the Charitable Corporation had a very close business relationship involving the operation of lotteries. See B.L., Add. MSS. 33226, fols. 221–8, General Court Minutes of the York Buildings Company, September 21–October 2, 1724.

The York Buildings Company, meanwhile, continued to fret over the entrails of its engine. The "dragon" came into effective operation in 1726 and from a technical perspective seems to have worked well enough. The difficulty was that, as with all the engines of this type, it was exceedingly "chargeable." It soon became obvious that the York Buildings Company could not expect much advantage from an engine that consumed coals at a prodigious rate – even when the coals came from the Company's own mines in the north.

It may be for this reason that John Payne created an engine that he claimed, from his own experience of the immense heat of glasshouses, could operate from air pressure caused by the rarification or elasticity of air in the vicinity of furnaces. Payne's petition for a patent in 1728 made the usual claims for engines "of sufficient force to turn or drive a wheel something like a Water Wheel that will grind Corn, or move large Hammers, raise water, or perform any other work that is done by the force of Wind, Water or horses" – but the patent was not granted until 1736. By this point, a detailed explanation had been provided to the Royal Society of his "new and more advantageous Method of Expanding Fluids . . . Sufficient to give Motion to Hydralopneumatical and other Engines & Machines."[63] The apparatus was later described in the *Philosophical Transactions* along with a list of experiments performed by Payne by which he professed to show that his method of producing steam could move an engine of a cylinder of 24 inches diameter. The structure involved a series of large globes connected to a reverberatory furnace so that they could be kept at a red heat. Into the globes was then injected hot water that, by means of a disperser with spouts rotating internally in the globes, enabled the rapid production of steam that could then impel the cylinder of an engine.[64] Payne's invention was intended to take advantage of the great waste of heat associated with many industrial processes. This was easily noticed in the manufacture of glass, and in a large operation, there would have been a continual draft into the factory.

[63] PRO, SP (Dom) 36/7/78, John Payne petition; B.L., Sloane MSS. 4056, fols. 315–16, "A Narrative of John Payne's New Invention of Expanding Fluids . . . letters Patent . . . bearing date at Westminster ye 15th day of November, 1736"; also B.L, Add. MSS. 4436, fols. 222–8. "A Narrative of a New Invention of Expanding Fluids," October 23, 1740.

[64] John Payne, "A Narrative of a New Invention of expanding Fluids, by their being conveyed into certain ignified Vessels, where they are immediately rareified into an elastic impelling Force, sufficient to give Motion to Hydraulo-pneumatical and other Engines, for raising Water, and other Uses, etc.," *Philosophical Transactions*, XLI, no. 461 (Aug.–Dec. 1741), 821–8. See Robert Stuart, *A Descriptive History of the Steam Engine* (London, 1824), pp. 85–6.

The proposal came too late to save the York Buildings engine. By the autumn of 1731, it was found necessary to shut down the "dragon" because of its expensive appetite for coal.[65] But the plight of the York Buildings Company had focused the attention of others on the resurgent promises of these engines. By February 1732, John Billingsley applied for a patent for an engine. His request was passed on through the Duke of Newcastle and granted in March.[66] Whether anyone took Billingsley seriously or not is difficult to tell. What is sure is that the Company had decided not to attempt to revive their engine, and in 1732, they sold two cylinders. One went to Sir James Lowther to be fitted to an engine in his mines at Whitehaven and a second to the London Lead Company.[67]

By this point, the attention of the engine makers came to focus increasingly on the mines where the likes of Sir James Lowther and Sir George Beaumont were interested in the application of new technology. Entrepreneurs, patentees, improving landlords, and natural philosophers found a future in each others' company. Perhaps this had been the most important consequence of scientific lectures in the coffeehouses. It was, of course, in these circumstances that the philosophers encountered great excitement over their designs to clear mines of foul air.[68] As a result, natural philosophy was drawn even more directly into the web of the joint-stock companies.

[65] Dickinson, *Water Supply*, pp. 62–4.

[66] PRO, SP (Dom) 36/26/94, John Billingsley petition, February 25, 1731/2; B.L., Add. MSS. 36142, fols. 268–270, report on the Billingsley petition, March 22, 1731[/2]. In April, Case Billingsley revived his longitude proposal.

[67] Rolt and Allen, *The Steam Engine of Thomas Newcomen*, p. 84; A. Raistrick, "The London Lead Company, 1692–1905," *Transactions of the Newcomen Society*, XIV (1933–4), 141, and J. N. Rhodes, "Early Steam Engines in Flintshire," *Transactions of the Newcomen Society*, XLI (1968–9), 218–19.

[68] A. Stanley Davies, "The Coalbrookdale Company and the Newcomen Engine, 1717–1769," *Transactions of the Newcomen Society*, XX (1939–40), 47; and Colin Owen, *The Leicestershire and South Derbyshire Coalfields* (Ashbourne, 1984), pp. 111–13. Hunt. Lib., Stowe MSS. 57, XLI, fols. 213–14, Chandos to [Lewis?] Baden, March 24, 1732/3; XLVII, fols. 171–2, Chandos to Sir George Beaumont, March 26, 1736.

12

The engines of providence

Natural philosophy and commerce were locked together in the embrace of the incipient industrial ventures of Hanoverian England. Natural philosophers – indeed, all those laying claim to special technical knowledge – were much sought after by investors and improvers throughout Britain and parts of the Continent. This cannot be surprising. The rhetoric of philosophers had increasingly deemed as irrelevant the profundities of the gospel-swilling clerics of the Burnet wars after the Revolution and of Convocation during Anne's reign. Providence – once so essential to natural theologians and to Newtonians as well – was to find deeper expression in the exploitation of nature. In Britain and on the Continent, philosophical lecturers stressed the utility of knowledge – and had obtained authority from the enthusiasm of their audiences in return. Thus, the rhetoric of natural philosophy had meaning primarily within the economy of the early eighteenth century. The companies that employed the philosophers were the great engines that would establish the providence of nature.

Natural philosophers were anxious to take advantage of the numerous opportunities the new technology brought their way. Such was obviously the case with Martin Triewald, the Swedish engineer, who arrived in England in 1716 and who quickly became part of the coterie of Desaguliers at a time when, as we know, Desaguliers was much immersed in hydraulics.[1] In the following year, Triewald became employed in the mines of Nicholas and Richard Ridley of Newcastle and came to assist in the

[1] See Margaret E. Rowbottom, "John Theophilus Desaguliers (1683–1744)," *Proceedings of the Huguenot Society of London*, XXI (1968), 213–14. On Triewald's link to Desaguliers, see Svante Lindqvist, *Technology on Trial. The Introduction of Steam Power Technology into Sweden, 1715–1736* (Stockholm, 1984), pp. 197–9.

construction of engines to drain the coal mines in his charge.[2] He remained on the northern moors for virtually ten years.

Desaguliers and Martin Triewald

When Triewald returned to Sweden in the spring of 1726, he did so at an especially propitious time. He had been planning to do so for some years even while there was more than enough demand for services such as he could provide in the coal fields of the north. As Svante Lindqvist has shown, in 1724, the *Newcastle Courant* reported that the Swedish government was especially anxious to encourage foreign Calvinists to settle if they held any special expertise in trade or manufactures. Even more directly, the Swedes had examined the promises of several engine builders, including the elusive John O'Kelly.[3] Within months of his return, Triewald had obtained a patent and entered into negotiations with mine owners in Sweden for the reopening of mines that had flooded out. As with the case of English engineers, Triewald was to supply the engine with its copper boiler, cylinder, and pumps and the mine partners at Dannemora would construct the engine house and supply the necessary timber. Although the engine was ultimately a failure, Triewald's venture was remarkably similar to those he had met with in Newcastle.[4]

Martin Triewald was as least as competent as most of the engineers erecting Newcomen engines. But when Triewald emigrated to Sweden, he had taken with him instruments in order to set himself up as a lecturer. The model was suggested by his attendance at Desaguliers' experiments in London in 1719. By 1724 and 1725, he was giving lectures with John Thorold in Newcastle at the "New Academy" and in Edinburgh.[5] Within a year of his return to Sweden, he presented lectures on natural philosophy in the Palace of Nobility in Stockholm.[6] His prospects of success, thus,

[2] This appears to have been an early family connection with the Ridleys. Lindqvist, *Technology on Trial*, p. 192. See also Lindqvist, "The Work of Martin Triewald in England," *Transactions of the Newcomen Society*, L (1978–9), 165–6; Are Waerland, "Marten Triewald and the First Steam Engine in Sweden," *Transactions of the Newcomen Society*, VII (1926–7), 30–2; B.L., MS. Rawl. D. 871, fols. 78–80, "An Account of a small Treatise lately transmitted to the Royal Society by its Author. It is entitled: A short Description of the Fire and Air Machine at the Mines of Dannemore . . . (1734)." On Ridley, see Edward Hughes, *North Country Life in the Eighteenth Century. The North-East, 1700–1750* (Oxford, 1952), Chapter V; and P. Cromar, "The Coal Industry on Tyneside, 1715–1750," *Northern History*, XIV (1978), 201–7.

[3] Lindqvist, *Technology on Trial*, pp. 212–14, 350n160, and Chapter 9.

[4] Lindqvist, *Technology on Trial*, pp. 239, 250–3.

[5] Lindqvist, *Technology on Trial*, pp. 205, 208–10; *Newcastle Courant*, no. 213 (Saturday, July 18, 1724), p. 11.

[6] Lindqvist, *Technology on Trial*, pp. 252–53.

did not hinge on the smooth functioning of the Dannemora engine. Triewald had not missed Desaguliers' exploitation of the rage for natural philosophy. And as the engineer John Smeaton a generation later, Triewald considered himself a natural philosopher.[7]

It was as though Martin Triewald set out to duplicate the adventures of Desaguliers – not only by posing as a technological adept, but also as a public lecturer. Desaguliers had not been satisfied merely to make public demonstrations, and so, too, did Triewald attempt to make his own contribution to the advance of natural philosophy. Toward the end of 1728, Triewald offered the Royal Society reports on some experiments he had performed on the phenomena of cohesion and electricity. Triewald's suggestion of a theory on electrical attraction and repulsion, although not published in the *Philosophical Transactions* with his other contributions, occasioned a correspondence with Desaguliers.

This exchange affords us with some insight into Desaguliers' views of the state of experimental knowledge. He first complained that Triewald was wrong "for want of being acquainted with several electrical Experiments that have been made." The difficulty arose, in part, from the number of experiments that were being done and from the ability to compare results, "so we ought not to acquiesce in probable solutions founded upon Hypotheses, and rather illustrated than prov'd by Experiments." Here was the strict Newtonian party line and Triewald would have difficulty disagreeing with it. Increasingly crucial, however, was access to the trials conducted by the community of philosophers. Indeed, in 1729, Desaguliers struck an important note in the evolution of experimental method that underlined the disadvantage of the isolated experimentalist:

> So odd are the Phaenomena in electrical Experiments; that it were to be wish'd that a History was made of such Experiments with all their Circumstances; that by comparing them with each other, and such Experiments might be made hereafter, without being too hasty to make Hypotheses, one might have at length by such comparisons to come at the Knowledge of the Cause. And if in the same manner all that have made new Experiments with the magnet wou'd communicate them, and a History shou'd be made of them and the Experiments already known; the many Matters of Fact wou'd, no doubt, be of use for the Improvement of natural Knowledge, and might at length come to give some Light in so intricate a Matter.[8]

7 Margaret C. Jacob, *The Cultural Meaning of the Scientific Revolution* (Philadelphia, 1987), pp. 161–2.

8 B.L., Add. MSS. 4432, fols. 75–6, Desaguliers, "Remarks upon Mr. Triewald's Attempt to give an Account of the Reasons of some Electrical Phenomena," February 27, 1728/9; fols. 70–4, Marten Triewald to the Secretary of the Royal Society, December 21, 1728.

The program of experimentation envisaged by Desaguliers was clearly founded on sharing results. This was the foundation of epistemological validity. And Martin Triewald appears to have been perfectly in accord with this view. Throughout the next few years, he communicated various experiments to the Royal Society. He undertook optical experiments as well as some curious observations on the sudden freezing of water, which he duly reported to Hans Sloane.[9] These reports were the result of his own experimental program even if his technical responsibilities frequently interfered.

But like Desaguliers, Triewald was also determined that natural philosophy should have a wider public than that confined to the universities. What emerged from the enlarged curiosity about the new science had now become a creed that was linked firmly to an incipient notion of testability in experimental method – and, significantly, to presumed practical benefits. Such a view, of course, coincided not only with the practice of the Swedish government to encourage arts and manufactures, but also with the interests of those merchants who appear to have attended his lectures.[10] It was this attitude that came to predominate amongst the eighteenth century lectures on science. As Svante Lindqvist has pointed out, in 1697, the Swedish government provided funds for the establishment of a *Laboratorium mechanicum* by Christopher Polhem, the aim of which was to provide an experimental basis for the improvement of machines. This notion was not without influence, and Triewald adopted Polhem's apparatus for his lectures.[11] The predominance of practicality amongst public lecturers by midcentury may well be counted as the legacy of the English experience – which was vigorously reinforced by the rapid industrialization that soon overtook a large part of Western Europe.[12]

Natural philosophers among the northern mines

The experience of the engineers and philosophers in the mines of Europe would have provided them not only with opportunity, but also with a

Triewald's reply after repeating some of his experiments is in fols. 77–8. Triewald to the Royal Society, August 4, 1729. See also Lindqvist, *Technology on Trial*, pp. 259–60; and J. L. Heilbron, *Electricity in the 17th and 18th Centuries. A Study of Early Modern Physics* (Berkeley and Los Angeles, 1979), pp. 290–2.

9 B.L., Sloane MSS. 4025, fols. 293–4, Triewald to Sloane, 1730; and Lindqvist, *Technology on Trial*, p. 261.

10 Lindqvist, *Technology on Trial*, p. 256.

11 B.L., Add. MSS. 4433, fols. 138–145, Triewald to Hans Sloane, April 24, 1735; and Lindqvist, *Technology on Trial*, pp. 67, 78.

12 Jacob, *The Cultural Meaning of the Scientific Revolution*, pp. 195–8.

powerful justification for natural philosophy. Therefore, the attitude of philosophers employed in the coal mines of the north of England was of great importance in the foundation of the practical side of natural philosophy. Because of flooding in the early eighteenth century, the coal fields of the north were ripe for the introduction of Newcomen engines. It is for this reason that negotiations were undertaken with the London Proprietors who held the patent. The Ridleys' rival for control of the coal trade, William Cotesworth, clearly had some interest in using fire engines as early as 1715, when he and his partners had leased Gateshead Park colliery. In 1719, Cotesworth purchased some of the estates of Lord Widdrington that had been confiscated by the Crown after the Jacobite rebellion of 1715. Cotesworth soon entered into extended negotiations with the Proprietors of the Newcomen patent in order to construct engines in his collieries.[13]

The little we know of the links between mine owners and the London entrepreneurs threatens to mask the significance of the natural philosophers in the adoption of the new technology. Martin Triewald's conception of the usefulness of natural philosophy was typical. It is evident that those who had a strong interest in the dissemination of natural philosophy were eager to adopt recent changes in mining technology. It was not the rigors of damp, fog-wrapped moors nor the rancor of disputes over way leaves that drove mine owners to distraction. Those obstacles were to be expected. But flooded pits did emphasize the opportunity that natural philosophers might represent if their new technology really could make mines profitable.

It is instructive that through his friend Colonel George Liddell, Cotesworth was connected with the antiquary, lecturer on natural philosophy, and Fellow of the Royal Society, John Horsley. Horsley seems to have played some role in Cotesworth's dealings with the Newcastle Common Council, which were often bitter and the result of northern rivalries in the coal trade.[14] Horsley, it has been argued, acted as agent for the York Buildings Company in its early dealings at Widdrington and supervised its Winton saltworks at Cockenzie.[15] But it was as one of the

[13] Joyce M. Ellis, *A Study of the Business Fortunes of William Cotesworth, c. 1688–1726* (New York, 1981), pp. 7–8, 61–2; and Edward Hughes, "The First Steam Engines in the Durham Coalfields," *Archaeologia Aeliana*, XXVII (1949), 32ff.

[14] Gateshead Public Library, Cotesworth MSS, CP1/21. Liddell to Cotesworth, March 5, 1717/18.

[15] Hughes, *North County Life*, pp. 22–3; and Christopher A. Whatley, *The Scottish Salt Industry 1570–1850. An Economic and Social History* (Aberdeen, 1987), p. 47. For Horsley's links with the Proprietors of the Fire Engine Patent, see J. S. Allen, "Bromsgrove and the Newcomen Engine," *Transactions of the Newcomen Society*, XLIII (1970–1),

early provincial lecturers that Horsley made his mark. Here he fit into the role defined most clearly by Desaguliers, whom we know to have lectured in provincial towns like Birmingham in 1739. Horsley was hardly alone in seeing the benefits that industrial areas of Britain might assume from the adoption of an interest in natural philosophy. That was clearly the intention of Martin Triewald's efforts. So, too, was it that of John Thorold or of Horsley's assistant, George Mark in his lectures in Morpeth and Newcastle. With Horsley's death in 1731, Mark went on to assist Isaac Thompson in his lectures in the North, which were specifically designed for those engaged in coal mines.[16]

The engineers prepared to give lectures were urgently representing themselves as experts to those who had the capability of patronizing them. It matters not at all that those who evidently appeared for the lectures were mine owners rather than artisans. It seems as though there was a widespread conviction propagated by Desaguliers and by the engineer, John Grundy, senior, that a knowledge of mathematics and natural philosophy was the best protection that an engineer might provide to an investor. But this is not to say that it was also necessary for those who attended the lectures to have a working knowledge of mathematics. Indeed, Horsley expressly recognized that such limitations had to be ignored.[17] When preparing an advertisement for the *Newcastle Courant*,

190. There is, admittedly, some doubt that the lecturer and antiquary John Horsley was the same as the land agent for the York Buildings Company, although it does seem clear that the lecturer was the author of the paper published in the *Philosophical Transactions* on rainfall at the Widdrington estate. The distinction, however, seems based on a very flimsy argument that I have been unable to confirm with any certainty. See Sir George MacDonald, "John Horsley, Scholar and Gentleman," *Archaeologia Aeliana*, fourth series, VIII (1933), 11. Among other reasons for being cautious about accepting the existence of two John Horsleys with an interest in Widdrington is the fact that the antiquary had an extensive interest in the estates of the north – even more so in the aftermath of the Jacobite debacle in 1715 and the confiscation of estates, Widdrington included, by the Crown. This is clearly demonstrated in the material he amassed for a projected *History of Northumberland*. See R. C. Bosanquet, "John Horsley and His Times," *Archaeologia Aeliana*, fourth series, VIII (1933), 59–81.

[16] Peter J. Wallis, *An Index of British Mathematicians, Part II (1701–1760)* (Newcastle Upon Tyne, 1976). Wallis, "Dissemination of Mathematics in the Eighteenth Century. The Evidence of Subscription Lists," Project for Historical Bio-Bibliography, no. 314 (December 1982). On Thompson, see E. G. R. Taylor, *The Mathematical Practitioners of Hanoverian England* (Cambridge, 1966), p. 247. On occasion, such engineers as Triewald might actually establish a claim to an invention useful to the mine owners as was the case in 1722, when he came to an agreement with Nicholas Ridley, Jr., Samuel Calley, who was engaged in building Newcomen engines, and William Prior, a mathematical and musical instrument maker from Gateshead. The partners obtained a patent for an atmospheric engine. Northumberland Record Office, Ridley (Blagdon) MSS, ZR1/23/2. Articles of Agreement, June 1, 1722; Lindqvist, *Technology on Trial*, pp. 201–2.

[17] Cf. Raymond James Evans, "The Diffusion of Science: The Geographical Transmission

with the help of the manufacturer Robert Cay, Horsley purposefully added, "All possible pains will be taken in these lectures to render everything plain & intelligible even to those who have no previous acquaintance wth mathematical learning."[18]

Although Horsley is now best known for his antiquarian activities, they have served to mask the range of his interests. Assuming our lecturer, who was undoubtedly well known to Cotesworth and Liddell, was the same Horsley engaged at Widdrington, we soon find that the estate was presenting Horsley with difficulties. As we know, the prospects of the York Buildings Company had not met expectations. By 1725, the Company had sufficient difficulties in its London operations to make the success of its northern estates much more urgent. But by 1725, Widdrington failed to show dividends and Horsley's efforts there, which he evidently exercised through an agent named Brown, caused much disappointment.[19] The true state of Horsley's affairs at Widdrington remains obscure, although his antiquarian manuscripts reveal a familiarity with the confiscated estate.[20]

Whether the interests of a land agent or of an antiquary, Horsley's own efforts display the problems that the eighteenth-century mine engineers often confronted. For example, his notes on natural history for 1729 indicate that he had begun to examine the damps to be found in coal mines, and in the company of Richard Peck, he had descended a pit

of Natural Philosophy into the English Provinces, 1660–1760" (Ph.D. thesis, Cambridge University, 1981), pp. 42, 138. On Grundy, see A. E. Musson and Eric Robinson, *Science and Technology in the Industrial Revolution* (Toronto, 1969), pp. 140–1; and Jacob, *The Cultural Meaning of the Scientific Revolution*, pp. 158–63.

18 Northumberland County Record Office, Society of Antiquaries MSS, M13/D19. Rev. John Horsley. MS notes and correspondence, ca. 1726–29. John Horsley to Robert Cay, n.d.. Robert Cay was evidently a great salt manufacturer of the north east and an old friend of Horsley. On the Cay family, see MacDonald, "John Horsley, Scholar and Gentleman," pp.29–30; and Bosanquet, "John Horsley and His Times," pp. 67ff.

19 Gateshead Public Library, CP1/176. Liddell to Cotesworth, May 4, 1725; CP1/239. Liddell to Cotesworth, April 19, 1726.

20 J. Hodgson Hinde, ed., *Inedited Contributions to the History of Northumberland. Materials for the History of Northumberland by the Rev. John Horsley, 1729–1730* (Newcastle, n.d.), pp. 25–6. These are taken from NCRO, Society of Antiquaries MSS., M13/D19. By 1729, he had evidently become known to the Scottish mine owner Sir John Clerk of Penicuik, who wrote to Roger Gale that Horsley had "acquired a great reputation for the mathematicks and his knowledge in all parts of philosophy." William Stukeley, "The Family Memoirs of the Rev. William Stukeley, III," *Surtees Society Publications*, LXXX (Durham, 1887), 390. Sir John Clerk to Roger Gale, December 22, 1729. In 1724, Sir John Clerk had visited Morpeth and seemingly had met Martin Triewald in his effort to learn more of mining technology. There is no evidence, however, that Clerk met Horsley at that time. Bosanquet, "John Horsley and His Times," p. 71; and, Lindqvist, *Technology on Trial*, pp. 203–4.

to observe the effects. Peck reported that an experiment to burn off bad air had resulted in an explosion that collapsed the timbers. Richard Peck was a mining engineer engaged by the London Lead Company in the construction of steam engines at collieries at Bushblades and Jesmond in the North-East.[21] Experiments on mine damps were conducted at some peril and Horsley came to realize, as had Desaguliers, that there were at least two forms of bad air: one a "fulminating damp" that was inflammatory and dangerous, and a suffocating damp that at least extinguished a candle before it was fatal and thus, at least, gave a warning.[22] The trick in mines was not just to rid the pits of the flood waters that had rendered them useless, but also to make them workable by detecting and expelling foul and dangerous gases that could easily kill the miners. Various attempts had been made to burn off foul air, but this could be dangerous, and vents from deep shafts were not always very practical, especially where the specific gravity of the gas caused it to settle. Numerous mechanical contrivances were proposed, not the least of which were those by Martin Triewald in 1724 and by Desaguliers in 1727. By 1729, Horsley was merely reporting from personal experience what others were actively investigating.[23]

The natural philosophers, especially those who not only gave public lectures, but who also claimed a special expertise in technical matters, were in increasing demand amongst investors in mines. It was widely believed they could spell the difference between fortune and failure. The substantial cost involved in the construction of the Newcomen engines, for example, explains why Chandos and the York Buildings Company came to rely so heavily on the reputation and expertise of Desaguliers.

[21] North of England Institute of Mining and Mechanical Engineers, Newcastle upon Tyne, Shelf 18. Richard Peck's View Book, 1710–35. fols. 42, 73; Shelf 49, vol. 5. T. E. Forster Collection. View Book, probably of Richard Peck, 1710–57, fols. 65, 76. The London Lead Company, it will be recalled, obtained parts of the York Buildings engine when it was sold. It is interesting that one of the long term directors of the company, and, in 1731, its Governor when the York Buildings engine became available, was Charles Du Bois, the merchant patron of scientific authors. NEIME, Shelf 1, London Lead Company, Court Minutes, vol. 4 (1711–20), pp. 262–4, 267, 294–7; vol. 6 (1727–32), p. 262.

[22] NCRO, Society of Antiquaries MSS., M13/D19, Rev. John Horsley, MS notes and correspondence, ca. 1726–9, "Natural History."

[23] NCRO, Society of Antiquaries MSS., M13/D19. John Horsley on Natural History. This MS is printed, in part, in J. Hodgson Hinde, ed., *Inedited Contributions to the History of Northumberland. Materials for the History of Northumberland by the Rev. John Horsley, 1729–1730* (Newcastle, n.d.), pp. 16–18. See also NEIME, Bell Collection, *History of Coal Mining*, vol. 4, pp. 113–17; Desaguliers, "An Attempt made before the Royal Society, to shew how Damps, or foul Air, may be drawn out of any Sort of Mines, &c. by and Engine contrived by . . . ," *Philosophical Transactions*, no. 399, XXXV (July-Sept. 1727), 353.

The introduction of steam engines became increasingly critical in the early industrial ventures of the eighteenth century. There can be little doubt, for example, that the growing fortunes of the North-East coal mines rested largely on the success in draining the pits. Likewise, the link between the engines and the efforts of the York Buildings Company indicates how crucial they became in the future of that complicated venture.

York Buildings in the north

The tenure of the second Sir John Meres as Governor of York Buildings lasted until 1726. It was a turbulent time for the Company. Meres was faced with the task of trying to secure sufficient capital to keep the Company afloat until the Widdrington estate, amongst others, could bring further returns. This proved increasingly futile and led to an abortive attempt to link the Company to the fortunes of the Charitable Corporation.[24] Late in 1725, rumors began to circulate about mismanagement in the Company's affairs. Buffeted by repeated problems with its water supply attempts and the failure of its engine, the timing of these accusations against Meres sent the Directors into a frenzy of activity. The original insinuations from James Marye of the English Copper Company were not easily dismissed.[25] The result of the investigation into Meres' activities was his replacement by Chandos's own agent, Samuel Horsey.

The accusations against the Governor could not have come at a worse time for a large investor. The "dragons" were problem enough, and Desaguliers was unable to complete the reservoirs in a manner satisfactory to Chandos. The York Buildings Company had been at the hub of a network of investment that had been created by the fertile mind of Case Billingsley. But the center was no longer holding. Chandos now had to

[24] B.L., Add. MSS. 36226, fols. 221–8, General Court Minutes of the York Buildings Company, September 21, 1724–October 2, 1724; and Armand B. DuBois, *The English Business Company after the Bubble Act, 1720–1800* (New York, 1938), pp. 327–30, 369–70, 424. There was possibly a Meres family connection with mining through the Beaumonts as early as the reign of William. Colin Owen, *The Leicestershire and South Derbyshire Coalfields* (Ashbourne, 1984), p. 99.

[25] Public Record Office, T1/258/98–106, Narrative and Account of Several Proceedings relating to the Affairs of the York Buildings Company; Hunt. Lib., St. 57, XXVII, fols. 125–7, Chandos to Samuel Horsey, November 30, 1725; f. 159, Chandos to Westfalling, December 14, 1725; f. 187, Chandos to James Marye, January 7, 1725/6. James Marye was also closely connected with the English Copper Company with which Billingsley had been involved. B.L., Add. MSS. 36142, fols. 119–20, Philip Yorke, report on petition. Chandos was purchasing English Copper stock though Thomas Watts at this time. Hunt. Lib., St. 57, XXVI, f. 260 Chandos to Watts, August 21, 1725.

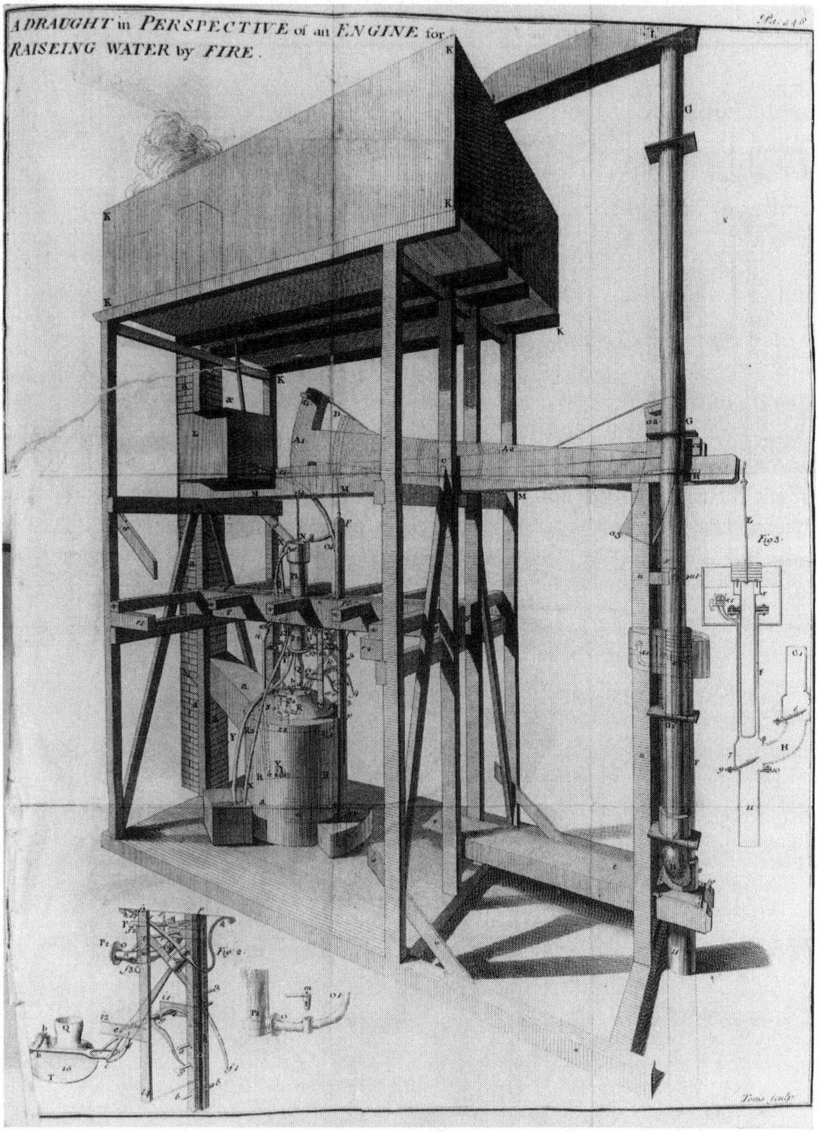

Figure 15. The design of the Newcomen engine at York Buildings, according to Stephen Switzer. By courtesy of Thomas Fisher Rare Book Library, University of Toronto.

take charge. Desaguliers and "Gun-Jones" had not proved enough insurance; Samuel Horsey and the genius of Thomas Watts of the Academy would now emerge to rescue York Buildings – or so it was hoped.

Samuel Horsey was to be one of the generation of unlucky men to come into contact with the York Buildings Company. Horsey's efforts on behalf of the Duke of Chandos placed him at the focus of the various attempts to rescue the fortunes of the Company in the mines of the north. Horsey, lately a Colonel in the Fourth Troop of Guards, had first come to the attention of Chandos during dealings with the African Company over abortive potash and soap manufacturing ventures.[26] Colonel Horsey was soon drawn into the Chandos circle and became one of his nominees to the Directors of the York Buildings Company in 1724.[27] The method was by now the usual one of the transfer of shares in trust. At this critical stage, Horsey became one of the key figures in the management of the York Buildings estates in Scotland and periodically made visits to Scotland to assess the likelihood of the success of the mines. It was through this means that Chandos was kept fully informed of York Buildings affairs, including the consistent failure of the engines on the Strand to be made to work.[28] By this point, of course, the accusations of fraud and mismanagement were surfacing at meetings of the Court of Directors and Meres emerged with little support. The result, ultimately, was that the Company unanimously chose Horsey to take Meres' place as Governor in 1726 – on the recommendation and support of Chandos.[29]

When Samuel Horsey first came into the management of York Buildings, he did so in the baggage train of the Duke of Chandos. But, as we have seen, linking to the fortunes of powerful patrons was standard practice in the eighteenth century. For some, it might be the only means to advancement. The opportunities presented by the York Buildings Company were not especially attractive. Chandos once told Thomas Watts that "Of all the Societys I have ever known there is none who have been more unfortunate in their Choice of Managers, nor where the Proprietors have been more grossly abused."[30] Horsey, however, looked on his own role as a temporary measure.

[26] Very little is known of Horsey's early career. In 1710, as a Captain of the Guards, he had a role in suppressing the Sacheverell riots. B.L., Add. MSS. 61653, Ro. Pringle to Captain Horsey, March 6, 1709/10.

[27] Hunt. Lib., Stowe MSS. 57, XXV, f.77, Chandos to Col. Horsey, December 19, 1724; XXV, f. 78, Chandos to Westfalling, December 19, 1724.

[28] Hunt. Lib., Stowe MSS. 57, XXVI, f. 107, Chandos to Horsey, June 17, 1725; f. 301. Chandos to Horsey, September 7, 1725.

[29] [Horsey], *The Case of Samuel Horsey, Esq; Governor of the York Buildings Company* (London, 1733), p. 3.

[30] Hunt. Lib., Stowe MSS. 57, XXVIII, fols. 201–2, Chandos to Watts, July 8, 1726.

While he struggled with the affairs of the Company, he also sought the Governorship of South Carolina, which had been promised to him by George I. This was no mere effort to escape the allegations of mismanagement that were ultimately directed at him as well, for Horsey seems to have been involved in the purchase of the Carolina colony by the Crown.[31] Into these efforts, he succeeded in drawing the Duke of Chandos, who offered to speak to Newcastle and Walpole about the appointment. However, by the time Horsey retired from York Buildings in 1733, in the midst of a storm of protest over the affairs of the Company, he still had not achieved his Royal commission. It was not until August 1738 that the unlucky Horsey actually took his oath of office – only to die within days of a fit of apoplexy at his house in Whitehall.[32] Undoubtedly, his years of facing the obstacles of the York Buildings Company and political opposition in the Carolina affair had taken a harsh toll.

Chandos, Desaguliers, Samuel Horsey, and Thomas Watts engineered many of the speculative schemes of the York Buildings Company that provided opportunities for philosophers and engineers as well as stock-

[31] Public Record Office, SP (Dom.) 36/2/59–60. Petition of Samuel Horsey, June 20, 1727; [Horsey,] *Answer of Samuel Horsey, Esq; Governor of the York Buildings Company, to the Misrepresentations of his Conduct, Publish'd in some of the Daily News Papers* (London, 1733), p. 18; and W. Eugene Sirmans, *Colonial South Carolina. A Political History, 1663–1763* (Chapel Hill, 1966), p. 194. In the history of the Carolina affair, there was a curious chapter involving the inventive Case Billingsley. In 1735, Billingsley proposed a scheme before the Georgia Board to secure the future of the colony, but when his reputation was revealed, the Board "declared it would bring a disrepute on our Society to treat with him on any foot. . . ." *Manuscripts of the Earl of Egmont. Diary of the First Earl of Egmont*, vol. II (1734–8). Historical Manuscripts Commission (London, 1923), pp. 179, 181–3.

[32] Hunt. Lib., Stowe MSS. 57, XXVII, fols. 204–5, Chandos to Horsey, January 15, 1725/5; f. 301, Chandos to Horsey, February 23, 1725/6; Public Record Office, SP36/31/81. Thomas Lowndes to ?, March 16, 1733/4; and Guildhall Library, MS 205. Diary of Stephen Monteage, I, p. 157, June 26, 1733; II, p. 180, August 21, 1738. The final strain upon Horsey was likely a suit filed against him by Sir James Grant when it became common knowledge that Horsey would soon be leaving England to take up his post. This concerned sureties given by the York Buildings Company when they began their expansion in Scotland under the direction of Horsey. Scottish Record Office, GD 248/135/1/11. Sir James Grant v. Samuel Horsey & others, filed in Chancery, June 29, 1738. In 1737, the Earl of Egmont of the Georgia Board recognized Horsey as "a person undone by being Governor of the York Buildings Company, and suffering (they say unjustly) in his character on that account." Curiously, just after Horsey secured his appointment, the Earl dined with him in March and went off immediately following to watch Desaguliers at the Royal Society demonstrate the theory of the tides. The Earl does not say whether Horsey accompanied him, but the philosopher and the Governor appear to have crossed paths on many occasions. At least they had a common patron in Chandos. See HMC, *Egmont Diary*, II, pp. 418, 471.

jobbers.[33] Under Horsey's direction, influenced mainly by his connection with Chandos, the Company began a major expansion of its interests in Scotland.[34] These projects included efforts to remove large amounts of timber from the Jacobite estates, but most activity centered around the exploration for various minerals. For example, in 1729, the Company found deposits of alum and one of its employees, a Mr. Place, was active in the search for new ore bodies in Scotland. Chandos, it seems, was none to sanguine about Place's reliability once it was discovered that he had been engaged in the efforts of Wood to obtain a patent for refining iron ore with sea coal.[35]

There was great excitement over the discoveries of iron ore deposits in the north and it is clear now that Chandos, Horsey, and both Thomas and William Watts – still in control of the Academy in Little Tower Street – were positioning themselves to best advantage.[36] As a result of the sighting of promising ore, the York Buildings Company in 1728 had taken out a fifty-year indenture to mine the lands of Sir James Grant of Grant at Abernethy. Initial assays indicated that the iron ore was better

[33] One such project was the involvement of the York Buildings Company with the Charitable Corporation. The extent of the interlocking relationships among the directors is difficult to sort out, but there is little question that the rumor of York involvement in the Charitable Corporation had a great effect on York stock after it was discovered that the cashier and warehouse keeper of the Charitable Corporation had absconded to Rome. Evidence in the Chandos correspondence clearly shows that many of his friends were involved, especially Archibald Grant, who had been treating with York Buildings for rights to minerals in Scotland. The result, by 1731, was that this scandal made it very difficult for Horsey to meet York Buildings obligations to their leases in Scotland. Hunt. Lib., Stowe MSS. 57, XXXVIII, f. 241. Chandos to Mr. Arnold, September 15, 1731; fols. 279–81, Chandos to Horsey, November 8, 1731; fols. 301–2, Chandos to Stuart, November 13, 1731; XXXIX, fols. 129–30, Chandos to Thomas Watts, March 20, 1731/2; f. 140, Chandos to Westfalling, March 23, 1731/2; f. 146, Chandos to Sir Archibald Grant and to Sir Thomas Robinson, April 11, 1732; SRO, GD. 248/47/3/6, Samuel Horsey to James Grant of Castle Grant, November 20, 1731. See also DuBois, *The English Business Company*, pp. 293–4, 326n93; Cummings, "The York Buildings Company," II, pp. 394ff.; and Sedgwick, *House of Commons*, II, pp. 77–8, 388–9.

[34] Chandos periodically denied that he had any concern in the Company, but this was a half truth as he controlled blocks of stock through his own nominees like Horsey or Watts. Hunt. Lib., Stowe MSS. 57, XXXI, f. 74, Chandos to Meres, January 2, 1727. On York Buildings in Scotland, see J. M. Lindsay, "The Iron Industry in the Highlands: Charcoal Blast Furnaces," *Scottish Historical Review*, LVI (April 1977), 59; Hunt. Lib., Stowe MSS. 57, XXX, f. 109, Chandos to Watts, May 19, 1729; f. 118, Chandos to Horsey, May 26, 1729; and Sidney Pollard, *The Genesis of Modern Management. A Study of the Industrial Revolution in Great Britain* (London, 1965), pp. 16–17.

[35] Hunt. Lib., Stowe MSS. 57, XXXIII, f. 146, Chandos to Horsey, June 16, 1729; f. 152, Chandos to Horsey, June 18, 1729.

[36] Hunt. Lib., Stowe MSS. 57, XXXIII, f. 29, Chandos to Thomas Watts, February 14, 1728/9; f. 152, Chandos to Watts, June 25, 1729; f. 277, Chandos to Horsey, October 10, 1729.

than had been expected and the Company drafted some Bristol operatives to erect a furnace for the iron works.[37] One of those so employed was Benjamin Lund, a Bristol merchant. Lund was friend of Francis Hauksbee with whom he had received a patent in 1728 for the refining of copper ores and extracting silver.[38] Once again, we find the philosophers and their confederates in the universe of the adventurers.

Sun Fire and Scotch Mines

The possibility of being able to refine ores for silver transfixed the desperate York Buildings Company. All trails now led north, along the great north road past the old frontier on the Tweed at Berwick and into the forests of the Highlands. And it was here that the links to the purveyors of public lectures were even more firmly established. The connection runs through Hauksbee, Benjamin Lund, Thomas Watts, and the mathematician James Stirling, who had lectured at Watts's Academy. The York Buildings Company materialized in the north working mines and refining ores at Leadhills, Strontian, and Kulnakele. Most of these arrangements were based on extended leases worked out by John Billingsley and Samuel Horsey.[39] Especially prominent were Sir John Erskine and Sir Archibald Grant, who were prepared to allow their estates to be worked in return for a portion of the revenue.

Erskine and Grant were prime movers in the creation of a Scotch Mines Company. This, however, is really only one view of the many sides of the York Buildings and Sun Fire Companies behind which we find the

[37] J. M. Lindsay, "The Iron Industry in the Highlands," p. 59; Hunt. Lib., Stowe MSS. 57, XXXIII, f. 277, Chandos to Horsey, October 10, 1729; XXXIV, fols. 3–4, Chandos to Watts, November 28, 1729; Scottish Record Office, GD 248/167/1/10, Samuel Horsey to Sir James Grant, August 7, 1731.

[38] Public Record Office, SP (Dom.) 36/2/86–109, Report on the Petition of Lund and Hauksbee, January 6, 1727[/8?]. The warrant for the patent was issued on January 18, 1727/8. B.L., Add. MSS. 36123, fols. 82–3, note the objections of Nehemiah Champion of Bristol, who was one of those who obtained pig iron from the highlands. PRO, SP (Dom.) 36/2/91. Lindsay, "The Iron Industry in the Highlands," p. 58. See also SP (Dom.) 36/5/253–262. Petition of Benjamin Lund and Francis Hauksbee for "a method for the more advantageous manufacturing of Copper Oare & extracting Silver from Copper and mixing of Copper Calamy & charcoale together for making Brass," March 30, 1728.

[39] Scottish Record Office, GD 248/81/1/27, agreement between the York Buildings Company and Sir James Grant, signed John Billingsley, April 21, 1729; GD 248/167/1/2, Samuel Horsey to Sir James Grant, February 18, 1729/30; [Horsey,] *The Case of Samuel Horsey*, pp. 12, 27–8; and Baron F. Duckham, *A History of the Scottish Coal Industry*, vol. I (Newton Abbot, 1970), *passim*. On the workings at Leadhills, see T. C. Smout, ed., "Journal of Henry Kalmeter's Travels in Scotland 1719–1720," in *Scottish Industrial History*. Scottish History Society, series IV, vol. XIV (1978), pp. 27–34.

shadows of Chandos and Thomas Watts. In April 1729, the York Buildings Company had reached the first of its formal agreements with Sir James Grant. Shortly thereafter, the Scotch Mines Company received its Royal Charter and the roads into the highlands were worn with the carts of the adventurers from London. In November, Benjamin Lund and Mr. Place of York Buildings were on their way to examine ore bodies on the estates of Lord Hoptoun, and by the spring of 1730, the mood in London was ebullient. Chandos was enthused by Horsey's reports upon the Colonel's return to London in April. The Duke was shown the accounts of Lund and Place, which were such, Chandos reported, "as, if they are not the greatest Cheats on Earth, must render the Stock very valuable." Throughout all of this, Chandos was buying York stock and consulting with Sir Archibald Grant on the cost of leases to his lands.[40]

Sir Archibald Grant of Monymusk, M.P., was known to Chandos in London because of his activities in several joint-stock companies. Grant was an investor in York Buildings and it had not served him well. Through his London connections, Grant was one of the creators of the Scotch Mines Company, which was mainly manipulated by the same gang that controlled York Buildings. The Scotch Mines, evidently, had been the brain child of Sir John Erskine, who had been mining silver on his own estates. Nonetheless, as T. C. Smout has shown, the Scotch Mines were actually controlled from London by the gentlemen of the Sun Fire Office. Chief among them was Thomas Watts.[41] This, in part, explains the sudden appearance at Leadhills of the mathematician James Stirling when he replaced Sir John Erskine as the chief agent for the Scotch Mines.[42] The

[40] Hunt. Lib., Stowe MSS. 57, XXXIV, fols. 3–4, Chandos to Watts, November 28, 1729; f. 164, Chandos to James Marye, February 8, 1729/30; fols. 178–9, Chandos to Horsey, February 14, 1729/30; fols. 303–4. Chandos to Watts, April 27, 1730; f. 343, Chandos to Sir Archibald Grant, May 11, 1730.

[41] T. C. Smout, "Lead-mining in Scotland, 1650–1850," in *Studies in Scottish Business History*, ed. Peter L. Payne (Edinburgh, 1967), pp. 115–20; Romney Sedgwick, ed., *The House of Commons, 1715–1754*, II, pp. 77–8. On the formation of the Scotch Mines Company, see PRO, Privy Council Minutes, PC1/4/82/1, August 15, 1728; and PC1/4/82/12. Graham Cummings provided this reference to me.

[42] Smout, "Lead-mining in Scotland," pp. 119–20, 124–5; and Hunt. Lib., Stowe MSS. 57, XLV, f. 135, Chandos to Farquharson, October 28, 1734. The precise date when Stirling became engaged in the operations in Scotland is not certain, although it was clearly before 1735, as has been presumed. On Stirling, see Musson and Robinson, *Science and Technology in the Industrial Revolution*, pp. 41–2; and Charles Tweedie, *James Stirling* (London, 1922), which deals briefly with his scientific career. By the time he returned to Scotland, Stirling was a Fellow of the Royal Society and an established mathematician. B.L., Add. MSS. 4432, fols. 316–18, "Of the figure of the Earth and the Variation of gravity on the surface," Stirling to Machin, December 6, 1733. This was published in *Philosophical Transactions*, no. 438 (July-Sept. 1735), 98–105.

bonds between the mines and the lecturers of the Academy provide further assurance that the practical interests of the likes of Desaguliers, Harris or Grundy were increasingly typical of the early eighteenth century.

Thomas Watts was one of the great entrepreneurs of eighteenth-century natural philosophy. The Academy in Little Tower Street competed with, as well as employed, the numerous public lecturers on science with which the century abounds. But Watts, it now seems, also played a pivotal role in the transformation of Britain into the world of industrialism. Repeatedly, he surfaced in the operations of the Duke of Chandos. Watts was a wily financier for whom, like Case Billingsley, the opportunism of the Royal Exchange and the coffeehouses were a natural element. The clues to his significance have long been overlooked.

Watts had come a long way from a mere schoolmaster. Thomas Watts infiltrated the management of the Sun during its reorganization of the summer of 1720. He entered some remarkable company. By that time, the Sun had gone through several upheavals involving many of the more manipulative financiers of early Hanoverian England. In 1715, one of those who was active in the Sun was Thomas Burgess with whom Case Billingsley was to submit proposals on coinage to the Crown.[43] Throughout the fall of 1720, as the shock waves of the South Sea Bubble worked their way through the wreckage, shares in the Sun changed hands rapidly to consolidate the control by the new managers. In these waters, the sharks had gathered. This group now included not only the founder Charles Povey, but also William Watts, Dr. John Woodward of Gresham College and confederate in Royal Society wars of the recently deceased John Harris, James Marye of the English Copper Company, Case Billingsley's son John, Christian Cole, and Sir Fisher Tench, M.P. for Southwark, and later, with Thomas Watts, one of the Directors of the scandalous Charitable Corporation.[44]

[43] B.L., Add. MSS. 61605, fols. 21–4, Burgess and Billingsley to the King, June 21, 1716; Add. MSS. 61640, f. 110, Burgess and Billingsley to Lord Sunderland, June 21, 1716; Guildhall Library, MS. 11931, vol. II, f. 13, Sun Insurance Office, General Court, October 14, 1715. On Burgess, see Dickson, *The Sun Insurance Office*, p. 269; and Scott, *Constitution and Finance of English, Scottish and Irish Joint-Stock Companies*, III, p. 446. There is the possibility of a link between Watts and Billingsley over the lease to some lands near Hawkesbury, which became the subject of a suit between the engineers Sparrow and Parrott. See J. S. Allen, "Some Early Newcomen Engines and the Legal Disputes Surrounding Them," *Transactions of the Newcomen Society*, XLI (1968–9), 185–90.

[44] Guildhall MS. 12023, Sun Insurance Office, Share Transfers, Vol. I (1720–1); MS. 12019C, Share Ledger, and MS. 11931. General Committee Minutes, Vol. II, f. 160, August 24, 1720. On Tench, see Sedgwick, *House of Commons*, II, p. 465; and Cummings, "The York Buildings Company," Vol. II, pp. 414–16.

Throughout the 1720s, the control of the Sun Fire was solidified. Watts bought more stock, and his associates like Benjamin Worster of the Academy or friends of the Duke of Chandos like Dr. Charles Stuart and Samuel Robinson took possession of their shares.[45] There can be no doubt that Watts's activities were closely linked to the plans that Chandos had launched to protect his personal fortune against the collapse of York Buildings.

Pretenders among the philosophers

The activities of the Duke of Chandos provide numerous excellent examples of a great lord seeking to exploit the expertise his patronage might purchase. During the early 1730s, when the York Buildings Company was struggling to develop its mines in Scotland at Swinhart and Kulnakele, the Duke of Chandos was constantly searching for a magic elixir that would unlock the riches he believed might yet be exploited. At one point, he thought he had found it in a patent granted in 1730 to Elizabeth Coppin for "an Art of fluxing and fixing Mundick & reducing that Sulphureous and volatile Mineral into a Metal from wch Silver might be extracted." Elizabeth Coppin (or Copping) was regarded as an authority on mundick and Chandos had extensive dealings with her until Sir Thomas Mackworth – hardly a novice in such matters – told him that "he had essayed the Regulus but found no mettal in it, nor any thing but dross; he despises very much this pretended discovery."[46] Mrs. Coppin's patent had been granted as a means of extracting silver from lead ores, which were known to contain small amounts of the precious metal. Given his investments in mines, it was hardly any surprise that Chandos should chase the grail Mrs. Coppin claimed to have in her hand. Chandos was transfixed by the possibilities this suggested and he had sought the advice of "Gun-Jones," a friend of Desaguliers who recently had constructed a working model of a Newcomen engine in Desaguliers' garden in Channel-Row at the request of the King of Spain. Jones, who was employed by the York-Buildings Company, was another of those artisan-engineers who con-

[45] Guildhall, MS. 12023, Sun Insurance. Share Transfers, Vol. II (1721–6); MS 12019C, Share Ledger, Vol. I.

[46] Hunt. Lib., Stowe MSS. 57, XXXIX, f. 314, Chandos to Dr. West, August 1, 1732. See also XXXIX, f. 283, Chandos to Mrs. Coppin, July 11, 1732; f. 294, Chandos to Dr. West, July 22, 1732; fols. 297–8, Chandos to Dr. West, July 23, 1732; fols. 249–50, Chandos to Mr. [Thomas] Watts, July 20, 1732; f. 312, Chandos to Mr. Benjamin Robinson (of the English Copper Company), August 1, 1732; f. 327, Chandos to Dr. West, August 5, 1732; Public Record Office, SP (Dom.) 36/43/71. Eliz. Coppin patent, August 1730.

sorted with philosophers. He was relied upon to test the value of samples of ore procured by Chandos specifically for their silver content.[47] Mrs. Coppin's secret turned out to be a mirage.

Chandos, like many wealthy investors one suspects, had become inoculated as much against disappointment as against the incessant claims of inventive pretenders. In any case, at the moment he severed his connections with Mrs. Coppin, the Duke drew upon Dr. John Arbuthnot for an assessment of yet another scheme. In the summer of 1732, when Desaguliers was away in Holland, Chandos was approached by a Dr. Hardesway of Dover, who claimed to have invented a method to prevent the rotting of wood caused by worms. With Arbuthnot's approval, the Duke set up Hardesway and his crucibles in rooms in Cavendish Square. As the smoke and odors from the trials wafted across the square, Chandos was entranced with the possibilities for such a composition. As he told Arbuthnot, not only might it help to stop leaks in water pipes already under the strain of pressure near reservoirs, but it might likewise prove invaluable on ships' bottoms and, especially, in the dikes of Holland, about which the Duke had visions of a population in dread of a rampaging North Sea.[48]

Plans were therefore set on foot to establish a joint stock in which his longtime associate Thomas Watts was to pay a major role. Yet, Hardesway still sensibly declined to reveal his secret until after the articles of agreement were signed. Chandos had attempted to gain the assistance of a Mr. Neale, probably John Neale, the broker who was concerned in the Mine Adventurers Company with Sir Thomas Mackworth. Neale was unconvinced. The company, which was to be established upon a patent for the invention, consisted of Chandos, Sir Roger Meredith, Arbuthnot, and Ralph Allen. The venture soon became unglued when it was discovered that the Dutch had just as good a preservative at a cheaper price.[49]

[47] Hunt. Lib., Stowe MSS. 57, XXXVI, f. 333, Chandos to Mr. Jones of York Buildings, January 3, 1730/1; f. 334, Chandos to Jones, January 4, 1730/1; Desaguliers, *Course of Experimental Philosophy*, II, p. 489. This was William Jones who also testified before the House on the Uxbridge water proposals. *House of Commons Journal*, XX (March 5, 1723) 284.

[48] Hunt. Lib., Stowe MSS. 57, XXXIX, f. 319, Chandos to Dr. Hardesway, August 4, 1732; XL, fols. 80–1, Chandos to Mr. Wilkieson (his Amsterdam Broker), September 22, 1732; XL, f. 128, Chandos to Arbuthnot, October 6, 1732. See Baker and Baker, *Life and Circumstances of James Brydges*, pp. 236, 354–6. Note that a patent for such a process was held by Benjamin Robinson and Francis Hauksbee in 1728. PRO, SP (Dom.) 36/5/115–120, report on petition, February 8, 1727/8.

[49] Hunt. Lib., Stowe MSS. 57, XXXIX, fols. 347–9, Chandos to Watts, August 14, 1732; XL, fols. 57–8, Chandos to Mr. Neale, September 12, 1732. See Baker and Baker, *Life and Circumstances of James Brydges*, pp. 353–6; and Charles de Labelye, "A Description

Like so many of the projects upon which Chandos had fixed his unlucky gaze, the prospects of the York Buildings Company in Scotland were to prove the dangers of enthusiasm. In some measure, the awareness of how easily investors could be misled was the basis for the urgency in Chandos's requests to "Gun-Jones" to evaluate the sample of alum and ores that were extracted from Scotland.[50] He arranged that engineers of his own acquaintance should examine the mines. Captain Samuel Robinson was to view the works in Scotland, and Chandos also took to taking advice from Sir Thomas Mackworth on the best means of draining the mines.[51] But equally important in the failure of many of their ventures was the obvious fact of continuous disagreements among the partners, landowners, and lessees.[52] By 1731, a new dispute had erupted between York Buildings and the Duke of Argyle over rights to some mines.[53]

In the end, one cannot say the mines in the north were unworkable; far from it. Between the York Buildings Company ventures at Swinhart and Kulnakele and the Scotch Mines at Leadhills great effort was made by investors like Thomas and William Watts and inventors like Lund to make the projects pay. Lund built furnaces throughout the highlands, at Kulnakele and Inverness, while sending glowing reports to Samuel Horsey in London.[54]

Chandos could not be satisfied with constant assurances of success; he

of the Carriages mde use of by Ralph Allen, Esq; to carry Stone from his Quarries," in Desaguliers, *Course of Experimental Philosophy*, II, pp. 283ff. For the connections between Ralph Allen and Chandos, see R. S. Neale, *Bath 1680–1850. A Social History OR A Valley of Pleasure, Yet a Sink of Iniquity* (London, 1981), Chapter 5. The John Neale with whom Chandos was involved was very likely the instrument maker, and friend of William Whiston, who obtained a patent in 1745 for a "quadrantal planetary machine" and lectured on globes and the longitude. He appeared in the bankruptcy list in *The Universal Magazine* in September 1758 as "watchmaker, broker, and chapman" of Leadenhall-Street near the Exchange. See also, E. G. R. Taylor, *The Mathematical Practitioners of Hanoverian England* (Cambridge, 1966), p. 214; and John Neale, *The Description of the Planetary Machine, for which His Majesty has granted his Royal Patent* (London, 1745).

[50] Hunt. Lib., Stowe MSS. 57, XXXVI, f. 309, Chandos to Mr. Jones of York Buildings, December 23, 1730; f. 333, Chandos to Jones, January 3, 1730/1; f. 334, Chandos to Jones, January 4, 1730/1.

[51] Hunt. Lib., Stowe MSS. 57, XXXV, fols. 140–1, Chandos to Mr. Watts, July 7, 1730; f. 141, Chandos to Mr. Robinson, July 8, 1730; f. 161, Chandos to Horsey, July 16, 1730; f. 166, Chandos to James Marye, July 18, 1730; f. 182, Chandos to Horsey, July 25, 1730.

[52] Scottish Record Office, GD 248/167/1/2, Samuel Horsey to Sir James Grant, February 18, 1729–30; GD 248/167/1/5, Samuel Horsey to Sir James Grant, April 5, 1730; GD 248/47/3/6, Samuel Horsey to James Grant, Castle Grant, November 20, 1731.

[53] Hunt. Lib., Stowe MSS. 57, XXXVII. f. 18, Chandos to Watts, January 20, 1730/1.

[54] Hunt. Lib., Stowe MSS. 57, XXXV, f.133, Chandos to Horsey, July 5, 1730; f. 225, Chandos to Horsey, August 28, 1730.

Figure 16. Desaguliers, tortured by gout and disappointment, shortly before his death in 1744. By permission of The Huntington Library, San Marino, California.

had heard them all before. The innumerable ventures of the early eighteenth century had revealed one unfortunate truth. For mine owners, entrepreneurs, and for the engineers and philosophers they employed, the success of their undertakings was often an illusion. But philosophers were also drawn to projectors as flies to the flame. It was often a fatal attraction. Financiers like Chandos or even Thomas Watts could, and often did, transfer their investments from one company to another as prospects changed. For the large investors the risk was as much as the nerve would allow. There was always the opportunity to blow up a bubble and escape before the carnage. The crisis of the South Sea Bubble, and the slightly less catastrophic circumstances of York Buildings or Mine Adventurers, provided ample evidence as far as Desaguliers was concerned of how greed had too frequently overcome the potential of natural philosophers. Here was the root of the bitterness Desaguliers expressed shortly before his death about "Persons of Fortune, who being covetous, but unwilling to own it, full of the Notion of the difference between Theory and Practice (or excus'd by it) have hearken'd to the most ignorant Pretenders, who have call'd themselves *Men of Practice*, in order to save Money."[55] In the contradictions of mechanization and modernity, there were too many disappointments.

[55] Desaguliers, *Course of Experimental Philosophy*, II, p. 416.

Conclusion

The transformation of Britain was to depend on the unification of opportunity with invention. This was the view of Desaguliers, Henry Beighton, and more inventive geniuses in the world of finance like Billingsley or Thomas Watts. For respectable philosophers like Desaguliers, and perhaps as well for many others whose lives in the provinces have left only the faintest trace, the transition into a technological future depended on the propagation of the true principles of natural philosophy. The rhetoric and practice of scientific lecturers in the early eighteenth century were crucial to the legitimation of improvement. The experimentalism of the public lecturers, whatever its foundation as a method of scientific investigation, was to be the symbol of their claims to expertise. Newtonianism was essential; it was the vehicle that lent them legitimacy.

Legitimation – effectively a process of attaining public approval – gave force to the meaning of experiment. It was one reason replication and verification became so crucial to the award of authority. In one sense, this was rather like Thomas Sprat's claim that Christ's miracles were "Divine Experiments of his Godhead."[1] Witnessing might provide a broader sanction than private vision. In this sense, seeing truly was believing. This formalism of the confirmation of Christian doctrine became part of the foundation of public knowledge. Clearly, erudition alone could not ensure salvation; by the mideighteenth century, accessibility was all.[2]

[1] Thomas Sprat, *The History of the Royal Society of London, For the Improving of Natural Knowledge* (London, 1667), p. 352. See also Peter Dear, "Miracles, Experiments, and the Ordinary Course of Nature," *Isis*, LXXXI (December 1990), 664–6.

[2] This may have been the most significant legacy of the humanists. See Barbara Shapiro, "Early Modern Intellectual Life: Humanism, Religion and Science in Seventeenth Century England," *History of Science* (March 1991), 53.

Because knowledge was to circulate like currency, then the authenticity of any event or fact might not be challenged without some social cost. This issue was especially critical to the apologists of the fledgling Royal Society; while unresolved, it became even more essential to the method of producing and repeating experiments. The means of the production of scientific facts by way of experiment had to be public in order for it to be socially legitimate – the more accessible, the more the credibility.

The issue here is the distinction between what is seen and what is read – or, in some cases, said. If I might adopt a notion from Michel Foucault, what became increasingly important in the early modern world was observation, credibility, and accessibility.[3] It was in this association that there emerged the kind of reliability that was fundamental to a discursive natural philosophy. There is some irony in suggesting that the very public Newton and others so actively disparaged as vulgar would turn out to be so important to a lasting faith in natural philosophy. Experimental philosophy soon *depended* on public acceptance.

But I also wish to suggest a continuity – not necessarily a novel one, but an important one nonetheless – between Thomas Sprat's intention that the early Royal Society be "free from all suspicion of men, or private interest," to Bishop Richard Cumberland's pantheon to the public good a decade later, to the promotion of public natural philosophy in the eighteenth century. This was the foundation of the unity of intellect and industry. This became the business of the apostles of Newton.

Social authority would increasingly be founded on the doctrine of utility. The community of experimenters, the instrument makers, and self-styled engineers with their varying degrees of dependence on Newton's principles, and the devotees of the public lectures, constructed a broad bottom for natural philosophy. This did not, as some feared it might, necessitate a world turned upside down. A revolution sensed by Swift and his friends was far more subtle. The rhetoric of confident experiment made seeming virtue of the renovation of social and intellectual relations that had once seemed inviolable, steeped as they were in the inertia of tradition or political might. Thus, the adoption of the machine, like the cultivation of Enlightenment, constituted an increasingly political act – largely unconscious even to those most responsible for the surge of industrialism in the eighteenth century.[4] But it would be a mistake to

[3] Michel Foucault, *The Order of Things. An Archaeology of the Human Sciences* (London, 1970), pp. 39–41. See also Gary Gutting, *Michel Foucault's Archaeology of Scientific Reason* (Cambridge, 1991), pp. 144ff.

[4] Cf. Michael W. McCahill, "Peers, Patronage, and the Industrial Revolution, 1760–1800," *Journal of British Studies*, XVI (Fall 1976), 84–107; Roy Porter, "Science, Provincial

assume this transformation to be a mere mechanical one determined, essentially, by the power machines could be forced to generate. The true power of the machine lies in its acceptance. And this acceptance is justified by the view that nature can, and should, be tamed.

To see the world this way is obviously to admit of a culture where power is determined less by custom or legacy than by the evidence of experiment and the command of nature. This seems to me to be the most crucial consequence of the variety of factors in the eighteenth century that transferred authority from theological disputation to natural law. The quest for authority was, therefore, both a social issue and a philosophical one – of recognizing that natural theology had once deflected the focus of experiment and led far too often to disputes whose social impact was diffused through a superficial veil of metaphysics.

Providence

At the beginning of the eighteenth century, the Boyle Lectures drifted into all too troublesome tendencies very far afield from the nettle in Whiston's garden. Despite the efforts to slay the deists, metaphysical and scriptural obscurites ultimately gave leave for freethinkers like Toland to voice "unanswerable doubts and difficulties, which, over their cups or their coffee, they pretend to have against Christianity."[5] Neither Newton's "General Scholium" nor Leibniz's *Theodicy* were likely to resolve this problem. By 1748, Colin MacLaurin regretted Newton's expression about the existence of a deity that had left him exposed to criticism and misinterpretation. MacLaurin pleaded that "There is no need of nice or subtle reasonings in this matter: a manifest contrivance immediately suggests a contriver. It strikes us like a sensation: and artful reasonings against it may puzzle us, but it is without shaking our belief."[6]

But Newton, in 1713, had also concluded: "And thus much concerning God, to discourse of whom from the appearances of things does certainly

Culture and Public Opinion in Enlightenment England," *British Journal for Eighteenth-Century Studies*, III (1980), 28–31; and Margaret C. Jacob, *The Cultural Meaning of the Scientific Revolution* (Philadelphia, 1987), pp. 163–9. For the metaphysical implications of mechanization, see Otto Mayr, *Authority, Liberty & Automatic Machinery in Early Modern Europe* (Baltimore and London), pp. 81–101, esp. Chap. 8.

5 Robert South, *Sermons*, I, pp. 130–1, quoted in John Spurr, "'Rational Religion' in Restoration England," *Journal of the History of Ideas*, XLIX (Oct.–Dec. 1988), 580. For the problem of rhetorical authority, see Thomas F. Gieryn, "Distancing Science from Religion in Seventeenth-Century England," *Isis*, LXXIX (December 1988), 592.

6 Colin MacLaurin, *An Account of Sir Isaac Newton's Philosophical Discoveries* (London, 1748), p. 381.

belong to natural philosophy."[7] This addition to the "General Scholium" was, ironically perhaps, a statement of some consequence for those struggling with public knowledge. Newton had elevated the rhetoric of experimentalism into a forceful epistemology upon which spiritual certainty might be constructed. This was a crucial point. In one neat stroke, he had assaulted the Restoration opponents of the sensory epistemology *and* asserted that religious enthusiasm might be rectified by appeal to the evidence of experiment. From Sprat to Boyle, no proponent of the mechanical philosophy had made the authority of the experimentalist seem so compelling. Newton's disciples took this one step further. They understood that experimental evidence made public was the first step in ensuring that utility was more convincing than the pestilence of enthusiasm. The Newtonian assertion of providence made legitimate the conquest of nature.

Providence and epistemology provided the keys to the new philosophy. The search for knowledge, the achievement of new methods of establishing the structure and laws of the natural world, and the possibility of improvement of creation could be derived from a notion of God's providence, which permitted an activist role for Christian men.[8] This was the legacy of the seventeenth century that was appropriated in a brilliant fashion by Newton and his followers. Such a providence was the soul of Newton's philosophy. Active powers in a dynamic universe demanded a providentialism evident in the structure and in the laws of the universe.[9] Moreover (to return to Thomas Sprat's notion of the importance of philosophers in the improvement of the world), such a view also emerged as a kind of economic providentialism that legitimized the mechanical projects in which natural philosophers were to have a major role. In the aftermath of the Great Fire of 1666, which delayed the completion of the *History*, Sprat made just such an argument to induce industrial activity.[10]

As it turned out, it proved impossible for Newton's acolytes to avoid answering the hostility of those who felt their world crumbling under a

[7] "Et haec de Deo; de quo utique ex Phaenomenis differe, ad Philosophiam Experimentalem pertinent." Newton, *Philosophiae Naturalis Principia Mathematica*, second edition (London, 1713), p. 483.

[8] Cf. Roy Porter, "Creation and Credence: The Career of Theories of the Earth in Britain, 1660–1820," in Barry Barnes and Steven Shapin, eds., *Natural Order. Historical Studies of Scientific Culture* (Beverley Hills and London, 1979), pp. 105–6.

[9] Cf. J. E. McGuire, "Force, Active Principles, and Newton's Invisible Realm," *Ambix*, XV (October 1968), 154–208; and, P. M. Heimann and J. E. McGuire, "Newtonian Forces and Lockean Powers: Concepts of Matter in Eighteenth-Century Thought," *Historical Studies in the Physical Sciences*, III (1971), 233–306.

[10] Sprat, *History*, pp. 122–3.

variety of political and economic pressures. Newtonian natural philosophy not only transformed the intellectual atmosphere of what has been called the "long eighteenth century."[11] It was itself altered by forces that, superficially, had little evidently to do with the physics of the universe. Yet, it is naive to assume Newton's natural philosophy could merely exist on the fringes of the polity that had given it birth. It was drawn back to the center of controversy over natural philosophy that Newton had not provoked but to which his physical principles spoke profoundly. Therefore, there evolved a Newtonianism as multivaried and as diverse as the interests of his apostles and their world cared to make it.

Had Newton's philosophy merely been an intellectual curiosity for mathematicians, this would have mattered little. But the natural phenomena examined by philosophers were not merely objects or effects; they were part of a grand design, the operation of which was revealed by the discovery of natural laws. In the emphasis on the providential in natural phenomena, and in the explanations of them, there existed a volatile issue that went to the very heart of the Protestant existence: to what extent was the Christian faith of God's role in creating and governing the universe actually undermined by the new philosophy, and particularly by that of Newton and his aggressive followers? It was an alarming proposition that the evidence derived from the practice of experimental philosophy might undermine the purity of the true Scripture. And Newton's effort to evade public demonstrations, and therefore controversy among the vulgar, was hopelessly misdirected. Undoubtedly, there was some protection in preserving the exclusive commonwealth of learning that reached from the Universities to the Royal Society.[12] Yet, in the eighteenth century, the Newtonian philosophy came to have such a public presence that it could hardly escape close scrutiny however much Newton may have wished otherwise. This partly explains Newton's later aversion to those of his converts like William Whiston or Nicolas Fatio de Duillier who were unable to restrain their own attractions for controversy.

This ultimate transformation away from the deep thickets of natural theology came about, ironically perhaps, first in the acknowledgment of the providence of nature. Providence was as ubiquitous a notion as it was frequent in Boyle's apology for natural philosophy and its usefulness.[13]

[11] Roy Porter, *Mind Forg'd Manacles. A History of Madness in England from the Restoration to the Regency* (Cambridge, Mass., 1987), p. 19.

[12] See Robert Charles Iliffe, "'The Idols of the Temple': Isaac Newton and the Private Life of Anti-Idolatry" (Ph.D. thesis, University of Cambridge, 1989), pp. 153, 252.

[13] Robert Boyle, *The Excellency of Theology* (London, 1674), p. 127; and Boyle, *The Christian Virtuoso* (London, 1690), p. 26.

It had also been central to Glanvill's defense in 1665 of his fellow philosophers so that "Solomon's House in the NEW ATLANTIS was a Prophetick Scheam of the Royal Society."[14] But providence was a complex notion locked in the history of political symbolism and the metaphysics of natural laws. By the early eighteenth century, the providentialism out of which was formed the canon of the Protestant Cause had taken a shape that linked politics with natural philosophy. Such a belief was the foundation of Newton's view of religion, of his defense of the Protestant Succession, and of the metaphysics behind his science. Within this nexus of experimentalism, nature's laws, and assertions of a Protestant destiny by Newton's disciples lay the seeds of a popular acceptance of a new Reformation that would be revealed in human history and in the uses of God's Providence of nature. What was needed was a social and economic framework that would establish a means to open the door to the utilization of the natural world that landed privilege had once kept closely guarded.

Improvement

The emergence in the late seventeenth century of joint-stock companies provided an atmosphere of opportunity ripe for the exploitation both of nature and of the gullible. This was erected on the long-established notion that the further improvement of nature was the proper mission of landowners in search of wealth as well as of survival. In 1677, Andrew Yarranton had made the argument in *England's Improvement* for the development of agriculture, the dredging of rivers, as well as for the support of industry. The same sentiments were to be promoted by Nehemiah Grew in Anne's reign and formed the ground tilled frequently by the fashionable writers on estates and their improvement throughout the entire spectrum from enclosure to the newest water engines.[15] Such an

[14] Joseph Glanvill, *Scepsis scientifica: Or, Confest Ignorance, the way to Science; In an Essay of the Vanity of Dogmatizing, And Confident Opinion. With a Reply to the Exceptions of the learned Thomas Albius* (London, 1665), "Dedication."

[15] Andrew Yarranton, *England's Improvement By Sea and Land* (London, 1677), pp. 40, 53, 64–6, 156ff.; Hunt. Lib., MS. HM 1264, Nehemiah Grew, "The Means of a Most Ample Encrease of the Wealth & Strength of England," pp. 87–8. For example, see also Stephen Switzer, *An Introduction To a General System of Hydrostaticks and Hydraulicks, Philosophical and Practical.... Containing in General A Physico-mechanical Enquiry into the Original and Rise of Springs, and of all the Hypotheses relating thereto; as also the Principles of Water-works,... Deduc'd from the Theory of Archimedes, Gallileo, Torricelli, Boyle, Wallis, Plot, Hook, Marriotte, Desaguliers, Derham, Hawksbee, and others* (London, 1729), esp. pp. 402ff.; Batty Langley, *A Sure Method of Improving Estates* (London, 1728), p. 62; Langley, *The City and Country Builder's, and Workman's Treasury of Designs* (London, 1740), "Advertisement"; Edward Laurence, *The Duty of a Steward to his Lord* (London, 1727), p. 49.

atmosphere permit projectors and inventors – those great improvers of providence – to find their way to monopolies and patents that the crown liberally granted. The employment of surveyors and engineers was one minor reflection of the opportunities glorified by John Houghton in the 1680s and later by Richard Bradley.[16] Houghton and Bradley, amongst countless others, operated on the fringes of the improvements and investments that legitimized the emergent money economy of the late seventeenth and early eighteenth centuries. An incipient middle class embraced the new possibilities in the nexus of money and land; it did not necessarily create them.

The survival of Baconianism in the eighteenth century owed more to the optimism and opportunism of the landed gentry than has generally been appreciated. The existence of a large group of technological adepts in their employ in the early eighteenth century cannot be dismissed. Nor can it be reduced to a pale reflection of what some may perceive as the idolatry of the bourgeosie in our present century.[17] In fact, the landed classes, out of a necessity magnified by war and high taxes, had already staked out practical territory. Thus, the efforts of natural philosophers to improve on available technological opportunities required no strategic conversion to Baconianism; the constant praise of the Lord Verulam is evidence of his stature even amongst the adherents of Newton.[18]

There were many landed families whose fortunes were founded on the ballast of resources nature provided and man exploited. This was as true of the Lowthers' deepening coal shafts under the Irish Sea, as it was of the seams of the Wortley–Monatagus on Tyneside, or the mines of Richard Beaumont of Whitley and Grizzlehurst in Yorkshire.[19] It ought to be apparent that such large landowners were hardly immune to the adventures of those London companies that took an active interest in the development of mineral resources. It does not require crude economic determinism to explain simple greed or desperation.[20] Land and commerce were not so remote as later to demand the coming of the bourgeoisie as

16 John Houghton, *A Collection of Letters for the Improvement of Husbandry & Trade* (London, 1681–3); and R. Bradley, ed., *Husbandry and trade improv'd: being a collection of many valuable materials* (London, 1728).

17 Cf. Michael Fores, "Newton on a Horse: A Critique of the Historiographies of 'Technology' and 'Modernity,' " *History of Science*, XXIII (1985), 370–1.

18 See, for example, Henry Pemberton, *A View of Sir Isaac Newton's Philosophy* (London, 1728), pp. 5–12; and Colin MacLaurin, *An Account of Sir Isaac Newton's Philosophical Discoveries* (London, 1748), pp. xviii, 56–63.

19 Peter Roebuck, *Yorkshire Baronets, 1640–1760. Families, Estates and Fortunes* (Oxford, 1980), pp. 38–40, 146–7, 324–5.

20 Cf., B. Hessen, *The Social and Economic Roots of Newton's "Principia,"* ed. Robert S. Cohen (New York, 1971).

the saviors of British innovation. Even in our age of born-again capitalism, the shopkeepers have never held the secrets of salvation.

Prometheus and Plutus were two faces of a very old, and perplexing, problem. Long before the tide of mechanization at the end of the eighteenth century, magnates like James Lowther and the Duke of Chandos perched above the hives of the London adventurers in Exchange Alley watching carefully for the slightest hint of advantage and opportunity. Lowther's experiments with early steam engines, Sir Richard Newdigate's arrangements with Henry Beighton, and Chandos's anxious demands on Hans Sloane for assayers reveal that the world of patents and projects was familiar territory for wealthy landowners.[21] And it was frequently to the Fellows of the Royal Society that a potential investor might look to evaluate the unique claims of one patentee or another. Sloane, Arbuthnot, and Desaguliers were continually harassed for advice. And with good reason.

The numerous abortive ventures were the factual confirmation of all of the suspicions that had justifiably surfaced during the South Sea Bubble. We should not assume that these were new fears; they were not.[22] But they did confirm a view that Desaguliers held dear – that natural philosophers were the best guide through the shoals of disappointment. During the Restoration, this had been anticipated in R. D'Acres' *The Art of Water Drawing* in which it was argued that those most familiar with a new technology were best equipped to reveal "the delusions and fallacies of water-machines." In an analogy that rings true of an era inundated with Longitudinarians, it was hoped to establish "experiments, whether reall or fallacious, that they may be as Sea-marks to keep those that come after from Ship-wrack."[23] Early warnings are often ignored, although these were later echoed by Desaguliers and those of his audience who took up the cause. From Robert Boyle to the popular lecturers, the limitations of vulgar knowledge were repeatedly asserted.[24] Yet the ref-

[21] B.L., Sloane MSS 4053, f. 316, Chandos to Sloane, November 10, 1734. See the large number of patents granted for procedures to improve the refining and working of ores noted in the Hardwicke papers. These are exceeded only by the number of patents for engines to raise water. B.L., Add. MSS. 36123, Warrants for Patents, 1720–8.

[22] Joan Thirsk, *Economic Policy and Projects. The Development of Consumer Society in Early Modern England* (Oxford, 1978), Chaps. III–IV.

[23] R. D'Acres, *The Art of Water-Drawing* (1659), ed. Rhys Jenkins (Cambridge, 1930), Preface; quoted in G. Hollister-Short, "Leads and Lags in late Seventeenth Century English Technology," *History of Technology*, V (1976), 165. D'Acres was likely a pseudonym for Robert Thornton of Brockall; see Jenkins, "The Art of Water Drawing," *Transactions of the Newcomen Society*, X (1929–30), 35.

[24] Boyle, *Some Considerations Touching the Usefulness of Experimental Natural Philosophy*, second edition (Oxford, 1664), p. 267.

ormation of vulgar opinion would lie, it was hoped, in the production of public philosophy.

Such early fears were frequently forgotten in the scarring and rape of the early industrial landscape. But enthusiasm could not disguise experience. D'Acres' views were obviously corroborated by the grand projects and delusions of late Stuart and early Hanoverian England. There proved considerable truth in his notion that it was not easy for a man to recognize "the tricks of Tumblers, matters of strangeness without worthynesse, neither can he know Sophisters nor deluding cheating Imposters (for such he shall often meet with)."[25] Warnings against perpetual motion men may seem comical, but Desaguliers more than once had to respond to such claims. So, too, evidently, was s'Gravesande to inspect such a claim for the Land-grave of Hesse in the summer of 1721.[26] The only things perpetual about motion were the incessant claims of the philosophically ignorant and the sport they provided for satirists.

Neither enthusiasm nor philosophers were confined by national boundaries. Amsterdam, Harburg, Leige, and London all suffered from the same dementia. Not incidentally, if natural philosophers and engineers – which at that time was a distinction far less rigorous than subsequently adopted – were the best preservative against technical and commercial disaster, then their claim to reduce uncertainty would likely fall on receptive ears. The early eighteenth century was a time transfixed by claims of innovation, and it was also one of rapid technological development. This was most especially obvious with the swift adoption of steam technology in mining and in water supply.[27]

A superficial view of the eighteenth century might suggest few links between science and technology. Yet so many of the disciples of Newton were engaged in projects that the role of the natural philosopher underwent a profound metamorphosis. The seams that connected science and industry ran deep into the heart of the community of natural philosophers and entrepreneurs, whether landed magnates or mine owners. Such re-

[25] D'Acres, *The Art of Water-Drawing*, Preface.

[26] At this point, the Dutchman may have come into contact with Fischer von Erlach, who was involved with the Liege engine and who was already in touch with Desaguliers in London. On the disputed role of s'Gravesande and Fischer von Erlach in the eventual construction of a Newcomen engine in the Hapsburg Empire, see Mikulas Teich, "The Early History of the Newcomen Engine at Nova Bana (Konigsberg): Isaac Potter's Negotiations with the Hofkammer and the Signing of the Agreement of 19 August 1721," *East-Central Europe*, XI (1982), 24–37. s'Gravesande's initiation into steam technology probably took place in London under the influence of Desaguliers. See Desaguliers, *Course of Experimental Philosophy*, II, pp. 484–5.

[27] See John Kanefsky and John Robey, "Steam Engines in 18th-Century Britain: A Quantitative Assessment," *Technology and Culture*, XXI (April 1980), 179–82.

lations had an obvious consequence to those who faced the great expense of the erection of the latest fire engines. The evidence suggests that the breadth of the community of interest in natural philosophy was linked closely to the desire for power over nature – the very aim vigorously justified in the seventeenth century by Bacon, Boyle, Glanvill, and Sprat, and by the natural theologians who followed them.

Public knowledge

Attracted by the market in natural philosophy, the disciples of Newton found their way into the industrial and commercial transformation of Britain. By trading on their expertise of the natural world, they were able to find patronage and opportunity among the adventuresome and the desperate. Polite society was not to be the final arbiter of popular science. As advisors to the wealthy, philosophers were also part of the transition from the intimate world of private patronage of the seventeenth century into patronage and employment by the public companies of the eighteenth century. Between the private and the public lay the balance of the prestige originally of the Royal Society, but more recently of natural philosophy in general.[28]

Crucial to this transition were the itinerant lecturers and the coffeehouse philosophers. Natural philosophy soon supplied theater and self-improvement. Desaguliers and his competitors delivered both – often through a wealth of increasingly portable apparatus designed precisely for that purpose. The key lay in the *presence* of the experiment; the public lecturers could achieve the dramatic success that Newton's earlier rhetorical flourishes only imagined. In the breadth of their travels, they brought to public discourse dramatic demonstrations of the force of nature's laws. Rhetoric had thus been transformed; nature had come to appear the arbiter of true philosophy.[29]

Above all, it was the *practice of experimental philosophy* that overwhelmed the power of the ideology that existed in the minds of Newton's early acolytes, whether Tory or Whig, Court or Country. It was the market for philosophy that forced their ideological committment to adapt to rapidly changing conditions. Like other promoters, the shackle of

[28] Peter Dear, "'Totius in verba.' Rhetoric and authority in the Early Royal Society," *Isis*, LXXVI (1985), 156; and J. V. Golinski, "A Noble Spectacle. Phosphorous and the Public Cultures of Science in the Early Royal Society," *Isis*, LXXX (1989), 11–15.

[29] See Alan G. Gross, *The Rhetoric of Science* (Cambridge, Mass., 1990), pp. 119–28, esp. 122–5; and Bruno Latour, *Science in Action. How to Follow Scientists and Engineers Through Society* (Cambridge, Mass., 1987), pp. 94–100.

corruption hobbled the claims of the Newtonian entrepreneurs. So Desaguliers had to represent natural philosophy – much as Sprat also had done – as a means of improving the human lot and protecting it from the piracy of ill-meaning projectors. In the early eighteenth century, John Grundy, senior, self-styled land surveyor and teacher of mathematics, in his introduction to the Adventurers of the Chester Navigation, argued for the significance of natural philosophy. His was a remarkable manifesto that required every engineer, in dealing with the laws of motion, of friction and falling bodies, "to understand Natural Philosophy in order to make his Enquiries just."[30] Here was the language of public science. It would also be the rhetoric that linked theory with practice, thereby, providing the moral foundation that ultimately made industrialism possible.

On the edge of the Industrial Revolution, the distinctions of private and public were incapable of so fine a definition that restricted natural philosophers to clients of the great landed patrons and aristocrats of Hanoverian Britain. On the contrary. This world of public philosophy had made virtue of experimentalism, but even more forcefully of opportunity and utility. The notion of public knowledge eroded the private paternalism of the client economy. Natural philosophy had become part of the social transformation of a prescient industrialism of which the landed soon proved enthusiastic supporters. It was in their own interest. It was precisely this social reality that gave credence to Desaguliers' caution "to those that have any Concerns in Mines, Water-works, Mills, or other Manufactures; that they might not be impos'd upon by Engine-makers, that pretend to (and often fancy they can) by some new invented Engine out-do all others."[31] While projectors misrepresented, natural philosophers might mark a surer path.

The suspicions of the projector disguised an essential contradiction in the dialectic of mechanization. The authority of experimentalism promoted by Hartlib and Boyle had been affirmed in the public demonstrations of Newton's apostles. Yet those very disciples depended on the farce of the bubbles to convince their audience to follow philosophers through the chaos. In a sense, the tension between scientist and technologist was founded on this fundamental conflict. When projects were proposed, then natural philosophers discovered that Thomas Sprat had not been entirely wide of the mark; defense of religion was one thing,

[30] Spalding Gentlemen's Society MSS., Drawer 2, John Grundy, *Chester Navigation consider'd* (n.d., ca. 1736).

[31] Desaguliers, *A Course of Experimental Philosophy*, second edition (London, 1745), I, p. 70, esp. 138.

but the promotion of technology quite another. The protection of civic virtue against the self-deceived or against Desaguliers' "Projectors [who] have been altogether knaves" made natural philosophy a commodity in a seller's market.[32] The industrial transformation at the end of the century was the epilogue to a long intellectual revolution. Bacon's New Atlantis had proved to be a Newtonian legacy.

[32] Desaguliers, *Course of Experimental Philosophy*, I (1744), 138.

Bibliography

1. Manuscript sources

British Library, London

Sloane MSS.	
1968, 3399, 3984, 4025–58	Hans Sloane Correspondence
Additional MSS.	
4432–36	Royal Society Papers
4478D	Thomas Birch, 1725–6
6116–17	Wake letters
6209	John Ward correspondence
6406	"Diploma Societatis Harburgensis"
22675	Cases and Opinions of Sir Thomas Pengelly
24197	Bishop Lloyd correspondence
30094	Benjamin Wilson correspondence
32548	Roger North papers
32702	Newcastle papers
36030, 36123, 36142, 36276A	Hardwicke papers
43741	Elizabeth Beighton correspondence
47009A–47031	Egmont papers, Percivale correspondence
61605, 61640	Case Billingsley correspondence
Loans	
Loan16	Mines Royal and the Mineral and Battery Works
Loan 29/2	New River Company

Lansdowne MSS.

841, 1056	Desaguliers papers, 1723–5.
1024	White Kennett's journal

House of Lords Record Office

Petitions, Box 195B, n.s. X, 3123.	Longitude Bill, 1714

Guildhall Library, London

173/1	Water Companies, 1719–1819
205	Diary of Stephen Monteage
537	London Bridge Water Works
5445, 5449, 5449a	Brewers' Company
11316	Land tax Assessment Books
11931, Vols. I–III.	Sun Insurance Office. General Committee Minutes
12019C, 12023	Ledgers and Transfers
Index to Freemen	Merchant Taylors' Company

Public Record Office, London

Admiralty Papers	
ADM 6	Warrant Books
ADM 7	Chatham, Salary and Pension
ADM 8	Disposal of Ships and Vessels in Sea Pay
ADM 33	Pay Book
ADM 36	Muster Book
ADM 42	Chatham Yard Books. Ordinary
Bankruptcy Commission	
B4/2–6	Docket Books
Chancery	
C11	Bills and Answers
C38/352/Part I	Order Books (1728)
C38/380/1726 A–C.	Queries
Privy Council Minutes	
PC1/4/106	Wood's patent and disputes
PC1/4	Scotch Mines Company
State Papers (Domestic)	
35 George I.	
36 George II.	
MPF 259	Drayton Water Supply, 1721

Treasury Papers	
T1/255	Treasury Board Papers
T1/258	York Buildings Company
T70/7–116	African Company
Forfeited Estates Commission	
FEC1/Y.1	1721. "Interrogatories"
Probate	
PROB. 3/48/40	Case Billingsley

Corporation of London Record Office

Court of Aldermen	Brokers Admissions, Vol. I.
Court of Aldermen	Minutes, Vol. 141 (1736–7).

Lambeth Palace Library

MS. 1770	Wake Diary
Misc. MSS. 1029	
Edmund Gibson MSS. 1742	

Royal Astronomical Society, London

Additional MSS.	
4	Byrdall letters
88	*Astronomical Scrapbook*

Royal Society, London

Certificates, I–II	1732–56
Classified Papers, III–XVIII	1660–1740
Council Minutes, II–III	1682–1747
Journal Book (C), X–XVI	1702–39
Letter Book (C), XII–XXI	
Letter Books, A, B.3, D.1–2, P.1	
Letters and Papers. Decade I, vols. I–V	1741–4
Register Book (C), XII–XXI	

University of London Library

Autograph Letters, 5	Billingsley correspondence

Bodleian Library, Oxford

Rawlinson MSS.

A. 241	Savery patent
D. 871	Triewald, Desaguliers papers
D. 916	Mine Adventurers minutes
J. fol.3	Desaguliers biography
Ballard	
12	Arthur Charlett papers
Gough (Somersetshire)	
7	Chandos papers
Rigaud	
39	Stephen Rigaud notebook

Cambridge University Library

Additional MSS.

3972	Newton on Longitude
3985, 4007(B)	Newton correspondence
Cholmondely (Houghton) MSS.	
1701, 1771, 1772, 2648	Billingsley to Walpole, 1730–6

Trinity College Library, Cambridge

MS. R.4.42	Roger Cotes correspondence
MS. R.4.47	John Craig letters

Carlisle Record Office

Lonsdale MSS.	Lowther correspondence

Leicestershire Record Office

Field-Barker MSS. Conant Papers.	Whiston correspondence

Northumberland County Record Office, Gosforth Park, Newcastle

Society of Antiquaries MSS.

M13/D19	John Horsley, ca. 1726–9.
Cotesworth MSS.	
ZCE 10/5	William Cotesworth to Steam Engine Proprietors
ZDE/10/2	William Cotesworth letter books

Ridley (Blagdon) MSS.

ZR1/11, 23	Articles of Agreement, 1718–5.

North of England Institute of Mining and Mechanical Engineers, Newcastle

Shelf 1	London Lead Company, vols. 4–6 (1711–32)
Shelf 18	Richard Peck's View Book, Grand Allies Agreement, 1726.
Shelf 49	T. E. Forster Collection

Gateshead Public Library, Gateshead.

CK4	Cotesworth MSS.
CP1	Liddell-Cotesworth correspondence

Spalding Gentlemen's Society, Spalding, Lincolnshire

Maurice Johnson MSS.

Drawer 1	Lecture course proposals
Drawer 2	John Grundy, Junior, letters
Loose MSS, vol. II.	Beighton, Grundy letters

Scottish Record Office, Edinburgh

Gifts and Deposits

GD 1/170, vols. I–III	York Buildings Company, Directors' correspondence.
GD 248/135	Sir James Grant *v.* Samuel Horsey and others, 1738.

Private Collection

Stirling of Garden MSS.	James Stirling MSS. and correspondence

Clements Library, University of Michigan, Ann Arbor

Miscellaneous Collection	James Jurin correspondence

Huntington Library, San Marino, California

A Catalogue of the Large and Valuable Library of the most Noble and Learned James Duke of Channdos . . . which will be sold at Auction, by Mr. Cock, At the Great Room in Exeter-Exchange in the Strand, On Thursday, the 12th of March 1746–7, and The Twenty-nine following Evenings (Sundays excepted) [copy of British Library K.r.2a].

HM 1264, Nehemiah Grew, "The Means of a Most Ample Encrease of the Wealth & Strength of England In a few years Humbly Represented to Her Majestie In the 5th Year of Her Reign," (1707–8).

Stowe MSS.

9	"Papers Relating to African Affairs"
13	James Brydges, Observations out of ye Old & New Testament, also out of Lord Bacon's Essays & Fuller's Holy State. [Begun] Jan: 16: 1696/7.
26, 2 vols.	James Brydges, "A Journal of My Daily Actions, 1697–1702"
28	"A Collection of Some Remarkable Papers"
57	James Brydges, Out-Letters
59	"A Book of Strangers" dining at Canons (1721–2)
85	Tradesmen accounts at Canons (1721–3)
329	Catalogue of Library, ca. 1761
Stowe Brydges MSS., Legal, Box 1.	
Stowe Nugent Financial MSS., No. 2	John Knight Accounts, Policies (1719–21)

Spencer Library, University of Kansas, Lawrence, Kansas

MS. 83/3	Arthur Moore correspondence
MS. G.3	Trenchard-Simpson correspondence
MSS. Q8	Accounts of Arthur Onslow

William Andrews Clarke Memorial Library, UCLA, Los Angeles

Isaac Newton, MS. "Out of Cudworth," ca. 1680s, from Cudworth, *The True Intellectual System of the Universe* (1678).

Isaac Newton, MS. "Paradoxical Questions concerning ye morals & actions of Athanasius & his followers," ca. 1690s.

Primary sources

Agricola, G.A., *The Experimental Husbandman and Gardener. Containing a New Method of Improving Estates and Gardens...*, ed. R. Bradley, second edition (London, 1726).

[Anon.,] *The Battle of the Bubbles. Shewing their Several Constitutions, Alliances, Policies, and Wars; From their first Suddain Rise, to their late Speedy Decay. By a Stander-By* (London, 1720).

[Anon.,] *The Broken Stock-Jobbers: Or, A Work for the Bailiffs. A New Farce. As it was lately Acted in Exchange-Alley* (London, 1720).

[Anon.,] *A description of the great burning-glass made by Mr. Villette and his two sons* (London, 1718).

[Anon.,] *A Letter from a Livery-Man of the City of London, to a Member of the Honourable House of Commons* (n.d.) [B.L., Shelf Mark 712.m.1 (74)].

[Anon.,] *A List of the Names of the Subscribers for Raising the Summe of One Million Sterling, as a Fund for Insuring Ships and Merchandise at Sea...* (London, 1718) [B.L., Shelf Mark 8225.a.38].

[Anon.,] "The Opinion of several Eminent Physicians of the Colledge, and others, concerning Sea-Water made Fresh," in *A Collection of Medical Advertisements*, no. 157 [B.L., Shelf Mark C.112. f. 7].

[Anon.,] *Proposals for Subscription to a New Invention for Raising Water, and working several useful Machines* [1724] [B.L., Shelf Mark 816.m.10 (110)].

[Anon.,] *The York-Buildings Dragons* (London, 1726).

[Anon.,] *The York Buildings Dragons... The second edition, Augmented by almost Half, Containing Several more wonderful Discoveries, and concluding with Orlando Furioso's Speech, and Prudentio's Answer* (London, 1726).

[Arbuthnot, John,] *An Essay on the Usefulness of Mathematical Learning, in A Letter from a Gentleman in the City to his Friend at Oxford*, second edition (Oxford, 1721).

The Longitude Examined in *Miscellaneous Works of the Late Dr. Ar-*

buthnot. With an Account of the Author's Life (London, 1770), II, pp. 295–310.

Articles of Agreement Made, Concluded, and Agreed upon the twentieth Day of July, in the Year of our Lord One thousand, seven hundred and twenty... (London, 1725) [B.L., Shelf Mark 522.m.12 (3)].

Atkinson, James, *Epitome of Navigation: or, A Short and Easy Methodical Way to become a Compleat Navigator*, fifth edition (London, 1704).

B., *Further Reasons Most humbly offer'd, for Increasing our Money; putting an immediate Stop to its being Melted Down, and sent Abroad* (London, 1716).

Some Proposals Humbly offer'd to the Parliament of Great Britain, For Easing the Nation of the Old Navy-Debt, (Now the South-Sea-Debt) of Ten Millions Sterling, Principal Money... (London, 1716).

B./B., *Reasons Humbly Submitted to the Consideration of the Honourable House of Commons; against the further lowering of Guineas; And against giving a Liberty to Export our Coin. With some Remarks upon Sir Isaac Newton's Representation of the 21st September last, to the Lords of the Treasury* (London, January 13, 1717–18).

Some Proposals, Humbly Offer'd, For the Increase and Preservation of our Currant-Coin, Particularly Silver, which is now become very scarce, And (if a speedy Remedy be not apply'd) will all be gone in a little time; For the further Payment of the Publick-Debts, So as to give every one their full Due, And instead of laying on any New Tax to accomplish this Work, To make the doing it an Advantage to all concern'd, And an Encouragement to Trade (London, n.d.).

Bentley, Richard, *Eight Sermons Preach'd at the Honourable Robert Boyle's Lecture... The Fifth edition. To which is added a Sermon Preach'd at the Publick-Commencement at Cambridge July V, MDCXCI. when he proceeded Doctor of Divinity* (Cambridge, 1724).

The Folly and Unreasonableness of Atheism Demonstrated from the Advantage and Pleasure of a Religious Life, The Faculties of Humane Souls, the Structure of Animate Bodies, & The Origin and Frame of the World: In Eight Sermons Preached at the Lecture Founded by the Honourable Robert Boyle, Esquire; in the First Year, MDCXCII, fourth edition (London, 1699).

A Bill for Incorporating several Undertakers, for the Better Accommodating the Inhabitants in and about the Cities of London and Westminster, and the Parts adjacent, with Water... in the Name of the Governor and Company of St. Albon's Waters (n.d.) [B.L., Shelf Mark 816.m.14 (32)].

Billingsley, Case, *A Letter to the Commissioners Appointed by the Parliament of Great Britain, for Discovering the Longitude at Sea, (from Case Billingsley) Further to Explain his Late Proposal for that Discovery, by the Sun, Moon, and Stars; and a True Time-keeper* (London, 1714).

Birch, Thomas, ed., *The Works of the Honourable Robert Boyle* (London, 1772).

Bissett, William, *The Modern Fanatick. Part II. Containing What is Necessary to Clear all the Matters of Fact in the First Part; and To Confute what has been Printed in the Pretended Vindication of Dr. Sacheverell, relating to myself*... (London, 1710).

[Booth, John,] *Course of Experimental Phylosophy, Consisting of Five Parts*... [1740] [Spalding Gentlemen's Society, Maurice Johnson MSS., Drawer 1].

Boyle, Robert, *The Christian Virtuoso* (London, 1690).

The Excellency of Theology, Compared with Natural Philosophy (London, 1674).

Some Considerations Touching the Usefulness of Experimental Natural Philosophy: Proposed in a familiar Discourse to a Friend, By Way of Invitation to the Study of it, second edition (Oxford, 1664).

Some Considerations Touching the Usefulness of Experimental Natural Philosophy, second tome (Oxford, 1671).

Bradford, Samuel, *A Sermon Preach'd in the Parish-Church of St. Mary-le-Bow. On Sunday, November 20, 1720* (London, 1720).

Bradley, Richard. *A General Treatise of Husbandry and Gardening; Containing a New System of Vegetation* (London, 1726).

ed., *Husbandry and trade improv'd: being a collection of many valuable materials* (London, 1728).

Browne, Robert. *Methods, Propositions and Problems, For Finding the Longitude; with the Degree and Minute of the equator upon the Meridean. And the Longitude at Sea, by Coelestial Observations only. And also by Watches, Clocks, etc. and to correct them and know their Alterations* (London, 1714).

A system of Theology Revealed from God by the Angels in the British Language herein is disclosed the mysteries of the Holy Scriptures, and the true Fundamentals of Religion, being a Lesson to the whole World. With any Curios and Occult Things in Church, State and Philosophy, All designed for the Public and Universal Good of Mankind, and a general Reformation of the present Times (London, 1729).

[Browne, Robert?] R.B., *Longitude To be found out with A new Invented Instrument, Both By Sea and Land*... *with a better Method for dis-*

covering Longitude, than that lately propos'd by Mr. Whiston and Mr. Ditton (London, 1714).

[Bumpstead, Robert,] *The Humble Petition of Robert Bumpstead, In Behalf of himself and several others. To the House of Commons* (n.d.) [B.L., Shelf Mark 357.b.6 (93)].

Burnet, Thomas, *The Theory of the Earth; Containing an Account of the Original of the Earth, and all the Changes which it hath already undergone, or is to undergo Till the Consummation of all Things*, second edition (London, 1691; reprint edn. with introduction by Basil Willey, London, 1965).

Chambers, Ephraim, *Cyclopaedia: Or, an Universal Dictionary of Arts and Sciences*..., fourth edition, corrected (London, 1741).

Chapman, Richard, *Publick Peace Ascertain'd; with some Cursory Reflections upon Dr. Sacheverel's Two Late Sermons. In a Sermon Preach'd On Tuesday, Nov. 22. 1709* (London, 1709).

Clarke, James, *An Essay, Wherein a Method is Humbly Propos'd for Measuring equal Time with the utmost Exactness; Without the Necessity of being confin'd to Clocks; Watches, or any other Horological Movements; in order to discover the Longitude at Sea* (London, 1714).

Clarke, John, ed., *Sermons... Published from the Author's Manuscript. Vol. I. With a Preface, giving some Account of the Life, Writings, and Character, of the Author: By Benjamin, Lord Bishop of Salisbury*, second edition (London, 1730).

Clarke, Samuel. *A Demonstration of the Being and Attributes of God: More Particularly in Answer to Mr. Hobbs, Spinoza, And their Followers... Being the Substance of Eight Sermons Preach'd at the Cathedral-Church of St. Paul, in the Year 1704. at the Lecture Founded by the Honourable Robert Boyle* (London, 1705).

A Discourse Concerning the Unchangeable Obligations of Natural Religion, and the Truth and Certainty of the Christian Revelation. Being Eight Sermons Preach'd at the Cathedral Church of St. Paul, in the Year 1705 at the Lecture founded by the Honourable Robert Boyle, Esq;, third edition, corrected (London, 1711).

A Letter Concerning the Argument a priori In the Proof of the Being of God... [added to] *A Discourse Concerning the Connexion of the Prophecies in the Old Testament, and the Application of them to Christ. Being an Extract from the Sixth Edition of a Demonstration of the Being and Attributes of God*, second edition (London, 1725).

Cumberland, Richard, *A Treatise of the Laws of Nature. Made English from the Latin by John Maxwell, M.A. Prebendary of Connor, and*

Chaplain to his Excellency the Lord Carteret, Lord Lieutenant of Ireland (London, 1727).

D'Acres, R., *The Art of Water-Drawing* (1659), ed. Rhys Jenkins (Cambridge, 1930).

Defoe, Daniel, *The Complete English Tradesman in Familiar Letters* (London, 1727; repr. New York, 1969).

An Essay on Projects (London, 1697).

The History Of the Principal Discoveries and Improvements, In the Several Arts and Sciences: Particularly the great Branches of Commerce, Navigation, and Plantation, In all Parts of the known World (London, 1727).

de Hatzfeld, Johann Conrad Francis, *The Case of the Learned represented, according to the merit of the ill progress hitherto made in Arts and Sciences, shewing I. The cause of gravity and attraction. II. What Nature is and the effects if is capable of* (London, 1724).

Derham, William. *Physico-Theology: Or, A Demonstration of the Being and Attributes of God, from his Works of Creation. Being the Substance of XVI Sermons Preached in St. Mary le Bow Church, London at the Honble Mr Boyle's Lectures, in the Years 1711 and 1712*, second edition, with additions (London, 1714).

Desaguliers, J. T., *A Course of Experimental Philosophy*, 2 vols. (London, 1734, 1744).

A Course of Experimental Philosophy, second edition, vol. I (London, 1745).

A Course of Mechanical and Experimental Philosophy, Whereby any one, although unskill'd in Mathematical Sciences, may be able to understand, all those Phaenomena of Nature, which have been discovered by Geometrical Principles, or accounted for by Experiments; and Mathematicians may be diverted in seeing those Machines used, and Physical Operations perform'd, concerning which they have read (London, 1725) [Spalding Gentlemen's Society, Maurice Johnson MSS., Drawer 1].

trans., *Fires Improv'd: Being a New Method of Building Chimneys, So as to prevent their Smoaking: in which A Small Fire, shall warm a Room better than a much Larger made the Common Way. With the manner of altering such Chimneys as are already Built, so that they shall perform the same Effects*, written in French, by Monsieur Gauger (London, 1715).

Lectures of Experimental Philosophy. Wherein The Principles of Mechanicks, Hydrostaticks, and Opticks, are Demonstrated and Explained at large, by a great Number of curious Experiments: With a

Description of the Air-Pump, and the several Experiments thereon: of the Condensing-Engine; as also of the different Species of Barometers, Thermometers, and Hygrometers; with several Experiments to prove and explain Sir Isaac Newton's Theory of Light and Colours, as performed in a Course of Mechanical and Experimental Philosophy. To which is added, A Description of Mr. Rowley's Machine, called the Orrery, which represents the Motion of the Moon about the Earth, Venus, and Mercury about the Sun, according to the Copernican System: All carefully Examined and Corrected by Mr. Desaguliers, second edition (London, 1719).

trans., *The Motion of Water, and other Fluids. Being a Treatise of Hydrostaticks, Written originally in French, by the late Monsieur MARRIOTTE, Member of the Royal Academy of Sciences at Paris* (London, 1718).

Physico-Mechanical Lectures. OR, An Account of What is Explain'd and Demonstrated in the Course of Mechanical and Experimental Philosophy... Wherin the Principles of Mechanics, Hydrostatics and Opticks, are demonstrated and explain'd by a great Number of Experiments. Design'd for the Use of all such as have seen, or may see Courses of Experimental Philosophy (London, 1717).

A Sermon Preach'd before the King, at Hampton-Court; on Sunday, Sept. 29, 1717 (London, 1717).

[Desaguliers, J.T.?,] *A Catalogue of the Experiments in Mr. Desaguliers's Course* (dated December 14, 1713), in *A Collection of Advertisements* [B.L., Shelf Mark C112.f.7].

Ditton, Humphry, *General Laws of Matter and Motion* (London, 1705).

An Institution of Fluxions (London, 1706).

The New Law of Fluids (London, 1714).

[Eachard, J.,] *The Grounds & Occasions of the Contempt of the Clergy and Religion Enquired into* (London, 1672).

Evelyn, John, *Navigation and Commerce, Their Original and Progress* (London, 1674).

Public Employment, and An Active Life, With all its Appanages, Such as Fame, Command, Riches, Conversation, &c. Preferr'd to Solitude. in William Upcott, ed., *The Miscellaneous Writings of John Evelyn, Esq. F.R.S.* (London, 1825).

Fatio de Duillier, Nicholas, *Navigation improved: Being chiefly the method for finding the latitude at sea as well as land, by taking any proper altitudes, Together with the time between observations: The Whole made easy by a new Trigonometric Rule...* (London, 1728).

[Gay, John,] "God's Revenge against Punning" (1716), in J. Swift and A. Pope, *Miscellanies*, vol. III (London, 1732).

"A True and Faithful Narrative of What pass'd in London during the general Consternation of all Ranks and Degrees of Mankind . . . ," in Swift and Pope, *Miscellanies*, vol. III (London, 1732).

Glanvill, Joseph, *Essays on Several Subjects in Philosophy and Religion* (London, 1676).

Essays on Several Important Subjects in Philosophy and Religion, ed. Richard H. Popkin (New York and London, 1970).

Philosophia Pia, Or, A Discourse of the Religious Temper, and Tendencies of the Experimental Philosophy Which is posest By the Royal Society (London, 1671).

Plus Ultra: Or, the Progress and Advancement of Knowledge Since the Days of Aristotle. In an Account of some of the most Remarkable Late Improvements of Practical, Useful Learning (London, 1668).

Scepsis scientifica: Or, Confest Ignorance, the way to Science; In an Essay of the Vanity of Dogmatizing, And Confident Opinion. With a Reply to the Exceptions of the learned Thomas Albius (London, 1665).

Scirei tuum nihil est: Or, The Authors Defense of the Vanity of Dogmatizing (London, 1665).

Seasonable Reflections and Discourses In Order to the Conviction, & Cure Of the Scoffing, & Infidelity Of a Degenerate Age (London, 1676).

The Vanity of Dogmatizing: Or Confidence in Opinions Manifest in a Discourse of the Shortness and Uncertainty of our Knowledge, And its Causes: With some Reflexions on Peripateticism; and An Apology for Philosophy (London, 1661).

Grew, Nehemiah, *Cosmologia Sacra: OR A Discourse of the Universe As it is the Creature and Kingdom of God. Chiefly Written, To Demonstrate the Truth and Excellency of the Bible; which contains the Laws of his Kingdom in this Lower World* (London, 1701).

New Experiments and Useful Observations Concerning Sea-Water made Fresh, According to the Patentees Invention, ninth edition, September 29 (London, 1684).

Tractatus de Salis Cathartici Amari in Aquis Ebeshamensibus (London, 1695).

Griffis, Will, *A Short Account of a Course of Mechanical and Experimental Philosophy and Astronomy. In which will be Exhibited the Experiments Necessary for the Explanation of Natural Philosophy in general,*

Mechanics, Hydrostatics, Hydraulics, Pneumatics, Optics, and Astronomy [1748] [Spalding Gentlemen's Society, Maurice Johnson MSS., Drawer 1].

Grundy, John, *Chester Navigation consider'd* (n.d., ca. 1736) [Spalding Gentlemen's Society, Maurice Johnson MSS., Drawer 2].

A Description of the Improvements that have been made in the Barometers by Mr. Charles Orme of Ashby-de-la-zouch in the County of Leicester (1734).

Philosophical and Mathematical Reasons, Humbly offer'd to the Consideration of the Publick; To prove that the Present Works, executed at Chester, to recover and preserve the Navigation of the River Dee Must intirely Destroy the same... (London, 1736) [Spalding Gentlemen's Society, Maurice Johnson MSS., Bound Pamphlets].

Hales, Stephen, *Philosophical Experiments: Containing Useful and Necessary Instructions for Such as undertake Long Voyages at Sea...* (London, 1739).

Vegetable Staticks, ed. M. A. Hoskins (London, 1969).

Harris, John, *Astronomical Dialogues Between and Gentleman and a Lady; Wherein the Doctrine of the Sphere, Uses of the Globes, and the Elements of Astronomy and Geography are Explain'd; In a Pleasant, Easy, and Familiar Way. With a Description of the famous Instrument called the Orrery*, second edition by George Gordon (London, 1729).

"The Atheist's Objections against the Immaterial Nature of God, etc. Refuted," in S. Letsome and J. Nicoll, eds., *A Defense of Natural and Revealed Religion*, Vol. I (London, 1739).

A Collection of Voyages and Travels, Some now Printed from Original Manuscripts, Vol. I (London, 1704).

The Evil and Mischief of a Fiery Spirit. A Sermon Preach'd in the Parish Churches of St. Mildred Bread-Street and St. Matthew Fryday-Street, On February 27, 1709/10 (London, 1710).

Lexicon Technicum, 2 vols. (London, 1704, 1710).

A Sermon preach'd in the Parish-Church of Stroud, near Rochester in Kent, on August the 1st, 1715. Being the Day of Thanksgiving appointed for King George's Happy Accession to the Crown (London, 1715).

The Wickedness of the Pretence of Treason & Rebellion for God's Sake. A Sermon Preach'd on November 5th. 1715. In the Parish-Church of St. Dunstan in the East, in the City of London (London, 1715).

[Harris, John,] *The Picture of a Highflying Clergyman: or, A True State of the Case Between John Harris, D.D. Rector of St. Mildred's Bread Street London; and Charles Humphreys, LL.D. late lecturer of that Parish* (London, 1716).

Hatton, E., *A New View of London* (London, 1708).

Hauksbee, F., *Physico-Mechanical Experiments on Various Subjects. Containing An Account of Several Surprizing Phenomena Touching Light and Electricity, Producible on the Attrition of Bodies. With many other Remarkable Appearances, not before observ'd. Together with the Explanations of all the Machines* (London, 1709).

Hawkins, Iaaac, *An Essay for the Discovery of the Longitude at Sea, By several Methods fully and Particularly laid before the Publick* (London, 1714).

[Hickes, George,] *Some Discourses Upon Dr. Burnet and Dr. Tillotson; Occasioned by the late Funeral Sermon of the Former upon the Later* (London, 1695).

Hobbs, William, *A New Discovery for Finding the Longitude. Humbly Submitted to the Right Honourable of the Lords Spiritual and Temporal...* (London, 1714).

Hooke, Robert, *An Attempt To prove the Motion of the Earth from Observations* (London, 1674).

Micrographia; or, Some physiological descriptions of minute bodies made by magnifying glasses. With observations and inquiries thereupon (London, 1665).

Horne, George, *A Fair, Candid, and Impartial State of the Case Between Sir Isaac Newton and Mr. Hutchinson* (Oxford, 1753).

The theology and philosophy in Cicero's Somnium Scipionis explained (London, 1751).

[Horsey, Samuel,] *Answer of Samuel Horsey, Esq; Governor of the York Buildings Company, to the Misrepresentations of his Conduct, Publish'd in some of the Daily News Papers* (London, 1733).

The Case of Samuel Horsey, Esq; Governor of the York Buildings Company (London, 1733).

House of Commons, *The Report of the Committee, Appointed to Examine into the Project, Commonly Called the Harburg-Lottery* (London, 1723).

A Report from the Committee to whom the Petition of the Proprietors of the Stock of the Governor and Company for Raising the Thames Water in York-Buildings, is Referred (London, 1733).

Keill, John, *An Examination of Dr. Burnet's Theory of the Earth, together with some remarks on Mr. Whiston's New Theory of the Earth* (Oxford, 1698).

An Introduction to Natural Philosophy: or, Philosophical Lectures Read in the University of Oxford, Anno. Dom. 1700 (London, 1720).

An Introduction to the True Astronomy (London, 1721).

[Kennett, White], *The Wisdom of Looking Backward, To Judge the Better of One Side and T'Other By the Speeches, Writings, Actions And other Matters of Fact on Both Sides, For the Four Years Last Past* (London, 1715).

King, William, *The Original Works of William King*, I (London, 1776).

[King, William,] "A Satire on Learning, With Reference to the Royal Society" (1709), in Frederick George Stephens, *Catalogue of Political and Personal Satires* (London, 1978), II, pp. 296–301.

[King, William,] *The Transactioneer With some of his Philosophical Fancies: In two Dialogues* (London, 1700).

Langley, Batty, *The City and Country Builder's, and Workman's Treasury of Designs* (London, 1740).

A Sure Method of Improving Estates (London, 1728).

Laurence, Edward, *The Duty of a Steward to his Lord* (London, 1727).

The Young Surveyor's Guide: Or, a New Introduction to the Whole Art of Surveying Land, Both by the Chain, and all Instruments now in Use. Now first Publish'd from an Original M.S., second edition (London, 1717).

Laurence, John, *A New System of Agriculture. Being a Complete Body of Husbandry and Gardening ... in Five Books. Containing All the best and latest, as well as many new Improvements, useful to the Husbandman, Grazier, Planter, Gardener and Florist* (London, 1726).

MacLaurin, Colin, *An Account of Sir Isaac Newton's Philosophical Discoveries* (London, 1748).

Marsh, Richard, *The Vanity and Danger of Modern Theories. A Sermon Preach'd at St. Marys Church in Cambridge, On Sunday the 13th Day of August, 1699* (Cambridge, 1699).

Martin, Benjamin, *Bibliotheca Technologica: Or, A Philosophical Library of Literary Arts and Sciences* (London, 1737).

Philosophia Britannica: Or, A New and Comprehensive System of the Newtonian Philosophy, Astronomy and Geography, in a Course of Twelve Lectures, second edition (London, 1759).

The Philosophical Grammar; Being a View of the Present State of Experimental Physiology, or Natural Philosophy (London, 1735).

The Philosophical Grammar, third edition (London, 1748).

Maultby, Josias, *God the Hope, and Saviour of his People, Apply'd to the Case of the Protestant-Dissenters: in a Sermon Preach'd at a*

Solemn Fast, June 2, 1714. On Occasion of the Bill for preventing the Growth of Schism, etc. (London, 1714).

Moxon, Joseph, *Mechanick Exercises*, third edition (London, 1703).

Newton, Isaac, *Philosophiae Naturalis Principia Mathematica*, second edition (London, 1713).

Mathematical Principles of Natural Philosophy, ed. F. Cajori (repr., New York, 1962).

Niewentyt, Bernard, *The Religious Philosopher: Or, the Right Use of Contemplating the Works of the Creator...*, trans. John Chamberlayne. To which is Prefix'd, A letter to the Translator, by the Reverend J. T. Desaguliers [dated February 2, 1717/8] (London, 1718–19).

Norris, John, *An Account of Reason & Faith: In Relation to the Mysteries of Christianity* (London, 1697).

North, Roger, *Examen: Or, An Enquiry into the Credit and Veracity of A Pretended Complete History* (London, 1740).

The Lives of the Right Hon. Francis North, Baron Guilford; The Hon. Sir Dudley North; and the Hon. and Rev. Dr. John North, ed. Augustus Jessop, Vol. III (London, 1890).

Pardies, I., *Short, but yet plain elements of geometry and plain trigonometry*, ed. John Harris (London, 1701).

Parker, Samuel, *A Demonstration of the Divine Authority of the Law of Nature, And of the Christian Religion* (London, 1681).

Pemberton, Henry, *A View of Sir Isaac Newton's Philosophy* (London, 1728).

Ray, John, *A Collection of English Words*, second edition (London, 1691).

The Wisdom of God Manifested in the Works of Creation, seventh edition, corrected (London, 1717, reprint edn., New York, 1977).

[Rohault, J.,] *A Treatise of Mechanicks: Or, The Science of the Effects of Powers of Moving Forces, as apply'd to Machines, demonstrated from its first Principles. Done out of French. In this Translation are inserted Particular References to the several Propositions in Euclid, on which Demonstrations are built: With considerable Additions, whereby the Whole is more Compleat and Universal, and particularly the Rising and Falling of the Quicksilver in the Weather-Glass explain'd and accounted for. Useful for all Artificers, as well as Natural Philosophers and Mathematicians* (London, 1716).

Rowning, John, *A Compendious System of Philosophy: With Notes, Containing the Mathematical Demonstrations, and Some Occasional Remarks*, seventh edition (London, 1772).

[Parker, Samuel,] *A Brief Account of the new Sect of Latitude-Men To-*

gether with some Reflections upon the New Philosophy. In answer to a letter to a Friend at Oxford (London, 1662).

Sacheverell, Henry, *The Natural Mischief of Prejudice and Partiality, Stated in a Sermon Preach'd at St. Mary's in Oxford, At the Assizes Held There, March 9 1703/4*, second edition (London, 1708).

The Perils of False Brethren, both in Church and State: Set forth in a Sermon Preach'd Before the Right Honourable The Lord-Mayor, Alderman, and the Citizens of London, at the Cathedral-Church of St. Paul, on the 5th of November, 1709 (London, 1709).

Sault, Richard, *A Treatise of Algebra, According to the late Improvements...* (London, 1695).

Shaw, Peter, *Chemical Lectures, Publickly Read at London, In the Years 1731, and 1732; And since at Scarborough, in 1733; For the Improvement of Arts, Trades and Natural Philosophy* (London, [1734]).

Proposals for Printing, by Subscription, The Philosophical Works of Francis Bacon ... in Three Volumes [London, March 25, 1731] [B.L., Add. MSS. 4258].

Three Essays in Artificial Philosophy, Or Universal Chemistry.... (London, 1731).

Shaw, Peter, and Francis Hauksbee, *An Essay For Introducing a Portable Laboratory: By Means Whereof all the Chemical Operations Are Commodiously Perform'd, For the Purposes of Philosophy, Medicine, Metallurgy, and a Family* (London, 1731).

Proposals For a Course of Chemical Experiments: With a View to Practical Philosophy, Arts, Trades, and Business [London, May 12, 1731].

Shower, John, *Enoch's Translation. A Funeral Sermon Upon the Sudden Death of Dr. Nehemiah Grew, Fellow of the College of Physicians. Who died March 25, 1712* (London, 1712).

Practical Reflections On the late Earthquakes in Jamaica, England, Sicily, Malta, &c. Anno 1692. With a Particular, Historical Account of those, and divers other Earthquakes (London, 1693).

Winter Meditations: Or, A Sermon Concerning Frost, and Snow, and Winds, &c. and the Wonders of God Herein, third edition, corrected (London, 1710).

Smith, Robert, *A Compleat System of Opticks in Four Books, viz. A Popular, A Mathematical, a Mechanical, and a Philosophical Treatise* (Cambridge, 1738).

South, Robert, *Sermons Preached upon Several Occasions* (Oxford, 1679).

Sprat, Thomas. *The History of the Royal Society of London, For the Improving of Natural Knowledge* (London, 1667).

St. John, Henry., *A Letter to Sir William Windham* (London, 1753).

Steele, Richard, *A Letter to a Member of Parliament Concerning the Bill for Preventing the Growth of Schism* (Edinburgh, 1714).

Stubbe, Henry, *Legends No Histories: Or, A Specimen of Some Animadversions Upon the History of the Royal Society* (London, 1670).

The Plus Ultra reduced to a Non Plus: OR, A Specimen of some Animadversions upon the Plus Ultra of Mr. Glanvill, wherein sundry Errors of some Virtuosi are discovered, the Credit of the Aristotelians in part Re-advanced; and Enquiries made... (London, 1670).

Switzer, Stephen, *Ichnographia Rustica: Or, The Nobleman, Gentleman, and Gardener's Recreation. Containing Directions for the general Distribution of a country Seat, into Rural and Extensive Gardens, Parks, Paddocks, &c. And a General System of Agriculture* (London, 1718).

An Introduction To a General System of Hydrostaticks and Hydraulicks, Philosophical and Practical.... Containing in General A Physico-mechanical Enquiry into the Original and Rise of Springs, and of all the Hypotheses relating thereto; as also the Principles of Water-works,... Deduc'd from the Theory of Archimedes, Gallileo, Torricelli, Boyle, Wallis, Plot, Hook, Marriotte, Desaguliers, Derham, Hawksbee, and others (London, 1729).

The Practical Fruit-Gardener. Being the best and newest Method of raising, Planting, and Pruning all Sorts of Fruit-Trees, agreeably to the Experience and Practice of the most eminent Gardeners and Nursery-Men, revised and recommended by the Revd. Mr. Laurence and Mr. Bradley (London, 1724).

Thacker, Jeremy, *The Longitude's Examin'd. Beginning with a short Epistle to the Longitudinarians, and Ending with the Description of a smart, pretty machine of my own, which I am (almost) sure will do for the Longitude, and procure me the Twenty Thousand Pounds* (London, 1714).

The Tryal of Dr. Henry Sacheverell, Before the House of Peers, for High Crimes and Misdemeanours; upon an Impeachment By the Knights, Citizens and Burgesses Assembled, in the Name of Themselves, and of all the Commons of Great-Britain: Begun in Westminster-Hall the 27th Day of February 1709/10, and from March following (London, Published by Order of the House of Peers, 1710).

Thompson, Isaac, *Proposals For Establishing A Course of Mechanic, Hydrostatic, and Pneumatic Experiments, In the Town of Newcastle* (n.d.) [Carlisle Record Office, D/Lons./W7/1/324].

Toland, John, *Letters to Serena* (London, 1704; reprint edn. Stuttgart, 1964).

Voltaire, *Letters Concerning the English Nation* (London, 1733).

Vream, William, *A Description of the Air-Pump. According to the late Mr. Hawksbee's best and last Improvements; With the manner of making Fifty of the most curious Experiments upon it: The Figures of the Air-Pump-Glasses, and all the Machines belonging to it, being curiously engraven in Copper Plates* (London, 1717).

Ward, John, *The Lives of the Professors of Gresham College* (London, 1740).

Watts, Thomas, *An Essay on the Proper Method for Forming the Man of Business*, ed. Arthur H. Cole (Boston, 1946).

Watts, William, *A Course of Mechanical and Experimental Philosophy, Consisting of Seven Parts* (London, 1727).

Webster, John, *Metallographia: Or, an History of Metals* (London, 1671).

Wells, Edward, *Remarks on Dr Clarke's Introduction to his Scripture-Doctrin of the Trinity* (Oxford, 1713).

The Young Gentleman's Astronomy, second edition (London, 1718).

Whiston, William, *An Account of a Surprising Meteor, Seen in the Air, March the 6th, 1715/16 at Night*, second edition (London, 1716).

An Account of a Surprising Meteor, Seen in the Air March 19. 1718/19 at Night, second edition (London, 1719).

Account of Convocation's Proceedings in Relation to Mr. Whiston. With a Postscript, Containing a Reply to the Considerations on the Historical Preface, and the Premonition to the Reader. To which is added, A Supplement to the foregoing Account of the Convocation's Proceedings (London, 1711).

Astronomical Principles of Religion, Natural and Reveal'd... Together with a Preface, Of the Temper of Mind necessary for the Discovery of Divine Truth; and of the Degree of Evidence that ought to be expected in Divine Matters (London, 1717).

The Cause of the Deluge Demonstrated, bound with his *Astronomical Principles of Religion, Natural and Reveal'd* (London, 1717).

Historical Memoirs of the Life of Dr. Samuel Clarke. Being a Supplement to Dr. Sykes's and Bishop Hoadley's Accounts (London, 1730).

Memoirs of the Life and Writings of Mr. William Whiston (London, 1749).

Mr. Whiston's Account of Dr. Sacheverell's Proceedings In order to Exclude him from St. Andrew's Church in Holborn (London, 1719).

A New Method for Discovering the Longitude both at Sea and Land, second edition (London, 1715).

A New Theory of the Earth, From its Original, to the Consummation of all Things. Wherein the Creation of the World in Six Days, the

Universal Deluge, and the General Conflagration, As laid down in the Holy Scriptures, Are shewn to be perfectly agreeable to Reason and Philosophy (London, 1696).

Several Papers Relating to Mr. Whiston's Cause Before the Court of Delegates;.... (London, 1715).

Whiston, William, and Humphry Ditton, *A New Method For Discovering the Longitude Both at Sea and Land, Humbly Proposed to the Consideration of the Publick* (London, 1714).

Whiston, William, and Francis Hauksbee, *An Experimental Course of Astronomy; Proposed by Mr Whiston and Mr Hauksbee* (n.d.).

Wilkins, John, *Mathematicall Magick. Or, the Wonders That may be performed by Mechanicall Geometry* (London, 1648).

Williams, John, *The Characters of Divine Revelation. A Sermon Preached at St. Martins in the Fields, March 4, 1694/5. Being the Third of the Lecture For the Ensuing Year, Founded by the Honourable Robert Boyle, Esquire*, second edition (London, 1697).

Woodward, John, *An Essay toward a Natural History of the Earth: And Terrestrial Bodies, Especially Minerals: As also of the Sea, Rivers, and Springs. With an Account of the Universal Deluge: And of the Effects that it had upon the Earth* (London, 1695).

Worster, Benjamin, *A Compendious and Methodical Account of the Principles of Natural Philosophy: As they are Explain'd and Illustrated in the Course of Experiments, perform'd at the Academy in little Tower-Street* (London, 1722).

[Wotton, William], *A Letter to Eusabia: Occasioned by Mr. Toland's Letters to Serena* (London, 1704).

Yarranton, Andrew, *England's Improvement By Sea and Land. To Outdo the Dutch without Fighting, To Pay Debts Without Moneys, To set at Work all the Poor of England with the Growth of our Lands* (London, 1677).

Newspapers and collections

Daily Post (1719–26), *Daily Courant* (1712–25), *The Englishman* (1713), *Gentleman's Magazine*, Vols. IV–VIII (1734–38), *The Guardian* (1713), *The History of the Works of the Learned*, IV (July 1702), *Journals of the House of Commons*, XVII–XXIII (1711–38), *Newcastle Courant* (1724), *The Newcastle General Magazine* (1755), *The Spectator* (1712), *St. James's Journal* (1722–23), *The Supplement to the Freeholder's Journal* of Wednesday April 25 (Friday, April 27,

1722), *The Weekly Journal, OR, British Gazetteer* (1719), *The Weekly-Journal or Saturday's-Post* (1719).

Dictionary of National Biography, eds. Leslie Stephen et al., 22 vols. (London: Smith, Elder & Co., 1908–).

Dictionary of Scientific Biography, ed. C. C. Gillispie. 15 vols. (New York: Scribners, 1970–80).

Hearne, Thomas, *Remarks and Collections*, eds. C. E. Doble, D. W. Rannie, et al., 11 vols. (Oxford: Clarendon Press, 1885–1921).

Nichols, John, *Literary Anecdotes of the Eighteenth Century*, VI–IX (London, 1812–1815).

Sedgwick, Romney, ed., *The House of Commons, 1715–1754*, 2 vols. (New York: History of Parliament Trust by Oxford University Press, 1970).

Turnbull, H. W., et al., eds., *The Correspondence of Isaac Newton*, 7 vols. (Cambridge: Cambridge University Press, 1959–77).

Selected secondary sources

Allen, D. G. C., and R. E. Schofield, *Stephen Hales: Scientist and philanthropist* (London: Scolar Press, 1980).

Allen, J. S., "Addendum to the Introduction of the Newcomen Engine from 1710–1733," *Transactions of the Newcomen Society*, XLIII (1970–1), 199–202.

"Bromsgrove and the Newcomen Engine," *Transactions of the Newcomen Society*, XLIII (1970–1), 183–98.

"The Introduction of the Newcomen Engine, 1710–1733," *Transactions of the Newcomen Society*, XLII (1969–70), 169–90.

"Some Early Newcomen Engines and the Legal Disputes Surrounding Them," *Transactions of the Newcomen Society*, XLI (1968–9), 181–201.

"The 1712 and other Newcomen Engines of the Earls of Dudley," *Transactions of the Newcomen Society*, XXXVII (1964–5), 57–84.

"The 1715 and the Other Newcomen Engines at Whitehaven, Cumberland," *Transactions of the Newcomen Society*, XLV (1972–3), 237–68.

"Thomas Newcomen (1663/4–1729) and His Family," *Transactions of the Newcomen Society*, LI (1979–80), 11–24.

Altick, Richard D., *The Shows of London* (Cambridge, Mass., and London: Belknap Press, 1978).

Appleby, John H., "Moses Stringer (f. 1695–1713): Iatrochemist and Mineral Master General," *Ambix*, XXXIV (March 1987), 31–45.

Atkinson, A.D., "William Derham, F.R.S. (1657–1735)," *Annals of Science*, VIII (December 1952), 368–92.

Axtell, James L., "Locke, Newton and the Two Cultures," in John W. Yolton, ed., *John Locke: Problems and Perspectives* (London: Cambridge University Press, 1969), pp. 165–82.

Baker, C. H. Collins, and Muriel I. Baker, *The Life and Circumstances of James Brydges, First Duke of Chandos* (Oxford: Clarendon Press, 1949).

Bardwell, Avelina, "Arts and Science in English Encyclopedias: The Lexicon Technicum of John Harris and the Cyclopedia of Ephraim Chambers" (M.A. thesis, University of Toronto, 1976).

Bataille, Robert R., "Robert Heath, Thomas Cowper, and Eighteenth-Century Mathematical Journalism," *Papers of the Bibliographical Society of America*, LXXXI (September 1987), 339–43.

Baumer, Edward, *The Early Days of the Sun Fire Office* (London: Sir J. Causton and Sons, 1910).

Beattie, Lester M., *John Arbuthnot: Mathematician and Satirist* (Cambridge, Mass.: Harvard University Press, 1935).

Beckett, J. V., *The Aristocracy in England, 1660–1914* (Oxford and New York: Cambridge University Press, 1986).

"A Back-Bench M.P. in the Eighteenth Century: Sir James Lowther of Whitehaven," *Parliamentary History*, I (1982), 79–98.

Coal and Tobacco. The Lowthers and the Economic Development of West Cumberland, 1660–1760 (London and New York: Cambridge University Press, 1981)

"Dr William Brownrigg, F.R.S.: Physician Chemist and Country Gentleman," *Notes and Records of the Royal Society of London*, XXXI (January 1977), 255–71.

Bellewes, G. O., *Notes on the Life and Works of John Laurence, Rector of Bishop Wearmouth, 1721–1732* (Sunderland: Reprinted from the Antiques of Sunderland, 1905).

Ben-Chaim, Michael, "Social Mobility and Scientific Change: Stephen Gray's Contribution to Electrical Research," *British Journal for the History of Science*, XXIII (March 1990), 3–24.

Bennett, G. V., *The Tory Crisis in Church and State, 1688–1730* (Oxford: Clarendon Press, 1975).

White Kennett, 1660–1728. Bishop of Peterborough (London: S.P.C.K., 1957).

Bennett, J. A., "Hooke's Instruments for Astronomy and Navigation," in Michael Hunter and Simon Schaffer, eds., *Robert Hooke. New Studies* (Woodbridge: Boydell Press, 1989), pp. 20–32.

Bjorkbom, Carl, "A Proposal to Erect an Atmospheric Engine in Sweden in 1725," *Transactions of the Newcomen Society*, XVIII (1937–8), 75–85.

Bosanquet, R. C., "John Horsley and His Times," *Archaeologia Aeliana*, fourth series, VIII (1933), 59–81.

Bowles, Geoffrey, "The Place of Newtonian Explanation in English Popular Thought" (Ph.D. thesis, Oxford University, 1977).

Buck, Peter, "People Who Counted: Political Arithmetic in the Eighteenth Century," *Isis*, LXXIII (March 1982), 28–45.

"Seventeenth-Century Political Arithmetic: Civil Strife and Vital Statistics," *Isis*, LXVIII (1977), 67–84.

Calvert, H. R., *Scientific Trade Cards in the Science Museum Collection* (London: Science Museum, 1971).

Cantor, Geoffrey, *Optics after Newton: Theories of light in Britain and Ireland, 1704–1840* (Manchester: Manchester University Press, 1983).

"The Rhetoric of Experiment," in David Gooding, Trevor Pinch, and Simon Schaffer, eds., *The Uses of Experiment. Studies in the Natural Sciences* (Cambridge: Cambridge University Press, 1989), pp. 159–80.

Capp, Bernard, *Astrology and the Popular Press. English Almanacs 1500–1800* (Ithaca: Cornell University Press, 1979).

Carr-Harris, Grant, "The Ancestry of John Harris (1667–1719) – Father of the Encyclopaedia," *Notes and Queries* (February 1978), 19–23.

Carson, Patricia, "The Building of the First Bridge at Westminster, 1736–1750," *Journal of Transport History*, III (November 1957), 111–22.

Carswell, John, *South Sea Bubble* (London: The Cresset Press, 1960).

Cawthorne, Henry H., "The Spitalfield's Mathematical Society (1717–1845)," *Journal of Adult Education*, III (1929), 155–66.

Clapp, Sarah L. C., "Subscription Publishers Prior to Jacob Tonson," *The Library*, XIII (1933), 158–83.

Clark, Sir George, *Science and Social Welfare in the Age of Newton*, second edition (Oxford: Clarendon Press, 1970).

Cohen, I. Bernard, *Franklin and Newton: An Inquiry into Speculative Newtonian Experimental Science and Franklin's Work in Electricity as an Example Thereof* (Philadelphia: American Philosophical Society, 1956).

Introduction to Newton's Principia (Cambridge, Mass: Harvard University Press, 1978).

Coleman, William, "Providence, Capitalism, and Environmental Degradation," *Journal of the History of Ideas*, XXXVII (1976), 27–44.

Colvin, H.M., et al., eds., *The History of the King's Works*, V (1660–1780) (London: H.M. Stationery Office, 1976).
Cope, Jackson I., *Joseph Glanvill, Anglican Apologist* (St. Louis: Washington University Press, 1956).
Cowie, Leonard W., *Henry Newman: An American in London, 1708–1743* (London: S.P.C.K., 1956).
Crawford, M. A., "Evidence from Trade Cards for the Scientific Instrument Community," *Annals of Science*, XLII (1985), 453–554.
Crouzet, Francois, *The First Industrialists. The Problem of Origins* (Cambridge and New York: Cambridge University Press, 1985).
Cudworth, William, *Life and Correspondence of Abraham Sharp* (London: S. Low, Marston, Searle, and Rivington Ltd., 1889).
Cummings, A. J. G., "The Harburgh Company and its Lottery, 1716–1723," *Business History*, XXVII (July 1986), 1–18.
"The York Buildings Company. A Case Study in Eighteenth Century Corporation Mismanagement" (Ph.D. thesis, University of Strathclyde, 1980).
Cummings, A. J. G., and Larry Stewart, "The Case of the Eighteenth-Century Projector: Entrepreneurs, Engineers, and Legitimacy at the Hanoverian Court in Britian," in Bruce T. Moran, ed., *Patronage and Institutions. Science, Technology and Medicine at the European Court, 1500–1750* (Woodbridge and Rochester: Boydell Press, 1991), pp. 235–61.
Curtis, T. C., and W. A Speck, "The Societies for the Reformation of Manners. A Case Study in the Theory and Practise of Moral Reform," *Literature and History*, III (1976), 45–64.
Davies, A. Stanley, "The Coalbrookdale Company and the Newcomen Engine, 1717–1769," *Transactions of the Newcomen Society*, XX (1939–40), 45–8.
Davies, K. G., "Joint-Stock Investment in the Later Seventeenth Century," *Economic History Review*, second series, IV (1952), 283–301.
Davies, Tony, "The Ark in Flames: Science, Language and Education in Seventeenth-Century England," in Andrew E. Benjamin, Geoffrey N. Cantor, and John R. R. Christie, eds., *The Figural and the Literal. Problems of Language in the History of Science and Philosophy, 1630–1800* (Manchester: Manchester University Press, 1987), pp. 83–102.
Day, Joan, "The Costers: Copper-Smelters and Manufacturers," *Transactions of the Newcomen Society*, XLVII (1974–6), 47–58.
de Beer, E. S., ed., *The Diary of John Evelyn* (Oxford: M. W. Dunne, 1955).

de Morgan, Augustus, *A Budget of Paradoxes* (London: Longmans, Green & Co., 1872).

Dear, Peter, "Miracles, Experiments, and the Ordinary Course of Nature," *Isis*, LXXXI (December 1990), 663–83.

" 'Totius in verba.' Rhetoric and Authority in the Early Royal Society," *Isis*, LXXVI (June 1985), 145–61.

Delree, Pierre, "A propos de l'introduction des machines a vapeur dans la region liegeoise," *Chronique archaeologique de pays de Liege*, LIII (1962), 109–16.

Dickinson, H. W., *A Short History of the Steam Engine* (Cambridge: University Press, 1939).

The Water Supply of Greater London (London: Newcomen Society, 1954).

Dickson, P. G. M., *The Financial Revolution in England, 1688–1756* (London and Melbourne: St. Martin's Press, 1967).

The Sun Insurance Office, 1710–1960 (London: Oxford University Press, 1960).

Domson, Charles Andrew, "Nicolas Fatio de Duillier and the Prophets of London: An Essay in the Historical Interpretation of Natural Philosophy and Millennial Belief in the Age of Newton" (Ph.D. thesis, Yale University, 1972).

DuBois, Armand B., *The English Business Company after the Bubble Act, 1720–1800* (New York: Columbia School of Law, 1938).

Duckham, Baron F., *A History of the Scottish Coal Industry*, Vol. I (Newton Abbot: David & Charles, 1970).

Duffy, Eamon, "Primitive Christianity Revived; Religious Renewal in Augustan England," in Derek Baker, ed., *Renaissance and Renewal in Christian History* (Oxford: Blackwell, 1977), pp. 287–300.

" 'Whiston's Affair': The trials of a Primitive Christian, 1709–1714," *Journal of Ecclesiastical History*, XXVII (April 1976), 129–50.

Eisenstein, Elizabeth L., *The Printing Press as an Agent of Change. Communications and Cultural Transformations in Early-Modern Europe* (Cambridge and New York: Cambridge University Press, 1979).

Ellis, Aytoun, *The Penny Universities: A History of the Coffee-Houses* (London: Secker & Warburg, 1956).

Ellis, Joyce M., ed., *The Letters of Henry Liddell to William Cotesworth. Surtees Society Publications*, CXVII (1985).

A Study of the Business Fortunes of William Cotesworth c. 1688–1726 (New York: Arno Press, 1981).

'Espinasse, Margaret. "The Decline and Fall of Restoration Science," in Charles Webster, ed., *The Intellectual Revolution of the Seventeenth Century* (London and Boston: Routledge and Kegan Paul, 1974), pp. 347–68.

Evans, Raymond James, "The Diffusion of Science: The Geographical Transmission of Natural Philosophy in the English Provinces, 1660–1760" (Ph.D. thesis, Cambridge University, 1982).

Farrell, Maureen, "The Life and Work of William Whiston" (Ph.D. thesis, University of Manchester, 1973).

Feingold, Mordechai, *The Mathematician's Apprenticeship. Science, Universities and Society in England, 1560–1640* (Cambridge: Cambridge University Press, 1984).

Flinn, Michael W., *The Origins of the Industrial Revolution* (London: Longmans, 1966).

"William Wood and the Coke-Smelting Process," *Transactions of the Newcomen Society*, XXXIV (1961–2), 55–71.

Force, James E., *William Whiston. Honest Newtonian* (Cambridge and New York: Cambridge University Press, 1985).

Force, James E., and Richard H. Popkin, eds., *Essays on the Context, Nature, and Influence of Isaac Newton's Theology* (Dordrecht, Boston and London: Kluwer Academic Publishers, 1990).

Fores, Michael, "Constructed Science and the Seventeenth Century 'Revolution,' " *History of Science*, XXII (1984), 217–44.

"Francis Bacon and the Myth of Industrial Science," *History of Technology*, VII (1982), 57–76.

"Newton on a Horse: A Critique of the Historiographies of 'Technology' and 'Modernity,' " *History of Science*, XXIII (1985), 351–78.

Foucault, Michel, *The Order of Things. An Archaeology of the Human Sciences* (London: Tavistock, 1970).

Freudenthal, Gad, "Early Electricity Between Chemistry and Physics: The Simultaneous Itineraries of Francis Hauksbee, Samuel Wall, and Pierre Poliniere," *Historical Studies in the Physical Sciences*, XI (1982), 203–29.

Freudenthal, Gideon, *Atom and Individual in the Age of Newton. On the Genesis of the Mechanistic World View* (Dordrecht and Boston: D. Reidel Pub. Co., 1986).

Gascoigne, John, *Cambridge in the Age of the Enlightenment. Science, Religion and Politics from the Restoration to the French Revolution* (Cambridge: Cambridge University Press, 1989).

"From Bentley to the Victorians: The Rise and Fall of British Newtonian Natural Theology," *Science in Context*, II (Autumn 1988), 219–56.

"The Universities and the Scientific Revolution: The Case of Newton and Restoration Cambridge," *History of Science*, XXIII (1985), 391–434.

Gee, Brian, "Amusement Chests and Portable Laboratories: Practical Alternatives to the Regular Laboratory," in Frank A. J. L. James, ed., *The Development of the Laboratory. Essays on the Place of Experiment in Industrial Civilization* (London: Macmillan), 1989, pp. 37–59.

Gibbs, F. W. "Itinerant Lecturers in Natural Philosophy," *Ambix*, VI (1960), 111–17.

"Peter Shaw and the Revival of Chemistry," *Annals of Science*, VII (September 1951), 211–37.

Gieryn, Thomas F., "Distancing Science from Religion in Seventeenth-Century England," *Isis*, LXXIX (December 1988), 582–93.

Gillespie, Neal C., "Natural History, Natural Theology, and Social Order: John Ray and the 'Newtonian Ideology,' " *Journal of the History of Biology*, XX (Spring 1987), 1–49.

Goldgar, Bertrand A. *The Curse of Party: Swift's Relations with Addison and Steele* (Lincoln: University of Nebraska Press, 1961).

Golinski, Jan V., "A Noble Spectacle. Phosphorous and the Public Culture of Science in the Early Royal Society," *Isis*, LXXX (March 1989), 11–39.

"Peter Shaw: Chemistry and Communication in Augustan England," *Ambix*, XXX (March 1983), 19–29.

"Robert Boyle: Scepticism and Authority in Seventeenth-Century Chemical Discourse," in Andrew E. Benjamin, Geoffrey N. Cantor, and John R. R. Christie, eds., *The Figural and the Literal. Problems of Language in the History of Science and Philosophy, 1630–1800* (Manchester: Manchester University Press, 1987), pp. 58–81.

"Utility and Audience in Eighteenth-Century Chemistry: Case Studies of William Cullen and Joseph Priestley," *British Journal for the History of Science*, XXI (March 1988), 1–31.

Greene, John C., *The Death of Adam: Evolution and Its Impact on Western Thought* (New York: Oxford University Press, 1961).

Grinnell, George, "Newton's *Principia* as Whig Propaganda," in P. Fritz and D. Williams, eds., *City and Society in the 18th Century* (Toronto: Hakkert, 1973), pp. 181–92.

Gross, Alan, *The Rhetoric of Science* (Cambridge, Mass.: Harvard University Press, 1990).

Guerlac, Henry, "Francis Hauksbee: Experimentateur au profit de Newton," in *Essays and Papers in the History of Modern Science* (Baltimore: John Hopkins University Press, 1977), pp. 107–19.

Guerrini, Anita, "James Keill, George Cheyne, and Newtonian Physiology, 1690–1740," *Journal of the History of Biology*, XVIII (Summer 1985), 247–66.

"Newtonian Matter Theory, Chemistry, and Medicine, 1690–1713" (Ph.D. thesis, Indiana University, 1983).

"The Tory Newtonians: Gregory, Pitcairne and their Circle," paper presented to the William Andrews Clarke Library seminar, July 1985.

Gunn, J. A. W., *Politics and Public Interest in the Seventeenth Century* (London: Routeledge and Kegan Paul, 1969).

Gunther, Robert W. T., ed., *Further Correspondence of John Ray* (London: Printed for the Ray Society, 1928).

Hackmann, W. D., "The Growth of Science in the Netherlands in the Seventeenth and Eighteenth Centuries," in Maurice Crosland, ed., *The Emergence of Science in Western Europe* (New York: Science of History Publications, 1976), pp. 89–109.

"Scientific Instruments: Models of Brass and Aids to Discovery," in David Gooding, Trevor Pinch, and Simon Schaffer, eds., *The Uses of Experiment. Studies in the Natural Sciences* (Cambridge: Cambridge University Press, 1989), pp. 31–65.

Hall, A. Rupert, "Newton in France: A New View," *History of Science*, XIII (1975), 233–50.

"On Knowing, and Knowing how to . . . ," *History of Technology*, III (1978), 91–103.

Philosophers at War: The Quarrel between Newton and Leibniz (Cambridge: Cambridge University Press, 1980).

"What did the Industrial Revolution in Britain Owe to Science?" in Neil McKendrick, ed., *Historical Perspectives: Studies in English Thought and Society in Honour of J. H. Plumb* (London: Europa, 1974), pp. 129–51.

Hall, Marie Boas, "Oldenburg, the *Philosophical Transactions* and Technology," in John G. Burke, ed., *The Uses of Science in the Age of Newton* (Berkeley and Los Angeles: University of California Press, 1983), pp. 21–47.

"Solomon's House Emergent: The Early Royal Society and Cooperative Research," in Harry Woolf, ed., *The Analytic Spirit: Essays in the History of Science* (Ithaca: Cornell University Press, 1981), pp. 177–94.

Hamilton, Henry, *The English Brass and Copper Industries to 1800* (London: Cass, 1926).

Hans, Nicholas, *New Trends in Education in the Eighteenth Century* (London: Routledge and Kegan Paul, 1966).

Hansotte, G., "L'Introduction de la machine a vapeur du Pays de Liege (1720)," *La vie wallonne*, XXIV (1950), 47–55.

"Machines a vapeur dans la region Liegeoise au XVIIIe siecle," *Societe royale le vieux Liege* (1955), 492–93.

Hart, Cyril, *The Industrial History of Dean* (Newton Abbot: David and Charles, 1971).

Harwood, John T., "Rhetoric and Graphics in *Micrographia*," in Michael Hunter and Simon Schaffer, eds., *Robert Hooke. New Studies* (Woodbridge: Boydell Press, 1989), pp. 119–47.

Hasquin, Herve, *Une mutation. le 'Pays de Charleoi' aux XVIIe et XVIIIe siecles. Aux origines de la Revolution industrielle en Belgique* (Brussels: Editions de l'Institut de Sociologie, 1971).

Heilbron, J. L., *Electricity in the 17th and 18th Centuries. A Study of Early Modern Physics* (Berkeley: University of California Press, 1979).

Physics at the Royal Society during Newton's Presidency (Los Angeles: William Andrews Clark Memorial Library, 1983).

"Robert Symmer and the Two Electricities," *Isis*, LXVII (March 1976), 7–20.

Heimann, P. M., and J. E. McGuire, "Newtonian Forces and Lockean Powers: Concepts of Matter in Eighteenth-Century Thought," *Historical Studies in the Physical Sciences*, III (1971), 233–306.

Henrey, Blanche, *British Botanical and Horticultural Literature before 1800*, 3 vols. (London and New York: Oxford University Press, 1975).

Hessen, B., *The Social and Economic Roots of Newton's "Principia,"* ed. Robert S. Cohen (New York: Howard Fertig, 1971).

Hills, Richard L., *Power from Steam. A History of the Stationery Steam Engine* (Cambridge: Cambridge University Press, 1989).

Hinde, J. Hodgson, ed., *Inedited Contributions to the History of Northumberland. Materials for the History of Northumberland by the Rev. John Horsley, 1729–1730* (Newcastle: E. Walker, n.d.).

Hollister-Short, G. J., "The Introduction of the Newcomen Engine into Europe," *Transactions of the Newcomen Society*, XLVIII (1976–7), 11–24.

"Leads and Lags in late Seventeenth Century English Technology," *History of Technology*, V (1976), 159–83.

Holmes, Geoffrey, *British Politics in the Age of Anne* (London: Macmillan, 1967).

"Science, Reason and Religion in the Age of Newton," *British Journal for the History of Science*, XI (July 1978), 164–71.

The Trial of Doctor Sacheverell (London: Eyre Methuen, 1973).

Houghton, Walter E., Jr., "The History of Trades: Its Relation to Seventeenth-Century Thought," *Journal of the History of Ideas*, II (1941), 37–43.

Hughes, Edward, "The First Steam Engines in the Durham Coalfields," *Archaeologia Aeliana*, XXVII (1949), 29–45.

North Country Life in the Eighteenth Century. The North East, 1700–1750 (London and New York: Oxford University Press, 1952).

Hunter, Michael, "The Crown, the Public and the New Science, 1689–1702," *Notes and Records of the Royal Society*, XLIII (1989), 165–81.

Establishing the New Science. The Experience of the Early Royal Society (Woodbridge and Wolfeboro: Boydell and Brewer, 1989).

Science and Society in Restoration England (Cambridge and New York: Cambridge University Press, 1981).

Hunter, Michael, and Paul B. Wood, "Towards Solomon's House: Rival Strategies for Reforming the Early Royal Society," *History of Science*, XXIV (1986), 49–108.

Hurst, W. R., *An Outline of the Career of John Theophilus Desaguliers, M.A., LL.D, F.R.S.* (privately printed, 1928).

Iliffe, Robert Charles, " 'The Idols of the Temple': Isaac Newton and the Private Life of Anti-Idolatry" (Ph.D. thesis, University of Cambridge, 1989).

Isaacs, Tina, "The Anglican Hierarchy and the Reformation of Manners, 1688–1738," *Journal of Ecclesiastical History*, XXXIII (July 1982), 391–411.

Jacob, James R., *Henry Stubbe, Radical Protestantism and the Early Enlightenment* (Cambridge and New York: Cambridge University Press, 1983).

"Restoration Ideologies and the Royal Society," *History of Science*, XVIII (1980), 25–38.

Jacob, James R., and Margaret C. Jacob, "The Anglican Origins of Modern Science: The Metaphysical Foundations of the Whig Constitution," *Isis*, LXXI (June 1980), 251–67.

Jacob, Margaret C., "The Church and the Boyle Lectures: The Social

Context of Newtonian Natural Philosophy" (Ph.D. thesis, Cornell University, 1969).

"The Church and the Formulation of the Newtonian World-View," *Journal of European Studies*, II (1972), 1–21.

The Cultural Meaning of the Scientific Revolution (Philadelphia: Temple University Press, 1987).

"John Toland and the Newtonian Ideology," *Journal of the Warburg and Courtauld Institutes*, XXXII (1969), 307–31.

The Newtonians and the English Revolution, 1689–1720 (Ithaca: Cornell University Press, 1976).

The Radical Enlightenment: Pantheists, Freemasons, and Republicans (London: Allen and Unwin, 1981).

"Scientific Culture in the Early English Enlightenment: Mechanisms, Industry, and Gentlemanly Facts," in Alan Charles Kors and Paul J. Korshin, eds., *Anticipations of the Enlightenment in England, France and Germany* (Philadelphia: University of Pennsylvania Press, 1987), pp. 134–64.

Jacob, Margaret C., and Henry Guerlac, "Bentley, Newton and Providence," *Journal of the History of Ideas*, XXX (1969), 307–18.

Jenkins, Philip, *The Making of a Ruling Class. The Glamorgan Gentry, 1640–1790* (Cambridge and New York: Cambridge University Press, 1983).

Johnson, B. L. C., "The Foley Partnerships: The Iron Industry at the End of the Charcoal Era," *Economic History Review*, second series, IV (1952), 322–40.

Johnson, Francis R., "Gresham College: Precursor of the Royal Society," *Journal of the History of Ideas*, I (1940), 414–21.

Jones, Clyve, "James Brydges, Earl of Carnarvon, and the 1717 Hereford By-election: A Case Study in Aristocratic Management," *Huntington Library Quarterly*, XLVI (Autumn 1983), 310–17.

Jones, Clyve, and Geoffrey Holmes, eds., *The London Diaries of William Nicolson Bishop of Carlisle, 1702–1718* (Oxford: Clarendon Press, 1985).

Jubb, Michael John, "Fiscal Policy in England in the 1720s and 1730s" (Ph.D. thesis, University of Cambridge, 1978).

King, Henry C., *Geared to the Stars. The Evolution of Planetariums, Orreries and Astronomical Clocks* (Toronto: University of Toronto Press, 1978).

Kirk, Linda, *Richard Cumberland and Natural Law. Secularisation of Thought in Seventeenth-Century England* (Cambridge: J. Clark and Co., 1987).

Uses of Science in the Age of Newton (Berkeley and Los Angeles: University of California Press, 1983), pp. 171–204.

wen, Colin, *The Leicestershire and South Derbyshire Coalfields* (Ashbourne: Moorland Publishers, 1984).

·cock, J. G. A. "Early Modern Capitalism – The Augustan Perception," in Eugene Kamenka and R. S. Neale, eds., *Feudalism, Capitalism and Beyond* (Canberra: Australia National University Press, 1975), pp. 62–83.

The Machiavellian Moment. Florentine Political Thought and the Atlantic Republican Tradition (Princeton: Princeton University Press, 1975).

Virtue, Commerce, and History. Essays on Political Thought and History, Chiefly in the Eighteenth Century (Cambridge and New York: Cambridge University Press, 1985).

llard, Sidney, *The Genesis of Modern Management. A Study of the Industrial Revolution in Great Britain* (London: E. Arnold, 1965).

rter, Roy, "Creation and Credence: The Career of Theories of the Earth in Britain, 1660–1820," in Barry Barnes and Steven Shapin, eds., *Natural Order: Historical Studies in Scientific Culture* (Beverley Hills: Sage Publications, 1979), pp. 97–123.

"The Enlightenment in England," in Roy Porter and M. Teich, eds., *The Enlightenment in National Context* (Cambridge and New York: Cambridge University Press, 1981), pp. 5–14.

he Making of Geology. Earth Science in Britain, 1660–1815 (Cambridge and New York: Cambridge University Press, 1977).

Science, Provincial Culture and Public Opinion in Enlightenment England," *British Journal for Eighteenth-Century Studies*, III (1980), 20–46.

nfrey, Stephen, "Ideas Above his Station: A Social Study of Hooke's Curatorship of Experiments," *History of Science*, XXIX (March 1991), 1–44.

strick, A, "The London Lead Company, 1692–1905," *Transactions of the Newcomen Society*, XIV (1933–4), 119–62.

uakers in Science and Industry (Newton Abbot: Bonnisdale Press, 1950).

The Steam Engine on the Tyneside," *Transactions of the Newcomen Society*, XVII (1936–7), 131–63.

en, Charles E., *John Ray Naturalist. His Life and Works* (Cambridge and New York: Cambridge University Press, 1942).

wood, John, *Reason, Ridicule and Religion: The Age of Enlightenment in England, 1660–1750* (London: Thames and Hudson, 1976)

Knorr-Cetina, Karin D., "The Ethnographic Study of Scientific Work: Towards a Constructivist Interpretation of Science," in Karin D. Knorr-Cetina and Michael Mulkay, eds., *Science Observed. Perspectives on the Social Study of Science* (London and Beverly Hills: Sage Publications, 1983), pp. 115–40.

Korshin, Paul J., "Types of Eighteenth-Century Literary Patronage," *Eighteenth-Century Studies*, VII (Summer 1974), 452–73.

Kramnick, Isaac, *Bolingbroke and His Circle. The Politics of Nostalgia in the Age of Walpole* (Cambridge, Mass.: Harvard University Press, 1968).

Kubrin, David Charles, "Providence and the Mechanical Philosophy: The Creation and Dissolution of the World in Newtonian Thought. A Study of the Relations of Science and Religion in Seventeenth Century England" (Ph.D. thesis, Cornell University, 1969).

Laslett, Peter, "John Locke, the Great Recoinage, and the Origins of the Board of Trade: 1695–1698," in John W. Yolton, ed., *John Locke: Problems and Perspectives* (London: Cambridge University Press, 1969), pp. 137–64.

Latour, Bruno, "Give Me a Laboratory and I will Raise the World," in Karin D. Knorr-Cetina and Michael Mulkay, eds., *Science Observed. Perspectives on the Social Study of Science* (London and Beverly Hills: Sage Publications, 1983), pp. 141–70.

Science in Action (Cambridge, Mass.: Harvard University Press, 1987).

Levine, Joseph M., *Dr. Woodward's Shield. History, Science, and Satire in Augustan England* (Berkeley: University of California Press, 1977).

Lillywhite, Bryant, *London Coffee Houses* (London: George Allen and Unwin, 1963).

Lindqvist, Svante, *Technology on Trial. The Introduction of Steam Power Technology into Sweden, 1715–1736* (Stockholm: Almqvist and Wicksell International, 1984).

"The Work of Martin Triewald in England," *Transactions of the Newcomen Society*, L (1978–9), 165–71.

Lindsay, J. M. "The Iron Industry in the Highlands: Charcoal Blast Furnaces," *Scottish Historical Review*, LVI (April 1977), 49–63.

Loftis, John, "Richard Steele's Censorium," *Huntington Library Quarterly*, XIV (1950), 53–62.

Steele at Drury Lane (Berkeley: University of California Press, 1952).

MacDonald, Sir George, "John Horsley, Scholar and Gentleman," *Archaeologia Aeliana*, fourth series, VIII (1933), 1–57.

MacLeod, Christine, *Inventing the Industrial Revolution. The English*

Patent System, 1660–1800 (Cambridge and New York: Cambridge University Press, 1988).

"Patents for Invention and Technical Change in England, 1660–1753" (Ph.D. thesis, Cambridge University, 1982).

"The 1690s Patents Boom: Invention or Stock-Jobbing?" *Economic History Review*, second series, XXXIX (1986), 549–71.

Maddison, R. E. W., "Notes on Some Members of the Hanckwitz Family in England," *Annals of Science*, XI (March 1955), 64–72.

"Studies in the Life of Robert Boyle, F.R.S., Part V. Ambrose Godfrey Hanckwitz, F.R.S.," *Notes and Records of the Royal Society*, XI (1954), 159–88.

Major, Paul Richard, "The Physical Researches of J. T. Desaguliers" (M.Sc. thesis, University of London, 1962).

Marshall, John, "The Ecclesiology of the Latitude-men 1660–1689: Stillingfleet, Tillotson, and 'Hobbism,' " *Journal of Ecclesiastical History*, XXXVI (July 1985), 407–27.

Matthews, William, *Hydraulia; An Historical and Descriptive Account of The Water Works of London* (London: R. Hunter, 1835).

McGuire, J. E., "Force, Active Principles, and Newton's Invisible Realm," *Ambix*, XV (October 1968), 154–208.

McKendrick, Neil, John Brewer, and J. H. Plumb, *The Birth of a Consumer Society. The Commercialization of Eighteenth-Century England* (Bloomington: Indiana University Press, 1982).

McKie, D., "John Harris and His *Lexicon Technicum*," *Endeavour*, IV (1945), 53–7.

Merton, Robert K., *Science, Technology and Society in Seventeenth Century England*, second edition (New York and London: Harper, 1970).

Millard, Peter T., "An Edition of Roger North's Life of Dr. John North with a Critical Introduction" (Ph.D. thesis, Oxford University, 1969).

Millburn, John R., *Benjamin Martin: Author, Instrument-Maker and "Country Showman"* (Leyden: Noordhoff International Publications, 1976).

"Benjamin Martin and the Royal Society," *Notes and Records of the Royal Society of London*, XXVIII (June 1973), 15–23.

"The London Evening Courses of Benjamin Martin and James Ferguson, Eighteenth-Century Lecturers on Experimental Philosophy," *Annals of Science*, XL (1983), 437–55.

"William Stukeley and the Early History of the Orrery," *Annals of Science*, XXXI (1974), 511–28.

Morton, Alan, "Public Science in Mid-18th Centur George III Collection at the Science Museum" 1990).

Morton, A. Q., "Lectures on Natural Philosoph 1765: S. C. T. Demainbray (1710–1782) and th Countrymen," *British Journal for the History o* cember 1990), 411–34.

Mulligan, Lotte, "Anglicanism, Latitudinarianism enteenth Century England," *Annals of Scienc* 213–19.

"'Reason,' 'Right Reason,' and 'Revelation' Century England," in Brian Vickers, ed., *Occ talities in the Renaissance* (Cambridge and N University Press, 1984), pp. 375–401.

Musson, A. E., and Eric Robinson, *Science a Industrial Revolution* (Toronto: Universit 1969).

Namier, Sir Lewis, "Brice Fisher, M.P.: A M Merchant and His Connexions," *English H* (October 1927), 514–32.

Nicolson, Marjorie Hope, *Newton Demands the and the Eighteenth Century Poets* (Princeto Press, 1946).

Nicolson, Marjorie Hope, and G. S. Rousseau, *Life*," *Alexander Pope and the Sciences* (Pri versity Press, 1968).

Nicolson, Marjorie Hope, and Nora M. Mohle ground to Swift's *Voyage to Laputa*," in (Ithaca: Great Seal Books, 1956), pp. 110–

Ochs, Kathleen H., "The Failed Revolution in of Industry by Members of the Royal So 1688" (Ph.D. thesis, University of Toront

"The Royal Society of London's History of Early Episode in Applied Science," *Notes Society*, XXXIX (April 1985), 129–58.

Ogden, H. V. S., "Thomas Burnet's *Telluris Th* Scenery," *ELH, A Journal of English Lite* 139–50.

Olson, Richard G., "Tory-High Church O Scientism in the Eighteenth Century: The Jonathan Swift, and Samuel Johnson," in

Rees, William, *Industry Before the Industrial Revolution*, 2 vols. (Cardiff: University of Wales Press, 1968).

Reynolds, Terry S., *Stronger Than a Hundred Men: A History of the Vertical Water Wheel* (Baltimore: Johns Hopkins University Press, 1983).

Robinson, F. J. G., "A Philosophic War: An Episode in Eighteenth Century Scientific Lecturing in North-East England," *Transactions of the Architectural and Archaeological Society of Durham and Northumberland*, II (1970), 101–8.

Robinson, F. J. G., and P. J. Wallis, *Book Subscription Lists: A Revised Guide* (Newcastle Upon Tyne: H. Hill, 1975).

Robinson, Henry W., and Walter Adams, eds., *The Diary of Robert Hooke, M.A., M.D., F.R.S., 1672–1680* (London: Taylor and Francis, 1935).

Robischon, Mary M., "Scientific Instrument Makers in London During the Seventeenth and Eighteenth Centuries" (Ph.D. thesis, University of Michigan, 1983).

Rogers, Pat, "Gulliver and the Engineers," *Modern Language Review*, LXX (1975), 260–70.

Rolt, L. T. C., and J. S. Allen, *The Steam Engine of Thomas Newcomen* (New York: Science History Publications, 1977).

Rousseau, G. S., "Science Books and Their Readers in the Eighteenth Century," in Isabel Rivers, ed., *Books and their Readers in Eighteenth-Century England* (Leicester: Leicester University Press, 1982), pp. 208–17.

" 'Wicked Whiston' and the Scriblerians: Another Ancients-Modern Controversy," *Studies in Eighteenth-Century Culture*, XVII (1987), 17–44.

Rousseau, G. S., and Roy Porter, eds., *The Ferment of Knowledge. Studies in the Historiography of Eighteenth-Century Science* (Cambridge and New York: Cambridge University Press, 1980).

Rowbottom, Margaret E., "John Theophilus Desaguliers (1683–1744)," *Proceedings of the Huguenot Society of London*, XXI (1968), 196–218.

"The Teaching of Experimental Philosophy in England, 1700–1730," *XIe congres international d'histoire des sciences (1965)*, Actes IV (1968), 46–53.

Rowlands, Marie B., "Stonier Parrott and the Newcomen Engines," *Transactions of the Newcomen Society*, XLI (1968–9), 49–67.

Rupp, Gordon, *Religion in England, 1688–1791* (Oxford: Clarendon Press, 1986).

Schaffer, Simon, "Defoe's Natural Philosophy and the Worlds of Credit," in John Christie and Sally Shuttleworth, eds., *Nature Transfigured. Science and Literature, 1700–1900* (Manchester and New York: Manchester University Press, 1989), pp. 13–44.

"Glass Works: Newton's Prisms and the Uses of Experiment," in David Gooding, Trevor Pinch, and Simon Schaffer, eds., *The Uses of Experiment. Studies in the Natural Sciences* (Cambridge: Cambridge University Press, 1989), pp. 67–104.

"Natural Philosophy and Public Spectacle in the Eighteenth Century," *History of Science*, XXI (1983), 1–43.

"Newtonian Cosmology and the Steady State" (Ph.D. thesis, University of Cambridge, 1980).

Schofield, Robert E., *Mechanism and Materialism. British Natural Philosophy in An Age of Reason* (Princeton: Princeton University Press, 1970).

Scott, W. R., *The Constitution and Finance of English, Scottish and Irish Joint-Stock Companies to 1720*, 3 vols. (Cambridge: Cambridge University Press, 1912; reprint edn., Gloucester, Mass., 1968).

Shapin, Steven, "The House of Experiment in Seventeenth-Century England," *Isis*, LXXIX (1988), 373–404.

"Of Gods and Kings: Natural Philosophy and Politics in the Leibniz-Clarke Disputes," *Isis*, LXXII (June 1981), 187–215.

"Pump and Circumstance: Robert Boyle's Literary Technology," *Social Studies of Science*, XIV (November 1984), 481–520.

"Robert Boyle and Mathematics: Reality, Representation, and Experimental Practice," *Science in Context*, II (1988), 23–58.

"Social Uses of Science," in G. S. Rousseau and Roy Porter, *Ferment of Knowledge. Studies in the Historiography of Eighteenth-Century Science* (Cambridge and New York: Cambridge University Press, 1980), pp. 93–105.

"Who was Robert Hooke?" in Michael Hunter and Simon Schaffer, eds., *Robert Hooke. New Studies* (Woodbridge: Boydell Press, 1989), 253–83.

Shapin, Steven, and Simon Schaffer. *Leviathan and the Air Pump: Hobbes, Boyle, and the Experimental Life* (Princeton: Princeton University Press, 1985).

Shapin, Steven, and Simon Schaffer, "Making Newton: On Interpreting Scientific Texts" (unpublished typescript).

Shapiro, Barbara, "Early Modern Intellectual Life: Humanism, Religion and Science in Seventeenth Century England," *History of Science* (March 1991), 45–71.

John Wilkins, 1614–1652: An Intellectual Biography (Berkeley: University of California Press, 1969).

"Latitudinarianism and Science in Seventeenth-Century England," in Charles Webster, ed., *The Intellectual Revolution of the Seventeenth Century* (London and Boston: Routledge and Kegan Paul, 1974), pp. 286–316.

Probability and Certainty in Seventeenth Century England (Princeton: Princeton University Press, 1983).

Smith, Alan, "Edward Elliott: Fire Engine Proprietor," *Transactions of the Newcomen Society*, L (1978–9), 230.

"The Newcomen Engine at Passy, France, in 1725: A Transfer of Technology Which Did Not Take Place" (typescript, November 1979).

"Steam and the City: the Committee of Proprietors of the Invention for Raising Water by Fire, 1715–1735," *Transactions of the Newcomen Society*, XLIX (1977–8), 5–20.

Smout, T. C., ed., "Journal of Henry Kalmeter's Travels in Scotland 1719–1720," in *Scottish Industrial History*, Scottish History Society, series IV, Vol. XIV (1978), 1–52.

"Lead-mining in Scotland, 1650–1850," in *Studies in Scottish Business History*, ed. Peter L. Payne (London: Cass, 1967), pp. 115–20.

Speck, W. A. "Politicians, Peers and Publication by Subscription, 1700–1750," in Isabel Rivers, ed., *Books and their Readers in Eighteenth-Century England* (Leicester: Leicester University Press, 1982), pp. 56–64.

Reluctant Revolutionaries. Englishmen and the Revolution of 1688 (Oxford and New York: Oxford University Press, 1989).

Speck, W. A., and W. A. Gray, eds., *London Poll Books, 1713* (London: London Record Society, 1981).

Spurr, John, " 'Rational Religion' in Restoration England," *Journal of the History of Ideas*, XLIX (Oct.–Dec. 1988), 563–85.

Stewart, L., "The Edge of Utility: Slaves and Smallpox in the Early Eighteenth Century," *Medical History*, XXIX (1985), 54–70.

"Public Lectures and Private Patronage in Newtonian England," *Isis*, LXXVII (March 1986), 47–58.

"Samuel Clarke, Newtonianism, and the Factions of Post-Revolutionary England," *Journal of the History of Ideas*, XLII (Jan.–March 1981), 53–72.

"The Selling of Newton: Science and Technology in Early Eighteenth-Century England," *Journal of British Studies*, XXV (April 1986), 178–92.

Stimson, Dorothy, *Scientists and Amateurs. A History of the Royal Society* (New York: Greenwood Press, 1968).

Straka, Gerald M., *Anglican Reaction to the Revolution of 1688* (Madison: State Historical Society of Wisconsin, 1962).

Strong, E. W., "Newtonian Explications of Natural Philosophy," *Journal of the History of Ideas*, XVIII (1957), 49–83.

Stroup, Alice, *Royal Funding of the Parisian Academie Royale des Sciences During the 1690s. Transactions of the American Philosophical Society*, LXXVII, 4 (Philadelphia: American Philosophical Society, 1987).

Stubbs, Mayling, "John Beale, Philosophical Gardener of Herefordshire. Part II. The Improvement of Agriculture and Trade in the Royal Society (1663–1683)," *Annals of Science*, XLVI (1989), 323–63.

Sullivan, Robert E., *John Toland and the Deist Controversy: A Study in Adaptations* (Cambridge, Mass.: Harvard University Press, 1982).

Supple, Barry, *The Royal Exchange Assurance. A History of British Insurance, 1720–1970* (Cambridge: Cambridge University Press, 1970).

Taylor, E. G. R., *The Mathematical Practitioners of Hanoverian England* (Cambridge: Cambridge University Press, 1966).

The Mathematical Practitioners of Tudor and Stuart England (Cambridge: Cambridge University Press, 1954).

Teich, Mikulas, "The Early History of the Newcomen Engine at Nova Bana (Konigsberg): Isaac Potter's Negotiations with the Hofkammer and the Signing of the Agreement of 19 August 1721," *East-Central Europe*, XI (1982), 24–37.

Thackray, Arnold, *Atoms and Powers. An Essay on Newtonian Matter-Theory and the Development of Chemistry* (Cambridge: Harvard University Press, 1970).

" 'The Business of Experimental Philosophy,' The Early Newtonian Group at the Royal Society," *XIIe congres international d'histoire des sciences* (1968), Actes, IIIB (1971), 155–9.

"Science and Technology in the Industrial Revolution," *History of Science*, IX (1970), 76–89.

Thirsk, Joan, *Economic Policy and Projects. The Development of a Consumer Society in Early Modern England* (Oxford: Clarendon Press, 1978).

"Plough and Pen: Agricultural Writers in the Seventeenth Century," in T. H. Aston, P. R. Cross, Christopher Dyer, and Joan Thrisk, eds., *Social Relations and Ideas. Essays in Honour of R. H. Hilton* (Cambridge and New York: Cambridge University Press, 1983), pp. 295–318.

Treadwell, J. M., "Swift, William Wood, and the Factual Basis of Satire," *Journal of British Studies*, XV (Spring 1976), 76–91.

"William Wood and the Company of Iron Masters of Great Britain," *Business History*, XVI (1974), 93–112.

Tuveson, Ernest, "Swift and the World-Makers," *Journal of the History of Ideas*, XI (1950), 54–74.

Tweedie, Charles, *James Stirling: A Sketch of His Life and Works along with his Scientific Correspondence* (Oxford: Clarendon Press, 1922).

Tyacke, Sarah, *London Map-Sellers, 1660–1720* (Tring: Map Collector Publications Ltd., 1978).

"Map-Sellers and the London Map Trade c1650–1710," in Helen Wallis and Sarah Tyacke, eds., *My Head is a Map Essays and Memoirs in Honour of R. V. Tooley* (London: Francis Edwards and Carta Press, 1973), pp. 63–80.

Van Reijn, Johan A., "John Theophilus Desaguliers, 1683–1983," *Thoth*, V (1983), 165–204.

Vermeulen, Ben, "Theology and Science: The Case of Bernard Nieuwentijt's Theo-Logical Positivism," in Sergio Rossi, ed., *Science and Imagination in XVIIIth-Century British Culture* (Milan: 1987), pp. 379–90.

Vickers, Brian, *In Defence of Rhetoric* (Oxford: Clarendon Press, 1988).

"The Royal Society and English Prose Style: A Reassessment," in Brian Vickers and Nancy S. Struever, *Rhetoric and the Pursuit of Truth. Language and Change in the Seventeenth and Eighteenth Centuries* (Los Angeles: William Andrews Clark Memorial Library, 1985), pp. 3–76.

Viner, Jacob, *The Role of Providence in the Social Order. An Essay in Intellectual History* (Princeton: Princeton University Press, 1976).

Waerland, Are, "Marten Triewald and the First Steam Engine in Sweden," *Transactions of the Newcomen Society*, VII (1926–7), 24–41.

Wallis, Peter J., "Book Subscription Lists" (unpublished lecture to the Bibliographical Society, April 1973).

"British Mathematical Bibliography," *Journal of the Institute of Navigation*, XX (April 1967), 200–5.

"British Philomaths – Mid-Eighteenth Century and Earlier," *Centaurus*, XVII (1973), 301–14.

"Dissemination of Mathematics in the Eighteenth Century. The Evidence of Subscription Lists," *Project for Historical Biobiliography*, no. 314 (December 1982).

An Index of British Mathematicians. A Check-list. Part 2. 1701–1760 (Newcastle: Project for Historical Biobibliography, 1976).

"Thomas Watts. Academy Master, Freemason, Insurance Pioneer, MP," *History of Education Society Bulletin*, XXXII (Autumn 1983), 51–3.

Wallis, P. J., and F. J. G. Robinson, "Scientists and Subscription Lists," *British Journal for the History of Science*, VI (1972), 227–8.

Wallis, R. V., and Peter Wallis, *Biobibliography of British Mathematics and its Applications. Part II. 1701–1760* (Newcastle: Project for Historical Biobibliography, 1988).

"Female Philomaths," *Historia Mathematica*, VII (1980), 57–64.

Webster, Charles, *The Great Instauration. Science, Medicine and Reform, 1626–1660* (New York: Holmes and Meier Publishers, 1975).

ed., *The Intellectual Revolution of the Seventeenth Century* (London and Boston: Routledge and Kegan Paul, 1974).

"Puritanism, Separatism, and Science," in David C. Lindberg and Ronald L. Numbers, eds., *God and Nature. Historical Essays on the Encounter between Christianity and Science* (Berkeley and Los Angeles: University of California Press, 1986), pp. 192–217.

ed., *Samuel Hartlib and the Advancement of Learning* (Cambridge and London: Cambridge University Press, 1970).

Westfall, Richard S., *Never at Rest: A Biography of Isaac Newton* (Cambridge: Cambridge University Press, 1980).

"Robert Hooke, Mechanical Technology and Scientific Investigation," in John G. Burke, ed., *The Uses of Science in the Age of Newton* (Berkeley and Los Angeles: 1983), pp. 85–110.

Science and Religion in Seventeenth Century England (Ann Arbor: University of Michigan Press, 1973).

White, A. W. A., "Early Newcomen Engines on the Warwickshire Coalfield, 1714–1736," *Transactions of the Newcomen Society*, XLI (1968–9), 203–16.

Williamson, F., "George Sorocold, of Derby. A Pioneer of Water Supply," *Journal of the Derbyshire Archaeological and Natural History Society*, LVII (1937), 43–93.

Winter, H. J. J., "Scientific Associations of the Spalding Gentlemen's Society During the Period 1710–1750," *Archives internationale d'histoire des sciences*, III (1950), 77–88.

Winton, Calhoun, *Captain Steele: The Early Career of Richard Steele* (Baltimore: Johns Hopkins Press, 1964).

Sir Richard Steele, M.P. The Later Career (Baltimore and London: Johns Hopkins University Press, 1970).

Wood, P. B., "Methodology and Apologetics: Thomas Sprat's *History of the Royal Society*," *British Journal for the History of Science*, XIII (1980), 1–26.

Woodward, David A., "English Cartography, 1650–1750," in Norman J. W. Thrower, ed., *The Compleat Plattmaker* (Berkeley: University of California Press, 1978), pp. 159–93.

Wordie, J. R., *Estate Management in Eighteenth-Century England. The Building of the Leveson-Gower Fortune* (London: Royal Historical Society, 1982).

Index

Knorr-Cetina, Karin D., "The Ethnographic Study of Scientific Work: Towards a Constructivist Interpretation of Science," in Karin D. Knorr-Cetina and Michael Mulkay, eds., *Science Observed. Perspectives on the Social Study of Science* (London and Beverly Hills: Sage Publications, 1983), pp. 115–40.

Korshin, Paul J., "Types of Eighteenth-Century Literary Patronage," *Eighteenth-Century Studies*, VII (Summer 1974), 452–73.

Kramnick, Isaac, *Bolingbroke and His Circle. The Politics of Nostalgia in the Age of Walpole* (Cambridge, Mass.: Harvard University Press, 1968).

Kubrin, David Charles, "Providence and the Mechanical Philosophy: The Creation and Dissolution of the World in Newtonian Thought. A Study of the Relations of Science and Religion in Seventeenth Century England" (Ph.D. thesis, Cornell University, 1969).

Laslett, Peter, "John Locke, the Great Recoinage, and the Origins of the Board of Trade: 1695–1698," in John W. Yolton, ed., *John Locke: Problems and Perspectives* (London: Cambridge University Press, 1969), pp. 137–64.

Latour, Bruno, "Give Me a Laboratory and I will Raise the World," in Karin D. Knorr-Cetina and Michael Mulkay, eds., *Science Observed. Perspectives on the Social Study of Science* (London and Beverly Hills: Sage Publications, 1983), pp. 141–70.

Science in Action (Cambridge, Mass.: Harvard University Press, 1987).

Levine, Joseph M., *Dr. Woodward's Shield. History, Science, and Satire in Augustan England* (Berkeley: University of California Press, 1977).

Lillywhite, Bryant, *London Coffee Houses* (London: George Allen and Unwin, 1963).

Lindqvist, Svante, *Technology on Trial. The Introduction of Steam Power Technology into Sweden, 1715–1736* (Stockholm: Almqvist and Wicksell International, 1984).

"The Work of Martin Triewald in England," *Transactions of the Newcomen Society*, L (1978–9), 165–71.

Lindsay, J. M. "The Iron Industry in the Highlands: Charcoal Blast Furnaces," *Scottish Historical Review*, LVI (April 1977), 49–63.

Loftis, John, "Richard Steele's Censorium," *Huntington Library Quarterly*, XIV (1950), 53–62.

Steele at Drury Lane (Berkeley: University of California Press, 1952).

MacDonald, Sir George, "John Horsley, Scholar and Gentleman," *Archaeologia Aeliana*, fourth series, VIII (1933), 1–57.

MacLeod, Christine, *Inventing the Industrial Revolution. The English*

Patent System, 1660–1800 (Cambridge and New York: Cambridge University Press, 1988).

"Patents for Invention and Technical Change in England, 1660–1753" (Ph.D. thesis, Cambridge University, 1982).

"The 1690s Patents Boom: Invention or Stock-Jobbing?" *Economic History Review*, second series, XXXIX (1986), 549–71.

Maddison, R. E. W., "Notes on Some Members of the Hanckwitz Family in England," *Annals of Science*, XI (March 1955), 64–72.

"Studies in the Life of Robert Boyle, F.R.S., Part V. Ambrose Godfrey Hanckwitz, F.R.S.," *Notes and Records of the Royal Society*, XI (1954), 159–88.

Major, Paul Richard, "The Physical Researches of J. T. Desaguliers" (M.Sc. thesis, University of London, 1962).

Marshall, John, "The Ecclesiology of the Latitude-men 1660–1689: Stillingfleet, Tillotson, and 'Hobbism,' " *Journal of Ecclesiastical History*, XXXVI (July 1985), 407–27.

Matthews, William, *Hydraulia; An Historical and Descriptive Account of The Water Works of London* (London: R. Hunter, 1835).

McGuire, J. E., "Force, Active Principles, and Newton's Invisible Realm," *Ambix*, XV (October 1968), 154–208.

McKendrick, Neil, John Brewer, and J. H. Plumb, *The Birth of a Consumer Society. The Commercialization of Eighteenth-Century England* (Bloomington: Indiana University Press, 1982).

McKie, D., "John Harris and His *Lexicon Technicum*," *Endeavour*, IV (1945), 53–7.

Merton, Robert K., *Science, Technology and Society in Seventeenth Century England*, second edition (New York and London: Harper, 1970).

Millard, Peter T., "An Edition of Roger North's Life of Dr. John North with a Critical Introduction" (Ph.D. thesis, Oxford University, 1969).

Millburn, John R., *Benjamin Martin: Author, Instrument-Maker and "Country Showman"* (Leyden: Noordhoff International Publications, 1976).

"Benjamin Martin and the Royal Society," *Notes and Records of the Royal Society of London*, XXVIII (June 1973), 15–23.

"The London Evening Courses of Benjamin Martin and James Ferguson, Eighteenth-Century Lecturers on Experimental Philosophy," *Annals of Science*, XL (1983), 437–55.

"William Stukeley and the Early History of the Orrery," *Annals of Science*, XXXI (1974), 511–28.

Morton, Alan, "Public Science in Mid-18th Century London: The King George III Collection at the Science Museum" (unpublished paper, 1990).

Morton, A. Q., "Lectures on Natural Philosophy in London, 1750–1765: S. C. T. Demainbray (1710–1782) and the 'Inattention' of His Countrymen," *British Journal for the History of Science*, XXIII (December 1990), 411–34.

Mulligan, Lotte, "Anglicanism, Latitudinarianism and Science in Seventeenth Century England," *Annals of Science*, XXX (June 1973), 213–19.

"'Reason,' 'Right Reason,' and 'Revelation' in Mid-Seventeenth-Century England," in Brian Vickers, ed., *Occult and Scientific Mentalities in the Renaissance* (Cambridge and New York: Cambridge University Press, 1984), pp. 375–401.

Musson, A. E., and Eric Robinson, *Science and Technology in the Industrial Revolution* (Toronto: University of Toronto Press, 1969).

Namier, Sir Lewis, "Brice Fisher, M.P.: A Mid-Eighteenth-Century Merchant and His Connexions," *English Historical Review*, XLII (October 1927), 514–32.

Nicolson, Marjorie Hope, *Newton Demands the Muse. Newton's Opticks and the Eighteenth Century Poets* (Princeton: Princeton University Press, 1946).

Nicolson, Marjorie Hope, and G. S. Rousseau, *"This Long Disease, My Life," Alexander Pope and the Sciences* (Princeton: Princeton University Press, 1968).

Nicolson, Marjorie Hope, and Nora M. Mohler, "The Scientific Background to Swift's *Voyage to Laputa*," in *Science and Imagination* (Ithaca: Great Seal Books, 1956), pp. 110–54.

Ochs, Kathleen H., "The Failed Revolution in Applied Science: Studies of Industry by Members of the Royal Society of London, 1660–1688" (Ph.D. thesis, University of Toronto, 1981).

"The Royal Society of London's History of Trades Programme: An Early Episode in Applied Science," *Notes and Records of the Royal Society*, XXXIX (April 1985), 129–58.

Ogden, H. V. S., "Thomas Burnet's *Telluris Theoria Sacra* and Mountain Scenery," *ELH, A Journal of English Literary History*, XIV (1947), 139–50.

Olson, Richard G., "Tory-High Church Opposition to Science and Scientism in the Eighteenth Century: The Works of John Arbuthnot, Jonathan Swift, and Samuel Johnson," in John G. Burke, ed., *The*

Uses of Science in the Age of Newton (Berkeley and Los Angeles: University of California Press, 1983), pp. 171–204.

Owen, Colin, *The Leicestershire and South Derbyshire Coalfields* (Ashbourne: Moorland Publishers, 1984).

Pocock, J. G. A. "Early Modern Capitalism – The Augustan Perception," in Eugene Kamenka and R. S. Neale, eds., *Feudalism, Capitalism and Beyond* (Canberra: Australia National University Press, 1975), pp. 62–83.

The Machiavellian Moment. Florentine Political Thought and the Atlantic Republican Tradition (Princeton: Princeton University Press, 1975).

Virtue, Commerce, and History. Essays on Political Thought and History, Chiefly in the Eighteenth Century (Cambridge and New York: Cambridge University Press, 1985).

Pollard, Sidney, *The Genesis of Modern Management. A Study of the Industrial Revolution in Great Britain* (London: E. Arnold, 1965).

Porter, Roy, "Creation and Credence: The Career of Theories of the Earth in Britain, 1660–1820," in Barry Barnes and Steven Shapin, eds., *Natural Order: Historical Studies in Scientific Culture* (Beverley Hills: Sage Publications, 1979), pp. 97–123.

"The Enlightenment in England," in Roy Porter and M. Teich, eds., *The Enlightenment in National Context* (Cambridge and New York: Cambridge University Press, 1981), pp. 5–14.

The Making of Geology. Earth Science in Britain, 1660–1815 (Cambridge and New York: Cambridge University Press, 1977).

"Science, Provincial Culture and Public Opinion in Enlightenment England," *British Journal for Eighteenth-Century Studies*, III (1980), 20–46.

Pumfrey, Stephen, "Ideas Above his Station: A Social Study of Hooke's Curatorship of Experiments," *History of Science*, XXIX (March 1991), 1–44.

Raistrick, A, "The London Lead Company, 1692–1905," *Transactions of the Newcomen Society*, XIV (1933–4), 119–62.

Quakers in Science and Industry (Newton Abbot: Bonnisdale Press, 1950).

"The Steam Engine on the Tyneside," *Transactions of the Newcomen Society*, XVII (1936–7), 131–63.

Raven, Charles E., *John Ray Naturalist. His Life and Works* (Cambridge and New York: Cambridge University Press, 1942).

Redwood, John, *Reason, Ridicule and Religion: The Age of Enlightenment in England, 1660–1750* (London: Thames and Hudson, 1976).

Rees, William, *Industry Before the Industrial Revolution*, 2 vols. (Cardiff: University of Wales Press, 1968).

Reynolds, Terry S., *Stronger Than a Hundred Men: A History of the Vertical Water Wheel* (Baltimore: Johns Hopkins University Press, 1983).

Robinson, F. J. G., "A Philosophic War: An Episode in Eighteenth Century Scientific Lecturing in North-East England," *Transactions of the Architectural and Archaeological Society of Durham and Northumberland*, II (1970), 101–8.

Robinson, F. J. G., and P. J. Wallis, *Book Subscription Lists: A Revised Guide* (Newcastle Upon Tyne: H. Hill, 1975).

Robinson, Henry W., and Walter Adams, eds., *The Diary of Robert Hooke, M.A., M.D., F.R.S., 1672–1680* (London: Taylor and Francis, 1935).

Robischon, Mary M., "Scientific Instrument Makers in London During the Seventeenth and Eighteenth Centuries" (Ph.D. thesis, University of Michigan, 1983).

Rogers, Pat, "Gulliver and the Engineers," *Modern Language Review*, LXX (1975), 260–70.

Rolt, L. T. C., and J. S. Allen, *The Steam Engine of Thomas Newcomen* (New York: Science History Publications, 1977).

Rousseau, G. S., "Science Books and Their Readers in the Eighteenth Century," in Isabel Rivers, ed., *Books and their Readers in Eighteenth-Century England* (Leicester: Leicester University Press, 1982), pp. 208–17.

" 'Wicked Whiston' and the Scriblerians: Another Ancients-Modern Controversy," *Studies in Eighteenth-Century Culture*, XVII (1987), 17–44.

Rousseau, G. S., and Roy Porter, eds., *The Ferment of Knowledge. Studies in the Historiography of Eighteenth-Century Science* (Cambridge and New York: Cambridge University Press, 1980).

Rowbottom, Margaret E., "John Theophilus Desaguliers (1683–1744)," *Proceedings of the Huguenot Society of London*, XXI (1968), 196–218.

"The Teaching of Experimental Philosophy in England, 1700–1730," *XIe congres international d'histoire des sciences (1965)*, Actes IV (1968), 46–53.

Rowlands, Marie B., "Stonier Parrott and the Newcomen Engines," *Transactions of the Newcomen Society*, XLI (1968–9), 49–67.

Rupp, Gordon, *Religion in England, 1688–1791* (Oxford: Clarendon Press, 1986).

Schaffer, Simon, "Defoe's Natural Philosophy and the Worlds of Credit," in John Christie and Sally Shuttleworth, eds., *Nature Transfigured. Science and Literature, 1700–1900* (Manchester and New York: Manchester University Press, 1989), pp. 13–44.

"Glass Works: Newton's Prisms and the Uses of Experiment," in David Gooding, Trevor Pinch, and Simon Schaffer, eds., *The Uses of Experiment. Studies in the Natural Sciences* (Cambridge: Cambridge University Press, 1989), pp. 67–104.

"Natural Philosophy and Public Spectacle in the Eighteenth Century," *History of Science*, XXI (1983), 1–43.

"Newtonian Cosmology and the Steady State" (Ph.D. thesis, University of Cambridge, 1980).

Schofield, Robert E., *Mechanism and Materialism. British Natural Philosophy in An Age of Reason* (Princeton: Princeton University Press, 1970).

Scott, W. R., *The Constitution and Finance of English, Scottish and Irish Joint-Stock Companies to 1720*, 3 vols. (Cambridge: Cambridge University Press, 1912; reprint edn., Gloucester, Mass., 1968).

Shapin, Steven, "The House of Experiment in Seventeenth-Century England," *Isis*, LXXIX (1988), 373–404.

"Of Gods and Kings: Natural Philosophy and Politics in the Leibniz-Clarke Disputes," *Isis*, LXXII (June 1981), 187–215.

"Pump and Circumstance: Robert Boyle's Literary Technology," *Social Studies of Science*, XIV (November 1984), 481–520.

"Robert Boyle and Mathematics: Reality, Representation, and Experimental Practice," *Science in Context*, II (1988), 23–58.

"Social Uses of Science," in G. S. Rousseau and Roy Porter, *Ferment of Knowledge. Studies in the Historiography of Eighteenth-Century Science* (Cambridge and New York: Cambridge University Press, 1980), pp. 93–105.

"Who was Robert Hooke?" in Michael Hunter and Simon Schaffer, eds., *Robert Hooke. New Studies* (Woodbridge: Boydell Press, 1989), 253–83.

Shapin, Steven, and Simon Schaffer. *Leviathan and the Air Pump: Hobbes, Boyle, and the Experimental Life* (Princeton: Princeton University Press, 1985).

Shapin, Steven, and Simon Schaffer, "Making Newton: On Interpreting Scientific Texts" (unpublished typescript).

Shapiro, Barbara, "Early Modern Intellectual Life: Humanism, Religion